Main Group Chemistry: From Synthesis to Applications— In Honor of the Great Contributions of Prof. J. Derek Woollins

Main Group Chemistry: From Synthesis to Applications— In Honor of the Great Contributions of Prof. J. Derek Woollins

Guest Editor

Petr Kilián

Basel • Beijing • Wuhan • Barcelona • Belgrade • Novi Sad • Cluj • Manchester

Guest Editor
Petr Kilián
School of Chemistry
University of St Andrews
St Andrews
United Kingdom

Editorial Office
MDPI AG
Grosspeteranlage 5
4052 Basel, Switzerland

This is a reprint of the Special Issue, published open access by the journal *Molecules* (ISSN 1420-3049), freely accessible at: www.mdpi.com/journal/molecules/special_issues/BLQABNA176.

For citation purposes, cite each article independently as indicated on the article page online and using the guide below:

Lastname, A.A.; Lastname, B.B. Article Title. *Journal Name* **Year**, *Volume Number*, Page Range.

ISBN 978-3-7258-3312-2 (Hbk)
ISBN 978-3-7258-3311-5 (PDF)
https://doi.org/10.3390/books978-3-7258-3311-5

© 2025 by the authors. Articles in this book are Open Access and distributed under the Creative Commons Attribution (CC BY) license. The book as a whole is distributed by MDPI under the terms and conditions of the Creative Commons Attribution-NonCommercial-NoDerivs (CC BY-NC-ND) license (https://creativecommons.org/licenses/by-nc-nd/4.0/).

Contents

About the Editor vii

Preface ix

Tristram Chivers and Richard T. Oakley
Structures and Spectroscopic Properties of Polysulfide Radical Anions: A Theoretical Perspective
Reprinted from: *Molecules* 2023, 28, 5654, https://doi.org/10.3390/molecules28155654 1

Kyzgaldak Ramazanova, Soumyadeep Chakrabortty, Fabian Kallmeier, Nadja Kretzschmar, Sergey Tin and Peter Lönnecke et al.
Access to Enantiomerically Pure *P*-Chiral 1-Phosphanorbornane Silyl Ethers
Reprinted from: *Molecules* 2023, 28, 6210, https://doi.org/10.3390/molecules28176210 22

Michael A. Beckett, Peter N. Horton, Michael B. Hursthouse and James L. Timmis
Synthesis and Thermal Studies of Two Phosphonium Tetrahydroxidohexaoxidopentaborate(1-) Salts: Single-Crystal XRD Characterization of [^{i}PrPPh$_3$][B$_5$O$_6$(OH)$_4$]·3.5H$_2$O and [MePPh$_3$][B$_5$O$_6$(OH)$_4$]·B(OH)$_3$·0.5H$_2$O †
Reprinted from: *Molecules* 2023, 28, 6867, https://doi.org/10.3390/molecules28196867 33

Anna E. Tarcza, Alexandra M. Z. Slawin, Cameron L. Carpenter-Warren, Michael Bühl, Petr Kilian and Brian A. Chalmers
Constrained Phosphine Chalcogenide Selenoethers Supported by *peri*-Substitution
Reprinted from: *Molecules* 2023, 28, 7297, https://doi.org/10.3390/molecules28217297 45

René T. Boeré
Hydrogen Bonds Stabilize Chloroselenite Anions: Crystal Structure of a New Salt and Donor-Acceptor Bonding to SeO$_2$
Reprinted from: *Molecules* 2023, 28, 7489, https://doi.org/10.3390/molecules28227489 62

Marko Rodewald, J. Mikko Rautiainen, Helmar Görls, Raija Oilunkaniemi, Wolfgang Weigand and Risto S. Laitinen
Formation, Characterization, and Bonding of *cis*- and *trans*-[PtCl$_2${Te(CH$_2$)$_6$}$_2$], *cis-trans*-[Pt$_3$Cl$_6${Te(CH$_2$)$_6$}$_4$], and *cis-trans*-[Pt$_4$Cl$_8${Te(CH$_2$)$_6$}$_4$]: Experimental and DFT Study †
Reprinted from: *Molecules* 2023, 28, 7551, https://doi.org/10.3390/molecules28227551 79

Nikita Demidov, Mateus Grebogi, Connor Bourne, Aidan P. McKay, David B. Cordes and Andreas Stasch
A Convenient One-Pot Synthesis of a Sterically Demanding Aniline from Aryllithium Using Trimethylsilyl Azide, Conversion to *β*-Diketimines and Synthesis of a *β*-Diketiminate Magnesium Hydride Complex
Reprinted from: *Molecules* 2023, 28, 7569, https://doi.org/10.3390/molecules28227569 95

Callum M. Inglis, Richard A. Manzano, Ryan M. Kirk, Manab Sharma, Madeleine D. Stewart and Lachlan J. Watson et al.
Poly(imidazolyliden-yl)borato Complexes of Tungsten: Mapping Steric vs. Electronic Features of *Facially* Coordinating Ligands
Reprinted from: *Molecules* 2023, 28, 7761, https://doi.org/10.3390/molecules28237761 118

Daniel J. Davidson, Aidan P. McKay, David B. Cordes, J. Derek Woollins and Nicholas J. Westwood
The Covalent Linking of Organophosphorus Heterocycles to Date Palm Wood-Derived Lignin: Hunting for New Materials with Flame-Retardant Potential
Reprinted from: *Molecules* **2023**, *28*, 7885, https://doi.org/10.3390/molecules28237885 147

R. Alan Aitken, Graham Dawson, Neil S. Keddie, Helmut Kraus, Heather L. Milton and Alexandra M. Z. Slawin et al.
Thermal Rearrangement of Thiocarbonyl-Stabilised Triphenylphosphonium Ylides Leading to (Z)-1-Diphenylphosphino-2-(phenylsulfenyl)alkenes and Their Coordination Chemistry
Reprinted from: *Molecules* **2023**, *29*, 221, https://doi.org/10.3390/molecules29010221 162

José A. Fuentes, Mesfin E. Janka, Aidan P. McKay, David B. Cordes, Alexandra M. Z. Slawin and Tomas Lebl et al.
Ligand Hydrogenation during Hydroformylation Catalysis Detected by In Situ High-Pressure Infra-Red Spectroscopic Analysis of a Rhodium/Phospholene-Phosphite Catalyst [†]
Reprinted from: *Molecules* **2024**, *29*, 845, https://doi.org/10.3390/molecules29040845 175

Laurence J. Taylor, Emma E. Lawson, David B. Cordes, Kasun S. Athukorala Arachchige, Alexandra M. Z. Slawin and Brian A. Chalmers et al.
Synthesis and Structural Studies of *peri*-Substituted Acenaphthenes with Tertiary Phosphine and Stibine Groups [†]
Reprinted from: *Molecules* **2024**, *29*, 1841, https://doi.org/10.3390/molecules29081841 194

Mehdi Elsayed Moussa, Susanne Bauer, Christian Graßl, Christoph Riesinger, Gábor Balázs and Manfred Scheer
Synthesis and Characterization of Novel Cobalt Carbonyl Phosphorus and Arsenic Clusters
Reprinted from: *Molecules* **2024**, *29*, 2025, https://doi.org/10.3390/molecules29092025 212

Alexandra M. Miles-Hobbs, Paul G. Pringle, J. Derek Woollins and Daniel Good
Monofluorophos–Metal Complexes: Ripe for Future Discoveries in Homogeneous Catalysis
Reprinted from: *Molecules* **2024**, *29*, 2368, https://doi.org/10.3390/molecules29102368 228

About the Editor

Petr Kilián

Dr. Petr Kilian obtained his first degree and a Ph.D. at Masaryk University, Brno, Czech Republic. After working as an RS/NATO Fellow at Loughborough University, UK, he moved to St Andrews. He became an EaStChem Fellow in 2005 and was appointed as a Lecturer at St Andrews in 2009. He has served as the Head of the Inorganic Section from 2017 till now and became a Fellow of the Royal Society of Chemistry (FRSC) in 2020. His scientific interests revolve around the synthesis, bonding, reactivity, and mechanistic studies of organoelement Group 15 compounds. His group has been one of the pioneers in using clamping frameworks, particularly peri-substitution, to stabilize fleeting species and create forced interactions, eliciting unusual structures, bonding, and reactivity.

Preface

Professor J. Derek Woollins is an outstanding British inorganic chemist who, in particular, made key contributions to the field of main group synthetic chemistry. After completing his B.Sc. and Ph.D. at the University of East Anglia, Derek completed his postdoctoral research in Canada, the US, and the UK before taking a lectureship position at Imperial College, where he stayed for twelve years. Derek then moved to Loughborough University as the Chair of Inorganic Chemistry, and in 1999, he moved to a Chair at St Andrews University. At St Andrews, Derek served as the Head of School for eight years and as a Vice Principal for Research and Innovation for six years. In 2019, Derek took a position as a Provost of Khalifa University, Abu Dhabi, from which he retired in 2020.

Derek is extremely well known in the academic community, in particular the inorganic and main group communities, both for his scientific contributions and for his brilliant personality. His scientific interests revolve around Group 16 and 15 chemistry, in particular S/N, Se, Te, and P chemistry. Derek's group established $Ph_2P_2Se_4$, now named Woollins reagent, as an accessible and powerful selenating reagent in a range of chemistries. Derek's efficiency is legendary; the Web of Science (January 2025) shows Derek to have 609 publications, 13,464 citations, and an H-index of 54.

Derek's hard-working nature, witty humor, jokes, and general ability to tell a good story about anything (from science to DIY) made him a legendary lecturer amongst students as well as an extremely valuable and likeable colleague.

In spring 2023, I agreed to be Guest Editor of a Special Issue of *Molecules* to celebrate Derek's retirement. The title of this Special Issue was formulated to enable all aspects of main group chemistry to be considered. Manuscripts were received throughout late 2023 and 2024. I was delighted by the response, which highlights the high esteem in which Derek is held by his colleagues and friends. I wish to thank all the authors for contributing their papers to this special issue. The quality of their submissions made my job quite easy and pleasurable!

Petr Kilián
Guest Editor

Structures and Spectroscopic Properties of Polysulfide Radical Anions: A Theoretical Perspective

Tristram Chivers [1,*] and Richard T. Oakley [2,*]

[1] Department of Chemistry, University of Calgary, Calgary, AB T2N 1N4, Canada
[2] Department of Chemistry, University of Waterloo, Waterloo, ON N2L 3G1, Canada
* Correspondence: chivers@ucalgary.ca (T.C.); oakley@uwaterloo.ca (R.T.O.)

Abstract: The potential involvement of polysulfide radical anions $S_n^{\bullet-}$ is a recurring theme in discussions of the basic and applied chemistry of elemental sulfur. However, while the spectroscopic features for n = 2 and 3 are well-established, information on the structures and optical characteristics of the larger congeners (n = 4–8) is sparse. To aid identification of these ephemeral species we have performed PCM-corrected DFT calculations to establish the preferred geometries for $S_n^{\bullet-}$ (n = 4–8) in the polar media in which they are typically generated. TD-DFT calculations were then used to determine the number, nature and energies of the electronic excitations possible for these species. Numerical reliability of the approach was tested by comparison of the predicted and experimental excitation energies found for $S_2^{\bullet-}$ and $S_3^{\bullet-}$. The low-energy (near-IR) transitions found for the two acyclic isomers of $S_4^{\bullet-}$ (C_{2h} and C_{2v} symmetry) and for $S_5^{\bullet-}$ (C_s symmetry) can be understood by extension of the simple HMO π-only chain model that serves for $S_2^{\bullet-}$ and $S_3^{\bullet-}$. By contrast, the excitations predicted for the *quasi*-cyclic structures $S_n^{\bullet-}$ (n = 6–8) are better described in terms of $\sigma \to \sigma^*$ processes within a localized 2c-3e manifold.

Keywords: polysulfide chemistry; radical anions; structures; spectroscopic properties; time-dependent density functional theory

1. Introduction

Polysulfide radical anions $S_n^{\bullet-}$ (n = 2–8) play a pivotal role as intermediates in the sulfur ↔ sulfide redox cycle [1–4]. The influence of these short-lived species is frequently invoked in contemporary investigations of sulfur chemistry, including alkali-metal-sulfur batteries [5–7], organic syntheses [8], biological chemistry [9,10], geochemical processes involving metal transport [11–13] and quantum-dot sensitized solar cells [14,15]. In solution, polysulfide radical anions are readily oxidized by atmospheric oxygen, but the smaller members can be trapped in an aluminosilicate matrix and are known to be the chromophores in yellow ($S_2^{\bullet-}$), blue ($S_3^{\bullet-}$) and green (simultaneous presence of $S_2^{\bullet-}$ and $S_3^{\bullet-}$) ultramarines [16] and related sodalite-group minerals [17]. The diatomic $S_2^{\bullet-}$ and the triatomic $S_3^{\bullet-}$ (C_{2v}) radical anions are readily detected in solution or in the solid state by their characteristic UV-visible, Raman or EPR spectra [18]. Indeed, one or more of these techniques is commonly invoked to provide evidence for the role of $S_3^{\bullet-}$ as an in-situ generated reagent in organic synthesis [8].

In contrast to the well-established spectroscopic signatures of $S_2^{\bullet-}$ and $S_3^{\bullet-}$, evidence for the larger members of the family (n = 4–8) is fragmentary and often conflicting. Of these species, $S_4^{\bullet-}$ has a long but somewhat checkered history. In 1970, as part of his pioneering study on solutions of alkali-metal polysulfides in electron-pair donor solvents, e.g., DMF, HMPA, Seel attributed a visible absorption at ca. 515 nm to $S_4^{\bullet-}$ [19], but later the band was reassigned to a dimer [20], an example of which has recently been structurally characterized in a dinuclear Bi(III)Bi(III) complex [21]. In 1983 Clark et al. investigated the nature of the sulfur chromophore in ultramarine pink by Raman spectroscopy [22], but they were unable to distinguish between S_4 and $S_4^{\bullet-}$ (or even S_3Cl). The association of a 490 nm band with $S_4^{\bullet-}$ has nonetheless persisted [9,23], and Chiba and co-workers have recently invoked formation of the radical anion during the photolysis of the closed-shell dianion S_4^{2-} [24–27]. The use of other techniques to identify $S_4^{\bullet-}$ in solution, notably Raman and IR spectroscopy [28–30], has been pursued, but band assignments based on calculated vibrational frequencies have been questioned [3]. Likewise, an EPR signal observed in solutions of lithium polysulfide solutions in DMF was proposed to belong to $S_4^{\bullet-}$ [31], but the isotropic g-value (2.031) lies close to that of the dominant radical anion $S_3^{\bullet-}$ (2.029) [32]. In principle, high field EPR spectroscopy could be used to distinguish between these (and other) polysulfide radical anions, and on this basis Chukanov et al. recently suggested the presence of $S_4^{\bullet-}$ in various sodalite minerals [33].

The larger anions $S_n^{\bullet-}$ (n = 5–8) are also acknowledged as potentially important intermediates in the $S_8 \leftrightarrow S^{2-}$ redox processes, as in the electrochemical reduction of cyclo-S_8, redox transformations in alkali metal-sulfur batteries [27] and the formation of polysulfides from photoexcited quantum dots [14,15]. Exploration of the stepwise electrochemical reduction of cyclo-S_8 in non-aqueous solvents has been extensively pursued, with formation of S_8^{2-} generally accepted by the battery community to occur first (Scheme 1) [3]. Initially, in 1970, Merritt and Sawyer claimed the preliminary formation of the one-electron reduction product $S_8^{\bullet-}$ [34], then revised this interpretation to a two-electron transfer [35], in agreement with the work of Bonnaterre and Cauquis [36] and also supported by results obtained by Hardacre and coworkers using ionic liquids as the solvent medium [37]. However, in 2008, the results of a detailed cyclic voltammetric study of the reduction of S_8 in various solvents were consistent with the formation of S_8^{2-} via two consecutive one-electron steps [38]. The potential for the involvement of $S_4^{\bullet-}$ in the S_8 reduction process has been argued [39–42], but in the absence of a clear spectroscopic signature for the anion, the issue has not been resolved. Although symmetrical dissociation of S_8^{2-} to give two $S_4^{\bullet-}$ radical anions is calculated to be exergonic [43], and an absorption band at ca. 700 nm was assigned to $S_4^{\bullet-}$ [44,45] in spectrochemical studies of the reduction of sulfur in DMSO and DMF, others insist that it has never been detected [46].

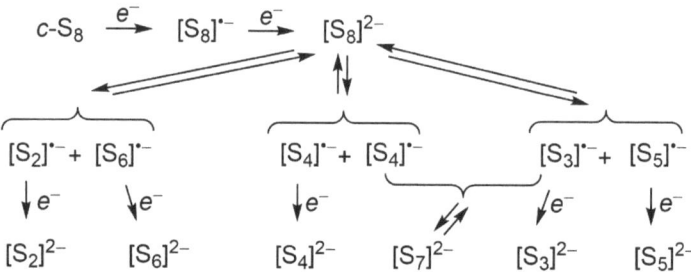

Scheme 1. Possible radical anion intermediates in the stepwise electrochemical reduction of c-S_8.

Surprisingly, although the preparation and structural characterization of several salts of the S_8^{2-} were reported more than 30 years ago [47,48], there is limited information on the behavior of ion-separated salts of S_8^{2-} in non-aqueous solvents. Formation of S_8^{2-} from S_8 has traditionally been interpreted to be followed by disproportionation to S_6^{2-} and $\frac{1}{4}S_8$ [39,43,49], the former dissociating to afford $S_3^{\bullet-}$ [48]. However, a disproportionation process simply represents a mass balance, and belies the reality that the formation of an eight-membered S_8 ring must involve the intermediacy of long chain polysulfide dianions S_n^{2-} with $n > 8$ and, possibly, polysulfide radical anions such as $S_n^{\bullet-}$ (n = 4, 6) [3]. Alternative fates for S_8^{2-} in dilute solution can be envisaged (Scheme 1) in terms of equilibria involving its symmetric and asymmetric dissociation to afford, in principle, the entire series of polysulfide radical anions $S_n^{\bullet-}$ (n = 2–6). Longer chain closed-shell dianions such as S_{10}^{2-} and S_{12}^{2-}, salts of which have recently been characterized [50,51], can then be viewed as arising from the reverse process, that is, symmetric coupling of the radical anions

$S_5{}^{\bullet-}$ and $S_6{}^{\bullet-}$, respectively, while association of $S_3{}^{\bullet-}$ and $S_4{}^{\bullet-}$ yields $S_7{}^{2-}$. Unfortunately, information on the electrochemical reduction of the cyclic allotropes S_6 and S_7 is lacking. That being said, in 2002 Dehnicke and coworkers isolated and structurally characterized crystals of the (orange/red) radical ion salt [Ph$_4$P][S$_6$], the only such characterization of an ion-separated salt of a polysulfide radical anion [52]. The importance of this result is discussed below.

As demonstrated in this brief survey, there are many unsettled questions regarding the basic chemistry of elemental sulfur, in particular relating to the stability, structure and properties of radical ion products that may be generated during the sequential reduction of cyclo-S_8 or the oxidation of the sulfide ion S^{2-} [1]. These questions have given rise to ongoing controversies, many relating to the colors of these ephemeral species—what do they look like, and do they exist if they cannot be seen?

While the colors of $S_2{}^{\bullet-}$ and $S_3{}^{\bullet-}$ are well characterized, the optical properties of the larger *putative* polysulfide radical anions have been explored only to a very limited extent. Fabian et al. used density functional theory (DFT) methods to probe the excited states of $S_4{}^{\bullet-}$, and predicted a strong absorption in the near-IR region, with a weaker band near 350 nm for cis $S_4{}^{\bullet-}$ (C_{2v}) isomer, which is slightly more stable than the trans (C_{2h}) isomer [53]. More recently, and using DFT and CASSCF methods, Rejmak confirmed that the cis $S_4{}^{\bullet-}$ radical anion could be identified by a strong absorption in the near-IR region [54] and proposed that the red chromophore in ultramarine red is *neutral* S_4 rather than the corresponding radical anion. Surprisingly, the excited state properties of the remaining radical ions in the series, that is $S_n{}^{\bullet-}$ (n = 5–8), have never been explored theoretically, perhaps because even their ground state geometries have remained somewhat of a puzzle.

The principal objective of the present article is to redress this issue, to fill in the blanks not only in regard to the spectroscopic signatures of these radical anions, that is, their excited state properties, but also to establish their ground state structures, particularly in solution in polar solvents, the media in which they are most likely to be generated.

2. Results

2.1. Structural Trends

In the following sections we describe the structural features and relative energies provided by spin unrestricted PBE0/D3/def2-QZVP calculations for the family of radical anions $S_n{}^{\bullet-}$ (n = 4–8). The results build upon the earlier systematic studies of Hunsicker et al. [55], Steudel [40] and Wong [56], but include several alternative shapes not previously considered. The possible effects of solvation are heavily stressed, as our overall aim has been to identify structures most likely to be present in solution in the polar solvents typically used for the spectroscopic observation of these species. To this end we performed not only standard "gas phase" geometry optimizations but also optimizations employing the polarized continuum model (PCM) to simulate solvation effects, with DMF (ε = 37.2) serving as a representative example. As observed by Steudel, the inclusion of solvation using the PCM approach leads to only minor geometrical adjustments, and for this reason only the "gas phase" structural parameters are presented in the main text (see Figure S1 for PCM-adjusted numbers). Solvation effects, however, have important energetic consequences, favoring structures with large molecular dipoles, and can play a pivotal role in adjusting the balance between structural alternatives which are otherwise closely matched energetically.

2.1.1. $S_2{}^{\bullet-}$ and $S_3{}^{\bullet-}$

Like molecular oxygen, the diatomic molecule S_2 possesses a triplet ground state [57]. Addition of an electron to one of the two half-filled π_g^* orbitals, to afford the $^2\Pi_g$ radical anion $S_2{}^{\bullet-}$, leads to an elongation of the S–S bond, calculated here = 1.996 Å. Attachment of a third sulfur introduces the possibility of structural options for $S_3{}^{\bullet-}$, namely linear ($D_{\infty h}$), equilateral and isosceles triangular (D_{3h} and C_{2v}, respectively); the last is established as the energetically preferred. Structural parameters calculated here for the $^2B_1$ state, an S–S bond distance of 1.984 Å and inter-bond angle of 115.7°, are consistent with previous estimates [58,59].

2.1.2. $S_4^{\bullet-}$

Based on both experimental and theoretical evidence [48,49,60] the geometry of a discrete neutral S_4 molecule displays C_{2v} symmetry, consisting of a planar broken-ring structure with one "long" S–S bond, calculated here = 3.222 Å. The most appealing option for the corresponding $S_4^{\bullet-}$ anion is also a planar C_{2v} structure (Figure 1a), akin to the neutral form but with the "long" S–S bond further stretched (calculated here = 3.505 Å). However, a C_{2h} isomer (Figure 1b), generated from the C_{2v} by a 180° rotation about the central S–S linkage, is also possible.

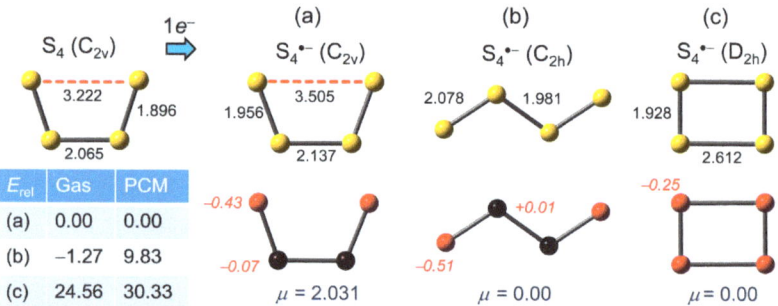

Figure 1. (U)PBE0/D3/def2-QZVP optimized geometries (distances in Å) for $S_4^{\bullet-}$, with Mulliken charges (in italics) and dipole moments μ (in Debye). Relative gas phase and PCM (=DMF) corrected total energies E_{rel} are in kJ mol^{-1}.

Frequency calculations confirm that both the C_{2v} (cis) and C_{2h} (trans) forms are true energetic minima, the latter being slightly more stable in the gas phase. But the energetic competition between the two ceases upon inclusion of PCM (=DMF) solvation, with the centrosymmetric ($\mu = 0$) C_{2h} isomer rising relative to the C_{2v} form by nearly 10 kJ mol^{-1}. The presence of both isomers in solvents with a low dielectric constant may nonetheless be possible. Other structures, based on closed rings, have been explored by previous workers and found to be energetically much more high-lying. Re-examination here of these variants, none of which represents a true energetic minimum, indicates the centrosymmetric ($\mu = 0$) D_{2h} modification (Figure 1c), formed by a Jahn–Teller distortion of a putative D_{4h} geometry [61], is the most stable of the closed-ring group, although it still lies well above the C_{2v} form and, with the inclusion of PCM, its relative energy rises even higher.

2.1.3. $S_5^{\bullet-}$

While the structure of neutral S_5 is unknown, an open envelope-like or chair shape with C_s symmetry has been predicted in previous studies [62,63], with the S–S bond bisected by the mirror plane slightly elongated. We concur with this result, and calculate the unique S–S distance = 2.157 Å. The apparent weakening may be attributed, in valence bond parlance, to lone-pair repulsion arising from the eclipsed alignment of the two neighboring S–S bonds. One-electron reduction to the radical anion $S_5^{\bullet-}$ leads to a variety of structural alternatives, the most obvious involving complete separation (to 4.095 Å) of the already weakened mirror-bisected linkage, to afford the distorted C_s chair illustrated in Figure 2a. Vibrational analysis confirms that the optimized structure represents a true energetic minimum and, as indicated by the associated Mulliken charge densities, negative charge is heavily localized on the two sulfurs associated with the "broken" bond. As expected, charge polarization, and its impact on the molecular dipole, increases with the inclusion of PCM (Table S1).

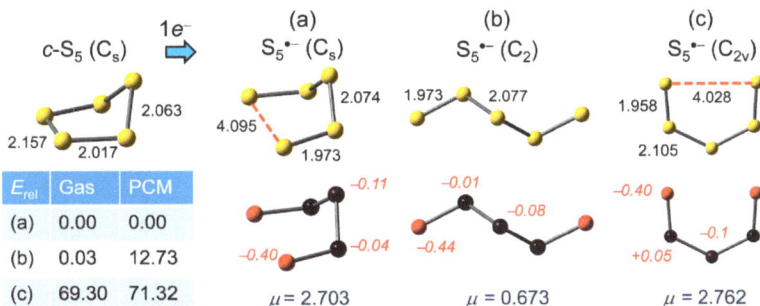

Figure 2. (U)PBE0/D3/def2-QZVP optimized geometries (distances in Å) for $S_5^{\bullet-}$, with Mulliken charges (in italics) and dipole moments μ (in Debye). Relative gas phase and PCM (=DMF) corrected total energies E_{rel} are in kJ mol^{-1}.

Open-chain structures for $S_5^{\bullet-}$ are also possible; several variants have been explored by previous workers, but in our hands these all gravitate on optimization towards the twisted chain (C_2 symmetry) minimum shown in Figure 2b. It is almost co-energetic with the C_s chair, perhaps not surprisingly as the two structures are interconvertible by a 180° rotation of one of the terminal bonds. However, by virtue of the lower dipole moment of the open-chain form, a substantial gap opens with the inclusion of PCM. In addition, we have considered two "forced" planar modifications, one being the cis-cis C_{2v} geometry shown in Figure 2c. While it does not represent a stable minimum, and its relative energy is substantially higher than the related C_s chair, its electronic structure provides a useful conceptual link (vide infra) to the shorter chain anions (n = 2–4). For the corresponding cis-trans isomer, which is isostructural with the closed shell SSNSS$^-$ anion [63,64], the energy gap is considerably less, both in the gas phase (26.1 kJ mol^{-1}) and in DMF (32.6 kJ mol^{-1}), but is still not a true minimum.

2.1.4. $S_6^{\bullet-}$

Here we have the unique advantage of experimental structural information on both the neutral molecule and its radical anion. The cyclic, chair-shaped structure of S_6, with D_{3d} symmetry and all neighboring bonds staggered, has been characterized crystallographically [65]; the observed S–S distance = 2.057(18) Å compares well with the value calculated here = 2.054 Å (Figure 3). In the corresponding radical anion $S_6^{\bullet-}$, identified in the crystal structure of the tetraphenylphosphonium salt [Ph$_4$P][S$_6$], the cyclic chair shape is retained (Figure 3a), despite some disorder, but with two elongated S–S bonds = 2.634(4) Å [47]. In their report, however, the authors cautioned that the apparently high molecular symmetry (C_{2h}) observed for the anion might be dictated by the high lattice symmetry (space group $C2/c$), and provided BP86/TZ2P results indicating that a distorted chair structure (Figure 3b) with C_2 symmetry was actually more stable.

From a theoretical perspective, one-electron reduction of the high-symmetry geometry of neutral S_6 gives rise to an orbitally degenerate ground state for the resulting radical anion $S_6^{\bullet-}$. Thus, when using D_{3h} symmetry constraints as a starting point for a geometry optimization, the symmetric chair immediately breaks symmetry and undergoes a first-order Jahn–Teller distortion [61] to C_{2h} symmetry, affording two elongated S–S bonds, calculated here = 2.357 Å (Figure 3a), somewhat shorter than that observed experimentally. However, as observed earlier, while this centrosymmetric C_{2h} structure represents a stationary point it is not an energy minimum. Upon release of symmetry constraints, it undergoes a second-order distortion to the C_2 modification (Figure 3b), in which one of the two elongated S–S bonds in the C_{2h} geometry stretches further to 2.823 Å, a result in accord with the earlier DFT work [47]. By our calculations the energy difference between the C_{2h} and C_2 structures is large (22.5 kJ mol^{-1}), even in the gas phase, and increases to 27.5 kJ mol^{-1} with the inclusion of PCM (μ = 0 in the C_{2h} form).

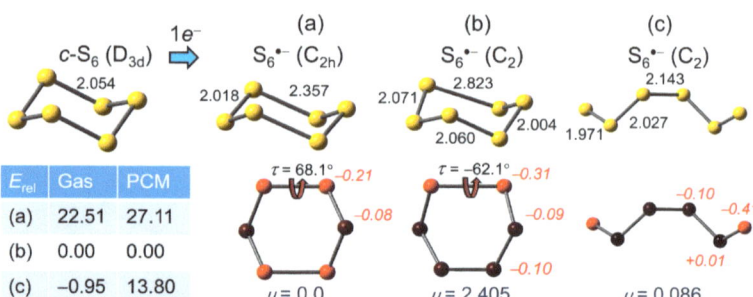

Figure 3. (U)PBE0/D3/def2-QZVP optimized geometries (distances in Å) for $S_6^{\bullet-}$, with Mulliken charges (in italics) and dipole moments (μ) in Debye. Relative gas phase and PCM (=DMF) corrected total energies E_{rel} are in kJ mol^{-1}.

In addition to the nominally closed-ring variants for $S_6^{\bullet-}$ several open chain options have been considered. Of these, we find the lowest energy C_2 structure (Figure 3c), which can be converted into the *quasi*-cyclic form by a ca. 180° rotation about the central S–S bond, constitutes a true minimum. Predictably, in the gas phase the total energies of the two rotamers are almost identical, but in accord with the low dipole moment of the open chain form the balance changes sharply in favor of the ring structure when the PCM is included.

2.1.5. $S_7^{\bullet-}$

Neutral S_7 possesses a chair-like structure with C_s symmetry [66,67], with the unique mirror-bisected bond lengthened to 2.18 Å (calculated here = 2.171 Å) by the effects of lone-pair repulsion occasioned by the eclipsed alignment of the neighboring bonds, as seen in c-S_5. In the structure of the global energetic minimum for $S_7^{\bullet-}$ the cyclic chair motif found in the neutral molecule is retained, but the already weakened mirror-bisected linkage is lengthened to 2.946 Å in the radical anion (Figure 4a), with the associated Mulliken charge densities heavily localized on the two sulfurs linked by the weakened bond.

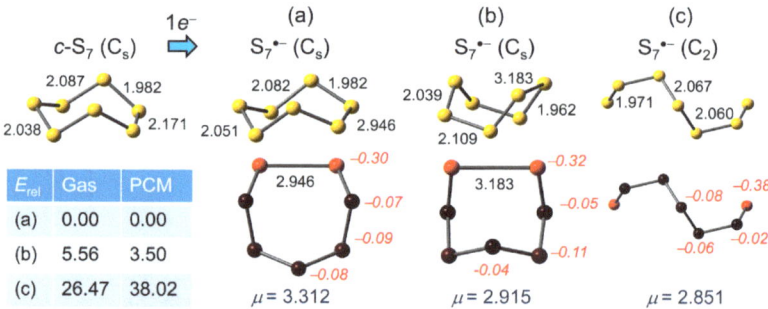

Figure 4. (U)PBE0/D3/def2-QZVP optimized geometries (distances in Å) for $S_7^{\bullet-}$, with Mulliken charges (in italics) and dipole moments (μ) in Debye. Relative gas phase and PCM (=DMF) corrected total energies E_{rel} are in kJ mol^{-1}.

Not surprisingly, a boat-shaped conformation (Figure 4b), in which the unique S–S bond is a little longer (3.183 Å) than in the chair, is also possible. This feature may be of relevance to optical properties, as its relative energy lies only slightly above that of the chair in both the gas phase and solution, so that the two conformers may coexist in equilibrium in solution. Outside of this pair of *quasi*-cyclic structures there is an open chain variant with C_2 symmetry (Figure 4c). It represents a local energy minimum, but is significantly less stable than the chair/boat structures in the gas phase, the gap increasing when PCM is invoked.

2.1.6. $S_8^{\bullet-}$

The eight-membered ring found in orthorhombic α-sulfur displays a classic crown conformation with D_{4d} symmetry, with all neighboring bonds (measured at 2.055(2) Å, calculated here = 2.044 Å) mutually staggered [68]. The structural changes accompanying formation of $S_8^{\bullet-}$ follow a similar pattern to that seen for $S_6^{\bullet-}$. Addition of an electron to the cyclo-S_8 in D_{4d} symmetry affords a degenerate ground state for the resulting radical anion, thereby setting up a first order Jahn–Teller distortion [61], which in this case affords the "squeezed" C_{2v} crown geometry shown in Figure 5a.

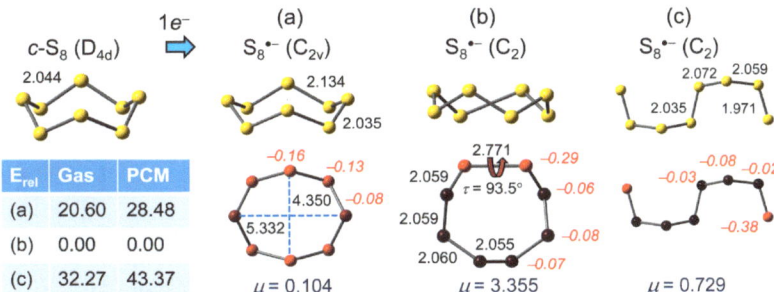

Figure 5. (U)PBE0/D3/def2-QZVP optimized geometries (distances in Å) for $S_8^{\bullet-}$, with Mulliken charges (in italics) and dipole moments (μ) in Debye. Relative gas phase and PCM (=DMF) corrected total energies E_{rel} are in kJ mol^{-1}.

While this high-symmetry structure is not an energy minimum, the possibility of trapping it in a crystal lattice, as in the case of the C_{2h} form of $S_6^{\bullet-}$, is worthy of consideration. That being said, upon release of all symmetry constraints the C_{2v} structure evolves into a C_2 variant (Figure 5b) which, like the C_2 structure of $S_6^{\bullet-}$, displays one elongated S–S bond, calculated here = 2.771 Å. Outside of distorted crown geometries, there are few energetically viable alternatives. Of these, the open-chain C_2-symmetry motif (Figure 5c) represents the only true minimum, but its energy lies well above that of the C_2 crown. Given its relatively low polarity, inclusion of PCM further widens the energy gap.

In summary, the smaller members (n = 3, 4) of the polysulfide radical anion family adopt open chain structures, in part because they have no choice, as there is too much ring strain in the cyclic alternatives. That being said, when alternatives exist, as in the C_{2v} (cis) and C_{2h} (trans) options for $S_4^{\bullet-}$, solvent effects may well dictate the outcome, with the non-centric cis isomer being preferred in polar solvents and the centric trans isomer possibly being viable in non-polar solvents. For medium-sized rings, i.e., n = 5, 6, closed or broken-ring structures compete with open-chain variants, and again the choice may depend upon the polarity of the solvent employed, with polar environments or lattice constraints (for n = 6)) favoring the cyclic or *quasi*-cyclic modifications. In the following sections we focus on the structures most likely preferred in the latter environments. Finally, when n = 7 and 8, the stability of the cyclic structures clearly outranks the open-chain alternatives, regardless of solvent effects. Dynamic equilibria between cyclic and acyclic forms are unlikely, a conclusion which may have consequences for the mechanism of formation of S_8^{2-} [39].

2.2. Electronic Spectra

Using the polar-medium preferred geometries afforded by the unrestricted DFT calculations described above, single point TD-DFT calculations were performed on the radical anions $S_n^{\bullet-}$ (n = 2–8), to explore the number, nature and energies of the possible electronic excitations. A compilation of the relevant states, dominant orbital transitions, frequencies ν, wavelengths λ and oscillator strengths f is provided in Table 1.

Table 1. TD-DFT electronic excitations for $S_n^{\bullet-}$ (n = 2–8).

n in $S_n^{\bullet-}$	State	Excitation [a]	State	ν, eV	λ, nm	f
2 ($D_{\infty h}$)	$^2\Sigma_g$	15β → 17β	$^2\Sigma_u$	3.177	390.3	0.0891
3 (C_{2v})	$^2B_1$	24β → 25β	$^2B_2$	2.059	602.1	0.0920
4 (C_{2v})	$^2A_2$	32β → 33β	$^2B_2$	1.242	998.3	0.0722
	$^2A_2$	29β → 33β	$^2A_1$	3.533	351.0	0.0286
4 (C_{2h})	^2B_g	32β → 33β	^2B_u	0.954	1299.5	0.0930
	^2B_g	29β → 33β	^2A_g	3.578	346.6	0.0000 [b]
5 (C_{2v})	$^2B_1$	40β → 41β	$^2B_2$	0.716	1732.3	0.0734
	$^2B_1$	38β → 41β	$^2A_1$	2.583	479.9	0.0332
	$^2B_1$	39β → 42β	$^2A_1$	3.519	352.4	0.1258
5 (C_s)	$^2A''$	40β → 41β	$^2A''$	0.685	1809.6	0.0589
	$^2A''$	38β → 41β	$^2A''$	1.940	638.2	0.0540
	$^2A''$	37β → 41β	$^2A''$	2.653	467.4	0.0158
6 (C_{2h})	B_g	49α → 50α	A_u	1.231	1007.3	0.0445
	B_g	48β → 49β	B_g	1.978	627.5	0.0000 [b]
6 (C_2)	^2B	48β → 49β	^2B	1.495	829.6	0.0611
	^2B	46β → 49β	^2B	2.283	543.1	0.0308
7 (C_s chair)	$^2A''$	56β → 57β	$^2A''$	1.749	708.7	0.1637
7 (C_s boat)	$^2A''$	56β → 57β	$^2A''$	1.435	863.9	0.1729
8 (C_2 crown)	^2B	64β → 65β	^2B	2.104	589.3	0.1551

[a] Dominant spin-orbital transitions from unrestricted TD-UωB97X-D/PCM/def2-QZVP calculations, with PCM = DMF. [b] Electric dipole forbidden.

As when dealing with geometrical trends, presentation and discussion of the results is developed according to the value of n, beginning with the three short-chain anions (n = 2–4), where the electronic excitations are all clearly $\pi \to \pi^*$. From there on (n = 5–8) the non-planar, distorted or broken-ring geometries militate against the use of conventional σ/π symmetry descriptors which usually aid with band assignments, but for n = 5 the calculated spectrum can still be rationalized by extension of the simple π-only model. Finally, the single elongated S–S linkages found in the *quasi*-cyclic structures (n = 6–8), which are broadly consistent with localized two-center three-electron (2c-3e) bonds, reminiscent of those found in transient organic disulfide radical anions (RS-SR)$^{\bullet-}$ [69,70], give rise to low energy excitations that are best described as $\sigma \to \sigma^*$ processes within the 2c-3e manifold.

2.2.1. $S_n^{\bullet-}$ (n = 2–4)

The origin of the electronic excitations in the short-chain radical anions $S_n^{\bullet-}$ (n = 2–4) can be readily understood with reference to the manifold of π-orbitals predicted by the classical Hückel molecular orbital (HMO) linear chain model [71,72], using linear arrays of overlapping sulfur 3p-orbitals as a basis set. For such systems the eigenvalues e_j are given by the analytical expression $e_j = \alpha + 2\beta \cos(j\pi/N + 1)$, where α and β are the respective Coulomb and resonance parameters, and N is the number of orbitals (atomic centers) in the chain. Schematic plots of the resulting π-energy levels and MOs are illustrated in Figure 6. Within this framework, a single $\pi \to \pi^*$ excitation ν_1 is expected for the diatomic anion $S_2^{\bullet-}$, with a slightly lower energy $n\pi \to \pi^*$ transition ν_1 anticipated for the triatomic chain $S_3^{\bullet-}$. Extrapolation to planar open chain $S_4^{\bullet-}$ systems suggests two excitations ν_1 and ν_2 are possible. Of these, ν_1 is predicted to occur at still lower energy, and its magnitude can be estimated by calibration against the known value of ν_1 for $S_3^{\bullet-}$ (λ_{max} = 615–620 nm in DMF or HMPA) [73]. Based on this simple model the first transition ν_1 in both isomers of $S_4^{\bullet-}$ is predicted to shift well beyond the visible region. For the cis (C_{2v}) isomer a second excitation ν_2 is anticipated towards the UV region, while for the trans (C_{2h}) form ν_2 should not be observed at all, as it is symmetry-forbidden ($g \to g$).

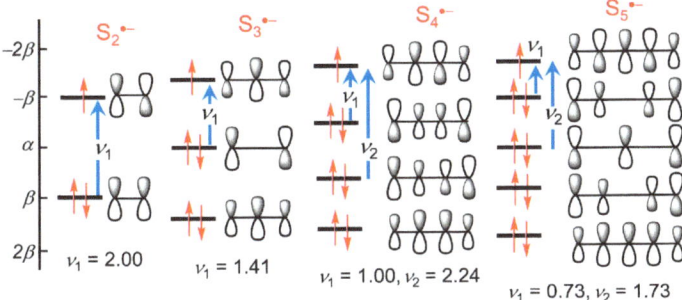

Figure 6. HMO energy levels for arrays of 3*p*-orbitals in model open chain $S_n^{\bullet-}$ (n = 2–5) radical anions; excitation energies ν_1, ν_2 are in units of β.

The TD-DFT calculations refine the qualitative predictions of the HMO model, confirming the nature of the expected transitions (Figure 7) and affording numerical estimates for the $\pi \to \pi^*$ excitation energies involved for both the cis (C_{2v}) and trans (C_{2h}) isomers. Calculated spectra for $S_n^{\bullet-}$ (n = 2–4) are shown in Figure 8.

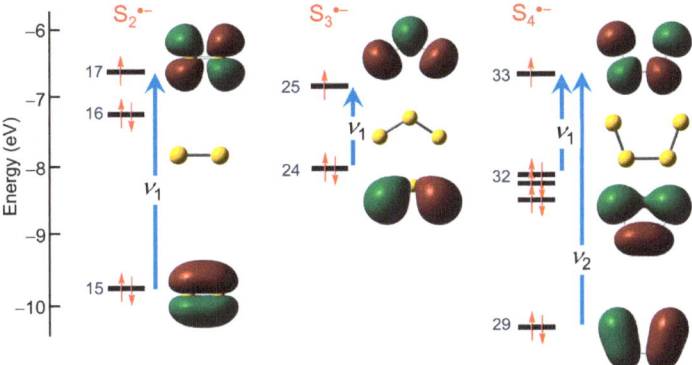

Figure 7. Spin-restricted TD-DFT frontier orbitals and excitations for $S_2^{\bullet-}$, $S_3^{\bullet-}$ and cis-$S_4^{\bullet-}$.

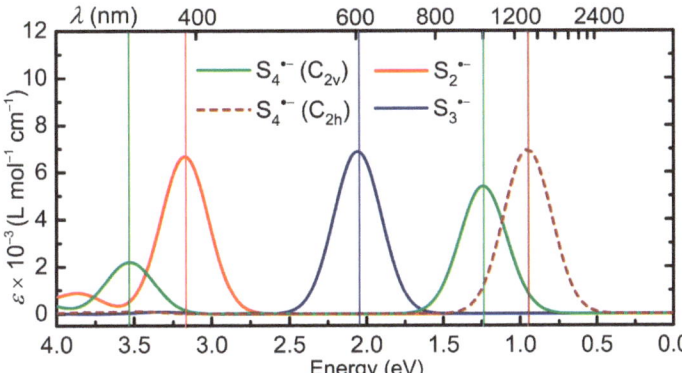

Figure 8. TD-DFT calculated electronic spectra in DMF of $S_n^{\bullet-}$ (n = 2, 3, 4), with HWHM = 0.18 eV; band assignments in Table 1.

The close correspondence between experimental λ_{max} values for $n = 2$ (~400 nm) [18] and $n = 3$ (615–620 nm in DMF or HMPA) [73] and those predicted by TD-DFT (Table 1) provide strong support for the choice of functional, and hence confidence in the calculated values for $n = 4$. The results for $n = 4$ are also in good qualitative agreement with those reported earlier [48,49], and thus help clarify some of the controversies surrounding the spectrophotometric identification of putative $S_4^{\bullet-}$ species. For both the cis and trans isomers, the first transition (ν_1, $32\beta \rightarrow 33\beta$) lies at or beyond the edge of the visible region (998 nm and 1300 nm, respectively), and for the cis isomer the second (ν_2, $29\beta \rightarrow 33\beta$) is predicted to have $\lambda_{max} = 351$ nm, placing it relatively close to $S_2^{\bullet-}$ and also many closed-shell dianionic species, e.g., S_3^{2-} [42], as well as other radical anions, e.g., $S_5^{\bullet-}$ (vide infra), from which it would be hard to distinguish. For the trans isomer, the second transition (ν_2, $29\beta \rightarrow 33\beta$) is symmetry-forbidden and has zero oscillator strength ($f = 0$). It will therefore display no signature at all in the visible region, regardless of its concentration in solution. In this light, assertions that $S_4^{\bullet-}$ has "never been observed" [39,43] by time-resolved spectroelectrochemistry perhaps deserve a second thought; absence of evidence is not evidence of absence.

2.2.2. $S_5^{\bullet-}$

TD-DFT analysis of the optical properties of $S_5^{\bullet-}$, using the coordinates of the chair-shaped C_s structure identified above as the most stable in polar media, affords an electronic spectrum (Figure 9) consisting of a series of bands spread across the entire visible and near-IR regions. However, in contrast to the three short-chain anions already discussed, the chair geometry of $S_5^{\bullet-}$ is not planar (although the molecule is bisected by a mirror plane), as a result of which rigorous characterization of individual orbitals and excitations between them according to their reflection symmetry in that plane, the classical σ/π classification, is no longer possible.

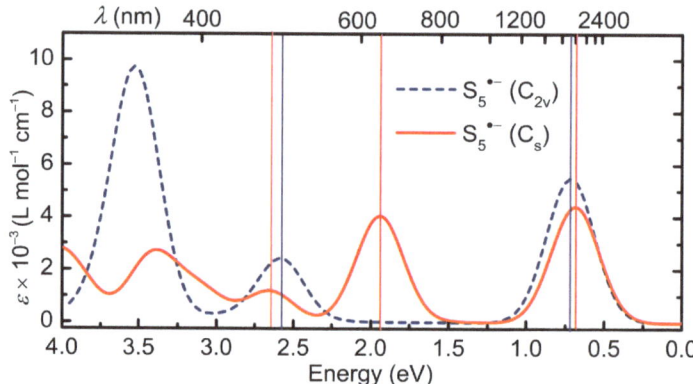

Figure 9. Calculated electronic spectrum of $S_5^{\bullet-}$ in DMF, in C_{2v} and C_s symmetry, with HWHM = 0.18 eV; band assignments in Table 1.

To resolve this difficulty, we examined the orbital make-up and electronic excitations found for the hypothetical planar variation with C_{2v} symmetry. While it is considerably less stable than the C_s form, by virtue of increased lone-pair repulsions, its higher symmetry allows for a clearer evaluation of its spectral signature, particularly in relation to the HMO open-chain model developed above. Indeed, it is immediately apparent by inspection of the frontier orbitals illustrated in Figure 10 that excitations $40\beta \rightarrow 41\beta$ and $38\beta \rightarrow 41\beta$ listed in Table 1 correspond to the two lowest energy $\pi \rightarrow \pi$ excitations ν_1 and ν_2 predicted by the HMO linear chain model with $N = 5$ (Figure 6). As expected, ν_1 lies deep into the near-IR ($\lambda_{max} = 1732$ nm), extending the shift to lower energy seen in cis and trans $S_4^{\bullet-}$ ($\lambda_{max} = 998$ and 1299 nm, respectively), with ν_2 likewise red-shifted to $\lambda_{max} = 478$ nm (from 351 nm in cis $S_4^{\bullet-}$). The third, very intense excitation, from $39\beta \rightarrow 42\beta$, with

λ_{max} = 352 nm, is not related to the chain model, nor even to a $\pi \rightarrow \pi$ transition, but is rather a lone-pair $\sigma \rightarrow \sigma$ process arising from the artificially enforced planarity of the structure.

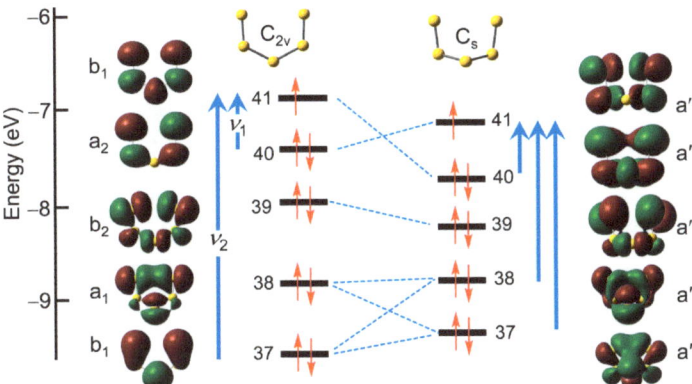

Figure 10. Correlation of spin-restricted TD-DFT frontier orbitals and electronic excitations for $S_5^{\bullet-}$ in C_{2v} and C_s symmetry.

With this information in hand, the origin of the optical signature of the C_s form emerges. The three lowest-lying states can each be described in terms a single dominant excitation from one of the doubly occupied molecular orbitals to the singly occupied molecular orbital (SOMO), that is, the lowest unoccupied molecular orbital (LUMO) for the unrestricted β-spins listed in Table 1. Moreover, correlation of the orbitals for the C_{2v} and C_s geometries confirms that the HMO chain model still applies, albeit more loosely because of the loss of planarity and consequent σ/π mixing. Thus, while the ordering of orbitals 40 and 41 is reversed, the first excitation, $40\beta \rightarrow 41\beta$ (λ_{max} = 1810 nm) can be considered a *quasi-* $\pi \rightarrow \pi$ transition related to ν_1 in the HMO model. The next two, $38\beta \rightarrow 41\beta$ (λ_{max} = 638 nm) and $37\beta \rightarrow 41\beta$ (λ_{max} = 467 nm), also involve heavily hybridized orbitals, but both are *quasi-* $\pi \rightarrow \pi$ processes that can be traced back to ν_2. The higher energy (>3 eV) absorptions comprise a series of less well-defined states arising from multiple excitations (see Table S2).

2.2.3. $S_6^{\bullet-}$

Addressing the optical properties of the $S_6^{\bullet-}$ anion presents a quandary. The crystallographic evidence indicates a symmetric chair structure with C_{2h} symmetry, while DFT optimizations point towards a distorted C_2 version. There are merits to both positions. In solution, and in the absence of environmental constraints, the lower-symmetry C_2 geometry is probably preferred, but the high space group symmetry of the [Ph_4P][S_6] salt appears to hold the chair in the higher-symmetry C_{2h} form. In that light we have performed TD-DFT calculations on both options, using geometries taken from the respective structural optimizations.

As a first step, however, we focus on a qualitative model for describing the two elongated bonds in the symmetric structure. Building on the ideas developed earlier by Dehnicke and coworkers [47], Figure 11a illustrates the two strongly coupled σ and σ^* orbitals arising from combinations of two S_3 fragments. A second-order Jahn–Teller distortion from C_{2h} to C_2 will give rise to mixing of the b_g SOMO and b_u LUMO, and a widening of the energy gap between them. Figure 11b refines this model, by showing the relevant spin-restricted Kohn–Sham orbitals and eigenvalues for the C_2 structure, that is, two heavily hybridized, but basically S–S σ-bonding, occupied orbitals (46 and 48), and a more localized σ^*-orbital (49).

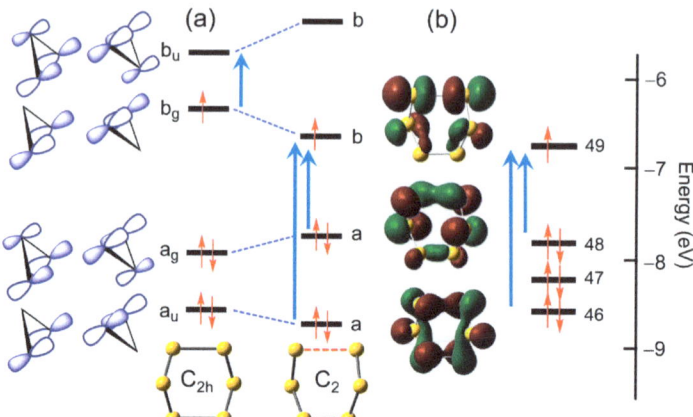

Figure 11. (a) Sketch of frontier orbital energies following distortion of $S_6^{\bullet-}$ from C_{2h} to C_2 symmetry. (b) Spin-restricted TD-DFT frontier orbitals and electronic excitations for $S_6^{\bullet-}$ in C_2 symmetry.

Given this conceptual framework, the optical signatures predicted for the two geometries are readily explained. As shown in Figure 12, the C_{2h} structure displays a single well-resolved band with λ_{max} = 1007 nm, which corresponds *not* to electron promotion from the HOMO to the SOMO, which is symmetry-forbidden ($g \rightarrow g$) in C_{2h}, but rather to the SOMO-to-LUMO excitation shown in Figure 11a ($49\alpha \rightarrow 50\alpha$, Table 1). In addition, there are a series of less well-defined states that give rise to a broad absorption that extends into the UV region.

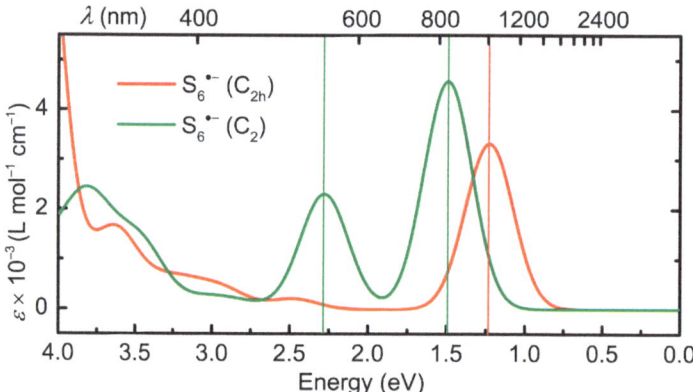

Figure 12. Calculated electronic spectrum of $S_6^{\bullet-}$ in DMF, in C_{2h} and C_2 symmetry, with HWHM = 0.18 eV; band assignments in Table 1.

For the distorted C_2 symmetry structure, the excited state manifold is quite different. Two bands are predicted in the visible and near-IR region (λ_{max} = 543 nm and 830), which arise primarily from the (now allowed) excitations, $46\beta \rightarrow 49\beta$ and $48\beta \rightarrow 49\beta$ (Table 1), from occupied orbitals to the SOMO, both of which are essentially $\sigma \rightarrow \sigma^*$ processes. As in the case of the C_{2h} geometry, there is a broad band extending into the UV region associated with a series of higher energy but less well-defined states.

In summary, the optical properties of both variations of the $S_6^{\bullet-}$ radical anion are associated with transitions associated with the lengthened S–S σ-bonds. In both its excited and ground states, the $S_6^{\bullet-}$ radical anion behaves like a cyclic molecule.

2.2.4. $S_7^{\bullet-}$ and $S_8^{\bullet-}$

The two largest radical anions (n = 7 and 8) are the easiest to analyze, as the structural perturbations occasioned by addition of an electron to the parent homocycles are small. Based on the structural parameters provided by the DFT optimizations of the chair and boat conformers of $S_7^{\bullet-}$, both of Cs symmetry, and of the C_2 distorted crown geometry of $S_8^{\bullet-}$, all three rings experience a lengthening of one of the S–S bonds, an effect which can best be described in terms of the formation of a largely localized 2c-3e σ-bond.

The TD-DFT calculations reinforce this picture, providing a description for the first excited state which involves promotion of an electron between the associated σ- and σ^*- orbitals of the 2c-3e manifold, that is, the β-spin HOMO and LUMO of the two conformers of $S_7^{\bullet-}$ (56β → 57β) and those of $S_8^{\bullet-}$ (64β → 65β) shown in Figure 13. These transitions give rise to single bands with large oscillator strength in the low-energy visible or near-IR region (Figure 14). As expected, there is a notable difference between the band maxima of the chair (λ_{max} = 709 nm) and boat (λ_{max} = 864 nm) conformations of $S_7^{\bullet-}$ which can be traced back to the longer S···S separation found in the latter (Figure 4).

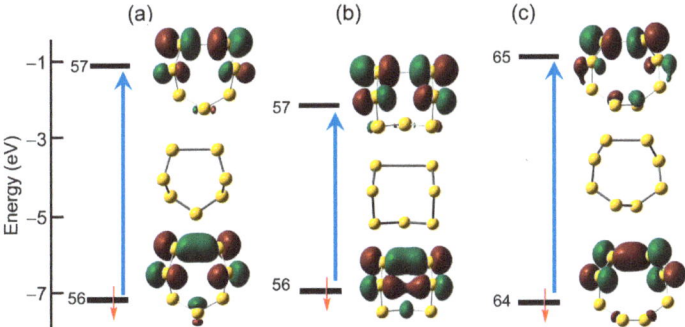

Figure 13. Unrestricted TD-DFT frontier orbitals and calculated $\sigma \to \sigma^*$ electronic excitations found for the (**a**) chair and (**b**) boat conformations of $S_7^{\bullet-}$ and (**c**) the C_2-distorted crown structure of $S_8^{\bullet-}$.

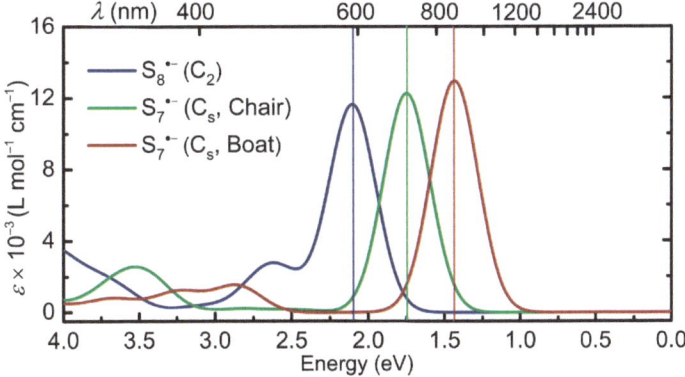

Figure 14. Calculated electronic spectra in DMF of $S_7^{\bullet-}$ (chair and boat conformations) and $S_8^{\bullet-}$, with HWHM = 0.18 eV; band assignments in Table 1 and Table S2.

The higher transition energy predicted for $S_8^{\bullet-}$ (λ_{max} = 589 nm) can be attributed to a similar effect, the shorter S···S separation stemming from the mutual staggering of the neighboring bonds and consequent relief from the effects of lone-pair repulsion. Secondary, less intense absorptions, with λ_{max} = 474 nm ($S_8^{\bullet-}$), 358 nm ($S_7^{\bullet-}$, chair) and 435 nm ($S_7^{\bullet-}$, boat), are also predicted. These are associated with poorly defined, higher-lying states (Table S2), but their presence may aid in identification. That being said, the strong low energy $\sigma \to \sigma^*$ excitation, which dominates the spectrum of all three species, represents the most distinguishing optical feature.

3. Discussion

As indicated in the introduction, there are considerable differences of opinion as to when and where the radical anions $S_n^{\bullet-}$ ($n = 4$–8) might be found. While their participation in the multistep redox equilibria associated with the operation of sulfide/polysulfide photoconductor cells, alkali-metal/sulfur batteries, as well as in many biological and organic transformations, is frequently implied, their identification in these complex systems has, not surprisingly, proved elusive. Equally, extensive and detailed spectroscopic and spectroelectrochemical studies, utilizing a wide range of techniques (optical, IR, Raman, EPR spectra), have been unable to provide decisive answers.

The purpose of the present work has been two-fold. Firstly, the most stable structures of the putative radical anions $S_n^{\bullet-}$ ($n = 4$–8) in polar solvents have been identified using high-level DFT methods. Secondly, TD-DFT calculations, performed on the most stable structural candidates, have been used to map out the number, nature and energies of the photochemically accessible excited states for these species. Critical to the validity of this latter step was the ability to assess the numerical reliability of the methods used (choice of LC-functional for TD-DFT work) by comparison of the predicted excitation energies with experimental values for the well-known radical anions $S_2^{\bullet-}$ and $S_3^{\bullet-}$. Even so, we neither expect nor claim that the present TD-DFT calculated transition energies will provide a perfect match with experiment for the larger members of the series, especially for the low-energy (near-IR and beyond) excitations. Taken together, however, the results on the entire series of anions $S_n^{\bullet-}$ ($n = 2$–8) provide a frame of reference for distinguishing between different members of the family. Equally important, from an interpretational viewpoint, has been the use of the classical one-electron HMO chain model [74,75] to anticipate both the number and *approximate* energies of $\pi \to \pi$ transitions, again using $S_2^{\bullet-}$ and $S_3^{\bullet-}$ as reference points. The TD-DFT results suggest that the calculated spectra for $S_4^{\bullet-}$, and even $S_5^{\bullet-}$, can be effectively rationalized by this approach. By contrast, the low energy excitations predicted for the essentially cyclic structures of $S_n^{\bullet-}$ ($n = 6$–8) are best described in terms of $\sigma \to \sigma^*$ processes within a relatively localized 2c-3e manifold.

The availability of this information opens the door to the design of experimental strategies for the generation, observation and perhaps even isolation of the radical anions $S_n^{\bullet-}$ ($n = 4$–8). We begin by considering the seminal 1991 report by Rauchfuss and coworkers on the structure and spectroscopic properties of the open-chain octasulfide dianion S_8^{2-} in the absence of counterion pairing effects [48]. As noted earlier, these authors attributed the strong band with $\lambda_{max} = 618$ nm that emerged upon dilution of a solution of [Mn(N-MeIm)$_6$][S$_8$] (N-MeIm = N-methylimidazole) in N-MeIm to the presence of the radical anion $S_3^{\bullet-}$. At the time they rationalized the generation of $S_3^{\bullet-}$ in terms of a disproportionation of S_8^{2-} to $\frac{1}{4}S_8$ and S_6^{2-}, and subsequent dissociation of the latter, following the conventional interpretation of the electrochemistry community [35,49]. Other radical anions, notably $S_n^{\bullet-}$ ($n = 4, 5$), were not included in the analysis, in part because their optical signatures were unknown. Given the present TD-DFT results, however, the potential involvement of these seemingly missing radical anions can be examined. In particular, we consider the possibility that both might be formed, along with $S_3^{\bullet-}$, by either symmetric or asymmetric dissociation of the S_8^{2-} dianion, as indicated in Scheme 1. In addition to the overall thermodynamics of such processes [43], mechanistic considerations may also be important—how easy is it to rupture the distinct S–S bonds along the chain? In response to this question, we suggest that dissociation may proceed via four-center intermediates, as illustrated in Scheme 2. Indeed, in the case of symmetric dissociation, an example of a four-center π-dimer has been characterized crystallographically [21].

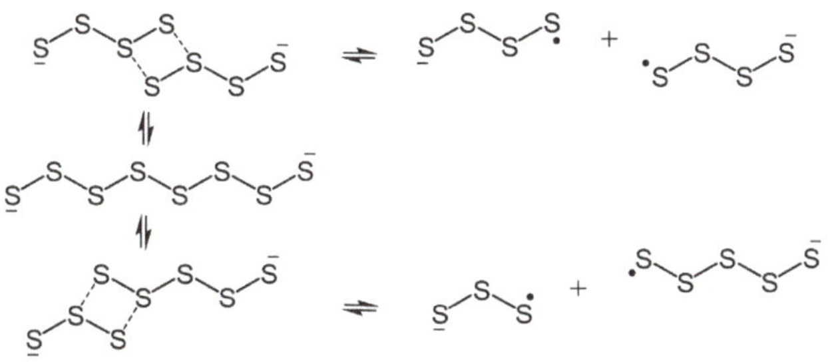

Scheme 2. Symmetric and asymmetric dissociation of S_8^{2-} to the radical anions $S_n^{\bullet -}$ ($n = 3, 4, 5$).

As an example of the spectroscopic ramifications of this interpretation, we compare in Figure 15 the experimental spectrum for the highly diluted solution of S_8^{2-}, as reported by Rauchfuss, with an equally weighted composite of the TD-DFT calculated spectra for $S_3^{\bullet -}$ and $S_5^{\bullet -}$, the two products of an *asymmetric* dissociation. In the mid-range visible region, the correspondence is remarkable, not only in terms of the overlap and coalescence of the two bands calculated for $n = 3, 5$, but also the presence of the weaker band near 480 nm which, on the present basis, may be assigned to $n = 5$ (calculated $\lambda_{max} = 467$ nm). Below 400 nm the match is less than ideal, but could be improved by inclusion into the composite of cis $S_4^{\bullet -}$ (calculated $\lambda_{max} = 351$ nm), the unique *symmetric* dissociation product. Alternatively, these higher energy absorptions may arise from *un*dissociated S_6^{2-}. That being said, the absence of bands attributable to $S_2^{\bullet -}$ or $S_6^{\bullet -}$ suggests dissociation of S_8^{2-} into these species does not occur to any great extent.

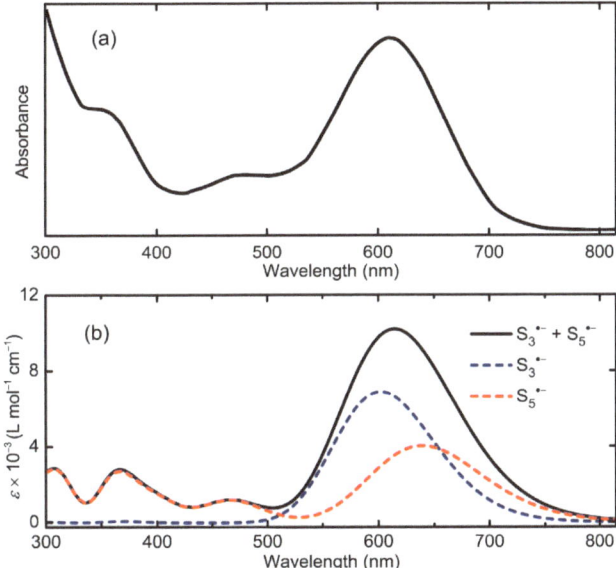

Figure 15. (a) Optical spectrum of 0.001 M solution of [Mn(*N*-MeIm)$_6$][S$_8$] in *N*-MeIm, reproduced from [48]; copyright American Chemical Society. (b) Calculated composite spectrum, with HWHM = 0.18 eV, for equally weighted mixture of $S_3^{\bullet -}$ and $S_5^{\bullet -}$ in DMF.

By itself, this single spectral deconvolution exercise does not constitute proof for multiple dissociation pathways, but critical support for the concept could be readily achieved by inspection of the near-IR region of dilute solutions of S_8^{2-}, where one or both of the low-energy (ν_1) bands of $S_4^{\bullet -}$ and $S_5^{\bullet -}$ should be present. Moreover, similar spectroscopic analysis at high dilution of solutions of salts of the known hepta- and hexasulfide dianions

S_7^{2-} and S_6^{2-} should reveal predicable patterns of radical anions, the former affording $S_3^{\bullet-}$ and $S_4^{\bullet-}$ (but possibly not $S_5^{\bullet-}$) and the latter specifically $S_3^{\bullet-}$; indeed, for S_6^{2-} this result has already been confirmed [48]. In the same way, observation of the cyclic radical anions $S_n^{\bullet-}$ (n = 6–8) may be possible, by examination of highly diluted solutions of long chain dianions such S_{10}^{2-} [49] and S_{12}^{2-} [50].

In addition to routes to the radical anions which rely solely on dissociative equilibria between radical anions and closed-shell dianions, direct chemical synthesis may be possible, as in the case of the radical anion salt $[Ph_4P][S_6]$, which was prepared by the rather unusual reaction of H_2S and Me_3SiN_3 in the presence of $[Ph_4P][N_3]$ [52]. If this procedure could be adapted to incorporate the use of other bulky cations, e.g., PPN^+, different crystal morphologies might be generated. That being the case, would the structure of the resulting anion be constrained to C_{2h} symmetry, as in $[Ph_4P][S_6]$, or display the more stable distorted C_2 shape? Even in the solid state, the expected optical signatures (Figure 12) are predicted to be quite different. Alternatively, not only $S_6^{\bullet-}$ but also $S_7^{\bullet-}$ and $S_8^{\bullet-}$ could be accessible by electrochemical reduction of the appropriate neutral allotrope [76–78]. The latter two anions have very distinct optical profiles, although we add the caveat that the footprint of $S_8^{\bullet-}$ may easily be confused with that of $S_3^{\bullet-}$.

Chemical reduction methods, for example using organometallic reducing agents such as cobaltocene, which is known to afford salts of closed [79] and open shell anions [80] with sulfur-based electron acceptors, may also provide access to salts of $S_8^{\bullet-}$. In this connection Woollins et al. obtained the salt $[Cp_2Co][S_3N_3]$ from the reaction of S_4N_4 with cobaltocene in THF [81]. This transformation probably involves the initial generation of the radical anion $S_4N_4^{\bullet-}$, known from electrochemical studies to be formed by one-electron reduction of S_4N_4, which then undergoes ring contraction to produce $S_3N_3^-$ [77].

Just as chemical oxidation of cyclo-S_8 has afforded the radical cation $S_8^{\bullet+}$ [78], treatment of salts of the dianions S_7^{2-} [82] and S_8^{2-} [47,48] with mild oxidants, e.g., iodine or N-bromosuccinimide, might well yield the corresponding radical anions $S_7^{\bullet-}$ and $S_8^{\bullet-}$. An alternative to chemical oxidation is the use of photolysis to generate polysulfide radical anions from the corresponding dianions. This approach is based on the recent work of Chiba and co-workers on the production of polysulfide radical anions $S_n^{\bullet-}$ (n = 3, 4) [25–27] via photolysis of the polysulfide dianion S_4^{2-}, as well as on insights provided by investigations into the photoelectrochemical oxidation of S^{2-} by metal-sulfide quantum dots [14,15].

Last, but not least, we acknowledge the role that serendipity has played in the advancement of the chemistry of polysulfide radical anions. For example, the procedures used to achieve the isolation and characterization of $S_6^{\bullet-}$ [52] and of the π-dimer of $S_4^{\bullet-}$ [21] would have been difficult to predict a priori, but their somewhat fortuitous discovery strengthens the conviction that continued exploration will yield new insights. Towards that end the present results may prove useful.

4. Computational Methods

Unrestricted density functional theory (DFT) calculations were performed with the Gaussian 16 suite of programs [74], using the default ultrafine integration grids. Geometry optimizations employed the hybrid-adapted Perdew–Burke–Ernzerhof functional (PBE0) [75,83] and Ahlrichs' quadruple-ζ valence def2-QZVP basis set [84], without additional diffuse functions [85] but with Grimme's empirical correction (D3) [81,86] included to account for possible dispersion effects [87]. For most anions several geometries, both cyclic and acyclic, were considered, and wherever a stationary point was located a full vibrational analysis was performed to determine whether or not it corresponded to a true energy minimum. Preferred geometries were further optimized with the inclusion of the polarized continuum model (PCM) [88] to account for the effects of solvation, dimethylformamide (DMF) being set as a representative polar solvent. Listings of total energies, vibrational frequencies and cartesian coordinates, with and without PCM, are provided in the SI.

The optical properties of the polysulfide radical anions were explored using single point unrestricted time-dependent (TD) DFT calculations, with the same def2-QZVP basis set and PCM included. The use of several long range corrected (LC) functionals, which are known to provide reasonable estimates of low energy (charge transfer, Rydberg-like) excitations in molecular species [89,90], including radicals [91] and sulfur-containing radical anions [92], was explored. The best results, reported here, employed the empirical dispersion-corrected density functional ωB97XD [93], which is well recognized for its overall performance [94]. All tabulated excitation energies refer to spin-unrestricted calculations, but for ease of visualization some of the orbital energy diagrams are based on spin-restricted wavefunctions. All spectral plots, prepared using Gaussview 6 [95], employed Gaussian band shapes with the half-width at half maximum (HWHM) value set at 0.18 eV. The associated extinction coefficients were derived using routines available within Gaussview. Kohn–Sham wavefunctions were also plotted using Gaussview.

5. Conclusions

The DFT and TD-DFT calculations reported here represent the first comprehensive attempt both to predict and to rationalize the optical properties of the entire family of polysulfide radical anions $S_n^{\bullet-}$ (n = 2–8). Our results confirm earlier predictions [52,53] that the first $\pi \to \pi$ transition for both the cis (C_{2v}) and slightly less stable trans (C_{2h}) isomers of $S_4^{\bullet-}$ should occur in the near-IR region. However, a second $\pi \to \pi$ transition at around 350 nm is expected for the cis isomer of $S_4^{\bullet-}$. Based on Seel's early results [19,20], a band near 490 nm has often been attributed to this species, but these conclusions are questionable [3]. At least in dilute solution this band may originate from $S_5^{\bullet-}$, the most stable form of which possesses an acyclic structure with C_s symmetry, and is predicted to display three optical absorption bands, two in the visible and one in the near-IR region.

The $S_6^{\bullet-}$ radical anion is an interesting and unique example of a polysulfide radical anion that has been structurally characterized in the solid state. In the ion-separated salt [Ph$_4$P][S$_6$] the anion displays a cyclic structure (C_{2h} symmetry) with two long S–S bonds [51], while DFT geometry optimization points to a distorted cyclic structure (C_2 symmetry) with one long S–S bond as a more stable arrangement. The predicted electronic spectra for these two forms are very different, with λ_{max} = 1007 nm vs. 830 and 543 nm, respectively.

To date the heptasulfide radical anion $S_7^{\bullet-}$ has received scant attention, but the present DFT results point to a cyclic structure with two energetically similar conformers, chair and boat (Cs symmetry), both displaying one long S–S bond best described in terms of a localized 2c-3e σ-interaction. Electronic excitation within this manifold gives rise to a strong visible/near-IR absorption with calculated values of λ_{max} = 709 nm and 864 nm for chair and boat, respectively. The octasulfide radical anion $S_8^{\bullet-}$, which carries particular significance as the initial product of the electrochemical reduction of cyclo-S$_8$ [3], is also predicted to possess a distorted cyclic structure (C_2 symmetry) exhibiting, like $S_7^{\bullet-}$, a single elongated 2c-3e S–S bond. The associated $\sigma \to \sigma^*$ excitation generates a strong visible absorption band with a calculated λ_{max} = 589 nm.

Supplementary Materials: The following supporting information (21 pages total) can be downloaded at: https://www.mdpi.com/article/10.3390/molecules28155654/s1, Figure S1: Optimized geometrical parameters; Table S1: Total electronic energies; Table S2: Excitation energies, oscillator strengths and orbital contributions; Tables S3–S11: Gaussian archive entries; Tables S12–S17: Frequency calculations.

Author Contributions: The two authors contributed equally to this work. All authors have read and agreed to the published version of the manuscript.

Funding: Financial support from NSERC (Canada).

Data Availability Statement: Details of the computational data are in the Supplementary Materials.

Conflicts of Interest: The authors declare no conflict of interest.

References

1. Chivers, T. Ubiquitous trisulphur radical ion $S_3^{\bullet-}$. *Nature* **1974**, *252*, 32–33. [CrossRef]
2. Steudel, R. Inorganic polysulfides S_n^{2-} and radical anions $S_n^{\bullet-}$. In *Elemental Sulfur and Sulfur-Rich compounds II. Topics in Current Chemistry*; Steudel, R., Ed.; Springer: Berlin/Heidelberg, Germany, 2003; Volume 231, pp. 127–152.
3. Steudel, R.; Chivers, T. The role of polysulfide dianions and radical anions in the chemical, physical and biological sciences, including sulfur-based batteries. *Chem. Soc. Rev.* **2019**, *48*, 3279–3319. [CrossRef] [PubMed]
4. Laitinen, R.S.; Oilunkaniemi, R.; Chivers, T.; McGeachie, L.; Kelly, P.F.; King, R.S.P. Polychalcogen molecules, ligands, and ions Part 2: Catenated acyclic molecules, ions, and p-block element derivatives. In *Reference Module in Chemistry, Molecular Sciences and Chemical Engineering*; Elsevier: Amsterdam, The Netherlands, 2023; pp. 970–1020.
5. Zheng, D.; Wang, G.; Liu, D.; Si, J.; Ding, T.; Qu, D.; Yang, X.; Qu, D. The progress of Li−S batteries—Understanding of the sulfur redox mechanism: Dissolved polysulfide ions in the electrolytes. *Adv. Mater. Technol.* **2018**, *3*, 1700233. [CrossRef]
6. Zhao, E.; Nie, K.; Yu, X.; Hu, Y.-S.; Wang, F.; Xiao, J.; Li, H.; Huang, X. Advanced characterization techniques in promoting mechanism understanding for lithium-sulfur batteries. *Adv. Func. Mater.* **2018**, *28*, 1707543. [CrossRef]
7. Zhang, G.; Zhang, Z.-W.; Peng, H.-J.; Huang, J.-Q.; Zhang, Q. A toolbox for lithium-sulfur research: Methods and protocols. *Small Methods* **2017**, *1*, 1700134. [CrossRef]
8. Song, P.; Rao, W.; Chivers, T.; Wang, S.-Y. Applications of trisulfide radical anion $S_3^{\bullet-}$ in organic chemistry. *Org. Chem. Front.* **2023**, *10*, 3378–3401. [CrossRef]
9. Bogdándi, V.; Ida, T.; Sutton, T.R.; Bianco, C.; Ditrói, T.; Koster, G.; Henthorn, H.A.; Minnion, M.; Toscano, J.P.; van der Vliet, A.; et al. Speciation of reactive sulfur species and their reactions with alkylating agents: Do we have any clue about what is present in the cell? *Br. J. Pharmacol.* **2019**, *176*, 646–670. [CrossRef] [PubMed]
10. Cortese-Krott, M.M.; Kuhnle, G.G.C.; Dyson, A.; Fernandez, B.O.; Grman, M.; DuMond, J.F.; Barrow, M.P.; McLeod, G.; Nakagawa, H.; Ondrias, K.; et al. Key bioactive reaction products of the NO/H_2S interaction are S/N hybrid species, polysulfides, and nitroxyl. *Proc. Natl. Acad. Sci. USA* **2015**, *112*, E4651–E4660. [CrossRef] [PubMed]
11. Pokrovski, G.S.; Dubrovinsky, L.S. The $S_3^{\bullet-}$ ion is stable in geological fluids at elevated temperatures and pressures. *Science* **2011**, *331*, 1052–1054. [CrossRef] [PubMed]
12. Pokrovski, G.S.; Kokh, M.A.; Guillaume, D.; Borisova, A.Y.; Gisquet, P.; Hazemann, J.-L.; Lahera, E.; Del Net, W.; Proux, O.; Testemale, D.; et al. Sulfur radical species form gold deposits on earth. *Proc. Nat. Acad. Sci. USA* **2015**, *112*, 13484–13489. [CrossRef]
13. Pokrovski, G.S.; Kokh, M.A.; Desmaele, E.; Laskar, C.; Bazarkina, E.F.; Borisova, A.Y.; Testemale, D.; Hazemann, J.-L.; Vuilleumier, R.; Ferlat, G.; et al. The trisulfur radical ion $S_3^{\bullet-}$ controls platinum transport by hydrothermal fluids. *Proc. Nat. Acad. Sci. USA* **2021**, *118*, e2109768118. [CrossRef] [PubMed]
14. Li, X.; McNaughter, P.D.; O'Brien, P.; Minamimoto, H.; Murakoshi, K. Photoelectrochemical formation of polysulfide at PbS QD-sensitized plasmonic electrodes. *J. Phys. Chem. Lett.* **2019**, *10*, 5357–5363. [CrossRef] [PubMed]
15. Chakrapani, V.; Baker, D.; Kamat, P.V. Understanding the role of the sulfide redox couple (S^{2-}/S_n^{2-}) in quantum dot-sensitized solar cells. *J. Am. Chem. Soc.* **2011**, *133*, 9607–9615. [CrossRef]
16. Reinen, D.; Lindner, G.-G. The nature of the chalcogen colour centres in ultramarine type solids. *Chem. Soc. Rev.* **1999**, *28*, 75–84. [CrossRef]
17. Chukanov, N.V.; Sapozhnikov, A.N.; Shendrik, R.Y.; Vigasina, M.F.; Steudel, R. Spectroscopic and crystal-chemical features of sodalite-group minerals from gem lazurite deposits. *Minerals* **2020**, *10*, 1042. [CrossRef]
18. Chivers, T.; Elder, P.J.W. Ubiquitous trisulfur radical anion: Fundamentals and applications in materials science, electrochemistry, analytical chemistry, and geochemistry. *Chem. Soc. Rev.* **2013**, *42*, 5996–6005. [CrossRef] [PubMed]
19. Seel, F. Polysulfide radical anions. *Angew. Chem. Int. Ed.* **1973**, *12*, 420–421. [CrossRef]
20. Seel, F.; Guttler, H.-J.; Simon, G.; Wieckowski, A. Colored sulfur species. *Pure Appl. Chem.* **1977**, *49*, 45–54. [CrossRef]
21. Schwamm, R.J.; Lein, M.; Coles, M.P.; Fitchett, C.M. Bismuth (III) complex of the $S_4^{\bullet-}$ radical anion: Dimer formation via pancake bonds. *J. Am. Chem. Soc.* **2017**, *139*, 16490–16493. [CrossRef] [PubMed]
22. Clark, R.J.H.; Dines, T.J.; Kurmoo, M. The nature of the sulfur chromophore in ultramarine blue, green, violent, and pink and of the selenium chromophore in ultramarine selenium. Characterization of radical anions by electronic and resonance Raman spectroscopy and the determination of their excited-state geometries. *Inorg. Chem.* **1983**, *22*, 2766–2772.
23. Rauh, R.D.; Shuker, F.S.; Marston, J.M.; Brummer, S.B. Formation of lithium polysulfides in aprotic media. *J. Inorg. Nucl. Chem.* **1977**, *39*, 1761–1766. [CrossRef]
24. Li, H.; Tang, X.; Pang, J.H.; Wu, X.; Yeow, E.K.L.; Wu, J.; Chiba, S. Polysulfide anions as visible light photoredox catalysts for aryl cross-couplings. *J. Am. Chem. Soc.* **2021**, *143*, 481–487. [CrossRef] [PubMed]
25. Li, H.; Liu, Y.; Chiba, S. Leveraging of sulfur anions in photoinduced molecular transformations. *JACS Au* **2021**, *1*, 2121–2129. [CrossRef] [PubMed]
26. Li, H.; Liu, Y.; Chiba, S. Anti-Markovnikov hydroarylation of alkenes via polysulfide anion photocatalysis. *Chem. Commun* **2021**, *57*, 6264–6267. [CrossRef] [PubMed]
27. Li, H.; Chiba, S. Synthesis of α-tertiary amines by polysulfide anions photocatalysis via single-electron transfer and hydrogen atom transfer in relays. *Chem. Catal.* **2022**, *2*, 1128–1142. [CrossRef]

28. Hagen, M.; Schiffels, P.; Hammer, M.; Dörfler, S.; Tübke, J.; Hoffmann, M.; Althues, H.; Kaskel, S. In-situ Raman investigation of polysulfide formation in Li-S cells. *J. Electrochem. Soc.* 2013, *160*, A1205–A1215. [CrossRef]
29. Clark, R.J.H.; Cobbold, D.G. Characterization of sulfur radical anions in solutions of alkali polysulfides in DMF and HMPA and in the solid state in ultramarine blue, green, and red. *Inorg. Chem.* 1978, *17*, 3169–3174. [CrossRef]
30. Wu, H.-L.; Huff, L.A.; Gewirth, A.A. In-situ Raman spectroscopy of sulfur speciation in Li-S batteries. *Appl. Mater. Interfaces* 2015, *7*, 1709–1719. [CrossRef]
31. Levillain, E.; Leghié, P.; Gobeltz, N.; Lelieur, J.-P. Identification of the $S_4^{\bullet-}$ radical anion in solution. *New J. Chem.* 1997, *21*, 335–341.
32. Wujcik, K.H.; Wang, D.R.; Raghunathan, A.; Drake, M.; Pascal, T.A.; Prendergast, D.; Balsara, N.P. Lithium polysulfide radical anions in ether-based solvents. *J. Phys. Chem. C* 2016, *120*, 18403–18410, and references cited therein. [CrossRef]
33. Chukanov, N.V.; Shendrik, R.Y.; Vigasina, M.F.; Pekov, I.V.; Sapozhnikov, A.N.; Shcherbakov, V.D.; Varlamov, D.A. Crystal chemistry, isomorphism, and thermal conversions of extra-framework components in sodalite-group minerals. *Minerals* 2022, *12*, 887. [CrossRef]
34. Merritt, M.V.; Sawyer, D.T. Electrochemical reduction of elemental sulfur in aprotic solvents. Formation of a stable $S_8^{\bullet-}$ species. *Inorg. Chem.* 1970, *9*, 211–215. [CrossRef]
35. Martin, R.P.; Doub, W.H.; Roberts, J.L.; Sawyer, D.T. Further studies of the electrochemical reduction of sulfur in aprotic solvents. *Inorg. Chem.* 1973, *12*, 1921–1924. [CrossRef]
36. Bonnaterre, R.; Cauquis, G. Spectrophotometric study of the electrochemical reduction of sulphur in organic media. *J. Chem. Soc. Chem. Commun.* 1972, 293–294. [CrossRef]
37. Manan, N.S.A.; Aldous, L.; Alias, Y.; Murray, P.; Yellowlees, L.J.; Lagunas, M.C.; Hardacre, C. Electrochemistry of sulfur and polysulfides in ionic liquids. *J. Phys. Chem. B* 2011, *115*, 13873–13879. [CrossRef]
38. Jung, Y.; Kim, S.; Kim, B.S.; Han, D.H.; Park, S.M.; Kwak, J. Effect of organic solvents and electrode materials on electrochemical reduction of sulfur. *J. Electrochem. Soc.* 2008, *3*, 566–577. [CrossRef]
39. Gaillard, F.; Levillain, E.; Lelieur, J.-P. Polysulfides in DMF: Only the radical anions $S_3^{\bullet-}$ and $S_4^{\bullet-}$ are reducible. *J. Electroanal. Chem.* 1997, *432*, 129–138. [CrossRef]
40. Evans, A.; Montenegro, M.I.; Pletcher, D. The mechanism for the cathodic reduction of sulphur in DMF: Low temperature voltammetry. *Electrochem. Commun.* 2001, *3*, 514–518. [CrossRef]
41. Levillain, E.; Gaillard, F.; Leghié, P.; Demortier, A.; Lelieur, J.-P. On the understanding of the reduction of sulfur (S_8) in dimethylformamide (DMF). *J. Electroanal. Chem.* 1997, *420*, 167–177. [CrossRef]
42. Leghié, P.; Lelieur, J.-P.; Levillain, E. Final comment on Reply to "Comments on the mechanism of the electrochemical reduction of sulphur in DMF". *Electrochem. Commun.* 2002, *4*, 406–411. [CrossRef]
43. Steudel, R.; Steudel, Y. Polysulfide chemistry in sodium-sulfur batteries and related systems: A computational study by G3X(MP2) and PCM calculations. *Chem. Eur. J.* 2013, *19*, 3162–3176. [CrossRef] [PubMed]
44. Kim, B.-S.; Park, S.-M. In situ spectroelectrochemical studies on the reduction of sulfur in dimethyl sulfoxide solutions. *J. Electrochem. Soc.* 1993, *140*, 115–122. [CrossRef]
45. Han, D.-H.; Kim, B.-S.; Choi, S.-J.; Jung, Y.; Kwak, J.; Park, S.-M. Time-resolved in situ spectroelectrochemical study on reduction of sulfur in DMF. *J. Electrochem. Soc.* 2004, *151*, E283–E290. [CrossRef]
46. Cuisinier, M.; Hart, C.; Balusubramanian, M.; Garsuch, A.; Nazar, L.F. Radical or not radical: Revisiting lithium-sulfur electrochemistry in non-aqueous solvents. *Adv. Energy Mater.* 2015, *5*, 1401801. [CrossRef]
47. Schliephke, A.; Falius, H.; Buchkremer-Hermanns, H.; Bottcher, P. Preparation and crystal structure of the bis(triethylammonium) octasulfide, [HN(C_2H_5)$_3$]$_2S_8$. *Z. Naturforsch.* 1988, *43b*, 21–24.
48. Dev, S.; Ramli, E.; Rauchfuss, T.B.; Wilson, S.R. Synthesis and structure of [M(N-methylimidazole)$_6$]S_8 (M = Mn, Fe, Ni, Mg). Polysulfide salts prepared by the reaction of N-methylimidazole + metal powder + sulfur. *Inorg. Chem.* 1991, *30*, 2514–2519. [CrossRef]
49. Fujinaga, T.; Kuwamoto, T.; Okazaki, S.; Hojo, M. Electrochemical Reduction of Elemental Sulfur in Acetonitrile. *Bull. Chem. Soc. Jpn.* 1980, *53*, 2851–2855. [CrossRef]
50. Mondal, M.K.; Zhang, L.; Feng, Z.; Tang, S.; Feng, R.; Zhao, Y.; Tan, G.; Ruan, H.; Wang, X. Tricoordinate pnictogen-centered radical anions: Isolation, characterization, and reactivity. *Angew. Chem. Int. Ed.* 2019, *58*, 15829–15833. [CrossRef]
51. Liebing, P.; Kühling, M.; Swanson, C.; Feneberg, M.; Hilfert, L.; Goldhahn, R.; Chivers, T.; Edelmann, F.T. Catenated and spirocyclic polychalcogenides from potassium carbonate and elemental chalcogens. *Chem. Commun.* 2019, *55*, 14965–14967. [CrossRef] [PubMed]
52. Neumüller, B.; Schmock, F.; Kirmse, R.; Voigt, A.; Diefenbach, A.; Bickelhaupt, F.M.; Dehnicke, K. (Ph$_4$P)S_6—A compound containing the cyclic radical anion $S_6^{\bullet-}$. *Angew. Chem. Int. Ed.* 2002, *39*, 4580–4582. [CrossRef]
53. Fabian, J.; Komiha, N.; Linguerri, R.; Rosmus, P. The absorption wavelengths of sulfur chromophores of ultramarines calculated by TD-DFT. *J. Mol. Struct. THEOCHEM* 2006, *801*, 63. [CrossRef]
54. Rejmak, P. Computational refinement of the puzzling red tetrasulfur chromophore in ultramarine pigments. *PhysChemChemPhys* 2020, *22*, 22684–22698. [CrossRef]
55. Hunsicker, S.; Jones, R.O.; Ganteför, G. Rings and chains in sulfur cluster anions S^- to S_9^-: Theory (simulated annealing) and experiment (photoelectron detachment). *J. Chem. Phys.* 1995, *102*, 5917–5936. [CrossRef]

56. Wong, M.W. *Quantum-chemical calculations of sulfur-rich compounds. Elemental Sulfur and Sulfur-Rich compounds II. Topics in Current Chemistry*; Steudel, R., Ed.; Springer: Berlin/Heidelberg, Germany, 2003; Volume 231, pp. 1–29.
57. Swope, W.C.; Lee, Y.-P.; Schaefer, H.F. Diatomic sulfur: Low lying bound molecular electronic states of S_2. *J. Chem. Phys.* **1979**, *70*, 947–953. [CrossRef]
58. Zakrzewski, V.G.; von Niessen, W. Structures, stabilities and adiabatic ionization and electron affinity energies of small sulfur clusters. *Theor. Chim. Acta* **1994**, *88*, 75–96. [CrossRef]
59. Koch, W.; Natterer, J.; Heinemann, C. Quantum chemical study on the equilibrium geometries of S_3 and S_3^-, The electron affinity of S_3 and the low lying electronic states of S_3^-. *J. Chem. Phys.* **1995**, *102*, 6159–6167. [CrossRef]
60. Wong, M.W.; Steudel, R. Structure and spectra of tetrasulfur S_4 — an ab initio MO study. *Chem. Phys. Lett.* **2003**, *379*, 162–169. [CrossRef]
61. Pearson, R.G. *Symmetry Rules for Chemical Reactions: Orbital Topology and Elementary Processes*; John Wiley and Sons: New York, NY, USA, 1976; pp. 78–79.
62. Jones, R.O.; Ballone, P. Density functional and Monte Carlo studies of sulfur. I. Structure and bonding in S_n rings and chains ($n = 2$–18). *J. Chem. Phys.* **2003**, *118*, 9257–9265. [CrossRef]
63. Chivers, T.; Laidlaw, W.G.; Oakley, R.T.; Trsic, M. Synthesis, crystal and molecular structure of $[(Ph_3P)_2N^+][S_4N^-]$, and the electronic structure of the acyclic anion, S_4N^-. *J. Am. Chem. Soc.* **1980**, *102*, 5773–5781. [CrossRef]
64. Burford, N.; Chivers, T.; Cordes, A.W.; Oakley, R.T.; Pennington, W.T.; Swepston, P.N. Variable geometry of the S_4N^- anion: Crystal and molecular structure of $Ph_4As^+S_4N^-$ and a refinement of the structure of PPN^+ S_4N^- (PPN = $[Ph_3P]_2N^+$). *Inorg. Chem.* **1981**, *20*, 4430–4432. [CrossRef]
65. Donohue, J.; Caron, A.; Goldish, E. The crystal and molecular structure of S_6 (sulfur-6). *J. Am. Chem. Soc.* **1961**, *83*, 3748–3751. [CrossRef]
66. Steudel, R.; Steidel, J.; Pickardt, J.; Schuster, F.; Reinhardt, R. X-ray structural analyses of two allotropes of cycloheptasulfur (γ and δ-S_7). *Z. Naturforsch.* **1980**, *35b*, 1378–1383. [CrossRef]
67. Schmidt, M.; Block, B.; Block, H.H.; Köpf, H.; Wilhelm, E. Cycloheptasulfur, S_7, and cyclododecasulfur, S_{12}: Two new sulfur rings. *Angew. Chem. Int. Ed.* **1968**, *7*, 632–633. [CrossRef]
68. Rettig, S.R.; Trotter, J. Refinement of the structure of orthorhombic sulfur, α-S_8. *Acta Crystallogr.* **1987**, *C43*, 2260–2262. [CrossRef]
69. Bonini, M.G.; Augusto, O. Carbon dioxide stimulates the production of thiyl, sulfinyl, and disulfide radical anion from thiol oxidation by peroxynitrite. *J. Biol. Chem.* **2001**, *276*, 9749–9754. [CrossRef]
70. Yamaji, M.; Tojo, S.; Takehira, K.; Tobita, S.; Fujitsuka, M.; Majima, T. S-S bond mesolysis in α,α'-dinaphthyl disulfide radical anion generated during γ-radiolysis and pulse radiolysis in organic solution. *J. Phys. Chem. A* **2006**, *110*, 13487–13491. [CrossRef]
71. Albright, T.A.; Burdett, J.K.; Whangbo, M.-H. *Orbital Interactions in Chemistry*; John Wiley and Sons: New York, NY, USA, 1985; pp. 212–213.
72. Heilbronner, E.; Bock, H. *The HMO Model and its Application. 1. Basis and Manipulation*; John Wiley and Sons: New York, NY, USA, 1976; pp. 131–132.
73. Chivers, T.; Drummond, I. Characterization of the trisulfur radical anion $S_3^{\bullet-}$ in blue solutions of alkali polysulfides in hexamethylphosphoramide. *Inorg. Chem.* **1972**, *11*, 2525–2527. [CrossRef]
74. Frisch, M.J.; Trucks, G.W.; Schlegel, H.B.; Scuseria, G.E.; Robb, M.A.; Cheeseman, J.R.; Scalmani, G.; Barone, V.; Petersson, G.A.; Nakatsuji, H.; et al. (Eds.) *Gaussian 16*; Revision B.01; Gaussian, Inc.: Wallingford, CT, USA, 2016.
75. Perdew, J.P.; Burke, K.; Ernzerhof, M. Generalized gradient approximation made simple. Generalized gradient approximation made simple. *Phys. Rev. Lett.* **1996**, *77*, 3865–3868. [CrossRef]
76. Berry, D.E.; Fawkes, K.L.; Chivers, T. Student-designed experiment: Preparation and mass spectrum of cyclohexasulfur. *Chem. Educ.* **2001**, *6*, 109–111. [CrossRef]
77. Boeré, R.T.; Chivers, T.; Roemmele, T.L.; Tuononen, H.M. An electrochemical and electronic structure investigation of the $[S_3N_3]^-$ radical and kinetic modeling of the $[S_4N_4]_n/[S_3N_3]_n$ ($n = 0, -1$) interconversion. *Inorg. Chem.* **2009**, *48*, 7294–7306. [CrossRef]
78. Derendorf, J.; Jenne, C.; Keßler, M. The first step of the oxidation of elemental sulfur: Crystal structure of the homopolyatomic sulfur radical cation $[S_8]^{\bullet+}$. *Angew. Chem. Int. Ed.* **2017**, *56*, 8281–8284. [CrossRef]
79. Jagg, P.N.; Kelly, P.F.; Rzepa, H.S.; Williams, D.J.; Woollins, J.D.; Wylie, W. The preparation, x-ray crystal structure and theoretical study of $[CoCp_2][S_3N_3]$, (Cp = cyclopentadienyl), a novel stacking compound incorporating multiple C-H—N(p_π) interactions. *J. Chem. Soc. Chem. Commun.* **1991**, 942–944. [CrossRef]
80. Konchenko, S.N.; Gritsan, N.P.; Lonchakov, A.V.; Irtegova, I.G.; Mews, R.; Ovcharenko, V.I.; Radius, U.; Zibarev, A.V. Cobaltocenium [1,2,5]thiadiazolo[3,4-c][1,2,5]thiadiazolidyl: Synthesis, structure, and magnetic properties. *Eur. J. Inorg. Chem.* **2008**, *2008*, 3833–3838. [CrossRef]
81. Grimme, S.; Antony, J.; Ehrlich, S.; Krieg, H. A consistent and accurate ab initio parametrization of density functional dispersion correction (DFT-D) for the 94 elements H-Pu. *J. Chem. Phys.* **2010**, *132*, 154104. [CrossRef] [PubMed]
82. Chivers, T.; Edelmann, F.; Richardson, J.F.; Schmidt, K.J. A convenient synthesis, X-ray crystal structure and Raman spectrum of the heptasulfide ion, S_7^{2-}, in $[PPN]_2S_7\cdot 2EtOH$. *Can. J. Chem.* **1986**, *64*, 145–151. [CrossRef]
83. Adamo, C.; Barone, V.J. Toward reliable density functional methods without adjustable parameters: The PBE0 model. *Chem. Phys.* **1999**, *110*, 6158–6169. [CrossRef]

84. Weigend, F.; Ahlrichs, R. Balanced basis sets of split valence, triple zeta valence and quadruple zeta valence quality for H to Rn: Design and assessment of accuracy. *Phys. Chem. Chem. Phys.* **2005**, *7*, 3297–3305. [CrossRef] [PubMed]
85. Treitel, N.; Shenhar, R.; Aprahamian, I.; Sheradsky, T.; Rabinovitz, M. Calculations of PAH anions: When are diffuse functions necessary? *Phys. Chem. Chem. Phys.* **2004**, *6*, 1113–1121. [CrossRef]
86. Grimme, S. Supramolecular binding thermodynamics by dispersion-corrected density functional theory. *Chem. Eur. J.* **2012**, *18*, 9955–9965. [CrossRef] [PubMed]
87. Grimme, S.; Hansen, A.; Brandenburg, J.G.; Bannwarth, C. Dispersion-corrected mean-field electronic structure methods. *Chem. Rev.* **2016**, *116*, 5105–5154. [CrossRef]
88. Tomasi, J.; Mennucci, B.; Cammi, R. Quantum mechanical continuum solvation models. *Chem. Rev.* **2005**, *105*, 2999–3093. [CrossRef]
89. Wang, C.-W.; Hui, K.; Chai, J.-D. Short- and long-range corrected hybrid density functionals with the D3 dispersion corrections. *J. Chem. Phys.* **2016**, *145*, 204101. [CrossRef]
90. Li, S.L.; Truhlar, D.G. Improving Rydberg excitations within time-dependent density functional theory with generalized gradient approximations: The exchange-enhancement-for-large-gradient scheme. *J. Chem. Theory Comput.* **2015**, *11*, 3123–3130. [CrossRef] [PubMed]
91. Li, Z.; Liu, W. Critical assessment of TD-DFT for excited states of open-shell systems: I. doublet–doublet transitions. *J. Chem. Theory Comput.* **2016**, *12*, 238–260. [CrossRef]
92. Fedunov, R.G.; Pozdnyakov, I.P.; Isaeva, E.A.; Zherin, I.I.; Egorov, N.B.; Glebov, E.M. Sulfur-containing radical anions formed by photolysis of thiosulfate: Quantum-chemical analysis. *J. Phys. Chem. A* **2023**, *127*, 4704–4714. [CrossRef] [PubMed]
93. Chai, J.-D.; Head-Gordon, M. Long-range corrected hybrid density functionals with damped atom–atom dispersion corrections. *Phys. Chem. Chem. Phys.* **2008**, *10*, 6615–6620. [CrossRef]
94. Liang, J.; Feng, X.; Hait, D.; Head-Gordon, M. Revisiting the performance of time-dependent density functional theory for electronic excitations: Assessment of 43 popular and recently developed functionals from rungs one to four. *J. Chem. Theory Comput.* **2022**, *18*, 3460–3473. [CrossRef] [PubMed]
95. *GaussView, Version 6.0.16*; Dennington, R.; Keith, T.A.; Millam, J.M. (Eds.) Semichem Inc.: Shawnee Mission, KS, USA, 2016.

Disclaimer/Publisher's Note: The statements, opinions and data contained in all publications are solely those of the individual author(s) and contributor(s) and not of MDPI and/or the editor(s). MDPI and/or the editor(s) disclaim responsibility for any injury to people or property resulting from any ideas, methods, instructions or products referred to in the content.

Access to Enantiomerically Pure *P*-Chiral 1-Phosphanorbornane Silyl Ethers

Kyzgaldak Ramazanova [1], Soumyadeep Chakrabortty [2], Fabian Kallmeier [2], Nadja Kretzschmar [1], Sergey Tin [2], Peter Lönnecke [1], Johannes G. de Vries [2] and Evamarie Hey-Hawkins [1,*]

[1] Institute of Inorganic Chemistry, Faculty of Chemistry and Mineralogy, Leipzig University, Johannisallee 29, 04103 Leipzig, Germany; kyzgaldak.ramazanova@yahoo.com (K.R.); nadja-kretzschmar@gmx.de (N.K.); loenneck@rz.uni-leipzig.de (P.L.)

[2] Leibniz Institute for Catalysis (LIKAT), Albert-Einstein-Straße 29A, 18059 Rostock, Germany; soumyadeep.chakrabortty@catalysis.de (S.C.); fabian.kallmeier@anu.edu.au (F.K.); sergey.tin@catalysis.de (S.T.); johannes.devries@catalysis.de (J.G.V.)

* Correspondence: hey@uni-leipzig.de

Abstract: Sulfur-protected enantiopure *P*-chiral 1-phosphanorbornane silyl ethers **5a,b** are obtained in high yields via the reaction of the hydroxy group of *P*-chiral 1-phosphanorbornane alcohol **4** with *tert*-butyldimethylsilyl chloride (TBDMSCl) and triphenylsilyl chloride (TPSCl). The corresponding optically pure silyl ethers **5a,b** are purified via crystallization and fully structurally characterized. Desulfurization with excess Raney nickel gives access to bulky monodentate enantiopure phosphorus(III) 1-phosphanorbornane silyl ethers **6a,b** which are subsequently applied as ligands in iridium-catalyzed asymmetric hydrogenation of a prochiral ketone and enamide. Better activity and selectivity were observed in the latter case.

Keywords: asymmetric hydrogenation; enantiopure; *P*-chiral phosphines; silylation

1. Introduction

Chiral phosphines play a pivotal role in asymmetric homogeneous catalysis [1–8]. *P*-stereogenic phosphines, a special class of chiral phosphines, have been well established in catalysis ever since the pioneering work on asymmetric hydrogenation (AH) employing a *P*-chiral ligand was introduced by Horner et al. [9] and Knowles et al. [10]. The development of the privileged *P*-chiral ligand (ethane-1,2-diyl)bis[(2-methoxyphenyl)(phenyl)phosphane] (DIPAMP) by Knowles and coworkers [11] led to the first industrial asymmetric hydrogenation in the production of the drug L-3,4-dihydroxyphenylalanine (L-DOPA) used in the treatment of Parkinson's disease [12] (Figure 1). A few years later, Noyori, another Nobel prize winner, and his group developed the axially chiral phosphine 2,2′-bis(diphenylphosphino)-1,1′-binaphthyl (BINAP) [13] and showed that complexes with ruthenium were effective in asymmetric hydrogenations of a wide range of olefins and carbonyl compounds [14–17]. Nowadays, AH is considered one of the most important enantioselective syntheses that gives access to many important optically active compounds. Among the widely used metals in such reactions, the iridium-catalyzed hydrogenations have been extensively studied [18–22]. The symmetric Crabtree catalyst [23], the chiral P,N bidentate PHOX ligand developed by Pfaltz et al. [24–26] and BIPI ligands by Busacca et al. [27] are key examples of Ir-based catalysts in hydrogenation reactions. The majority of the developed procedures employ bidentate hetero-donor P,X (X = N, O) ligands as they were believed to considerably influence enantioselectivities due to better chirality transfer [28–31]. Thus, the reluctance to employ monodentate ligands in AH is understandable, especially given the proven history of success with chelate ligands. However, evidence of high enantiomeric excess (*ee*) values in AH achieved using monodentate ligands has been reported [32–35]. Despite the success of the published compounds in enantioselective catalysis, industry and academia are still searching for better, more efficient and sustainable catalysts, resulting in a number of new *P*-chiral compounds being reported regularly [36–38].

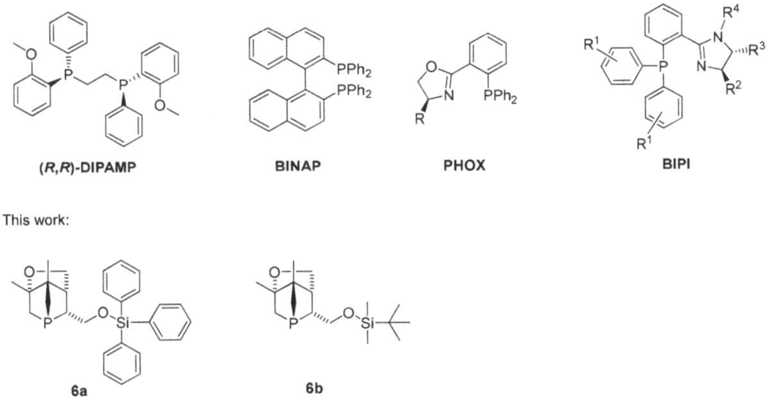

Figure 1. Previously reported chiral phosphines (**top row**) and the *P*-chiral 1-phosphanorbornane silyl ethers **6a,b** (**bottom row**) reported in this work.

Phospholes are known to undergo hetero-Diels–Alder (HDA) reactions with various dienophiles to afford P-heterocyclic compounds [39], and recently, we reported the unprecedented phospha-aza-Diels–Alder reaction using an N-sulfonyl α-imino ester to produce 1-phospha-2-azanorbornenes (PANs) [40]. We also showed that the reactive P–N bond of PANs can be cleaved by both achiral and enantiopure nucleophiles to yield racemic 2,3-dihydrophosphole and optically pure 1-alkoxy-2,3-dihydrophosphole derivatives, respectively [40,41]. Moreover, the reduction of PAN with lithium aluminum hydride (LAH) resulted in a seven-membered P-heterocycle [42]. Previously, we reported the first stereoselective HDA reaction between (5*R*)-(L-menthyloxy)-2(5*H*)-furanone (MOxF) and 2*H*-phospholes (Scheme 1) to produce *P*-chiral 1-phosphanorbornenes (**2**) [43] as well as *P*-chiral 7-phosphanorbornenes [44] in high yields. Moreover, the reduction of 1-phosphanorbornenes yields access to 1-phosphanorbornene diol **3**. The latter undergoes an intramolecular Michael addition to afford 1-phosphanorbornane alcohol **4**, which can be converted into enantiopure 1-phosphanorbornane bromide for subsequent functionalization [45].

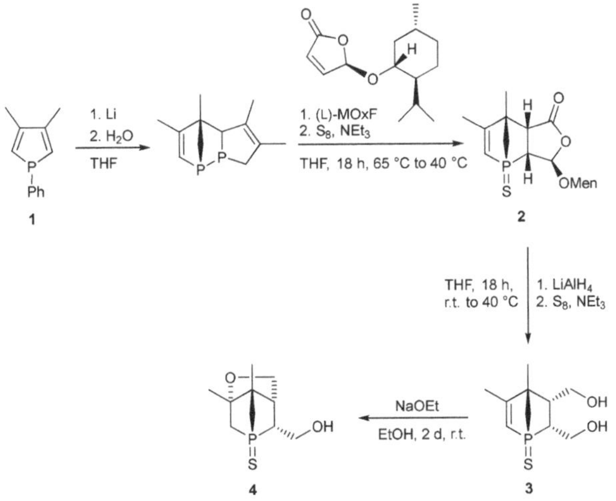

Scheme 1. Preparation of 1-phosphanorbornane alcohol **4**.

Herein, we report the one-step synthesis of enantiomerically pure P-stereogenic 1-phosphanorbornane silyl ethers obtained via reaction of the hydroxy group in **4** with chlorosilanes followed by desulfurization. The application of these ligands in iridium-catalyzed AH of prochiral enamides, namely methyl-(Z)-α-acetamidocinnamate (MAC), was studied. To our knowledge, such ligands have not yet been tested in AH nor any other enantioselective homogeneous catalysis.

2. Results and Discussion
2.1. Synthesis and Characterization of 5a,b

The enantiopure 1-phosphanorbornane alcohol **4** (PNA) is readily prepared in very good yields [45]. The P-chiral 1-phosphanorbornane silyl ethers **5a,b** are obtained by reaction of PNA **4** with chlorosilanes in dimethylformamide (DMF) in the presence of base and catalyst (Scheme 2). The formation of silyl ethers is widely exploited for the protection of alcohols, and numerous suitable silylation reagents have been reported [46–49]. We selected *tert*-butyldimethylsilyl chloride (TBDMSCl) and triphenylsilyl chloride (TPSCl), as the corresponding bulky siloxy groups provide high stability in acidic and basic media compared to the less sterically demanding trimethylsilyl or triethylsilyl ethers [49,50].

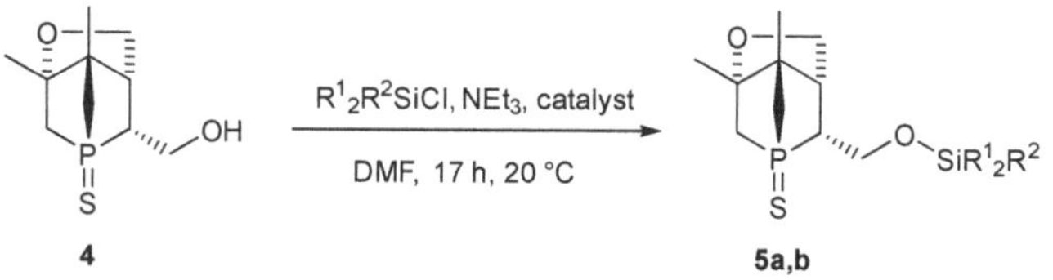

5a: $R^1 = R^2 = Ph$; catalyst = 4-dimethylaminopyridine
5b: $R^1 = Me$; $R^2 = {}^tBu$; catalyst = imidazole

Scheme 2. Synthesis of sulfur-protected 1-phosphanorbornane silyl ethers **5a,b**.

In this kind of established reaction, the choice of catalyst, solvent and base is important. Initially, when CH_2Cl_2 was used as the solvent, the reaction of **4** with TPSCl was much slower compared to DMF as the solvent. This supports the reported evidence of DMF acting as a catalyst itself in silylation reactions of alcohols [51]. Consistent with the classical procedure developed by Corey et al. [52], imidazole was employed as catalyst to afford **5b**, while for the reaction with TPSCl, 4-dimethylaminopyridine (DMAP) was used as it was previously reported to be a successful catalyst.

Stirring at 20 °C overnight resulted in full consumption of PNA as confirmed by $^{31}P\{^1H\}$ NMR spectroscopy (CDCl$_3$, singlet at 43.4 ppm for **5a** and 43.6 ppm for **5b**). Thus, this one-step procedure gives access to **5a,b** in very good yields under mild conditions. Pure **5a,b** were isolated by crystallization; single crystals suitable for X-ray crystallography (Supplementary Materials, Section S3) were obtained by dissolving **5a,b** in a hot $^iPrOH/n$-hexane mixture and cooling to −25 °C for 17 h. High chemical (98%) and optical purity of the UV-active compound **5a** were confirmed by HPLC using a chiral column (Supplementary Materials, Figure S13), while the chemical purity of **5b** was verified by elemental analysis. High-resolution mass spectrometry (HRMS) showed the presence of the expected ions, namely [**5a** + H]$^+$ (m/z 491.1617), [**5a** + NH$_4$]$^+$ (m/z 508.1878), and [**5a** + Na]$^+$ (m/z 513.1447) or [**5b** + H]$^+$ (m/z 347.1631) and [**5b** + Na]$^+$ (m/z 369.1451), respectively. The molecular structures of **5a,b** were also confirmed by 2D NMR spectroscopy.

The enantiopure compounds crystallize in the triclinic space group $P1$ with two independent molecules in the unit cell (**5a**) or in the monoclinic space group $P2_1$ with $Z = 2$ (**5b**), respectively. The phosphorus atom has a distorted tetrahedral environment (Figure 2). The Si–O bond lengths are in the range of 164.1(2) to 165.4(2) pm, which is in agreement with the literature [53,54].

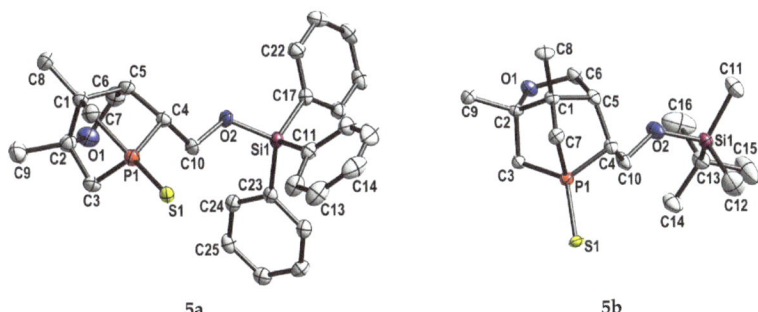

Figure 2. Molecular structures of **5a** and **5b**. Hydrogen atoms were omitted for clarity. Only one of the two independent molecules of **5a** is shown. Displacement ellipsoids are drawn at the 50% probability level.

2.2. Desulfurization of Compounds 5a,b

The *P*-chiral 1-phosphanorbornane silyl ethers **5a,b** can be reduced (desulfurized) to the corresponding phosphorus(III) derivatives with excess of freshly activated Raney nickel at room temperature (Scheme 3). No further work up is required after the reaction is finished. Moreover, this method is mild and tolerates many other functional groups guaranteeing selective desulfurization of the phosphorus atom. In contrast, treating **5a,b** with the very strong base lithium aluminum hydride (LAH) at 50 °C requires further quenching and has a risk of side reactions. Nevertheless, $^{31}P\{^{1}H\}$ NMR spectra (CDCl$_3$) of the reaction mixtures of **5a** and both reducing agents revealed full conversion of the starting material and formation of **6a** (singlet at −45.9 ppm). In contrast, **5b** can only be reduced cleanly with excess Raney nickel (singlet at −46.3 ppm for **6b** in the $^{31}P\{^{1}H\}$ NMR spectrum (CDCl$_3$)), while the reduction of **5b** with LAH resulted in formation of side products, which are presumably formed by deprotection of the silyl group. Although the TBDMS and TPS groups are known to be stable in various media, examples of TBDMS ether cleavage by LAH have been reported previously [55–57]. Therefore, the reduction of both compounds was carried out with Raney nickel.

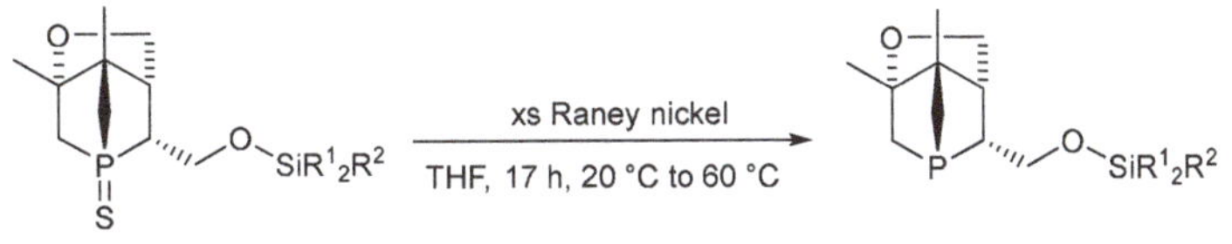

Scheme 3. Desulfurization of **5a,b** with excess Raney nickel.

The structures of **6a,b** were fully confirmed by 2D NMR spectroscopy. However, due to the high oxophilicity of the phosphorus atom, mainly the corresponding oxides were observed by HRMS ([**6a** + O + H]$^+$ (*m/z* 475.1795), [**6a** + O + Na]$^+$ (*m/z* 497.1644), [**6b** + Na]$^+$ (*m/z* 337.1722), [**6b** + O + Na]$^+$ (*m/z* 353.1668), and [**6b** + O + K]$^+$ (*m/z* 369.1420)).

3. Catalysis

Bidentate (mixed donor) chiral ligands developed by Pfaltz et al. [58] and Andersson et al. [59] are mostly used in Ir-catalyzed asymmetric hydrogenation of olefins. On the other hand, the use of chiral monodentate phosphines in Ir-catalyzed enantioselective hydrogenation is uncommon. Encouraged by the previous result on an Ir/phosphoramidite catalyst in AH [34], we evaluated the activity of the bulky monodentate *P*-chiral 1-phosphanorbornane silyl ethers **6a,b** in the asymmetric hydrogenation of carbonyl compounds and olefins. No or minor conversion was observed in the asymmetric hydrogenation of acetophenone (**S-1**) using [Ir(COD)Cl]$_2$/**6a** (1 mol%, M:L = 1:3) as catalyst in dichloromethane (Scheme 4). However, up to 20% conversion was obtained with potassium *tert*-butoxide as base (20 mol%), albeit with formation of racemic 1-phenylethan-1-ol (**P-1**) (Table 1).

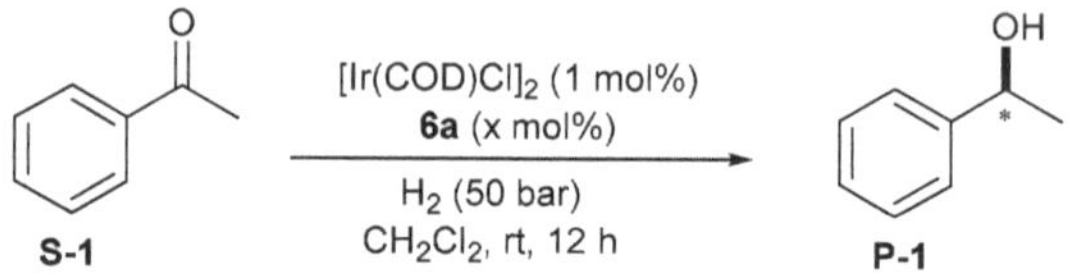

Scheme 4. Ir-catalyzed asymmetric hydrogenation of acetophenone using **6a** as ligand (* indicates a chiral center).

Table 1. Asymmetric hydrogenation of acetophenone (**S-1**) employing Ir/**6a** as catalyst.

Entry	M/L	Conversion	ee
1	1:3	-	-
2 [a]	1:2	-	-
3 [a]	1:3	6%	racemate
4 [b]	1:3	20%	racemate

[a] K$_2$CO$_3$ (20 mol%); [b] KO^{t}Bu (20 mol%).

Then, the catalytic activities of **6a,b** in the asymmetric hydrogenation of the functionalized olefin methyl (Z)-2-acetamido-3-phenylacrylate as benchmark substrate was studied. The catalytic experiments were performed by premixing the ligand (**6a** or **6b**) and the iridium complex (Scheme 5). The hydrogenation of **S-2** proceeds with 98% conversion using [Ir(COD)Cl]$_2$/**6a** (5 mol%, M:L = 1:1) as the catalyst, but with poor enantioselectivity (Table 2, Entry 1). A similar activity was observed when the catalyst loading was decreased to 0.5 mol% (Table 2, Entry 2) in dichloromethane. Changing the solvent to MeOH and THF did not improve the *ee*, but resulted in lower conversion (Table 2, Entry 3 and 4). The catalytic activity was not affected by altering the silyl substituent from SiPh$_3$ (**6a**) to SiMe$_2^t$Bu (**6b**) (Table 2, Entry 5). Apparently, the bulky silyl group is not in close proximity to the catalytically active iridium center.

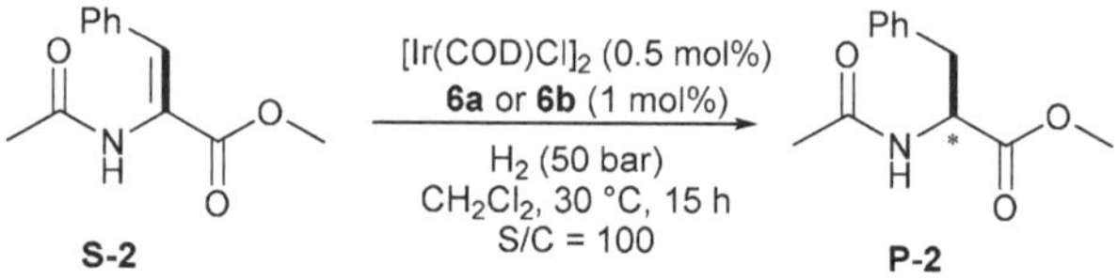

Scheme 5. Asymmetric hydrogenation of methyl (Z)-2-acetamido-3-phenylacrylate (* indicates a chiral center).

Table 2. Asymmetric hydrogenation of methyl (Z)-2-acetamido-3-phenylacrylate (**S-2**) using Ir/**6a** or **6b** as catalyst.

Entry	Solvent	Conversion	ee
1 [a]	CH$_2$Cl$_2$	98%	8%
2 [b]	CH$_2$Cl$_2$	>99%	9%
3 [b]	MeOH	90%	8%
4 [b]	THF	50%	-
5 [c]	CH$_2$Cl$_2$	95%	8%

Reaction conditions: [Ir(COD)Cl]$_2$/**6a** (1:2), substrate/catalyst (S/C) = 100, [substrate] = 0.5 mmol, H$_2$ (50 bar), solvent = dichloromethane, 30 °C, 15 h. [a] 5 mol% catalyst (M:**6a** = 1:2); [b] **6a** (1 mol%); [c] **6b** (1 mol%).

4. Conclusions

A highly efficient and facile synthesis of enantiomerically pure sulfur-protected *P*-stereogenic 1-phosphanorbornane silyl ethers **5a,b** via reaction of the alcohol function of **4** with chlorosilanes is described. Moreover, this method can be applied to prepare a variety of compounds with desired electronic and steric effects via the appropriate choice of the corresponding chlorosilane. The phosphorus(III) derivatives **6a,b** are readily accessible via desulfurization of **5a,b** with excess Raney nickel. The phosphines **6a,b** were tested as ligands in the Ir-catalyzed asymmetric hydrogenation of acetophenone and methyl (*Z*)-2-acetamido-3-phenylacrylate resulting in moderate to high conversions but poor *ee*. Further studies on different ligand variations based on the chiral phosphanorbornane motif and their application in enantioselective catalysis are underway.

5. Materials

5.1. General Information

All air-sensitive reactions were carried out under dry high purity nitrogen using standard Schlenk techniques. THF was degassed and distilled from potassium. DMF was degassed and dried under activated 4 Å molecular sieves. TBDMSCl and TPSCl were purchased from Carbolution (St. Ingbert, Germany) or Sigma Aldrich (St. Louis, MO, USA), respectively. The NMR spectra were recorded with a Bruker Avance DRX 400 spectrometer (^{1}H NMR 400.13 MHz, ^{13}C NMR 100.63 MHz, ^{31}P NMR 161.98 MHz) or a Bruker Fourier 300 spectrometer (^{1}H NMR 300.23 MHz, ^{13}C NMR 75.50 MHz). ^{13}C{^{1}H} NMR spectra were recorded as APT spectra. The assignment of the chemical shifts and configurations was performed using correlation spectroscopy (COSY) and heteronuclear single quantum coherence (HSQC) techniques. Tetramethylsilane (TMS) was used as the internal standard in the ^{1}H NMR spectra and all other nuclei spectra were referenced to TMS using the Ξ-scale [60]. The numbering scheme of **5a,b** and **6a,b** is given in the Supplementary Materials. High-resolution mass spectra (HRMS; electrospray ionization (ESI)) were measured using a Bruker Daltonics APEX II FT-ICR spectrometer (Billerica, MA, USA). IR spectra were obtained with an FTIR spectrometer (Nicolet iS5 FTIR by Thermo Scientific, Waltham, MA, USA) in the range of 400–4000 cm^{-1} in KBr. Column chromatography was performed using silica 60 (0.015–0.040 mm) purchased from Merck (Rahway, NJ, USA). UV light (389 nm) and iodine (saturated atmosphere) were used as staining reagents. The synthesis of the starting material PNA **4** and Raney nickel activation were carried out according to the literature [45].

5.2. Synthesis

5.2.1. Synthesis of **5a**

TPSCl (0.38 g, 1.28 mmol) was added to a solution of **4** (0.2 g, 0.86 mmol) and NEt$_3$ (0.18 mL, 1.28 mmol) in 12 mL DMF at room temperature. Further 20 mg of DMAP (0.017 mmol) was added and the reaction mixture was stirred for 17 h at 20 °C. The mixture was washed with sat. aq. NH$_4$Cl solution and the separated organic layer was further washed with 5 mL water 3 times. The combined organic phases were dried over MgSO$_4$. The solvent was removed under reduced pressure to give a white powder. The compound was dissolved in hot ^{i}PrOH/*n*-hexane and then cooled at −25 °C for 17 h. The formed white solid was isolated, washed with 3 mL cold *n*-hexane 3 times and dried in vacuo to afford 308 mg of **5a** as a white powder. Yield: 308 mg (73%). Single crystals of **5a** suitable for X-ray crystallographic studies were obtained by dissolving **5a** in a hot ^{i}PrOH/*n*-hexane mixture and cooling to −25 °C for 17 h (Supplementary Materials, Figure S14).

^{1}H NMR (400 MHz, CDCl$_3$): δ 7.61 (m, 5H), 7.50–7.25 (m, 10H), 4.33 (m, 1H, H-5 or H-6a), 4.16–4.01 (m, 2H), 3.94 (dd, *J* = 9.9, 5.7 Hz, 1H, H-5 or H-6a), 2.67–2.45 (m, 2H), 2.25 (m, 1H, H-6 or H-7 or H-2), 2.00 (m, 1H), 1.92–1.78 (m, 2H), 1.24 (s, 3H, H-3a or H-4a), 1.20 (s, 3H, H-3a or H-4a) ppm; ^{13}C{^{1}H} NMR (101 MHz, CDCl$_3$): δ 135.4 (s, C-aryl), 135.2 (s, C-aryl), 133.4 (s, C-aryl quart.), 130.3 (s, C-aryl), 129.8 (s, C-aryl), 128.0 (s, C-aryl), 127.7 (s,

C-aryl), 86.3 (s, C-quart.), 66.3 (s), 59.0 (d, $J_{C,P}$ = 6.1 Hz), 51.3 (d, $^2J_{C,P}$ = 19.5 Hz, C-quart.), 47.3 (d, $^2J_{C,P}$ = 2 Hz, C-5), 44.8 (d, $^1J_{C,P}$ = 46.7 Hz, C-6), 41.4 (d, $^1J_{C,P}$ = 44.8 Hz, C-2 or C-7), 40.3 (d, $^1J_{C,P}$ = 51.8 Hz, C-2 or C-7), 23.9 (d, $^3J_{C,P}$ = 7.2 Hz, C-3a or C-4a), 18.3 (d, $^3J_{C,P}$ = 15.9 Hz, C-3a or C-4a) ppm; ^{31}P{^{1}H} NMR (162 MHz, CDCl$_3$): δ 43.4 (s) ppm; HRMS (ESI, MeCN), m/z: found: 491.1617, calculated for [M + H]$^+$: 491.1624; found: 508.1878, calc. for [M + NH$_4$]$^+$: 508.1890; found: 513.1447, calc. for [M + Na]$^+$: 513.1444; found: 998.3474, calc. for [2M + NH$_4$]$^+$: 998.3441; found: 1003.3032, calc. for [2M + Na]$^+$: 1003.2996; IR (KBr, $\tilde{v}$/cm^{-1}): 3067 (w), 2975 (w), 2881 (w), 1588 (w), 1485 (w), 1427 (m), 1381 (w), 1369 (w), 1306 (w), 1250 (w), 1189 (w), 1114 (s), 1077 (s), 1053 (m), 1042 (m), 1012 (m), 996 (m), 958 (m), 928 (w), 881 (m), 862 (w), 841 (w), 800 (m), 775 (m), 738 (m), 709 (s), 697 (s), 675 (m), 619 (m), 609 (w), 582 (w), 506 (s), 481 (s), 448 (m), 435 (m).

5.2.2. Synthesis of **5b**

TBDMSCl (0.146 g, 0.97 mmol) was added to a solution of **4** (0.15 g, 0.65 mmol) and NEt$_3$ (0.135 mL, 0.97 mmol) in 10 mL DMF at room temperature. Further 13 mg of imidazole (0.19 mmol) were added and the reaction mixture was stirred for 17 h at 20 °C. The mixture was washed with sat. aq. NH$_4$Cl solution and the separated organic layer was further washed with 5 mL water 3 times. The combined organic phases were dried over MgSO$_4$. The solvent was removed under reduced pressure to give a white powder. The compound was dissolved in hot ^{i}PrOH/n-hexane and then cooled at −25 °C for 17 h. The resulting white solid was washed with 3 mL cold n-hexane 3 times and dried in vacuo to afford 154 mg of **5b** as a white powder. Yield: 154 mg (69%). Elemental analysis: C$_{16}$H$_{31}$O$_2$PSSi (346.54) calc. C 55.5%, H 9.0%; found C 55.7%, H 9.1%. Single crystals of **5b** suitable for X-ray crystallographic studies were obtained by dissolving **5b** in a hot ^{i}PrOH/n-hexane mixture and cooling to −25 °C for 17 h (Supplementary Materials, Figure S15).

^{1}H NMR (400 MHz, CDCl$_3$): δ 4.20–4.07 (m, 2H, H-5a/6a), 3.94 (m, 2H, H-5a/6a), 2.62–2.41 (m, 2H, H-5 or H-6), 2.31 (m, 1H, H-5 or H-6), 2.02 (m, 1H, H-2 or H-7), 1.96–1.83 (m, 2H, H-7 or H-2), 1.26 (s, 3H, H-3a or H-4a), 1.20 (s, 3H, H-3a or H-4a), 0.88 (s, 9H, H-9a), 0.09 (s, 6H, H-8) ppm; ^{13}C{^{1}H} NMR (101 MHz, CDCl$_3$): δ 86.3 (s, C-quart.), 66.2 (s, C-5a), 58.1 (d, $^2J_{C,P}$ = 5.6 Hz, C-6a), 51.3 (d, $^2J_{C,P}$ = 19.4 Hz, C-quart.), 47.3 (d, $^2J_{C,P}$ = 2.3 Hz, C-5), 44.8 (d, $^1J_{C,P}$ = 47.0 Hz, C-6), 41.6 (d, $^1J_{C,P}$ = 44.8 Hz, C-2 or C-7), 40.3 (d, $^1J_{C,P}$ = 51.8 Hz, C-2 or C-7), 25.8 (s, C-9a), 23.9 (d, $^3J_{C,P}$ = 7.3 Hz, C-3a or C-4a), 18.3 (d, $^3J_{C,P}$ = 16.0 Hz, C-3a or C-4a), 18.1 (s, C-9), −5.5 (d, J = 6.1 Hz, C-8) ppm; ^{31}P{^{1}H} NMR (162 MHz, CDCl$_3$) δ 43.6 (s) ppm; HRMS (ESI, MeCN), m/z: found: 347.1631, calc. for [M + H]$^+$: 347.1624; found: 369.1451, calc. for [M + Na]$^+$: 369.1444; IR (KBr, $\tilde{v}$/cm^{-1}): 2948 (m), 2925 (m), 2877 (m), 2853 (m), 1497 (w), 1468 (w), 1426 (w), 1383 (w), 1360 (w), 1311 (m), 1258 (m), 1245 (m), 1198 (w), 1162 (w), 1122 (m), 1077 (s), 1041 (m), 1029 (m), 1011 (m), 959 (m), 930 (w), 881 (s), 867 (s), 829 (m), 815 (m), 783 (s), 768 (s), 749 (m), 721 (s), 675 (s), 658 (s), 586 (w), 563 (w), 511 (m), 481 (w), 447 (m).

5.2.3. Synthesis of **6a**

Compound **5a** (200 mg, 0.41 mmol) was added to a suspension of freshly activated Raney nickel in THF (ca. 2 g, excess) and stirred for 17 h at room temperature. The clear solution was filtered and the black solid was washed four times with 5 mL THF each. The solution was concentrated to give 142 mg of **6a** as a white solid (76%). Yield: 142 mg (76%).

^{1}H NMR (400 MHz, THF-d$_8$): δ 7.54–7.44 (m, 3H), 7.39–7.33 (m, 3H), 7.33–7.19 (m, 6H), 7.13 (m, 3H), 3.86 (m, 2H, H-2 or H-7), 3.61–3.51 (m, 2H, H-5a or H-6a), 2.38 (m, 1H, H-6 or H-5), 2.13 (m, 1H, H-5 or H-6), 1.44–1.16 (m, 4H, H-2 or H-7, H-5a or H-6a), 0.99 (s, 3H, H-3a or H-4a), 0.97 (s, 3H, H-3a or H-4a) ppm; ^{13}C{^{1}H} NMR (101 MHz, THF-d$_8$): δ 135.1 (s, C-aryl), 134.9 (s, C-aryl), 134.2 (s, C-aryl quart.), 129.6 (s, C-aryl), 129.5 (s, C-aryl), 127.5 (s, C-aryl), 127.4 (s, C-aryl), 86.9 (s, C-quart.), 63.7 (s), 62 (d, $^1J_{C,P}$ = 14.9 Hz, C-2 or C-7), 47.5 (d, $^2J_{C,P}$ = 3.3 Hz, C-5), 45.1 (d, $^1J_{C,P}$ = 13.5 Hz, C-6), 38 (d, $^1J_{C,P}$ = 16.0 Hz, C-2 or C-7), 36.7 (d, J = 6.6 Hz, C-6a or C-5a), 23.7 (s, C-3a or C-4a), 17.5 (s, C-3a or C-4a) ppm; ^{31}P{^{1}H}

NMR (162 MHz, C$_6$D$_6$): δ −45.6 (s) ppm; HRMS (ESI, MeCN), m/z: found: 475.1795, calc. for [M + O + H]$^+$: 475.1863; found: 497.1644, calc for [M + O + Na]$^+$: 497.1672.

5.2.4. Synthesis of 6b

Compound **5b** (53 mg, 0.153 mmol) was added to a suspension of freshly activated Raney nickel in THF (ca. 0.44 g, excess) and stirred for 17 h at room temperature. The clear solution was filtered and the black solid was washed four times with 2 mL THF each. The solution was concentrated to give 31 mg of **6b** as a colorless oil (63%). Yield: 31 mg (63%).

^{1}H NMR (400 MHz, C$_6$D$_6$): δ 3.89–3.82 (m, 2H), 3.77–3.67 (m, 2H), 2.34 (m, 1H), 1.95–1.90 (m, 1H), 1.75 (dt, J = 15.4, 3.1 Hz, 1H, H-2 or H-7), 1.43–1.33 (m, 1H), 1.21–1.15 (m, 1H), 1.1 (s, 3H, H-3a or H-4a), 0.91 (s, 9H, H-8a), 0.89 (s, 3H, H-3a or H-4a), 0.29 (s, 6H, H-7) ppm; ^{13}C{^{1}H} NMR (101 MHz, C$_6$D$_6$): δ 64.1(s, C-5a), 61.3 (d, J = 15.0 Hz, C-6a), 47.6 (d, $^2J_{C,P}$ = 3.6 Hz, C-5), 45.2 (d, $^1J_{C,P}$ = 12.9 Hz, C-6), 38.52 (d, $^1J_{C,P}$ = 15.7 Hz, C-2 or C-7), 37.03 (d, $^1J_{C,P}$ = 6.4 Hz, C-2 or C-7), 25.7 (s, C-9a), 24.5 (s, C-3a or C-4a), 18.1 (s, C-3a or C-4a), −5.7 (d, J = 11.6 Hz, C-8) ppm; ^{31}P NMR (162 MHz, C$_6$D$_6$): δ −45.6 (s) ppm; HRMS (ESI, MeCN), m/z: found: 337.1722, calc. for [M + Na]$^+$: 337.1723; found: 353.1668, calc. for [M + O + Na]$^+$: 353.1672; found: 369.1420, calc. for [M + O + K]$^+$: 369.1412.

5.3. Catalysis

General Procedure for Hydrogenations

Ketone hydrogenation: The hydrogenation experiments were performed in stainless steel autoclaves charged with an insert suitable for up to 8 reaction vessels (4 mL) with teflon mini stirring bars. In a typical experiment, a reaction vessel was charged with [Ir(COD)Cl]$_2$ (1 mol%), ligand (1–3 mol%, as desired) and base (20 mol%) and stirred for 10–15 min in the dichloromethane (2 mL). Then, acetophenone (**S-1**, 0.5 mmol) was added to the reaction vials maintaining the inert atmosphere and the vessels were placed in a high pressure autoclave. The autoclave was purged two times with nitrogen and three times with hydrogen. Finally, it was pressurized at 50 bar H$_2$ at 25 °C for 12 h. Afterwards, the autoclave was depressurized and the contents of the reaction vessels were diluted with EtOAc and filtered through a short pad of silica. The conversion was determined by GC, GC-MS and NMR measurement and the enantiomeric excess was measured by chiral GC analysis.

Olefin hydrogenation: The hydrogenation experiments were performed in stainless steel autoclaves charged with an insert suitable for up to 8 reaction vessels (4 mL) with teflon mini stirring bars. In a typical experiment, a reaction vessel was charged with [Ir(COD)Cl]$_2$ (0.5 mol%), ligand (1 mol%) in the appropriate solvent (2 mL). Then, methyl (Z)-2-acetamido-3-phenylacrylate (**S-2**, 0.5 mmol) was added to the reaction vials maintaining the inert atmosphere and the vessels were placed in a high pressure autoclave. The autoclave was purged two times with nitrogen and three times with hydrogen gas. Finally, it was pressurized at 50 bar H$_2$ at 30 °C for 15 h. Afterwards, the autoclave was depressurized and the contents of the reaction vessels were diluted with EtOAc and filtered through a short pad of silica. The conversion was determined by GC, GC-MS and NMR measurements and the enantiomeric excess was measured by chiral GC analysis.

5.4. X-ray Crystallography Data

The data were collected on a Gemini diffractometer (Rigaku Oxford Diffraction) using Mo-Kα radiation and ω-scan rotation. Data reduction was performed with CrysAlisPro [61] including the program SCALE3 ABSPACK for empirical absorption correction. All structures were solved by dual space methods with SHELXT [62] and the refinement was performed with SHELXL [63]. For **5b**, hydrogen atoms were calculated on idealized positions using the riding model, whereas for **5a**, a difference-density Fourier map was used to locate hydrogen atoms. Structure figures were generated with DIAMOND-4 [64].

CCDC deposition numbers 2287331 for **5a** and 2287332 for **5b** contain the supplementary crystallographic data for this paper. These data can be obtained free of charge via https://www.ccdc.cam.ac.uk/structures/, accessed on 25 May 2023 (or from the Cambridge Crystallographic Data Centre, 12 Union Road, Cambridge CB2 1EZ, UK; fax: (+44)1223-336-033 or deposit@ccdc.cam.uk).

Supplementary Materials: The following supporting information can be downloaded at: https://www.mdpi.com/article/10.3390/molecules28176210/s1, NMR spectra of **5a,b** and **6a,b**, details for the crystallographic characterization, HPLC data of **5a** as well as chromatograms from catalytic tests are available in the Supplementary Materials.

Author Contributions: Conceptualization, E.H.-H. and K.R.; methodology, K.R.; formal analysis (spectroscopy, HRMS, HPLC of **5a**), K.R.; formal analysis (NMR of **6b**), N.K.; formal analysis (XRD), P.L.; catalytic tests, S.C., F.K. and S.T.; writing—original draft preparation, K.R. and S.C.; writing—review and editing, K.R., E.H.-H., J.G.d.V., S.C., P.L., N.K., F.K. and S.T.; supervision, E.H.-H., J.G.d.V.; project administration, E.H.-H. and J.G.d.V.; funding acquisition, E.H.-H. and J.G.d.V. All authors have read and agreed to the published version of the manuscript.

Funding: We thank the DFG (HE 1376/46-1 and Project number 411421782) and the Graduate School BuildMoNa for financial support.

Institutional Review Board Statement: Not applicable.

Informed Consent Statement: Not applicable.

Data Availability Statement: The data presented in this study are available in the Supplementary Materials.

Acknowledgments: Mara Wolniewicz is gratefully acknowledged for her help with chiral HPLC measurements. The analytical facilities at LIKAT are highly acknowledged.

Conflicts of Interest: The authors declare no conflict of interest.

Sample Availability: Not applicable.

References

1. Dutartre, M.; Bayardon, J.; Jugé, S. Applications and stereoselective syntheses of P-chirogenic phosphorus compounds. *Chem. Soc. Rev.* **2016**, *45*, 5771–5794. [CrossRef] [PubMed]
2. Imamoto, T. Synthesis and applications of high-performance P-chiral phosphine ligands. *Proc. Jpn. Acad. Ser. B Phys. Biol. Sci.* **2021**, *97*, 520–542. [CrossRef] [PubMed]
3. Cabré, A.; Verdaguer, X.; Riera, A. Recent Advances in the Enantioselective Synthesis of Chiral Amines via Transition Metal-Catalyzed Asymmetric Hydrogenation. *Chem. Rev.* **2022**, *122*, 269–339. [CrossRef] [PubMed]
4. Kamer, P.C.J.; van Leeuwen, P.W.N.M. *Phosphorus(III) Ligands in Homogeneous Catalysis*; John Wiley & Sons: Hoboken, NJ, USA, 2012; ISBN 9780470666272.
5. van Leeuwen, P.W.N.M.; Kamer, P.C.J.; Claver, C.; Pàmies, O.; Diéguez, M. Phosphite-containing ligands for asymmetric catalysis. *Chem. Rev.* **2011**, *111*, 2077–2118. [CrossRef]
6. Tang, W.; Zhang, X. New chiral phosphorus ligands for enantioselective hydrogenation. *Chem. Rev.* **2003**, *103*, 3029–3070. [CrossRef]
7. Xu, G.; Senanayake, C.H.; Tang, W. P-Chiral Phosphorus Ligands Based on a 2,3-Dihydrobenzo d1,3oxaphosphole Motif for Asymmetric Catalysis. *Acc. Chem. Res.* **2019**, *52*, 1101–1112. [CrossRef]
8. Fu, W.; Tang, W. Chiral Monophosphorus Ligands for Asymmetric Catalytic Reactions. *ACS Catal.* **2016**, *6*, 4814–4858. [CrossRef]
9. Horner, L.; Siegel, H.; Büthe, H. Asymmetric Catalytic Hydrogenation with an Optically Active Phosphinerhodium Complex in Homogeneous Solution. *Angew. Chem. Int. Ed.* **1968**, *7*, 942. [CrossRef]
10. Knowles, W.S.; Sabacky, M.J. Catalytic asymmetric hydrogenation employing a soluble, optically active, rhodium complex. *Chem. Commun.* **1968**, 1445–1446. [CrossRef]
11. Vineyard, B.D.; Knowles, W.S.; Sabacky, M.J.; Bachman, G.L.; Weinkauff, D.J. Asymmetric hydrogenation. Rhodium chiral bisphosphine catalyst. *J. Am. Chem. Soc.* **1977**, *99*, 5946–5952. [CrossRef]
12. Knowles, W.S. Asymmetric hydrogenations (Nobel lecture). *Angew. Chem. Int. Ed.* **2002**, *41*, 1999–2007. [PubMed]
13. Noyori, R. Asymmetric Catalysis: Science and Opportunities (Nobel Lecture). *Angew. Chem. Int. Ed.* **2002**, *41*, 2008–2022. [CrossRef]
14. Sandoval, C.A.; Ohkuma, T.; Muñiz, K.; Noyori, R. Mechanism of asymmetric hydrogenation of ketones catalyzed by BINAP/1,2-diamine-rutheniumII complexes. *J. Am. Chem. Soc.* **2003**, *125*, 13490–13503. [CrossRef] [PubMed]
15. Akutagawa, S. Asymmetric synthesis by metal BINAP catalysts. *Appl. Catal. A Gen.* **1995**, *128*, 171–207. [CrossRef]

16. Li, J.J. Noyori Asymmetric Hydrogenation. In *Name Reactions*; Li, J.J., Ed.; Springer International Publishing: Berlin/Heidelberg, Germany, 2021; pp. 399–402, ISBN 978-3-030-50864-7.
17. Seo, C.S.G.; Morris, R.H. Catalytic Homogeneous Asymmetric Hydrogenation: Successes and Opportunities. *Organometallics* 2019, 38, 47–65. [CrossRef]
18. Biosca, M.; Salomó, E.; de La Cruz-Sánchez, P.; Riera, A.; Verdaguer, X.; Pàmies, O.; Diéguez, M. Extending the Substrate Scope in the Hydrogenation of Unfunctionalized Tetrasubstituted Olefins with Ir-P Stereogenic Aminophosphine-Oxazoline Catalysts. *Org. Lett.* 2019, 21, 807–811. [CrossRef]
19. Andersson, P.G. *Iridium Catalysis*; Springer: Berlin/Heidelberg, Germany, 2011; ISBN 9783642153341.
20. Hu, X.-P.; Wang, D.-S.; Yu, C.-B.; Zhou, Y.-G.; Zheng, Z. Adventure in Asymmetric Hydrogenation: Synthesis of Chiral Phosphorus Ligands and Asymmetric Hydrogenation of Heteroaromatics. In *Asymmetric Catalysis from a Chinese Perspective*; Ma, S., Ed.; Scholars Portal: Berlin/Heidelberg, Germany, 2011; pp. 313–354, ISBN 978-3-642-19471-9.
21. Akiyama, T.; Ojima, I. *Catalytic Asymmetric Synthesis*; John Wiley & Sons, Ltd.: Hoboken, NJ, USA, 2022.
22. Mazuela, J.; Verendel, J.J.; Coll, M.; Schäffner, B.; Börner, A.; Andersson, P.G.; Pàmies, O.; Diéguez, M. Iridium phosphite-oxazoline catalysts for the highly enantioselective hydrogenation of terminal alkenes. *J. Am. Chem. Soc.* 2009, 131, 12344–12353. [CrossRef]
23. Crabtree, R. Iridium compounds in catalysis. *Acc. Chem. Res.* 1979, 12, 331–337. [CrossRef]
24. Roseblade, S.J.; Pfaltz, A. Iridium-catalyzed asymmetric hydrogenation of olefins. *Acc. Chem. Res.* 2007, 40, 1402–1411. [CrossRef]
25. Helmchen, G.; Pfaltz, A. Phosphinooxazolines—A new class of versatile, modular P,N-ligands for asymmetric catalysis. *Acc. Chem. Res.* 2000, 33, 336–345. [CrossRef]
26. von Matt, P.; Pfaltz, A. Chiral Phosphinoaryldihydrooxazoles as Ligands in Asymmetric Catalysis: Pd-Catalyzed Allylic Substitution. *Angew. Chem. Int. Ed.* 1993, 32, 566–568. [CrossRef]
27. Busacca, C.A.; Grossbach, D.; So, R.C.; O'Brien, E.M.; Spinelli, E.M. Probing electronic effects in the asymmetric Heck reaction with the BIPI ligands. *Org. Lett.* 2003, 5, 595–598. [CrossRef] [PubMed]
28. Carroll, M.P.; Guiry, P.J. P,N ligands in asymmetric catalysis. *Chem. Soc. Rev.* 2014, 43, 819–833. [CrossRef]
29. Elsevier, C.J.; de Vries, J.G. (Eds.) *The Handbook of Homogeneous Hydrogenation*; Wiley-VCH: Weinheim, Germany, 2007; ISBN 3527311610.
30. Helmchen, G.; Kudis, S.; Sennhenn, P.; Steinhagen, H. Enantioselective catalysis with complexes of asymmetric P,N-chelate ligands. *Pure Appl. Chem.* 1997, 69, 513–518. [CrossRef]
31. Peters, B.B.C.; Zheng, J.; Birke, N.; Singh, T.; Andersson, P.G. Iridium-catalyzed enantioconvergent hydrogenation of trisubstituted olefins. *Nat. Commun.* 2022, 13, 361. [CrossRef] [PubMed]
32. Frank, D.J.; Franzke, A.; Pfaltz, A. Asymmetric hydrogenation using rhodium complexes generated from mixtures of monodentate neutral and anionic phosphorus ligands. *Chem. Eur. J.* 2013, 19, 2405–2415. [CrossRef]
33. Minnaard, A.J.; Feringa, B.L.; Lefort, L.; de Vries, J.G. Asymmetric hydrogenation using monodentate phosphoramidite ligands. *Acc. Chem. Res.* 2007, 40, 1267–1277. [CrossRef] [PubMed]
34. Giacomina, F.; Meetsma, A.; Panella, L.; Lefort, L.; de Vries, A.H.M.; de Vries, J.G. High enantioselectivity is induced by a single monodentate phosphoramidite ligand in iridium-catalyzed asymmetric hydrogenation. *Angew. Chem. Int. Ed.* 2007, 46, 1497–1500. [CrossRef]
35. Komarov, I.V.; Börner, A. Highly Enantioselective or Not?—Chiral Monodentate Monophosphorus Ligands in the Asymmetric Hydrogenation. *Angew. Chem. Int. Ed.* 2001, 40, 1197–1200. [CrossRef]
36. Cabré, A.; Riera, A.; Verdaguer, X. P-Stereogenic Amino-Phosphines as Chiral Ligands: From Privileged Intermediates to Asymmetric Catalysis. *Acc. Chem. Res.* 2020, 53, 676–689. [CrossRef]
37. Xie, X.; Li, S.; Chen, Q.; Guo, H.; Yang, J.; Zhang, J. Synthesis and application of novel P-chiral monophosphorus ligands. *Org. Chem. Front.* 2022, 9, 1589–1592. [CrossRef]
38. Murai, T. Axis-to-Center Chirality Transfer Reactions of Phosphates with a Binaphthyl Group and their Congeners: New Synthetic Routes to P-Chirogenic Organophosphorus Compounds. *Chem. Lett.* 2023, advance publication. [CrossRef]
39. Mathey, F. The organic chemistry of phospholes. *Chem. Rev.* 1988, 88, 429–453. [CrossRef]
40. Wonneberger, P.; König, N.; Kraft, F.B.; Sárosi, M.B.; Hey-Hawkins, E. Access to 1-Phospha-2-azanorbornenes by Phospha-aza-Diels-Alder Reactions. *Angew. Chem. Int. Ed.* 2019, 58, 3208–3211. [CrossRef] [PubMed]
41. Ramazanova, K.; Lönnecke, P.; Hey-Hawkins, E. Facile Synthesis of Enantiomerically Pure P-Chiral 1-Alkoxy-2,3-dihydrophospholes via Nucleophilic P-N Bond Cleavage of a 1-Phospha-2-azanorbornene. *Chem. Eur. J.* 2023, 29, e202300790. [CrossRef]
42. Wonneberger, P.; König, N.; Sárosi, M.B.; Hey-Hawkins, E. Reductive Rearrangement of a 1-Phospha-2-azanorbornene. *Chem. Eur. J.* 2021, 27, 7847–7852. [CrossRef]
43. Möller, T.; Sárosi, M.B.; Hey-Hawkins, E. Asymmetric phospha-Diels-Alder reaction: A stereoselective approach towards P-chiral phosphanes through diastereotopic face differentiation. *Chem. Eur. J.* 2012, 18, 16604–16607. [CrossRef] [PubMed]
44. Möller, T.; Wonneberger, P.; Kretzschmar, N.; Hey-Hawkins, E. P-chiral phosphorus heterocycles: A straightforward synthesis. *Chem. Commun.* 2014, 50, 5826–5828. [CrossRef]
45. Möller, T.; Wonneberger, P.; Sárosi, M.B.; Coburger, P.; Hey-Hawkins, E. P-chiral 1-phosphanorbornenes: From asymmetric phospha-Diels-Alder reactions towards ligand design and functionalisation. *Dalton Trans.* 2016, 45, 1904–1917. [CrossRef]

46. Denmark, S.E.; Hammer, R.P.; Weber, E.J.; Habermas, K.L. Diphenylmethylsilyl ether (DPMS): A protecting group for alcohols. *J. Org. Chem.* **1987**, *52*, 165–168. [CrossRef]
47. Bols, M.; Pedersen, C.M. Silyl-protective groups influencing the reactivity and selectivity in glycosylations. *Beilstein J. Org. Chem.* **2017**, *13*, 93–105. [CrossRef] [PubMed]
48. Crouch, R.D. Recent Advances in Silyl Protection of Alcohols. *Synth. Commun.* **2013**, *43*, 2265–2279. [CrossRef]
49. Wuts, P.G.; Green, T.W. *Greene's Protective Groups in Organic Synthesis*, 4th ed.; John Wiley & Sons Inc.: Hoboken, NJ, USA, 2007; ISBN 0471697540.
50. Davies, J.S.; Higginbotham, C.L.; Tremeer, E.J.; Brown, C.; Treadgold, R.C. Protection of hydroxy groups by silylation: Use in peptide synthesis and as lipophilicity modifiers for peptides. *J. Chem. Soc. Perkin Trans. 1* **1992**, 3043–3048. [CrossRef]
51. Patschinski, P.; Zhang, C.; Zipse, H. The Lewis base-catalyzed silylation of alcohols—A mechanistic analysis. *J. Org. Chem.* **2014**, *79*, 8348–8357. [CrossRef] [PubMed]
52. Corey, E.J.; Venkateswarlu, A. Protection of hydroxyl groups as tert-butyldimethylsilyl derivatives. *J. Am. Chem. Soc.* **1972**, *94*, 6190–6191. [CrossRef]
53. Weinhold, F.; West, R. The Nature of the Silicon–Oxygen Bond. *Organometallics* **2011**, *30*, 5815–5824. [CrossRef]
54. Kaftory, M.; Kapon, M.; Botoshansky, M. The Structural Chemistry of Organosilicon Compounds. In *The Chemistry of Organic Silicon Compounds*; Patai, S., Rappoport, Z., Eds.; Wiley: Chichester, UK, 1998; pp. 181–265, ISBN 9780471967576.
55. Wender, P.A.; Bi, F.C.; Brodney, M.A.; Gosselin, F. Asymmetric synthesis of the tricyclic core of NGF-inducing cyathane diterpenes via a transition-metal-catalyzed 5 + 2 cycloaddition. *Org. Lett.* **2001**, *3*, 2105–2108. [CrossRef]
56. de Vries, E.F.J.; Brussee, J.; van der Gen, A. Intramolecular Reductive Cleavage of tert-Butyldimethylsilyl Ethers. Selective Mono-Deprotection of Bis-Silyl-Protected Diols. *J. Org. Chem.* **1994**, *59*, 7133–7137. [CrossRef]
57. Crouch, R.D. Selective deprotection of silyl ethers. *Tetrahedron* **2013**, *69*, 2383–2417. [CrossRef]
58. Pfaltz, A.; Blankenstein, J.; Hilgraf, R.; Hörmann, E.; McIntyre, S.; Menges, F.; Schönleber, M.; Smidt, S.P.; Wüstenberg, B.; Zimmermann, N. Iridium-Catalyzed Enantioselective Hydrogenation of Olefins. *Adv. Synth. Catal.* **2003**, *345*, 33–43. [CrossRef]
59. Peters, B.B.C.; Andersson, P.G. The Implications of the Brønsted Acidic Properties of Crabtree-Type Catalysts in the Asymmetric Hydrogenation of Olefins. *J. Am. Chem. Soc.* **2022**, *144*, 16252–16261. [CrossRef] [PubMed]
60. Harris, R.K.; Becker, E.D.; Cabral de Menezes, S.M.; Goodfellow, R.; Granger, P. NMR Nomenclature: Nuclear Spin Properties and Conventions for Chemical Shifts. IUPAC Recommendations 2001. *Solid State Nucl. Magn. Reson.* **2002**, *22*, 458–483. [CrossRef] [PubMed]
61. Rigaku Corporation. *CrysAlisPro Software System*; Rigaku Oxford Diffraction: Wroclaw, Poland, 1995–2023.
62. Sheldrick, G.M. SHELXT—Integrated space-group and crystal-structure determination. *Acta Cryst. Sect. A Found. Adv.* **2015**, *A71*, 3–8. [CrossRef] [PubMed]
63. Sheldrick, G.M. Crystal structure refinement with SHELXL. *Acta Cryst. Sect. C Struct. Chem.* **2015**, *71*, 3–8. [CrossRef]
64. Brandenburg, K. *Crystal Impact GbR*, version 4.6.8; DIAMOND 4: Bonn, Germany.

Disclaimer/Publisher's Note: The statements, opinions and data contained in all publications are solely those of the individual author(s) and contributor(s) and not of MDPI and/or the editor(s). MDPI and/or the editor(s) disclaim responsibility for any injury to people or property resulting from any ideas, methods, instructions or products referred to in the content.

Article

Synthesis and Thermal Studies of Two Phosphonium Tetrahydroxidohexaoxidopentaborate(1-) Salts: Single-Crystal XRD Characterization of [^{i}PrPPh$_3$][B$_5$O$_6$(OH)$_4$]·3.5H$_2$O and [MePPh$_3$][B$_5$O$_6$(OH)$_4$]·B(OH)$_3$·0.5H$_2$O [†]

Michael A. Beckett [1,*], Peter N. Horton [2], Michael B. Hursthouse [2] and James L. Timmis [1]

[1] School of Natural Sciences, Bangor University, Bangor LL57 2UW, UK
[2] Chemistry Department, University of Southampton, Southampton SO17 1BJ, UK
* Correspondence: m.a.beckett@bangor.ac.uk
[†] Dedicated to Professor J. Derek Woollins on the occasion of his retirement.

Abstract: Two substituted phosphonium tetrahydoxidohexaoxidopentaborate(1-) salts, [^{i}PrPPh$_3$][B$_5$O$_6$(OH)$_4$]·3.5H$_2$O (**1**) and [MePPh$_3$][B$_5$O$_6$(OH)$_4$]·B(OH)$_3$·0.5H$_2$O (**2**), were prepared by templated self-assembly processes with good yields by crystallization from basic methanolic aqueous solutions primed with B(OH)$_3$ and the appropriate phosphonium cation. Salts **1** and **2** were characterized by spectroscopic (NMR and IR) and thermal (TGA/DSC) analysis. Salts **1** and **2** were thermally decomposed in air at 800 °C to glassy solids via the anhydrous phosphonium polyborates that are formed at lower temperatures (<300 °C). BET analysis of the anhydrous and pyrolysed materials indicated they were non-porous with surface areas of 0.2–2.75 m^2/g. Rhe recrystallization of **1** and **2** from aqueous solution afforded crystals suitable for single-crystal XRD analyses. The structure of **1** comprises alternating cationic/anionic layers with the H$_2$O/pentaborate(1-) planes held together by H-bonds. The cationic planes have offset face-to-face (*off*) and vertex-to-face (*vf*) aromatic ring interactions with the ^{i}Pr groups oriented towards the pentaborate(1-)/H$_2$O layers. The anionic lattice in **2** is expanded by the inclusion of B(OH)$_3$ molecules to accommodate the large cations; this results in the formation of a stacked pentaborate(1-)/B(OH)$_3$ structure with channels occupied by the cations. The cations within the channels have *vf*, *ef* (edge-to-face), and *off* phenyl embraces. Both H-bonding and phenyl embrace interactions are important in stabilizing these two solid-state structures.

Keywords: organotriphenylphosphonium salts; π-interactions; pentaborate(1-); phenyl embraces; phosphonium salts; tetrahydroxidohexaoxidopentaborate(1-); X-ray structures

1. Introduction

Hydrated polyhydroxidooxidoborates and anhydrous polyoxidoborates are a well-known, naturally occurring classes of compounds [1–9] with many synthetic analogues [3,9–12]. Some of these compounds are industrially important bulk chemicals (e.g., Na$_2$B$_4$O$_5$(OH)$_4$·3H$_2$O, tincalconite and the synthetic borax pentahydrate largely produced from Na$_2$B$_4$O$_6$(OH)$_2$·3H$_2$O (kernite) and Na$_2$B$_4$O$_5$(OH)$_4$·8H$_2$O, borax (tincal)) with many applications [13–15], whilst others, such as β-BaB$_2$O$_4$ (BBO), have found more specialist niche applications in NLO materials [9,16]. Structurally, these polyoxidoborates are a diverse class of compounds with the polyoxidoborate moieties as discrete insular anions or as more highly condensed polymeric 1-D, 2-D or 3-D anionic networks, and the associated cations as simple *s*-, *p*-, *d*- or *f*-block element cations, cationic *p*- or *d*-block complexes or non-metal/organic based [1–16]. With some late transition metals (e.g., Cu^{II}, Zn^{II}, Cd^{II} and Ni^{II}), oxidoborates can also function as *O*-donor ligands [17]. Examples of hydroxidooxidopolyborate salts with phosphorus-containing cations are rare and are currently limited to [Ph$_3$PNPPh$_3$][B$_3$O$_3$(OH)$_4$]·2.5H$_2$O [18], [Ph$_3$PNPPh$_3$][B$_5$O$_6$(OH)$_4$]·1.5

H$_2$O [18] and [PPh$_4$][B$_5$O$_6$(OH)$_4$]·1.5H$_2$O [19]; the latter compound has been structurally characterized by a single-crystal X-ray diffraction (sc-XRD). We have previously published thermal studies and BET analysis on materials that were thermally obtained from pentaborate(1-) salts containing organic cations [20–22]. This manuscript extends our structural studies on phosphonium salts of pentaborate(1-) anions and examines them using BET analysis on materials derived thermally from these salts. A schematic drawing of the tetrahydroxidohexaoxidopentaborate(1-) anion, hereafter generally abbreviated to pentaborate(1-) [23], is shown in Figure 1.

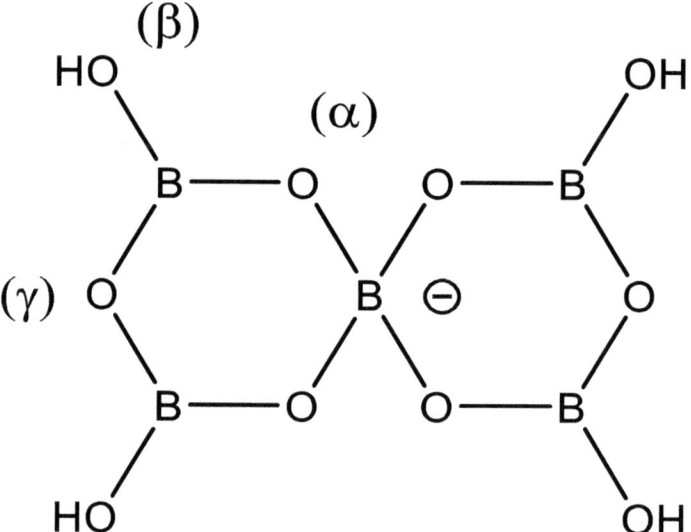

Figure 1. Schematic drawing of tetrahydroxidohexaoxidopentaborate(1-) as found in [^{i}PrPPh$_3$][B$_5$O$_6$(OH)$_4$]·3.5H$_2$O (**1**) and [MePPh$_3$][B$_5$O$_6$(OH)$_4$]·B(OH)$_3$·0.5H$_2$O (**2**). The oxygen H-bond acceptor sites are labelled as in Ref. [24].

2. Results and Discussion

2.1. Synthesis

The two new tetraorganophosphonium pentaborate(1-) salts [^{i}PrPPh$_3$][B$_5$O$_6$(OH)$_4$]·3.5H$_2$O (**1**) and [MePPh$_3$][B$_5$O$_6$(OH)$_4$]·B(OH)$_3$·0.5H$_2$O (**2**) were obtained by crystallization from basic aqueous solutions primed with B(OH)$_3$ and the appropriate substituted phosphonium cation, as shown in Scheme 1. The phosphonium iodide salts were converted to their hydroxide salts by use of an ion-exchange resin and B(OH)$_3$ was used as the boron source for the pentaborate(1-) salts. The boron-containing substrate B(OH)$_3$, is only present in aqueous solution as B(OH)$_3$ under acidic conditions [25–27]. At a higher pH, it is present as rapidly attained equilibrium concentrations of various hydroxidooxidopolyborate anions and [B(OH)$_4$]$^-$ [25–27]. The observed crystalline products arise through the cation templated self-assembly/crystallization processes [28–31], as are often observed in related systems involving non-metal cations derived from organic amines and B(OH)$_3$ [10,32].

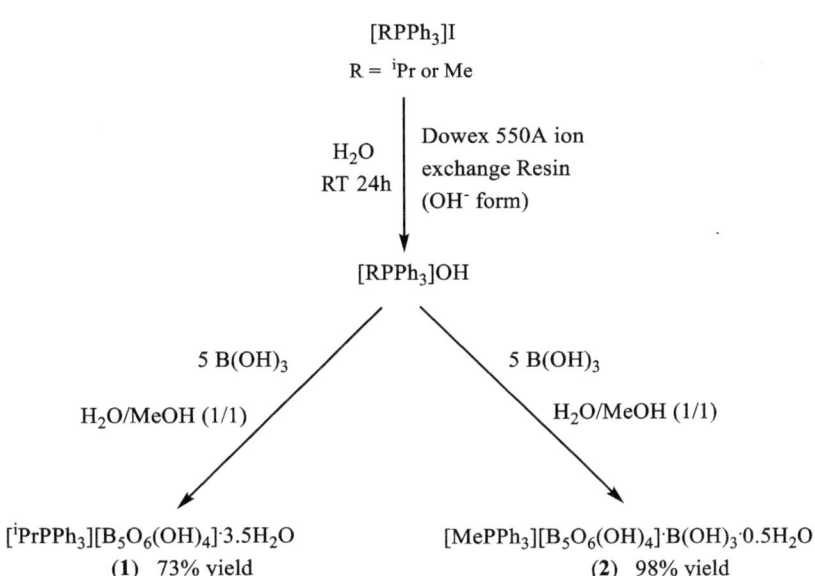

Scheme 1. Synthesis of compounds **1** and **2**.

Compounds **1** and **2** were prepared in excellent yields and were characterized spectroscopically (multinuclear NMR, IR) and thermally (TGA/DSC). Porosity data (BET) were also obtained on materials derived thermally in air from **1** and **2**. Crystallization of the crude products **1** and **2** from H_2O afforded crystals suitable for sc-XRD studies (Section 2.4). The co-crystallization of $B(OH)_3$ with pentaborate(1-) anions, as in **2**, has occasionally been observed in structures with large organic cations [18,33–36].

2.2. Thermal Studies

Organic cation polyborates are known to thermally decompose in air with the formation of glassy B_2O_3 at 800 °C [10,34]. The closely related tetraphenylphosphonium pentaborate salt, $[PPh_4][B_5O_6(OH)_4]\cdot 1.5H_2O$, is reported to be thermally decomposed in a similarly manner [19]. In previous studies, water is lost at lower temperatures with the formation of 'anhydrous' pentaborates and this is followed at higher temperatures by oxidation of the cation, gaseous evolution, and the formation of darkened intumesced solids. At higher temperatures again, these solids shrink down to form glassy residual materials of B_2O_3 [10,24]. The thermal decomposition of **1** and **2** was studied by TGA/DSC analysis in air over the temperature range 20–800 °C.

The data for **1** were consistent with the initial loss of $5.5 \times H_2O$ in the first stage (<275 °C) in an endothermic process to form anhydrous $[^iPrPPh_3][B_5O_8]$. This dehydration stage involved the loss of interstitial H_2O ($3.5 \times H_2O$) and condensation/cross-linking of the B-OH groups ($2.0 \times H_2O$) as one large continuous step (see Supplementary Information, Figure S10 for TGA plots). The TGA plot of **2** had a similar profile, with loss of $4.0 \times H_2O$ in the first stage (100–275 °C) and the formation of anhydrous $[MePPh_3][B_6O_{9.5}]$. Since **2** is a 1:1 $B(OH)_3$/pentaborate(1-) co-crystal, this material is formulated as an anhydrous hexaborate [18,33,34]. This endothermic dehydration step for **2** is a two-stage process involving the loss of interstitial H_2O and partial condensation/cross-linking of the pentaborate B-OH groups ($2.0 \times H_2O$, 100–150 °C), with further condensation/cross-linking of the pentaborate B-OH groups ($2.0 \times H_2O$, 150–275 °C). This two-stage water loss is qualitatively very similar to that observed for $[PPh_4][B_5O_6(OH)_4]\cdot 1.5H_2O$ [19].

It was anticipated that, upon further heating (275–800 °C), **1** and **2** would leave, after oxidation of the cations during the exothermic second stages, with glassy residues comprised of 2.5 or 3.0 equivalents of B_2O_3, respectively. However, the residual masses from **1** and **2** were higher than calculated, indicating that they both contained additional, non-B_2O_3 material. It has been noted that phosphonium salts, with simple non-polar substituents, generally decompose cleanly with little residue [37], and our previous studies on the thermal decomposition of phosphonium polyborates are consistent with this [18,19]. However, some phosphonium salts are also known to decompose with residual material [37]. The additional residual material from **1** and **2** possibly arises through the incorporation of phosphorus and/or an organic char slowing down the oxidation process.

Porosity data (BET analysis [38]) of organic pentaborates salts and their thermally derived anhydrous, pyrolysed and residual glasses have been reported and the results indicated that they were non-porous [20–22]. Compounds **1** and **2** both possess unusual solid-state pentaborate structures (see sc-XRD studies, Section 2.4), with **1** layered and **2** having its cations stacked in channels. We were, therefore, interested in obtaining porosity data on the thermally derived intermediate materials from these phosphonium pentaborates to see if these structural modifications are influential. Thus, samples of 'anhydrous' and 'pyrolysed' materials were obtained from **1** and **2** by heating ca. 0.5 g samples in a furnace in air for 24 h at 300 °C and 625 °C, respectively. These materials had surface areas of 0.2–2.75 m^2/g and were essentially non-porous, with similar values to those obtained for materials derived thermally from $[PPh_4][B_5O_6(OH)_4]\cdot 1.5H_2O$ [39] and organic pentaborates [20–22], again suggesting that the intumesced solids have 'foamlike' gas-encapsulated macroporous structures [40].

2.3. Spectroscopic Studies

IR and NMR (^{1}H, ^{13}C, ^{11}B and ^{31}P) data for **1** and **2** are reported in the experimental section. These spectroscopic data are in agreement with the expected data for the anions and cations found in **1** and **2**.

The IR spectra, obtained as KBr discs, show the expected broad H-bonded O-H stretches (ca. 3300 cm^{-1}) and strong B-O stretches/bends (1450–620 cm^{-1}) [41] associated with the pentaborate(1-) anions. Specifically, a diagnostic strong band (B_{trig}-O (sym.) at ca. 925 cm^{-1} [34]) for the anion was observed at 918 and 924 cm^{-1} for **1** and **2**, respectively, helping to confirm their identities.

Compounds **1** and **2** are insoluble in organic solvents but 'dissolve' in H_2O with decomposition of the pentaborate(1-) anion by the borate equilbria processess that are also involved in their formation [25–27]. The cations in **1** and **2** are not affected by this, and their presence was confirmed as substituted phosphonium cations by ^{1}H, ^{13}C and ^{31}P spectroscopic analysis. Thus, ^{31}P spectra of **1** and **2** both show only one signal at the expected chemical shift for their phosphonium cations [42]. The ^{11}B spectra of **1** and **2** (in D_2O) show three signals, corresponding to the tetrahedral boron centre of $[B_5O_6(OH)_4]^-$ (ca. +1 ppm), $[B_3O_3(OH)_4]^-$ (ca. +13 ppm) and $B(OH)_3/[[B(OH)_4]^-$ (ca. +18 ppm) in the form of 'signature spectra', as was previously observed [43]. These signals arise from the equilbrium concentrations of the borate anions present from the 'disolution' of the original pentaborate(1-) anion.

2.4. X-ray Crystallography

There are two independent isopropyltriphenylphosphonium(1+) cations, two independent tetrahydroxidohexaoxidopentaborate(1-) anions, and seven waters of crystallization within the unit cell of **1**. The asymmetric unit cell of **2** contains two independent tetrahydroxidohexaoxidopentaborate(1-) anions and two independent methyltriphenylphosphonium(1+) cations. Additionally, **2** also contains two independent $B(OH)_3$ molecules and a single disordered H_2O of crystallization. These crystallographic studies are in agreement with the formulation of **1** and **2** as ionic phosphonium(1+)/pentaborate(1-) salts, as indicated by their spectroscopic and thermal analysis. The co-crystallization of $B(OH)_3$ is not uncommon in recrystallized samples of pentaborate(1-) salts containing bulky cations [18,33–36]. Drawings of the structures of **1** and **2** showing atomic numbering are shown in Figures 2 and 3, respectively. Selected crystallographic information is available in the experimental section and full details can be found in the Supplementary Information.

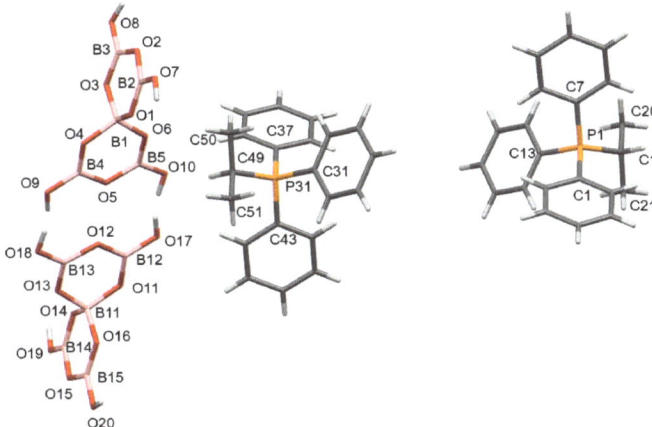

Figure 2. Drawing of the structure of [^{i}PrPPh$_3$][B$_5$O$_6$(OH)$_4$]·3.5H$_2$O (**1**), showing selected crystallographic atomic numbering schemes. The seven waters of crystallization have been omitted for clarity. The O atoms of these H$_2$O molecules are numbered O61–O67. Only the lowest numbered carbon in each aryl ring is labelled; the other five carbons are numbered sequentially. H atoms take the same label number as the heavy atoms to which they are attached.

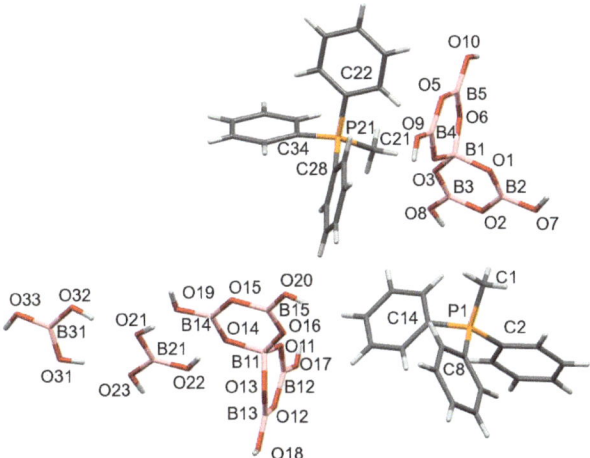

Figure 3. Drawing of the structure of [MePPh$_3$][B$_5$O$_6$(OH)$_4$]·B(OH)$_3$·0.5H$_2$O (**2**) showing the crystallographic atomic numbering scheme. The waters of crystallization have been omitted for clarity and are disordered over four sites. The O atoms of these waters are labelled O41–O43. Only the lowest numbered carbon in each phenyl ring is labelled; the other five carbons are numbered sequentially around the ring. H atoms take the same label number as the heavy atoms to which they are attached.

The tetrahydroxidohexaoxidopentaborate(1-) anion is crystallographically well-known [10] and has the gross structure of two fused, slightly puckered ('planar') boroxole (B$_3$O$_3$) rings sharing a spiro 4-coordinate boron centre; all other boron atoms are 3-coordinate, and are bound solely to oxygen atoms within the rings, or to *exo* hydroxido groups (see Figure 1). B-O bonds lengths and OBO and BOB angles within these compounds are within normal limits [10,20,24]. The B-O distances involving 4-coordinate boron centres range from 1.455(3) to 1.487(3) Å (*av.* 1.473(3) Å) and 1.434(4) to 1.481(4) Å (*av.* 1.462(4) Å), whilst B-O distances involving 3-coordinate boron centres are shorter and range from 1.344(2) to 1.401(3) Å (*av.* 1.369(3) Å) and 1.334(4) to 1.395(4) Å (*av.* 1.359(4) Å) in **1** and **2**, respectively. The OBO angles involving 4-coordinate (sp^3 hybridized, tetrahedral) boron centres range from 106.24(17) to 111.19(16)° (*av.* 109.5(2)°) and 107.5(2) to 110.9(2)° (*av.* 109.5(2)°), whilst OBO angles involving 3-coordinate (sp^2 hybridized, trigonal planar) borons are larger and range from 116.21(19) to 122.2(2)° (*av.* 120.0(2)°) and 114.3(3) to 123.7(3) (*av.* 120.0(3)°) for **1** and **2**, respectively. The BOB angles within the boroxole rings range from 118.40(17) to 123.39(18)° (*av.* 121.46(18)°) and 116.7(2) to 124.2(2)° (*av.* 121.4(3)°) for **1** and **2** respectively, indicative of these oxygens being sp^2 hybridised [44]. The B-O distances and OBO angles within the B(OH)$_3$ molecules of **2** are normal for B(OH)$_3$ and are also within the ranges found for the trigonal borons of the pentaborate(1-) rings in **1** and **2**.

The [^{i}PrPPh$_3$]$^+$ and [MePPh$_3$]$^+$ cations in **1** and **2** are also well-known crystallographically [45,46] with P-C distances ranging from 1.795(2) to 1.882(2) Å (av. 1.803(2) Å) and 1.757(4) to 1.785(3) (av. 1.776(5)A), respectively. Likewise, the CPC angles about the (sp^3) phosphorus centres range from 107.84(9) to 111.31(10)° (av. 109.47(10)°) and 106.82(17) to 110.8(2)° (av.109.47(19)°). These values are within previously observed ranges for these cations [45,46].

The closely related hydrated tetraphenylphosphonium pentaborate salt, [PPh$_4$][B$_5$O$_6$(OH)$_4$] ·1.5H$_2$O, has an interesting supramolecular giant structure composed of interpenetrating networks of complex H-bonded anion–anion interactions and cation–cation interactions involving multiple embraces of their aromatic rings [19]. Aromatic embraces are known to be strong stabilizing interactions [47,48] and are likely to be responsible (together with H-bonding interactions) for the crystallized self-assembly [28–31] of this compound. We examined the structures of **1** and **2** to see if similar aromatic interactions occur in these compounds, and details, together with their H-bonding interactions, are described below.

Compound **1** is a co-crystallised phosphonium pentaborate salt with 3.5 H$_2$O per cation/anion. These water molecules H-bond with the pentaborate(1-) anions and form a unique H$_2$O/pentaborate H-bonded anionic network. Unusually for pentaborate salts, this anionic network is arranged in layers (Figure 4). Each pentaborate(1-) has four H-bond donor sites and the pentaborate anions containing B1 forms donor H-bonds to two β-sites (see Figure 1 for acceptor site labels [24]) of two neighbouring pentaborate anions (O10H10···O7' and O8H8···O20") and two H$_2$O molecules (O7H7···O62 and O9H9···O63). The repeating O10-H-10···O7' interaction is part of an infinite chain that links the B1 containing pentaborates(1-) anions. This interaction is C(8) using Etter terminology [49]. Likewise, the pentaborates containing B11 are similarly linked into infinite chains by C(8) interactions involving O18-H18···O19'. The other three H-bond acceptor sites for the pentaborate containing B11 are water molecules: O17H17···O61, O19H19···O67, and O20H20···O66. These two pentaborate chains are linked through a complex series of H-bonds involving four H$_2$O molecules (containing O61, O62, O63 and O67) and a C$_5^5$(10) chain of three H$_2$O molecules (containing O66, O65, O64), linking with O8H8-O20' interborate interaction into layers (Figure 4).

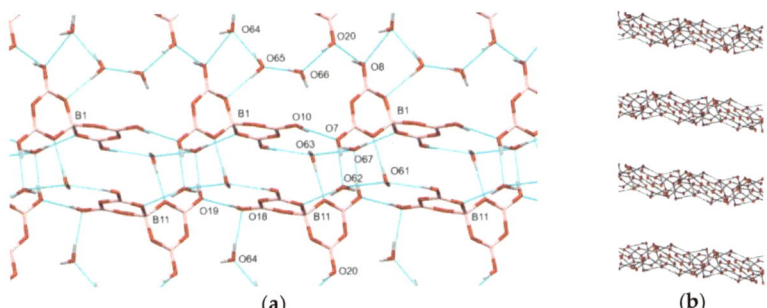

Figure 4. (a) A view along the c axis in compound **1** (perpendicular to a anionic pentaborate/H$_2$O layer) with selected atomic labels, illustrating the two C(8) chains associated with the pentaborate(1-) anions containing B5 and B11. A third $C_5^5(10)$ chain, involving three H$_2$O molecules (containing O64, O65, and O66) and pentaborate(1-) oxygens (O8 and O20), can also be seen. (b) The pentaborate(1-) anion/H$_2$O molecules in **1** can be viewed along the b axis, illustrating their layered structure.

The [^{i}PrPPh$_3$]$^+$ cations in **1** are also arranged in layers, and these layers alternate with the anionic/H$_2$O layers. Within these cationic layers, there are several cation–cation interactions involving their aromatic rings [47,48]. The two independent cations are both arranged as centrosymmetric pairs, with aromatic (phenyl) ring embrace interactions between each pair. Thus, the cation containing P1 forms an offset face-to-face (*off*) interaction between a phenyl ring (containing C1–C6) and the C1′–C6′ ring in its pair, with a centroid–centroid distance of 3.748(2) Å and a centroid-to-plane distance of 3.634(2) Å with a shift of 0.917(4) Å. These distances are indicative of a strong interaction and are considerably shorter than that found in [PPh$_4$][B$_5$O$_6$(OH)$_4$]·3/2H$_2$O (ca. 4.3 Å) [19]. These phenyl rings are also involved in vertex-to-face (*vf*) interactions between C4H4 and its paired C7′–C12′ phenyl ring, with the H4-to-plane distance at 2.538(1) Å (Figure 5). Dance and Scudder refer to this type of interaction in tetraphenylphosphonium salts as a parallel quadruple phenyl embrace (PQPE) and calculate this interaction energy as −41 kJ.mol^{-1} [48]. Similarly, the cation containing P31 in **1** is also involved in a PQPE interaction with paired *off* interactions involving the C37–C42/C37′–C42′ phenyl rings and paired *vf* interactions from C40H40 to the phenyl ring containing C43′–C48′. For these rings, the centroid–centroid distance is 3.712(2) Å; the centroid-to-plane distance is 3.528(2) Å with a shift of 1.155(4) Å, and the H40-to-plane distance is 2.636(1) Å. The P1···P1 and P31···P31 distances are 8.004(1) Å and 8.170(1) Å, with P1···P31 of 11.129(1) Å. The [^{i}PrPPh$_3$]$^+$ cations in **1** are oriented within the cationic layers with the ^{i}Pr groups towards the pentaborate(1-)/H$_2$O layers.

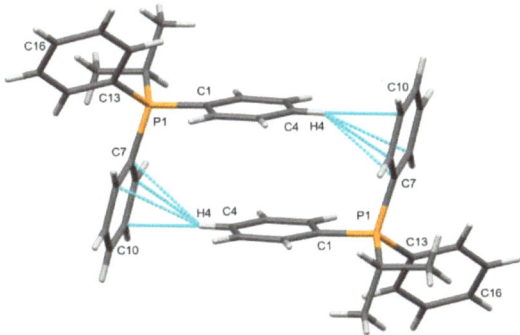

Figure 5. The centrosymmetric paired [^{i}PrPPh$_3$]$^+$ cations (containing P1) in **1** display vertex-to-face interactions phenyl ring interactions, in addition to an offset face-to-face interactions (not highlighted). Similar interactions also occur in centrosymmetric paired cations containing P31.

Compound **2** is a further example of a co-crystallized phosphonium pentaborate salt with one B(OH)$_3$ and 0.5 (disordered) H$_2$O per cation/anion. The supramolecular structure of **2** also displays anion–anion H-bond interactions and cation–cation aromatic embraces, but the details of these stabilizing interactions differ from those observed in **1** and [PPh$_4$][B$_5$O$_6$(OH)$_4$]·1.5H$_2$O and are described below.

All hydroxyl groups of the two independent B(OH)$_3$ and the two independent pentaborate(1-) anions are used as H-bond donor centres. The anion containing B1 forms two donor H-bonds to two α-sites (O9H9···O11′ and O10H10···O6′) of two adjacent pentaborates (one containing B11 and one containing B1) and both these interactions are R$_2^2$(8) [49] with the ring involving the O10H10 donor centrosymmetric (reciprocal). The anion containing B1 also forms two donor H-bonds to two adjacent B(OH)$_3$ molecules: O8H8···O32, and O7′H7′···O31. The anion containing B11 forms three donor H-bonds to three adjacent anions at two α-sites (O17H17···O4′ and O18H18···O13′, reciprocal) and one β-site, (O19H19···O7′). This O10H19···O7′ interaction is part of a two larger R$_4^4$(12) ring interactions with both these rings including both B(OH)$_3$ molecules (Figure 6). The fourth pentaborate donor interaction is to the disordered H$_2$O (O20H20···O44), and overall the anion can be represented as α,α,β,ω [21]. The hydroxido groups of the two B(OH)$_3$ molecules are arranged asymmetrically to maximise their acceptor/donor H-bond interactions. The B(OH)$_3$ containing B31 forms a R$_2^2$(8) 'pincer' ring with the B(OH)$_3$ containing B21, and likewise this B(OH)$_3$ forms a 'pincer' R$_2^2$(8) interaction with the pentaborate containing B11 (Figure 6a). These interactions allow for the two co-crystallised B(OH)$_3$ molecules to function as 'spacer' units to expand the lattice and replace what would otherwise be a simpler pentaborate/pentaborate R$_2^2$(8) interaction [18,33–36].

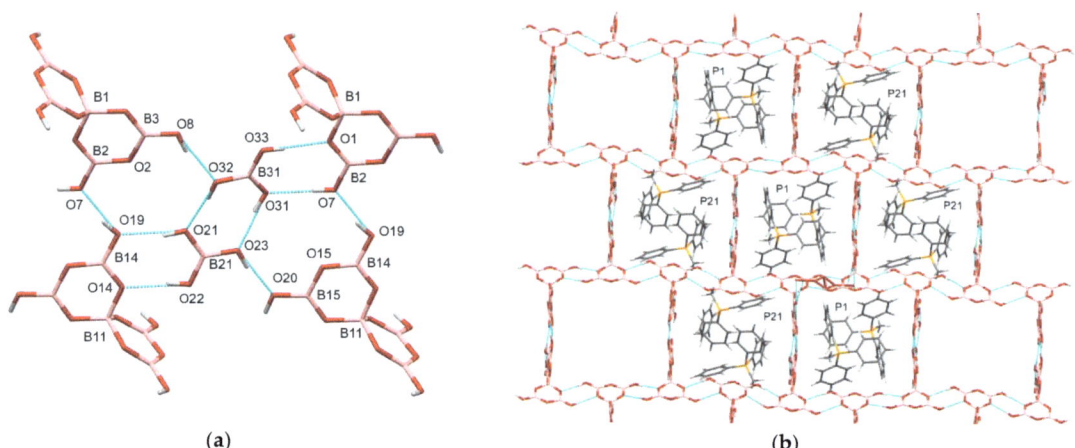

(a) (b)

Figure 6. (a) View along the *c* axis in [MePPh$_3$][B$_5$O$_6$(OH)$_4$]·B(OH)$_3$·0.5H$_2$O (**2**) showing H-bond interactions involving the two B(OH)$_3$ moieties and their three R$_2^2$(8) and the two R$_4^4$(12) ring motifs. (b) View along the *a* axis in **2**, illustrating the stacking pattern of the anionic network. The [MePPh$_3$]$^+$ cations (for clarity, only some are shown) occupy the channels as shown forming stacks (chains) with cations in the stacks only containing P atoms, as labelled. From this perspective, the B(OH)$_3$ units are side-on and are in the 'vertical' section of the borate channels.

A view along the *a* axis of **2** (along the plane shown in Figure 6a) is shown in Figure 6b. This view reveals a stacked anionic lattice (rectangular and honeycomb-like) with channels that are occupied by the cations; interestingly, each cationic stack is occupied by either cations containing solely P1 or P21 and adjacent cationic stacks in the arrangement, as shown in Figure 6b. Cations are arranged as centrosymmetric pairs within the stacks with P1···P1 and P21···P21 distances of 6.253(2) Å and 6.255(2) Å, respectively. The repeating P···P distances in both stacks are 10.1076(10) Å, but the interpair dimer interactions differ. These interactions involve aromatic embraces and are *vf* (C19H19···C6′) and an edge-to face (*ef*) (C6H6···C11′, C7H7A···C10′) for P1-containing cations and *vf* (C39H39···C23′/C24′ and C24H24···C31′/C32′) for P21-containing cations. The closest contacts between centrosymmetric pairs arise from Me···Ph (C1H1B···C10′/C11′) and Ph···Ph (C3H3···C12′) interactions for the P1-containing stack whilst the P21-containing stack has an *off* phenyl ring interaction (between C34-C39 and C34′-C39′) with a centroid–centroid distance of 4.889(3) Å, a centroid-to-plane distance of 3.326(6) Å and a plane-to-plane shift of 3.583(7) Å. The C38-to-plane distance is 3.323(8) Å.

3. Materials and Experimental Methods

3.1. General

Reagents were all obtained commercially. FTIR spectra were obtained as KBr pellets on a Perkin-Elmer 100FTIR spectrometer (Perkin-Elmer, Seer Green, UK). ^{1}H, ^{11}B and ^{13}C and ^{31}P NMR spectra were obtained on a Bruker Avance-500 spectrometer (Bruker, Coventry, UK) on samples dissolved in D_2O at 500, 160, 125 and 202 MHz, respectively. Chemical shifts are in ppm, with positive values to high frequency (downfield) of TMS (^{1}H, ^{13}C), $BF_3 \cdot OEt_2$ (^{11}B) or H_3PO_4 (^{31}P). TGA and DSC were performed on an SDT Q600 instrument (TA Instruments, New Castle, DE, USA) using Al_2O_3 crucibles with a temperature ramp-rate of 10 °C per minute (25 °C to 800 °C in air). BET measurements were performed on a Gemini 2375 analyser (Norcross, GA, USA) with N_2 gas as the adsorbent. Samples were analysed between partial pressures (P/P$_o$) of 0.05 and 0.30. X-ray crystallography was performed at the EPSRC national crystallography service centre at Southampton University. CHN analyses were obtained from OEA Laboratories (Callingham, Cornwall, UK).

3.2. X-ray Crystallography

Crystallographic data for **1** and **2** are given in the experimental section and in the Supplementary Data. Data collection for **1** and **2** was performed on a *Nonius KappaCCD* area detector (ϕ scans and ω scans to fill *asymmetric unit* sphere) diffractometer at 120(2)K. Unit cell parameters were determined using *DirAx* [50], with data collection using Collect [51]. Denzo [52] was used for data reduction and cell refinement and *SORTAV* [53,54] was used for absorption correction. *SHELXS97* [55] was used to solve the structure and was refined using *SHELXL 2018/3 97* [56]. *Olex2* [57] was used for graphics in the Supplementary Information.

3.3. Preparation of [^{i}PrPPh$_3$][B$_5$O$_6$(OH)$_4$]·3.5H$_2$O (1)

[^{i}PrPPh$_3$]I (3.0 g, 6.9 mmol) was dissolved in H_2O (50 mL). To this solution, excess Dowex 550A monosphere (OH$^-$ form) was added and the suspension was stirred for 24 h. The ion-exchange resin was removed by filtration and MeOH (50 mL) was added to the filtrate. B(OH)$_3$ (2.14 g, 34.6 mmol) was added to the resulting solution, which was then heated for 1 h. The solvent was removed by rotary evaporation to yield a solid, which was dried at 110 °C for 24 h to give a white crude product (2.94 g, 73%). NMR. (δ^1H/ppm): 1.30 (6*H*, dd, ^3J(HH) 6.9, ^2J (PH) 18.6 Hz), 3.73 (1*H*, dq, ^3J(HH) 6.9 Hz), 7.50 (6*H*, m), 7.63 (9H, m); (δ(^{13}C/ppm): 15.42 (2 × CH$_3$), 20.09 (CH, d, ^1J(CP) 48.9 Hz), 117.21 (3 × C, d, ^1J(CP) 83.9 Hz), 130.04 (6 × CH, d, ^2J(CP) 12.1 Hz), 133.62 (6 × CH, d, ^3J(CP) 9.2 Hz), 134.78 (3 × CH, d, ^4J(CP) 2.1 Hz); (δ ^{11}B/ppm): 1.1, 12.9, 18.5; (δ ^{31}P/ppm): 30.15. IR (KBr/cm^{-1}); 3292 (vs), 3149 (vs), 1425 (vs), 1404 (vs), 1311 (vs), 1148 (m), 1106 (s), 1095 (s), 1061 (s), 1015 (s), 924 (s), 897 (s), 811 (m), 709 (s). Elemental analysis, $C_{21}H_{33}B_5O_{13.5}P$ req. C 43.0%, H 5.7%, found C, 42.9%, H, 4.9%. TGA: 1st stage-loss of 5.5H$_2$O (<275 °C): 17% expt., 17% calc., 2nd stage-oxidation (275–800 °C) to glassy residue: 36% expt., 30% calc. for 2.5 × B$_2$O$_3$. BET: multi-point surface area (m^2/g) 0.7979 (**1**); 2.7599 (anhydrous); 0.2723 (pyrolysed). Crystals suitable for sc-XRD studies were obtained by crystallization from H$_2$O. sc-XRD: $C_{21}H_{33}B_5O_{13.5}P$, M_r = 586.49, triclinic, *P*−1, *a* = 9.1810(3) Å, *b* = 13.4812(5) Å, *c* = 23.8357(8) Å, α = 75.3130(10)°, β = 83.095(2)°, γ = 86.919(2)° V = 2832.24(17) Å^3,

$Z = 4$, $T = 120(2)$ K, $\lambda = 0.71073$ Å, $D_{(calc.)}$ 1.375 Mg/m, absorption coefficient 0.162 mm^{-1}, $F(000)$ 1228, 50,870 reflections measured, 99.7% complete to $\theta = 27.48°$, 12,947 unique [$R_{int} = 0.0578$], which were used in all calculations. The final $R1 = 0.0563$ ($I > 2\sigma(I)$) and $wR2 = 0.1571$ (all data).

3.4. Synthesis of [MePPh$_3$][B$_5$O$_6$(OH)$_4$]·B(OH)$_3$·0.5H$_2$O (2)

[MePPh$_3$]I (2.50 g, 6.2 mmol) was dissolved in H$_2$O (50 mL). To this solution, excess Dowex 550A monosphere (OH$^-$ form) was added and the suspension was stirred for 24 h. The ion-exchange resin was removed by filtration and MeOH (50 mL) was added to the filtrate. B(OH)$_3$ (1.91 g, 30.9 mmol) was added to the resulting solution, which was then heated for 1 h. The solvent was removed by rotary evaporation to yield an orange solid as the crude product, which was dried at 110 °C for 24 h (2.84 g, 98%). NMR. (δ^1H/ppm): 2.69 (3H, d, ^2J(HP) 13.8 Hz), 7.50 (12H, m), 7.68 (3H, m); (δ^{13}C/ppm): 7.96 (CH$_3$, d, ^1J(CP) 58.5 Hz), 119.06 (C, d, ^1J(CP) 89.2 Hz); (δ^{11}B/ppm) ppm: 1.2, 13.1, 18.6; (δ^{31}P/ppm): 21.05. IR (KBr/cm^{-1}): 3307 (s), 1417 (vs), 1296 (vs), 1118 (s), 1019 (m), 918 (s), 775 (m), 748 (m), 720 (m), 688 (m). Elemental analysis, C$_{19}$H$_{26}$B$_6$O$_{13.5}$P req. C 40.3%, H 4.6%, found C, 42.8%, H, 4.6%. TGA: 1st stage-loss of 4H$_2$O (<275 °C): 12% expt., 13% calc., 2nd-stage oxidation (275–800 °C) to glassy residue: 42% expt., 37% calc. for 3 × B$_2$O$_3$. BET: multi-point surface area (m^2/g) 0.5098 (2); 0.4134 (anhydrous); 0.2226 (pyrolysed). Crystals suitable for sc-XRD were obtained by crystallization from H$_2$O. sc-XRD: C$_{38}$H$_{52}$B$_{12}$O$_{27}$P$_2$, M_r = 1132.46, triclinic, $P-1$, $a = 10.1076(10)$ Å, $b = 13.0403(10)$ Å, $c = 20.6260(15)$ Å, $\alpha = 84.364(5)°$, $\beta = 82.811(4)°$, $\gamma = 76.492(4)°$, $V = 2615.9(4)$ Å^3, $Z = 2$, $T = 120(2)$ K, $\lambda = 0.71073$ Å, $D_{(calc.)}$ 1.438 Mg/m, absorption coefficient 0.172 mm^{-1}, $F(000)$ 1172, 35,616 reflections measured, 98.8% complete to $\theta = 27.48°$, 12,083 unique [$R_{int} = 0.0649$], which were used in all calculations. The final $R1 = 0.0742$ ($I > 2\sigma(I)$) and $wR2 = 0.1676$ (all data).

4. Conclusions

Two substituted aryl phosphonium pentaborate salts were synthesized by templated crystallization from aqueous solution primed with B(OH)$_3$ and appropriate aryl phosphonium cation and characterized by sc-XRD. Their structures are unusual for pentaborates in that [^{i}PrPPh$_3$][B$_5$O$_6$(OH)$_4$]·3.5H$_2$O has alternating layers of cations and anions whilst [MePPh$_3$][B$_5$O$_6$(OH)$_4$]·B(OH)$_3$·0.5H$_2$O has a rectangular honeycomb-like structure with cations stacked within channels. Despite their unusual structures, the materials derived by thermal oxidation of the cations are non-porous. The solid-state structures of both compounds are stabilized by multiple H-bonding and phenyl embrace interactions.

Supplementary Materials: The following supporting information can be downloaded at: https://www.mdpi.com/article/10.3390/molecules28196867/s1, Crystallographic data for **1** and **2** are available as Supplementary Materials. CCDC 2291682 (**1**) and 2291683 (**2**) also contain the supplementary crystallographic data for this paper. These CCDC data can be obtained free of charge via http://www.ccdc.cam.ac.uk/conts/retrieving.html (or from CCDC, 12 Union Road, Cambridge, CB2 1EZ. Fax: +44-1223-336033; E-mail: deposit@ccdc.cam.ac.uk).

Author Contributions: Conceptualization, M.A.B. and J.L.T.; experimental methodology, J.L.T. and P.N.H.; writing—original draft preparation, M.A.B. and J.L.T.; writing—review and editing, M.A.B. and P.N.H.; supervision, M.A.B.; funding acquisition, M.A.B. and M.B.H. All authors have read and agreed to the published version of the manuscript.

Funding: This research received no external funding.

Institutional Review Board Statement: Not applicable.

Informed Consent Statement: Not applicable.

Data Availability Statement: Crystallographic data available from CCDC, UK; see Supplementary Material for details.

Acknowledgments: We thank the EPSRC for the NCS X-ray crystallography service (Southampton).

Conflicts of Interest: The authors declare no conflict of interest.

Sample Availability: Samples of the compounds are not available from the authors.

References

1. Farmer, J.B. Metal borates. *Adv. Inorg. Chem Radiochem.* **1982**, *25*, 187–237.
2. Heller, G. A survey of structural types of borates and polyborates. *Top. Curr. Chem.* **1986**, *131*, 39–98.
3. Belokonova, E.L. Borate crystal chemistry in terms of extended OD theory and symmetry analysis. *Crystallogr. Rev.* **2005**, *11*, 151–198. [CrossRef]
4. Topnikova, A.P.; Belokoneva, E.L. The structure and classification of complex borates. *Russ. Chem. Rev.* **2019**, *88*, 204–228. [CrossRef]
5. Burns, P.C.; Grice, J.D.; Hawthorne, F.C. Borate minerals I. Polyhedral clusters and fundamental building blocks. *Can. Mineral.* **1995**, *33*, 1131–1151.
6. Grice, J.D.; Burns, P.C.; Hawthorne, F.C. Borate minerals II. A hierarchy of structures based upon the borate fundamental building block. *Can. Mineral.* **1999**, *37*, 731–762.
7. Christ, C.L.; Clark, J.R. A crystal-chemical classification of borate structures with emphasis on hydrated borates. *Phys. Chem. Miner.* **1977**, *2*, 59–87. [CrossRef]
8. Touboul, M.; Penin, N.; Nowogrocki, G. Borates: A survey of main trends concerning crystal chemistry, polymorphism and dehydration process of alkaline and pseudo-alkaline borates. *Solid State Sci.* **2003**, *5*, 1327–1342. [CrossRef]
9. Mutailipu, M.; Poeppelmeier, K.R.; Pan, S. Borates: A rich source for optical materials. *Chem. Rev.* **2021**, *121*, 1130–1202. [CrossRef]
10. Beckett, M.A. Recent Advances in crystalline hydrated borates with non-metal or transition-metal complex cations. *Coord. Chem. Rev.* **2016**, *323*, 2–14. [CrossRef]
11. Schubert, D.M.; Smith, R.A.; Visi, M.Z. Studies of crystalline non-metal borates. *Glass Technol.* **2003**, *44*, 63–70.
12. Schubert, D.M.; Knobler, C.B. Recent studies of polyborate anions. *Phys. Chem. Glasses Eur. J. Glass Sci. Technol. B* **2009**, *50*, 71–78.
13. Schubert, D.M. Borates in industrial use. *Struct. Bond.* **2003**, *105*, 1–40.
14. Schubert, D.M. Boron oxide, boric acid, and borates. In *Kirk-Othmer Encyclopedia of Chemical Technology*, 5th ed.; John Wiley & Sons: Hoboken, NJ, USA, 2011; pp. 1–68.
15. Schubert, D.M. Hydrated zinc borates and their industrial use. *Molecules* **2019**, *24*, 2419. [CrossRef]
16. Becker, P. Borate materials in nonlinear optics. *Adv. Mater.* **1998**, *10*, 979–992. [CrossRef]
17. Xin, S.-S.; Zhou, M.-H.; Beckett, M.A.; Pan, C.-Y. Recent advances in crystalline oxidopolyborate complexes of d-block or p-block metals: Structural aspects, synthesis, and physical properties. *Molecules* **2021**, *26*, 3815. [CrossRef]
18. Beckett, M.A.; Coles, S.J.; Horton, P.N.; Jones, C.L. Polyborate anions partnered with large non-metal cations: Triborate(1-), pentaborate(1-) and heptaborate(2-) salts. *Eur. J. Inorg. Chem.* **2017**, 4510–4518. [CrossRef]
19. Beckett, M.A.; Horton, P.N.; Hursthouse, M.B.; Timmis, J.L.; Varma, K.S. Synthesis, thermal properties and structural characterization of the tetraphenylphosphonium pentaborate salt, [PPh$_4$][B$_5$O$_6$(OH)$_4$]·1.5H$_2$O. *Inorg. Chim. Acta* **2012**, *383*, 199–203. [CrossRef]
20. Beckett, M.A.; Horton, P.N.; Hursthouse, M.B.; Knox, D.A.; Timmis, J.L. Structural (XRD) and thermal (DSC, TGA) and BET analysis of materials derived from non-metal cation pentaborate salts. *Dalton Trans.* **2010**, *39*, 3944–3951. [CrossRef]
21. Beckett, M.A.; Horton, P.N.; Hursthouse, M.B.; Timmis, J.L.; Varma, K.S. Templated heptaborate and pentaborate salts of cyclo-alkylammonium cations: Structural and thermal properties. *Dalton Trans.* **2012**, *41*, 4396–4403. [CrossRef]
22. Beckett, M.A.; Horton, P.N.; Hursthouse, M.B.; Timmis, J.L. Triborate and pentaborate salts of non-metal cations derived from N-substituted piperazines: Synthesis and structural (XRD) and thermal properties. *RSC Adv.* **2013**, *3*, 15181–15191. [CrossRef]
23. Beckett, M.A.; Brellocks, B.; Chizhevsky, I.T.; Damhus, T.; Hellwich, K.-H.; Kennedy, J.D.; Laitinen, R.; Powell, W.H.; Rabinovich, D.; Vinas, C.; et al. Nomenclature for boranes and related species (IUPAC Recommendations 2019). *Pure Appl. Chem.* **2020**, *92*, 355–381. [CrossRef]
24. Visi, M.Z.; Knobler, C.B.; Owen, J.J.; Khan, M.I.; Schubert, D.M. Structures of self-assembled nonmetal borates derived from α,ω-diaminoalkanes. *Cryst. Growth Des.* **2006**, *6*, 538–545. [CrossRef]
25. Anderson, J.L.; Eyring, E.M.; Whittaker, M.P. Temperature jump rate studies of polyborate formation in aqueous boric acid. *J. Phys. Chem.* **1964**, *68*, 1128–1132. [CrossRef]
26. Salentine, G. High-field ^{11}B NMR of alkali borate. Aqueous polyborate equilibria. *Inorg. Chem.* **1983**, *22*, 3920–3924. [CrossRef]
27. Liu, H.; Liu, Q.; Lan, Y.; Wang, D.; Zhang, L.; Tang, X.; Yang, S.; Luo, Z.; Tian, G. Speciation of borate in aqueous solution studied experimentally by potentiometry and Raman spectroscopy and computationally by DFT calculations. *New J. Chem.* **2023**, *47*, 8499–8506. [CrossRef]
28. Corbett, P.T.; Leclaire, J.; Vial, L.; West, K.R.; Wietor, J.-L.; Sanders, J.K.M.; Otto, S. Dynamic combinatorial chemistry. *Chem. Rev.* **2006**, *106*, 3652–3711. [CrossRef]
29. Sola, J.; Lafuente, M.; Atcher, J.; Alfonso, I. Constitutional self-selection from dynamic combinatorial libraries in aqueous solution through supramolecular interactions. *Chem. Commun.* **2014**, *50*, 4564–4566. [CrossRef]
30. Desiraju, G.R. Supramolecular synthons in crystal engineering—A new organic synthesis. *Angew. Chem. Int. Ed. Engl.* **1995**, *34*, 2311–2327. [CrossRef]

31. Dunitz, J.D.; Gavezzotti, A. Supramolecular synthons: Validation and ranking of intermolecular interaction energies. *Cryst. Growth Des.* **2012**, *12*, 5873–5877. [CrossRef]
32. Freyhardt, C.C.; Wiebcke, M.; Felsche, J.; Engelhardt, G. Clathrates and three dimensional host structures of hydrogen bonded pentaborate $[B_5O_6(OH)_4]^-$ ions: Pentaborates with cations NMe_4^+, NEt_4^+, $NPhMe_3^+$ and $pipH^+$ ($pipH^+$ = piperidinium). *Z. Naturforsch B.* **1993**, *48*, 978–985.
33. Beckett, M.A.; Coles, S.J.; Horton, P.N.; Rixon, T.A. Structural (XRD) characterization and an analysis of H-bonding motifs in some tetrahydroxidohexaoxidopentaborate(1-) salts of *N*-substituted guanidinium cations. *Molecules* **2023**, *28*, 3273. [CrossRef] [PubMed]
34. Beckett, M.A.; Horton, P.N.; Coles, S.J.; Kose, D.A.; Kreuziger, A.-M. Structural and thermal studies of non-metal cation pentaborate salts with cations derived from 1,5-diazobicyclo[4.3.0]non-5-ene, 1,8-diazobicyclo[5.4.0]undec-7-ene and 1,8-bis(dimethylamino)naphthalene. *Polyhedron* **2012**, *38*, 157–161. [CrossRef]
35. Yang, Y.; Fu, D.S.; Li, G.F.; Zhang, Y. Synthesis, crystal structure, and variable-temperature-luminescent property of the organically templated pentaborate $[C_{10}N_2H_9][B_5O_6(OH)_4]\cdot H_3BO_3\cdot H_2O$. *Z. Anorg. Chem.* **2013**, *639*, 722–727. [CrossRef]
36. Freyhardt, C.C.; Wiebcke, M.; Felsche, J.; Engelhardt, G. $N(^nPr_4)[B_5O_6(OH)_4][B(OH)_3]_2$ and $N(^nBu_4)[B_5O_6(OH)_4][B(OH)_3]_2$: Clathrates with a diamondoid arrangement of hydrogen bonded pentaborate anions. *J. Inclusion Phenom. Mol. Recogn. Chem.* **1994**, *18*, 161–175. [CrossRef]
37. Ferrillo, R.G.; Granzow, A. Thermogravimetric study of phosphonium halides. *Thermochem. Acta* **1981**, *45*, 177–187. [CrossRef]
38. Brauner, S.; Emmett, P.H.; Teller, E. Adsorption of gases in multimolecular layers. *J. Am. Chem. Soc.* **1938**, *60*, 309–319. [CrossRef]
39. Timmis, J.L. Characterization of Non-Metal Cation Polyborate Salts and Silicate Solutions. Ph.D. Thesis, Bangor University, Bangor, UK, 2011.
40. Schubert, U.; Husing, N. *Synthesis of Inorganic Materials*, 2nd ed.; Wiley VCH: Weinheim, Germany, 2007; Volume Ch 6, pp. 305–352.
41. Li, J.; Xia, S.; Gao, S. FT-IR and Raman spectroscopic study of hydrated borates. *Spectrochim. Acta* **1995**, *51*, 519–532.
42. Grim, S.O.; McFarlane, W.; Davidoff, E.F.; Marks, T.J. Phosphorus-31 chemical shifts of quaternary phosphonium salts. *J. Am. Chem. Soc.* **1966**, *70*, 581–584. [CrossRef]
43. Beckett, M.A.; Coles, S.J.; Davies, R.A.; Horton, P.N.; Jones, C.L. Pentaborate(1−) salts templated by substituted pyrrolidinium cations: Synthesis, structural characterization, and modelling of solid-state H-bond interactions by DFT calculations. *Dalton Trans.* **2015**, *44*, 7032–7040. [CrossRef]
44. Beckett, M.A.; Brassington, D.S.; Owen, P.; Hursthouse, M.B.; Light, M.E.; Malik, K.M.A.; Varma, K.S. π-Bonding in B-O ring species: Lewis acidity of $Me_3B_3O_3$, synthesis of $Me_3B_3O_3$ amine adducts, and the crystal and molecular structure of $Me_3B_3O_3\cdot NH_2^iBu\cdot MeB(OH)_2$. *J. Organomet. Chem.* **1999**, *585*, 7–11. [CrossRef]
45. Hosten, E.; Gerber, T.; Betz, R. Crystal structure of methyltriphenylphosphonium iodide, $C_{19}H_{18}IP$. *Z. Kristallogr. NCS* **2012**, *227*, 331–332.
46. Jaliliana, E.; Lidi, S. Bis(isopropyltriphenylphosphonium)di-µ-iodidobis[iodidocopper(I)]. *Acta Cryst.* **2010**, *E66*, m432–m433.
47. Hunter, C.A.; Sanders, J.K.M. The nature of π-π-interactions. *J. Am. Chem. Soc.* **1990**, *112*, 5525–5534. [CrossRef]
48. Dance, I.; Scudder, M. Supramolecular motifs: Concerted multiple phenyl embraces between PPh_4^+ cations are attractive and ubiquitous. *Chem. Eur. J.* **1996**, *2*, 481–486. [CrossRef]
49. Etter, M.C. Encoding and decoding hydrogen-bond patterns of organic chemistry. *Acc. Chem. Res.* **1990**, *23*, 120–126. [CrossRef]
50. Duisenberg, A.J.M. Indexing in single-crystal diffractometry with an obstinate list of reflections. *J. Appl. Cryst.* **1992**, *25*, 92–96. [CrossRef]
51. Hooft, R.; Nonius, B.V. COLLECT, Data Collection Software. 1998.
52. Otwinowski, Z.; Minor, W. Processing of X-ray diffraction data collected in oscillation mode. *Meth. Enzymol.* **1997**, *276*, 307–326.
53. Blessing, R.H. An empirical correction for absorption anisotropy. *Acta Cryst.* **1995**, *A51*, 33–37. [CrossRef]
54. Blessing, R.H. Outlier Treatment in Data Merging. *J. Appl. Cryst.* **1997**, *30*, 421–426. [CrossRef]
55. Sheldrick, G.M. A short history of ShelX. *Acta Cryst.* **2008**, *A64*, 339–341.
56. Sheldrick, G.M. Crystal structure refinement with ShelXL. *Acta Cryst.* **2015**, *C71*, 3–8.
57. Dolomanov, O.V.; Bourhis, L.J.; Gildea, R.J.; Howard, J.A.K.; Puschmann, H. Olex2: A complete structure solution, refinement and analysis program. *J. Appl. Cryst.* **2009**, *42*, 339–341. [CrossRef]

Disclaimer/Publisher's Note: The statements, opinions and data contained in all publications are solely those of the individual author(s) and contributor(s) and not of MDPI and/or the editor(s). MDPI and/or the editor(s) disclaim responsibility for any injury to people or property resulting from any ideas, methods, instructions or products referred to in the content.

Article

Constrained Phosphine Chalcogenide Selenoethers Supported by *peri*-Substitution

Anna E. Tarcza, Alexandra M. Z. Slawin, Cameron L. Carpenter-Warren, Michael Bühl, Petr Kilian and Brian A. Chalmers *

EaStCHEM School of Chemistry, University of St Andrews, North Haugh, St Andrews, Fife KY16 9ST, UK
* Correspondence: bac8@st-andrews.ac.uk; Tel.: +44-1334-463785

Abstract: A series of phosphorus and selenium *peri*-substituted acenaphthene species with the phosphino group oxidized by O, S, and Se has been isolated and fully characterized, including by single-crystal X-ray diffraction. The P(V) and Se(II) systems showed fluxional behavior in solution due to the presence of two major rotamers, as evidenced with solution NMR spectroscopy. Using Variable-Temperature NMR (VT NMR) and supported by DFT (Density Functional Theory) calculations and solid-state NMR, the major rotamers in the solid and in solution were identified. All compounds showed a loss of the through-space J_{PSe} coupling observed in the unoxidized P(III) and Se(II) systems due to the sequestration of the lone pair of the phosphine, which has been previously identified as the major contributor to the coupling pathway.

Keywords: *peri*-substitution; selenium; phosphorus; NMR; single-crystal X-ray structures; rotational conformation; DFT calculations

Citation: Tarcza, A.E.; Slawin, A.M.Z.; Carpenter-Warren, C.L.; Bühl, M.; Kilian, P.; Chalmers, B.A. Constrained Phosphine Chalcogenide Selenoethers Supported by *peri*-Substitution. *Molecules* **2023**, *28*, 7297. https://doi.org/10.3390/molecules28217297

Academic Editor: Yves Canac

Received: 9 October 2023
Revised: 24 October 2023
Accepted: 26 October 2023
Published: 27 October 2023

Copyright: © 2023 by the authors. Licensee MDPI, Basel, Switzerland. This article is an open access article distributed under the terms and conditions of the Creative Commons Attribution (CC BY) license (https://creativecommons.org/licenses/by/4.0/).

1. Dedication

This paper is dedicated to Professor J. Derek Woollins on the occasion of his well-earned retirement and for his outstanding contributions to main group chemistry.

2. Introduction

The selective, stepwise lithiation reaction of 5,6-dibromoacenaphthene allows synthetic access to heteroleptic bis(phosphino)acenaphthenes and has been used to synthesize bis(phosphine) **A** (Figure 1) [1]. Due to the inherent asymmetry of the heteroleptic phosphine groups, in the $^{31}P\{^{1}H\}$ NMR spectrum, **A** shows two doublets of an AB spin system at δ_P −11.3 and −12.8 ppm, with a remarkably large $^{4TS}J_{PP}$ of 180.0 Hz. This is attributed to the through-space coupling resulting from the overlap of the phosphorus lone pairs due to the constraints imposed by the rigid acenaphthene skeleton. Oxidation of the P(III) centers to P(V) with sulfur, atmospheric oxygen, or hydrogen peroxide results in a loss or significant decrease in the magnitude of the through-space J_{PP} coupling as the lone pairs of the phosphines are sequestered [1,2]. In only a handful of cases, where **A** acts as a bidentate ligand with MCl$_2$ (M = Zn, Cd, Hg), the magnitude of J_{PP} increases as the coupling is mediated by the large, diffuse *s*-character orbitals of the group 12 metals (e.g., **A**·HgCl$_2$ J_{PP} 309 Hz, Figure 1) [2].

Heteroleptic substitution is not limited to phosphorus substituents but can also involve other *p*-block and *d*-block heteroatoms (for some examples, see references [3–11]). Not only do heteroatoms present a challenging synthetic opportunity for *peri*-substitution, but they yield interesting NMR spectra when both nuclei are NMR active, as these nuclei can also experience through-space spin–spin coupling. When heavier nuclei are used, the orbitals are larger and more diffuse. As a result, through-space coupling can occur at longer *peri*-distances [12]. An excellent example of this is the series of phosphine–tin *peri*-substituted acenaphthene reported by Athukorala Arachchige et al. where $^{31}P \cdots ^{119}Sn$

J coupling can be observed [13]. The ^{119}Sn isotope has $I = \frac{1}{2}$ and a natural abundance of 8.6%, making it possible to observe J coupling with ^{31}P ($I = \frac{1}{2}$, 100%). In **B** (Figure 1), there is a direct P–Sn bond (2.815(3) Å) with $^1J_{PSn}$ 754 Hz, yet, in **C**, where there is no direct P–Sn bond but there is a sub-van der Waals P···Sn interaction (3.251(1) Å), a significant J_{PSn} of 373 Hz is still observed, demonstrating a clear 3c–4e type overlap of the phosphorus lone pair with the Sn–C$_{Ph}$ σ* orbital. Other P/Sn acenaphthenes have also been reported with diphenylphosphino groups instead of diisopropylphosphino groups [14].

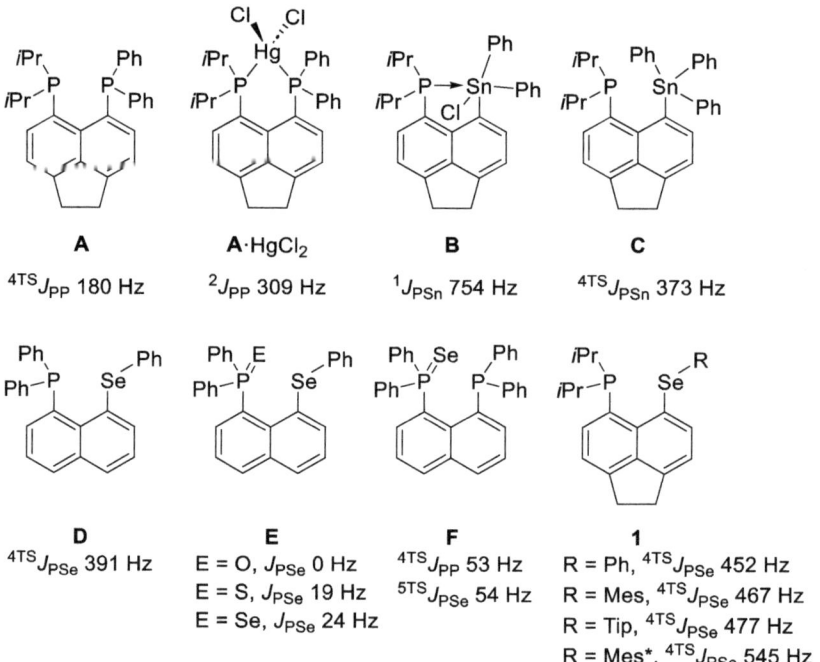

Figure 1. Compounds discussed in the introduction with selected J couplings shown (Mes = 2,4,6-trimethylphenyl; Tip = 2,4,5-triisopropylphenyl; Mes* = 2,4,6-tri-tert-butylphenyl).

Woollins et al. previously published a series of naphthalene-based phosphine selenoethers [15]. In **D**, there is an efficient transfer of spin information between P and Se, as indicated by the $^{4TS}J_{PSe}$ of 391 Hz (note, TS superscript indicates through-space coupling). When the P(III) center is oxidized with chalcogens to P(V) (compounds **E**), the magnitude of J_{PSe} diminishes to <24 Hz (Figure 1) [15]. An in-depth computational study has shown that the magnitude of J_{PP} and J_{PSe} in the related compound **F** has contributions from both through-space and through-bond pathways [16]. We recently reported a series of acenaphthene analogues (**1**) with various aryl groups bound to selenium [17]. As the electron-donating ability of the aryl group attached to selenium increases, so does the magnitude of J_{PSe} from 452 Hz, when R = phenyl, up to 545 Hz, when R = Mes* (2,4,6-tri-*tert*-butylphenyl).

3. Results and Discussion

3.1. Synthesis

Utilizing compound **1$_{Ph}$** as our workhorse, we herein report the synthesis and characterization of the P(V) chalcogen oxidized species **1-O**, **1-S**, and **1-Se** and the P(V)/Se(IV) species **1-O2** (Scheme 1).

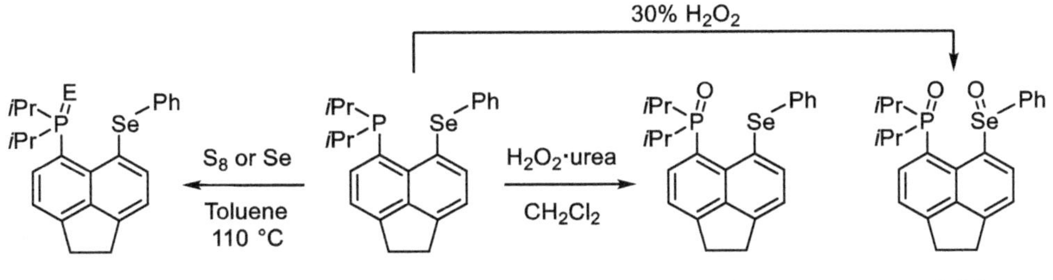

Scheme 1. The synthetic pathway for compounds **1-O**, **1-S**, **1-Se**, and **1-O2**.

Compound **1$_{Ph}$** showed a singlet in the ^{31}P{^{1}H} NMR spectrum at δ_P −6.5 ppm with ^{77}Se satellites giving J_{PSe} 452.2 Hz. This was complemented by a doublet at δ_{Se} 425.3 ppm, observed in the ^{77}Se{1H} NMR spectrum. Heating a solution of **1$_{Ph}$** under reflux in toluene with one equivalent of gray selenium for 15 h, followed by purification, afforded **1-Se** as a yellow microcrystalline powder (54% yield). **1-S** was prepared in a similar manner but only required heating under reflux for 6 h for complete consumption of **1$_{Ph}$**. After purification, **1-S** was afforded as beige microcrystalline powder (86% yield).

The first attempt to produce **1-O** using hydrogen peroxide resulted in the formation of a mixture of **1-O** and **1-O2**, as determined by solution ^{77}Se{^{1}H} NMR spectroscopy. Direct synthesis of **1-O2** was achieved by using an excess of hydrogen peroxide. After purification, **1-O2** was obtained as a white solid (45% yield). To control the oxidation of **1$_{Ph}$** to selectively oxidize the phosphine, we attempted air oxidation by leaving a vigorously stirring solution of **1$_{Ph}$** exposed to air; however, even after 24 h, no reaction had occurred, as judged by ^{31}P{^{1}H} NMR spectroscopy. Instead, **1$_{Ph}$** was stirred with one equivalent of H$_2$O$_2$·urea complex. The conversion was slow, but, as monitored by ^{31}P{^{1}H} NMR spectroscopy, complete consumption of **1$_{Ph}$** was observed after 72 h. After recrystallization, **1-O** was isolated in a 39% yield. The mechanism of the phosphine oxidation was not studied in the scope of this work, as P(III) to P(V) oxidations by peroxides, cyclooctasufur (S$_8$), and gray selenium are well established from early thermochemical and mechanistic studies [18–20]. All compounds reported herein were found to be air stable, in the solid state, with no signs of degradation after twelve months.

3.2. Crystallography

Crystals of **1-Se**, **1-S**, and **1-O** were grown from a solution of CH$_2$Cl$_2$:hexane (1:3 v/v), and crystals of **1-O2** were grown from evaporation of a solution in CH$_2$Cl$_2$. The structures of **1-O**, **1-S**, and **1-Se** are very similar with only minor differences due to the increased size of the chalcogen bound to phosphorus. The crystal structures are shown in Figure 2, and selected crystallographic data are presented in Table 1.

The most notable differences between the structures of the precursor (**1$_{Ph}$**) [17] and the oxidized P(V) species (**1-O**, **1-S**, and **1-Se**) are in the *peri*-region. In **1$_{Ph}$**, the P···Se distance is 3.055(1) Å; this increases to 3.322(2) Å in **1-O**, 3.4863(5) Å in **1-S**, and 3.5012(7) Å in **1-Se**. Similarly, there are large increases in the splay angles (12.6° in **1$_{Ph}$** to 32.0° in **1-Se**), the P–C···C–Se dihedral angles, and the out-of-plane displacements of the P and Se atoms from the mean C$_{12}$ acenaphthene plane (see Table 1). These changes are expected due to the new steric demands placed on the molecule caused by the addition of another atom into the *peri*-gap when the iPr$_2$P group is oxidized to iPr$_2$P = E (where E = O, S, Se). When compared to the crystal structures of the naphthalene analogues (**E**, Figure 1), there are no significant differences [15]. The only dissimilarity observed between the acenaphthene and naphthalene analogues is that the absence of the ethylene bridge in the naphthalene structures results in a slightly decreased P···Se distance and slightly smaller splay angles. For example, in the naphthalene compound **E** (where E = Se), the P⋯Se distance is 3.278(2) Å (vs. 3.5012(7) Å for **1-Se**), and the splay angle drops from 32.0° to 24.8°.

Table 1. Selected bond lengths (ångströms (Å)) and angles (degrees, °) for **1-O**, **1-S**, **1-Se**, and **1-O2**.

Compound	1-O	1-S	1-Se	1-O2 [b]
		peri-region bond distances		
P1⋯Se1	3.322(2)	3.4863(5)	3.5012(7)	3.578(1) [3.610(1)]
P1-E	1.491(6)	1.9657(5)	2.1219(7)	1.487(4) [1.492(4)]
Se1⋯E	2.825(6)	3.2272(5)	3.2829(6)	2.646(3) [2.625(3)]
Se1-O1	—	—	—	1.669(3) [1.671(4)]
		peri-region bond angles		
C9-P1-E	110.7(4)	112.39(5)	112.40(8)	112.7(2) [113.1(2)]
P1-E-Se1	95.7(3)	80.36(2)	77.47(2)	117.1(2) [118.5(2)]
E-Se1-C19	165.4(3)	164.53(5)	166.29(8)	84.6(2) [86.4(2)]
O2-Se1-O1	—	—	—	169.3(1) [169.2(2)]
C1-Se1-C19	99.3(3)	97.01(6)	96.8(1)	97.7(2) [95.4(2)]
Splay [a]	18.5	19.9	20.2	28.1 [29.1]
		dihedral angles		
C9-C10-C5-C4	174.8(8)	173.7(1)	173.2(2)	178.3(5) [177.4(5)]
P1-C9⋯C1-Se1	22.2(5)	31.87(8)	32.0(1)	1.8(3) [1.7(3)]
		out-of-plane displacements		
P1	0.508	0.605	0.593	0.065 [0.007]
Se1	−0.393	−0.700	−0.725	0.006 [0.128]

[a] splay angle = sum of the bay region angles—360. [b] values in square parentheses are for the 2nd molecule in the asymmetric unit.

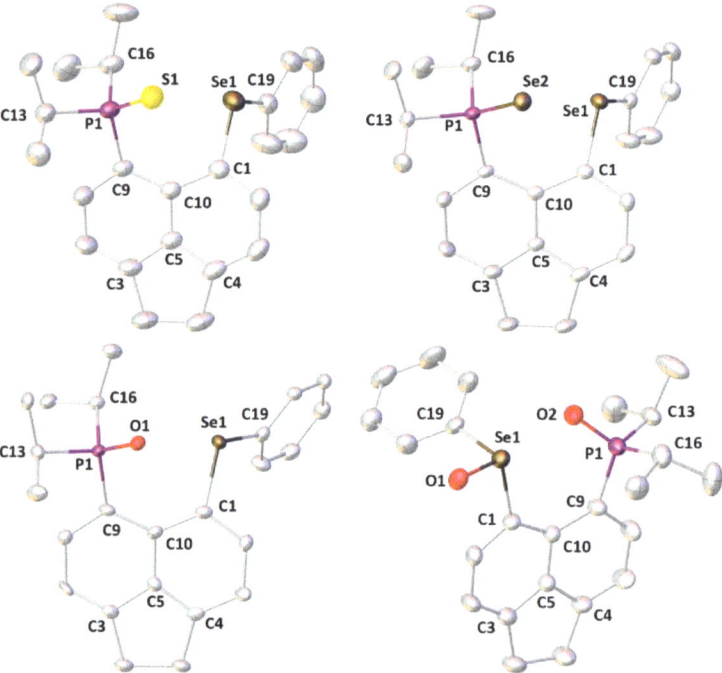

Figure 2. Molecular structures of **1-S**, **1-Se**, **1-O**, and **1-O2**. Hydrogen atoms and the second molecule in the asymmetric unit of **1-O2** are omitted for clarity. Anisotropic displacement ellipsoids are plotted at the 50% probability level.

The structure of **1-O2** is very different from that of the other three due to the Se(IV) group (selenoxide) being present. The trends are similar to those observed in the Se(II) complexes, but taken to a new extreme. There is a greater steric demand on the molecule, as evidenced by the much larger splay angle of 28° (29° in the second molecule of the asymmetric unit) and *peri*-distance between the P and Se atoms of 3.578(1) Å (3.610(1) Å). Somewhat unexpectedly, the dihedral angle is much smaller at 1.8(3)° (1.7(3)°); however, this arises from the rotation around the C9–Se1 bond, such that the Se=O group points away from the *peri*-gap, which significantly reduces the steric crowding and the need for any out-of-plane deformations to relieve the steric strain.

3.3. NMR Spectroscopy of 1-Se

In solution, the precursor **1$_{Ph}$** showed a sharp singlet in the ^{31}P{^{1}H} NMR spectrum at δ_P −6.5 ppm with ^{77}Se satellites (7.6% natural abundance), giving $^{4TS}J_{PSe}$ of 452.2 Hz. This was complemented by a doublet in the ^{77}Se{^{1}H} NMR spectrum centered at δ_{Se} 425.3 ppm ($^{4TS}J_{SeP}$ 452.8 Hz). Recently, we have shown that the large through-space coupling between ^{31}P and ^{77}Se arises from the overlap of the phosphorus lone pair with the orbitals localized on the Se–C$_{Ph}$ bond [17]. In this study, the lone pair of the phosphorus was sequestered by oxidation with a chalcogen atom in all complexes, and as expected, this significantly reduced the magnitude of J_{PSe} in all compounds. Based on the recent findings by Makina et al. [16], we assume the dominant pathway of coupling information being exchanged is through space in the unoxidized **1$_{Ph}$**; therefore, we attribute the drop in magnitude to J_{PSe} to the loss of this pathway.

3.3.1. Fluxionality in Solution

The ^{31}P{^{1}H} NMR spectrum of **1-Se** revealed two singlets, at δ_P 86.3 and 58.4 ppm, when only one was anticipated (Figure 3, top). The signal at δ_P 58.4 ppm was accompanied by a broadened set of ^{77}Se satellites with $^1J_{PSe}$ of ca. 690 Hz. This was indicative of a P = Se double bond and closely resembled those reported in the literature (cf. Ph$_3$PSe; $^1J_{PSe}$ 730 Hz) [21]. The signal at δ_P 86.3 ppm showed significant broadening and no resolvable ^{77}Se satellites. The presence of two rotational conformers was confirmed by ^{77}Se{^{1}H} NMR with the spectrum showing two broad singlets at δ_{Se} 426.2 and 419.0 ppm corresponding to the selenoether, as well as two doublets at δ_{Se} −358.5 and −451.0 ppm ($^1J_{SeP}$ ca. 696 and 693 Hz, respectively) corresponding to the phosphine selenide (Figure 3, bottom). Initially, the presence of two sets of peaks suggested the occurrence of a side reaction; however, upon further investigation using Variable-Temperature NMR, it was concluded they were due to the very large steric bulk around the *peri*-region exhibiting fluxional behavior in solution. Acquisition of a ^{1}H–^{31}P HMBC spectrum (Figure S1) showed a strong correlation between both ^{31}P signals and the hydrogen atoms in the isopropyl groups, strongly supporting the idea of fluxional behavior. This was unexpected as the naphthalene equivalent (**E**) was reported as showing one sharp singlet in the ^{31}P{^{1}H} NMR spectrum, which does not suggest fluxional behavior [15].

To obtain further evidence of the fluxional behavior, a one-dimensional exchange spectroscopy (EXSY) NMR experiment was performed on **1-Se**. This showed magnetization transfer within the NMR timescale at 253 K. This indicated an exchange between two magnetic environments in two different isomers and confirmed the presence of fluxional behavior in solution. Therefore, the two sets of signals observed in the ambient temperature (293 K) ^{31}P{^{1}H} and ^{77}Se{^{1}H} NMR spectra arose from two different rotational conformations present in solution. The notion of different rotational conformations in solution in *peri*-substituted naphthalenes has been reported by Woollins previously [22]. The species Nap(POCl$_2$)(PCl$_2$) (Nap = naphthalene-1,8-diyl), with a P(III)/P(V) *peri*-substitution, was demonstrated to have two rotamers in solution, with the ^{31}P{^{1}H} NMR spectrum at 233 K showing two similar signals for the PCl$_2$ group (δ_P 145.52 and 145.50 ppm) and one signal for the POCl$_2$ group (δ_P 42.9 ppm). At 298 K, the signals at δ_P 145.52 and 145.50 ppm were not observed. Kilian et al. reported that these results are interpreted as "the hindered rotation around the P-C$_{(Nap)}$ bonds, resulting the presence of two conformers whose interconversion is slow on the NMR time scale".

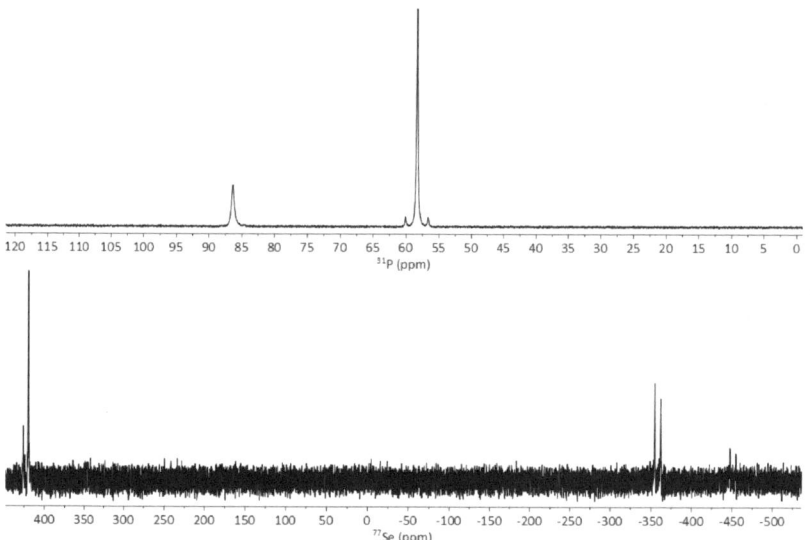

Figure 3. The ambient temperature ^{31}P{^{1}H} (**top**) and ^{77}Se{^{1}H} (**bottom**) NMR spectra of **1-Se**, acquired at 202.5 and 95.4 MHz, respectively.

Variable-temperature ^{31}P{^{1}H} and ^{77}Se{^{1}H} NMR experiments were carried out using **1-Se**; however, to overcome the coalescence point, a high-boiling solvent was needed. For the elevated-temperature experiments, d$_5$-bromobenzene (boiling point 156 °C, 429 K) was used as the NMR solvent. For low-temperature experiments, d-chloroform (melting point −64 °C, 209 K) was used. At 253 K, fully resolved signals of the two conformations with observable satellites were observed. The ^{31}P{^{1}H} NMR spectrum at 253 K showed two singlets at δ$_P$ 86.2 and 58.0 ppm, with $^1J_{PSe}$ of 682.9 and 681.3 Hz, respectively (Figure 4). The ^{77}Se{^{1}H} NMR spectrum at 255 K showed two singlets at δ$_{Se}$ 422.8 and 415.3 ppm corresponding to the selenoether and two doublets at δ$_{Se}$, −362.6 and −452.4 ppm, with $^1J_{SeP}$ 681.5 and 683.6 Hz, corresponding to the phosphine selenide (Figure 5). The singlet at δ$_{Se}$ 415.3 ppm also showed a $^{5TS}J_{SeSe}$ coupling of 182.0 Hz as ^{77}Se satellites (Figure 5). Even in the slow-motion regime at 253–255 K, no through-space coupling was present between the phosphorus and the selenoether. This was expected as the phosphorus lone pair was sequestered in the P = Se bond and, hence, was no longer available to overlap with the SePh orbitals. The ^{31}P{^{1}H} NMR spectrum acquired at 363 K showed one broad singlet at δ$_P$ 68.3 ppm as the energy barrier between the two conformations had been overcome, but the speed of the exchange was only marginally faster than the NMR timescale. At 363 K, the ^{77}Se{^{1}H} NMR spectrum showed one singlet at δ$_{Se}$ 433.0 ppm, corresponding to the selenoether. The upfield signal attributed to the P = Se group was not observed, likely due to the fact that 363 K is close to the coalescence temperature. Due to limitations with the equipment, we could not acquire any data at temperatures exceeding 368 K.

As the coalescence was observed in the ^{31}P{^{1}H} and the ^{77}Se{^{1}H} NMR spectra, the coalescence method could be used to estimate the rotational barrier (ΔG‡) of **1-Se**, assuming the coalescence followed typical Eyring behavior.

$$\Delta G^\ddagger = aT_C \left[9.972 + log\left(\frac{T_c}{\Delta \nu}\right) \right]$$

Using this equation, the temperature of coalescence (T_C = 363 K) and the largest separation between the signals of the two conformers obtained from the lowest-temperature $^{31}P\{^1H\}$ VT NMR spectra ($\Delta\nu$ = 5715 Hz); $\Delta G^{\ddagger}$ was estimated as 61 kJ mol^{-1}.

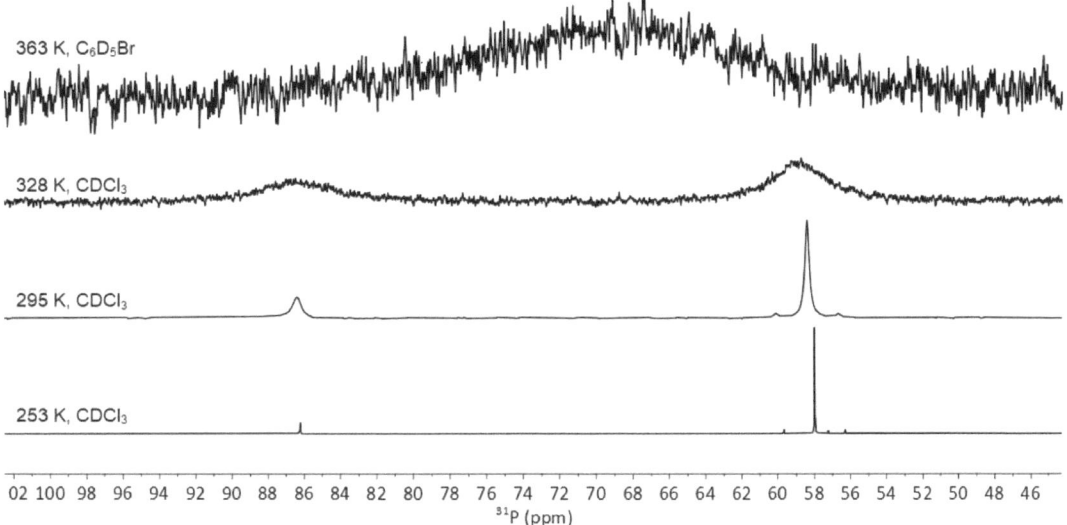

Figure 4. The $^{31}P\{^1H\}$ VT NMR spectra of **1-Se** with solvent and temperatures indicated (acquired at 202.5 MHz).

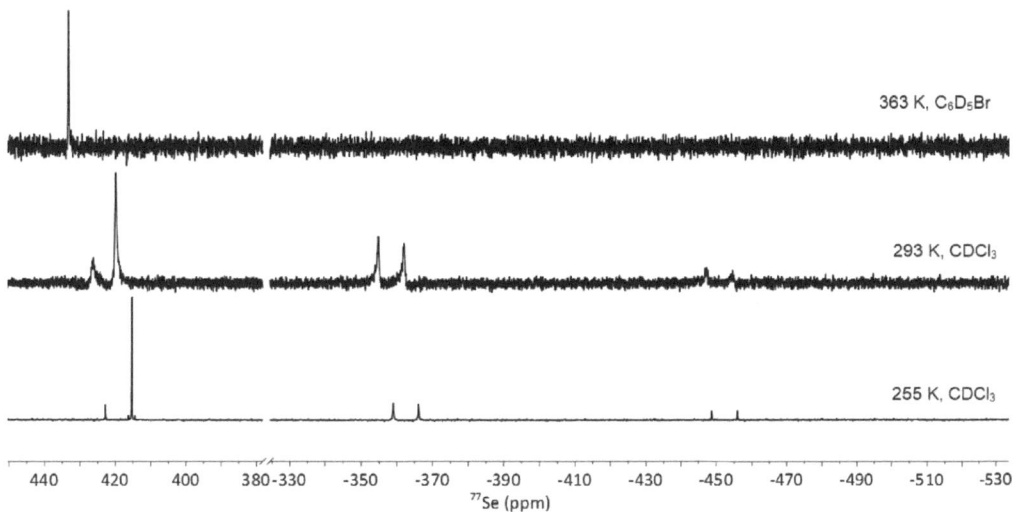

Figure 5. The $^{77}Se\{^1H\}$ VT NMR spectra of **1-Se** with solvent and temperatures indicated (acquired at 95.4 MHz).

3.3.2. Solid-State NMR of **1-Se**

To corroborate the large coupling values observed and to confirm the number of conformations in the solid state, $^{31}P\{^1H\}$ (Figure S2) and $^{77}Se\{^1H\}$ SS-MAS NMR spectra of **1-Se** were acquired (Figure 6). In the $^{31}P\{^1H\}$ SS-MAS NMR spectrum, a singlet at δ_P 60.2 (with ^{77}Se satellites giving $^1J_{PSe}$ = 699.2 Hz) was observed, with spinning sidebands. In the $^{77}Se\{^1H\}$

SS-MAS NMR spectrum, there were two signals: a singlet at δ_{Se} 431.6 ppm and a doublet at δ_{Se} −353.9 ppm, with $^1J_{SeP}$ of 700.8 Hz (Figure 6). The singlet corresponds to the selenoether and the doublet to the phosphine selenide environment. The large $^1J_{SeP}$ was still present in the solid state, albeit with a slightly larger magnitude than in the solution state.

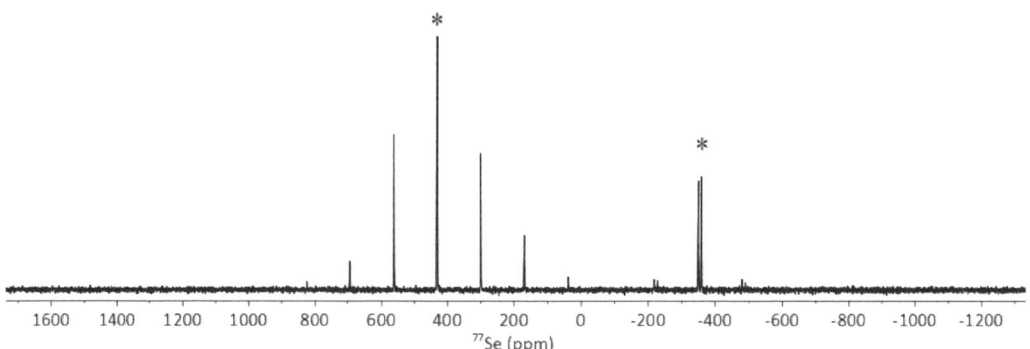

Figure 6. The ^{77}Se{^{1}H} SS-MAS NMR spectrum of **1-Se** recorded at 76.3 MHz. The isotropic peaks are located at δ_{Se} 431.6 and −353.9 ppm and denoted with *.

The key finding is that no other signals were present for each ^{31}P and ^{77}Se environment, indicating that while the bonding environments and connections were the same in both the solution and the solid state, only one conformer was present in the solid state. If two conformers were present in the solid state, two isotropic signals (with spinning side bands) would be expected in both the upfield and downfield regions of the ^{77}Se{^{1}H} MAS spectrum. It is likely that the dominant isomer corresponded to the conformation elucidated by the crystal structure (Figure 2). However, it is also possible that some solvates were formed, as demonstrated recently [23].

3.4. NMR Spectroscopy of 1-S and 1-O

For other chalcogen-oxidized compounds of **1$_{Ph}$**, the change in rotational barrier was expected to follow the trend **1-Se** > **1-S** > **1-O**, as the larger atomic radius of selenium provides a greater barrier to the rotation of the molecule (single-bond covalent radii Se 1.16 Å; S 1.03 Å; O 0.63 Å) [24].

The lighter congeners, **1-S** and **1-O** were prepared; as with **1-Se**, the solution state ^{31}P{^{1}H} NMR spectrum of **1-S** acquired at ambient conditions was notably broad with two signals at δ_P 82.8 and 65.1 ppm, neither of which showed any ^{77}Se satellites (Figure 7, left). The solution-state ^{77}Se{^{1}H} NMR spectrum mirrored the observations of the ^{31}P{^{1}H} spectrum with two broad singlets present at δ_{Se} 422.9 and 418.2 ppm (Figure 7, right). To determine the coalescence temperature and thus determine the rotational energy barrier, VT NMR experiments were performed on **1-S** (Figure 7). The ^{31}P{^{1}H} NMR spectra showed coalescence was reached at 368 K, although the signal at δ_P 71.9 ppm was still observed as a reasonably broad singlet, while completely resolved signals of the two rotamers in the slow-motion regime were observed at 253 K, showing two singlets at δ_P 82.6 and 64.7 ppm. Comparatively, the ^{77}Se{^{1}H} NMR spectra showed fast free rotation was achieved at 368 K with a sharp singlet observed at δ_{Se} 433.2 ppm and verified the full resolution of signals in the slow-motion regime at 255 K (δ_{Se} 418.9 and 413.2 ppm).

Similar observations were made for **1-O**. The ^{31}P{1H} NMR spectrum at ambient conditions revealed two broad singlets at δ_P 55.4 and 54.3 ppm, with the ^{77}Se{^{1}H} NMR spectrum showing two singlets at δ_{Se} 436.6 and 400.9 ppm (Figures S3 and S4). Additionally, VT NMR studies were carried out, with the coalescence observed at 323 K in the ^{31}P{^{1}H} NMR spectrum with complete sharpening of signals observed at 373 K. Two fully resolved singlets were observed in both the ^{31}P{^{1}H} and ^{77}Se{^{1}H} spectra at 255 K (δ_P 56.1 and

55.2 ppm; δ_{Se} 430.4 and 397.0 ppm). Using $\Delta \nu$ = 3617 Hz from the 255 K spectra for **1-S** and $\Delta \nu$ = 184 Hz from the 255 K spectra for **1-O**, $\Delta G^{\ddagger}$ was estimated to be ca. 62 kJ mol^{-1} for **1-S** and 63 kJ mol^{-1} for **1-O** (at 323 K), which, somehow contrary to expectations, was marginally higher than for **1-Se** (61 kJ mol^{-1}). However, one needs to realise that the coalescence method is only an approximation, and that the $\Delta G^{\ddagger}$ values were obtained for different temperatures. Full Erying analysis was not possible as the spectrometer could not exceed 373 K, meaning complete sharpening of the peaks was never observed. As the van der Waals radii of Se is larger than that of S, which is larger than that of O (1.93, 1.85, 1.37 Å, respectively) [25], one may expect the rotational barriers to follow this order; therefore, it is likely that other steric and electronic effects were dominant here.

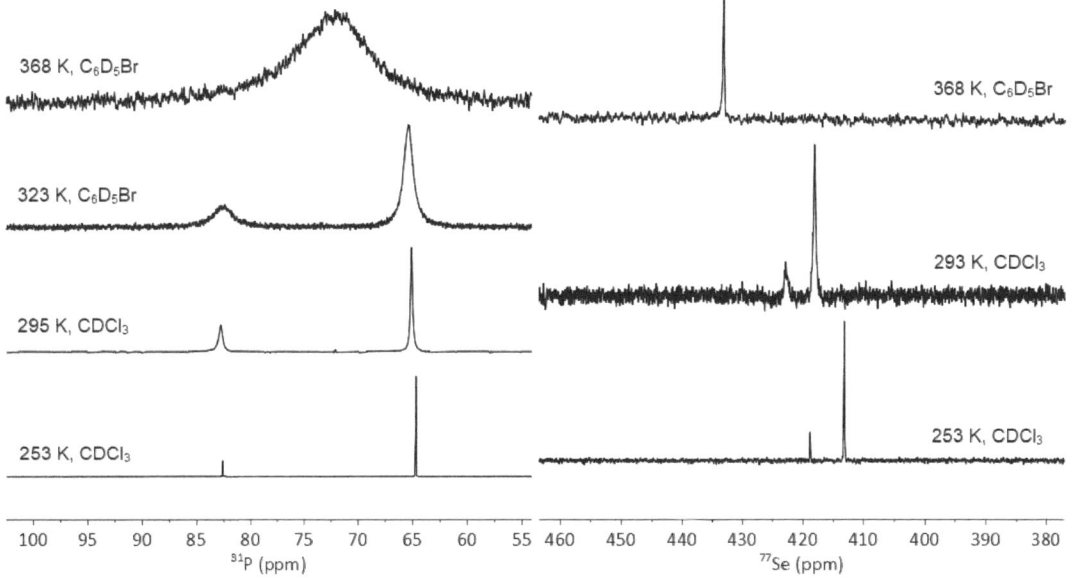

Figure 7. The ^{31}P{^{1}H} VT NMR spectra (**left**) and ^{77}Se{^{1}H} VT NMR spectra (**right**) of **1-S** with solvent and temperatures indicated (acquired at 202.5 and 95.4 MHz, respectively).

3.5. NMR Spectroscopy of 1-O2

The solution-state ^{31}P{^{1}H} NMR spectrum of **1-O2** showed one sharp downfield-shifted singlet at δ_P 56.4 ppm with the ^{77}Se{^{1}H} NMR spectrum also showing a sharp downfield-shifted singlet at δ_{Se} 896.4 ppm (cf. **1$_{Ph}$** δ_P −6.5 ppm; δ_{Se} 425.3 ppm) (Figure S5). The large downfield shift of both peaks was consistent with the oxidation of both the iPr$_2$P and SePh moieties to the P(V) and Se(IV) species, iPr$_2$P(O) and Se(O)Ph. No other signals were present in the NMR spectra, unlike for **1-O**, **1-S**, and **1-Se**, indicating that only one rotational conformation was present in solution, presumably due to the increased steric bulk in the *peri*-region causing extremely hindered rotation of the iPr$_2$P(O) and Se(O)Ph groups. Due to the phosphorus lone pair being sequestered, as well as one of the selenium lone pairs, no J_{PSe} couplings were observed in either spectra.

3.6. Computational Studies

To complement these findings, we performed calculations at the B3LYP-D3/6-311+G(d,p)/CPCM(C$_6$H$_5$Br)//B3LYP-D3/6-31+G(d,p) level of density functional theory (DFT). Starting from the conformation observed in the solid, selected rotamers were constructed by rotating the SePh and iPr$_2$P(Se) moieties about the C(acenaphthene)–E bonds (E = P, Se). The resulting optimized structures are shown in Figure 8, and computed relative energies are collected in Table 2.

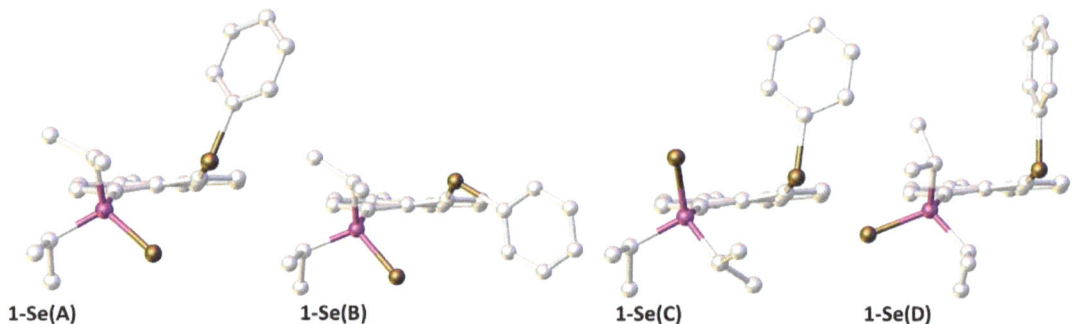

Figure 8. B3YP-D3 optimized rotamers of **1-Se** viewed along the central C–C bond of the acenaphthene moiety; 1-Se(A) is the conformation observed in the solid state. Color code: gray, purple, and bronze for C, P, and Se, respectively. Hydrogen atoms are omitted for clarity.

Table 2. Computed [a] relative energies ΔE, enthalpies ΔH, and free energies ΔG[a] for selected rotamers in kJ mol^{-1} relative to **1-Se(A)**, as well as computed [a] and experimental [b] (in italics) δ(^{31}P) and δ(^{77}Se) chemical shifts of **1-Se(A)** and **1-Se(D)**.

Molecule	ΔE	ΔH^{298}	ΔG^{298}	δ(P)	δ(Se = P)	δ(SePh)
1-Se(A)	0	0	0	61.7 *58.0* [b]	−389.9 *−362.6* [b]	385.1 *415.3* [b]
1-Se(B)	19.0	19.7	23.5			
1-Se(C)	32.8	33.8	34.3			
1-Se(D)	−2.6	−2.4	−2.4	89.1 *86.2* [b]	−555.7 *−452.4* [b]	433.8 *422.8* [b]

[a] Energies at B3LYP-D3/6-311+G(d,p)/CPCM(C$_6$H$_5$Br)//B3LYP-D3/6-31+G(d,p) level, thermodynamic corrections from B3LYP/6-31+G(d,p); chemical shifts at GIAO-B3LYP/ILGO-II'/CPM(CHCl$_3$)//B3LYP-D3/6-31+G(d,p) level [b] 235 K in CDCl$_3$ (this work).

In the conformer found in the solid (structure **1-Se(A)** in Figure 8), the two Se atoms displayed sub-van der Waals contact (Se···Se distance 3.29 Å and 3.28 Å from B3LYP-D3 and XRD, respectively), and the Se–Ph group was oriented along the Se···Se axis and anti with respect to the Se atom on the phosphine. Rotating either the SePh group or the iPr$_2$P(Se) group such that the Se atoms were still in contact but the SePh group was roughly perpendicular to the Se···Se axis afforded two minima (**1-Se(B)** and **1-Se(C)**, respectively, in Figure 8) which were significantly higher in energy than conformer **1-Se(A)** (by ca. 19–34 kJ mol^{-1}, see Table 2). Further rotating the iPr$_2$P(Se) moiety such that the Se atom on the phosphine was pointing away from the other Se atom in the SePh substituent afforded a new minimum (**1-Se(D)** in Figure 8) which was slightly more stable than conformer **1-Se(A)** (by ca. −2 to −3 kJ mol^{-1}, see Table 2). One of the isopropyl groups was also rotated to minimize steric clash between a methyl group and the Se(Ph) atom in rotamer **1-Se(D)**. These results are fully compatible with the observation of a mixture of two slowly interconverting isomers. Based on the comparison of computed and observed ^{31}P and ^{77}Se chemical shifts (see Table 2), we assigned the more deshielded ^{31}P resonance, and the more "extreme" ^{77}Se shifts (i.e., the most deshielded and the most shielded one), to rotamer **1-Se(D)**. From the observed relative intensities of these two sets of signals (Figures 4 and 5), it appears that it was indeed rotamer **1-Se(A)** that was more abundant, i.e., more stable. This assignment also agrees with the comparison of the ^{77}Se resonances observed in the solid (Figure 6), arguable arising from **1-Se(A)**, and those of the more abundant form in solution. In addition, only for **1-Se(A)**, a notable indirect J_{SeSe} spin–spin coupling constant was computed (145 Hz, with 182 Hz observed), whereas that in **1-Se(D)** was negligibly small. The reason why the computed relative stabilities of **1-Se(A)** and **1-Se(D)** were reversed is not clear at the moment. Indeed, switching the solvent model to CHCl$_3$, or the functional to M06-2X, which has performed very well for energetics in other related systems [5], did not change the relative sequence of both.

The reason for the apparent stability of rotamer **1-Se(D)** seems to be more the relief of Se···Se repulsion rather than Se···P bonding interactions; the optimized Se···P distance in **1-Se(D)** was 3.72 Å. This is close to the sum of the van der Waals radii of 4.09 Å; consequently, only a very small Wiberg bond index of 0.01 was obtained between these two atoms.

4. Materials and Methods

4.1. General Considerations

All synthetic manipulations were performed under an atmosphere of dry nitrogen using standard Schlenk techniques or under an argon atmosphere in a Saffron glove box. However, all compounds reported herein were found to be air stable, so repeated reactions were performed under air with no detrimental effects. All glass apparatus were stored in a drying oven (ca. 120 °C) prior to use. Dry solvents were collected from an MBraun Solvent Purification System and stored over appropriate molecular sieves. Water used in experiments was subject to nitrogen sparging and stored under nitrogen prior to use. Chemicals were taken from the laboratory inventory and used without further purification. Infrared Spectra were acquired using a Nicolet 308 FT-IR (Thermo Fisher Scientific, Oxford, UK) with Specac ATR attachment, recorded between 4000 and 500 cm^{-1}.

All solution-state NMR spectra were recorded using either a Bruker Avance III (500 MHz) or Bruker Avance III-HD (500 MHz) spectrometer operating at a magnetic field strength of 11.7 Tesla at 20 °C, unless otherwise specified. Assignments of ^{1}H and ^{13}C spectra were made in conjunction with appropriate 2D spectra. ^{13}C NMR spectra were recorded using the DEPTQ pulse sequence with broadband proton decoupling. The following external standards were used: ^{1}H and ^{13}C NMR, tetramethylsilane; ^{31}P NMR, 85% H$_3$PO$_4$ in D$_2$O; ^{77}Se NMR, dimethyldiselende (Me$_2$Se$_2$) and diphenyldiselenide (Ph$_2$Se$_2$) as a secondary reference at 463.0 ppm. Residual solvent peaks were also used for secondary calibration (CDCl$_3$ δ_H 7.260 ppm; δ_C 77.160 ppm; C$_6$D$_5$Br δ_H 7.300, 7.019, 6.946 ppm; δ_C 130.900, 129.339, 126.162, 122.181 ppm). Chemical shifts (δ) are given in parts per million (ppm) relative to the residual solvent peaks where possible. Coupling constants (*J*) are quoted in Hertz (Hz). The NMR numbering scheme for all compounds is shown in Figure 9.

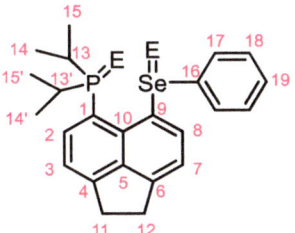

Figure 9. NMR numbering system for compounds reported (E = null, O, S, Se).

Solid-state ^{31}P{^{1}H} and ^{77}Se{^{1}H} NMR (SS-MAS NMR) measurements were performed using a Bruker Avance III 400 MHz spectrometer operating at a magnetic field strength of 9.4 T. Experiments were carried out using a conventional 4 mm MAS probe with a MAS rate of 14 KHz for ^{31}P{^{1}H} and 10 kHz for ^{77}Se{^{1}H}. The ^{77}Se{^{1}H} cross-polarization MAS experiments (using ramped contact pulse durations of 5 ms and TPPM 1H decoupling) were carried out with signal averaging for 2048 transients with a recycle interval of 3 s. Chemical shifts are reported in ppm, relative to Me$_2$Se at 0 ppm, using the isotropic resonance of solid H$_2$SeO$_3$ at 1288.1 ppm as a secondary reference. The position of the isotropic resonance within the spinning sideband patterns was unambiguously determined by recording a second spectrum at a different MAS rate.

Melting and decomposition points were determined by heating solid samples in sealed glass capillaries using a Stuart SMP30 Melting Point Apparatus. High-Resolution Mass Spectrometry of **1-S** and **1-Se** was performed by the EPSRC UK National Mass Spectrometry Facility (NMSF) at Swansea University using a Thermofisher LTQ Orbitrap XL (Atmospheric-Pressure Chemical Ionization). Mass Spectrometry on **1-O** and **1-O2** was performed at the University of St Andrews using a Micromass LCT (Electrospray Ionization) from solutions of the analyte in methanol or acetonitrile. Elemental Analysis was performed by the EA Service at London Metropolitan University.

4.2. Synthetic Procedures and Analytical Data

4.2.1. Synthesis of **1-O**

A solution of **1_Ph** (500 mg, 1.18 mmol) in dichloromethane (25 mL) was prepared. To this, a solution of hydrogen peroxide urea adduct (111 mg, 1.18 mmol) in water (100 mL) was added in one batch. The solution was stirred vigorously for three days. The organic layer was separated and dried over magnesium sulfate. The volatiles were removed in vacuo to afford the crude product. Recrystallization from dichloromethane:n-hexane (1:4 v/v) at $-20\,^\circ$C afforded analytically pure crystals of **1-O** (200 mg, 40%) (melting with decomp. 170–173 $^\circ$C). These crystals were suitable for single-crystal X-ray diffraction.

^{1}H NMR: (500.1 MHz, C$_6$D$_5$Br, 368 K) δ_H 7.84 (1H, d, $^3J_{HP}$ 7.3 Hz, H-2), 7.20 (1H, d, $^3J_{HH}$ 7.2 Hz, H-7), 7.19–7.13 (2H, m, H-17), 7.00–6.90 (4H, m, H-3, 18, 19), 3.15–3.03 (4H, m, H-11,12), 1.33 (6H, dd, $^3J_{HP}$ 15.2, $^3J_{HH}$ 6.9 Hz, H-14/15), 1.00 (6H, $^3J_{HP}$ 15.5, $^3J_{HH}$ 6.9 Hz, H14/15). **^{13}C DEPTQ NMR:** (125.8 MHz, C$_6$D$_5$Br, 368 K), δ_C 151.1 (d, $^4J_{CP}$ 2.4 Hz, qC-6), 147.4 (s, qC-4), 140.6 (d $^3J_{CP}$ 8.2 Hz, qC-5), 140.3 (s, C-2), 138.2 (s, qC-16), 135.9 (d, $^2J_{CP}$ 23.0 Hz, qC-10), 131.2 (s, C-18), 130.4 (s, qC-1/8), 128.8 (s, C-17), 126.1 (s, C-19), 125.3 (s, qC-9), 120.7 (s, C-3), 118.5 (d, $^3J_{CP}$ 11.6 Hz, C-7), 30.0 (s, C-11/12), 29.7 (s, C-11/12), 29.5 (d, $^1J_{CP}$ 67.0 Hz, C-13), 17.3 (d, $^2J_{CP}$ 3.4 Hz, C-14/15), 16.9–16.8 (m, C-14/15). **^{31}P{^{1}H} NMR:** (202.4 MHz, CDCl$_3$, 253 K) δ_P 56.1 (s), 55.2 (s). **^{31}P{^{1}H} NMR:** (202.4 MHz, CDCl$_3$, 295 K) δ_P 55.4 (br s), 54.3 (s). **^{31}P{^{1}H} NMR:** (202.4 MHz, C$_6$D$_5$Br, 373 K) δ_P 53.0 (s). **^{77}Se{^{1}H} NMR:** (95.4 MHz, CDCl$_3$, 253 K) δ_{Se} 430.4 (s), 397.0 (s). **^{77}Se{^{1}H} NMR:** (95.4 MHz, CDCl$_3$, 295 K) δ_{Se} 436.6 (s), 400.9 (s). **^{77}Se{^{1}H} NMR:** (95.4 MHz, C$_6$D$_5$Br, 363 K) no signals observed. **IR:** ν_{max} ATR/cm^{-1} 3067w (ν_{CH}), 2963w (ν_{CH}), 1576m ($\nu_{C=C}$), 1138s ($\nu_{P=O}$), 851m, 733s, 691s. **HRMS:** (ES+): m/z (%) Cacld. for C$_{24}$H$_{27}$POSeNa: 465.0857, found: 465.0842 (100) [M+Na].

4.2.2. Synthesis of **1-O2**

A solution of **1_Ph** (500 mg, 1.18 mmol) in dichloromethane (20 mL) was prepared. To this, 30% aqueous hydrogen peroxide (0.25 mL, 2.47 mmol) was added dropwise over five minutes with vigorous stirring. The solution was stirred at ambient conditions for a further six hours. The organic layer was separated and dried over magnesium sulfate. The volatiles were removed in vacuo to afford **1-O2** as a pale orange solid (240 mg, 45%) (melting with decomp. 208–214 $^\circ$C). The aqueous layer was quenched with aqueous sodium metabisulfite before disposal. Crystals of **1-O2** suitable for single-crystal X-ray diffraction were grown from a dichloromethane/n-hexane vapor diffusion set up at ambient conditions.

^{1}H NMR (500.1 MHz, CDCl$_3$) δ_H 8.45 (1H, dd, $^3J_{HH}$ 7.5 Hz, H-8), 8.05–7.98 (2H, m, H-18), 7.57 (1H, dd, $^3J_{HP}$ 13.7, $^3J_{HH}$ 7.3 Hz, H-2), 7.37 (1H, d, $^3J_{HH}$ 7.6 Hz, H-7), 7.32 (1H, d, $^3J_{HH}$ 7.3 Hz, H-3), 7.23 (3H, m, H-17,19), 3.34–3.21 (4H, m, H-11,12), 2.54–2.36 (2H, m, H-13,13′), 1.28–1.14 (9H, m, H-14/14′/15/15′, 3 × CH$_3$), 0.88 (3H, dd, $^3J_{HP}$ 15.3, $^3J_{HH}$ 7.1 Hz, H-14/14′/15/15′, 1 × CH$_3$). **^{13}C DEPTQ** (125.8 MHz, CDCl$_3$) δ_C 152.8 (s, qC-4), 151.0 (s, qC-6), 148.3 (s, $^1J_{CSe}$ 127.9 Hz, C-16), 140.3 (d, $^3J_{CP}$ 8.8 Hz, C-5), 139.8 (d, $^3J_{CP}$ 3.3 Hz, qC-9), 134.1 (d, $^2J_{CP}$ 11.7 Hz, C-1), 132.3 (s, C-8), 132.2 (d, $^2J_{CP}$ 4.6 Hz, qC-10), 129.5 (s, C-19), 128.5 (s, C-17), 127.7 (s, C-18), 121.5 (s, C-7), 119.2 (d, $^1J_{CP}$ 84.8 Hz, qC-1), 118.6 (d, $^3J_{CP}$ 13.0 Hz, C-3), 30.6 (s, C-11/12), 29.5 (s, C-11/12), 28.8 (d, $^1J_{CP}$ 64.1 Hz, C-13/13′), 26.5 ($^1J_{CP}$ 68.3 Hz, C-13/13′), 17.1 (s, C-14/15, 1 × CH$_3$), 16.1 (d, $^2J_{CP}$ 2.1 Hz, C-14/15, 1 × CH$_3$),

15.9 (d, $^2J_{CP}$ 3.5 Hz, C-14/15, 1 × CH$_3$), 15.6 (d, $^2J_{CP}$ 1.8 Hz, C-14/15, 1 × CH$_3$). ^{31}P{^{1}H} **NMR** (202.5 MHz, CDCl$_3$) δ_P 56.4 (s). ^{77}**Se{^{1}H} NMR** (95.4 MHz, CDCl$_3$) δ_{Se} 869.4 (s). **IR:** ν_{max} ATR/cm^{-1} 3047w (ν_{CH}), 2962w (ν_{CH}), 1597m ($\nu_{C=C}$), 1437m ($\nu_{C=C}$), 1149s ($\nu_{P=O}$), 818vs ($\nu_{Se=O}$), 752s, 690s. **HRMS:** (ES+): m/z (%) Cacld. for C$_{24}$H$_{28}$PO$_2$Se: 459.0987, found: 459.0973 (100) [M+H].

4.2.3. Synthesis of **1-S**

A suspension of **1$_{Ph}$** (1.50 g, 3.52 mmol) and sulfur (177 mg, 3.65 mmol) in toluene (30 mL) was heated under reflux for six hours. The solution was cooled to ambient conditions and all volatiles removed in vacuo to afford the crude product. Recrystallization from dichloromethane:n-hexane (1:3 v/v) at −20 °C afforded white analytically pure crystals of **1-S** (1.38 g, 86%) (melting with decomp. 232–237 °C). These crystals were suitable for single-crystal X-ray diffraction. **Elemental Analysis:** Cacld. (%) for C$_{24}$H$_{27}$PSSe: C 63.01, H 5.95, found: C 62.89, H 6.03. ^{1}**H NMR** (500.1 MHz, C$_6$D$_5$Br, 363 K) δ_H 7.98 (1H, d, $^3J_{HH}$ 7.2 Hz, H-8), 7.15 (1H, d, $^3J_{HH}$ 7.3 Hz, H-3), 7.03–6.98 (2H, m, H-18), 6.93 (1H, d, $^3J_{HH}$ 7.2 Hz, H-3), 6.96–6.87 (3H, m, H-17,19), 3.47 (2H, br s, H-13,13'), 3.09–3.01 (4H, m, H-11,12), 1.34 (6H, dd, $^3J_{HP}$ 17.1, $^3J_{HH}$ 6.9 Hz, H-14,14'), 1.04 (6H, dd, $^3J_{HP}$ 17.7, $^3J_{HH}$ 6.9 Hz, H-15/15'). ^{13}**C DEPTQ NMR** (125.8 MHz, C$_6$D$_5$Br, 368 K) δ_C 151.4 (d, $^4J_{CP}$ 2.5 Hz, qC-4), 148.2 (s, qC-6), 142.2 (s, C-8), 141.0 (d, $^3J_{CP}$ 8.6 Hz, qC-5), 135.8 (s, qC-10), 130.0 (s, C-18), 128.8 (s, C-17), 125.9 (s, C-19), 124.1 (d, $^1J_{CP}$ 64.2 Hz, C-1), 120.8 (s, C-7), 118.3 (d, $^3J_{CP}$ 12.3 Hz, C-3), 30.1 (d, $^1J_{CP}$ 50.2 Hz, C-13,13'), 29.8 (s, C-11/12), 29.8 (s, C-11/12), 17.8 (s, C-14,14'), 17.5 (s, C-15/15'). ^{31}**P{^{1}H} NMR** (202.4 MHz, CDCl$_3$, 253 K) δ_P 82.6 (s), 64.7 (s). ^{31}**P{^{1}H} NMR** (202.4 MHz, CDCl$_3$, 295 K) δ_P 82.8 (s), 65.1 (s). ^{31}**P{^{1}H} NMR** (202.4 MHz, C$_6$D$_5$Br, 363 K) δ_P 71.9 (s). ^{77}**Se{^{1}H} NMR** (95.4 MHz, CDCl$_3$, 253 K) δ_{Se} 418.9 (s), 413.2 (s). ^{77}**Se{^{1}H} NMR** (95.4 MHz, CDCl$_3$, 293 K) δ_{Se} 422.9 (br s), 418.2 (br s). ^{77}**Se{^{1}H} NMR** (95.4 MHz, C$_6$D$_5$Br, 368 K) δ_{Se} 433.2 (s). **IR:** ν_{max} ATR/cm^{-1} 3055w (ν_{CH}), 2958w (ν_{CH}), 1578m ($\nu_{C=C}$), 1477m ($\nu_{C=C}$), 1022m, 744s, 687vs ($\nu_{P=S}$). **HRMS** (APCI+): m/z (%) Cacld. for C$_{24}$H$_{28}$PSSe: 459.0815, found: 459.0814 (100) [M+H].

4.2.4. Synthesis of **1-Se**

A suspension of **1$_{Ph}$** (1.50 g, 3.52 mmol) and selenium (276 mg, 3.50 mmol) in toluene (30 mL) was heated under reflux for fifteen hours. The solution was cooled to ambient conditions and all volatiles removed in vacuo to afford the crude product. Recrystallization from dichloromethane:n-hexane (1:3 v/v) at −20 °C afforded yellow analytically pure crystals of **1-Se** (1.10 g, 62%) (melting with decomp. 237–242 °C). These crystals were suitable for single-crystal X-ray diffraction. **Elemental Analysis:** Cacld. (%) for C$_{24}$H$_{27}$PSe$_2$: C 57.15, H 5.40, found: C 56.93, H 5.36. ^{1}**H NMR** (500.1 MHz, C$_6$D$_5$Br, 363 K) δ_H 8.02 (1H, d, $^3J_{HH}$ 7.2 Hz, H-8), 7.12 (1H, d, $^3J_{HH}$ 7.4 Hz, H-3), 7.00–6.92 (3H, m, H-7,18), 6.92–6.85 (3H, m, H-17,19), 3.55 (2H, br s, H-13,13'), 3.16–2.90 (4H, m, H-11,12), 1.33 (6H, dd, $^3J_{HP}$ 17.5, $^3J_{HH}$ 6.8 Hz, H-14,14'), 1.04 (6H, dd, $^3J_{HP}$ 18.3, $^3J_{HH}$ 6.9 Hz, H-15,15'). ^{13}**C DEPTQ NMR** (126.8 MHz, C$_6$D$_5$Br, 363 K) δ_C 151.6 (d, $^4J_{CP}$ 2.6 Hz, qC-4), 148.3* (s, qC-6), 142.4 (s, C-8), 141.0 (d, $^3J_{CP}$ 8.2 Hz, qC-5), 135.8 (s, qC-10), 129.8 (s, C-18), 128.8 (s, C-17), 125.9 (s, C-18*), 123.4 (s, qC-9), 121.9* (s, qC-1), 120.9 (s, C-7), 118.3 (d, $^3J_{CP}$ 12.4 Hz, C-3), 29.8 (s, C-11/12), 29.6 (s, C-11/12), 29.3 (d, $^1J_{CP}$ 42.5 Hz, C-13,13'), 18.8* (s, C-14,14'), 18.7* (s, C-15,15'). ^{31}**P{^{1}H} NMR** (202.5 MHz, CDCl$_3$, 253 K), δ_P 86.2 (s, $^1J_{PSe}$ 682.9 Hz), 58.0 (s, $^1J_{PSe}$ 681.3 Hz). ^{31}**P{^{1}H} NMR** (202.5 MHz, CDCl$_3$, 295 K) δ_P 86.3 (br s), 58.4 (br s). ^{31}**P{^{1}H} NMR** (202.5 MHz, C$_6$D$_5$Br, 363 K), δ_P 68.3 (br s). ^{31}**P{^{1}H} SS-MAS NMR** (162.0 MHz) 60.2 (s, $^1J_{PSe}$ 699.2 Hz). ^{77}**Se{^{1}H} NMR** (95.4 MHz, CDCl$_3$, 255 K) δ_{Se} 422.8 (s), 415.3 (s, $^{5TS}J_{SeSe}$ 182.0 Hz), −362.6 (d, $^1J_{SeP}$ 681.5 Hz), −452.4 (d, $^1J_{SeP}$ 683.6 Hz). ^{77}**Se{^{1}H} NMR** (95.4 MHz, CDCl$_3$, 293 K) δ_{Se} 426.2 (br s), 419.0 (br s), −358.5 (d, $^1J_{SeP}$ 696.4 Hz), −451.0 (d, $^1J_{SeP}$ 693.4 Hz). ^{77}**Se{^{1}H} NMR** (95.4 MHz, C$_6$D$_5$Br, 363 K) δ_{Se} 433.0 (s). ^{77}**Se{^{1}H} SS-MAS NMR** (76.3 MHz) 431.6 (s), −353.9 (d, $^1J_{PSe}$ 700.8 Hz). **IR:** ν_{max} ATR/cm^{-1} 3047w (ν_{CH}), 2962w (ν_{CH}), 1601m ($\nu_{C=C}$), 1473m ($\nu_{C=C}$), 1018m, 744s, 636s. **HRMS** (APCI+): m/z (%) Cacld.

for $C_{24}H_{28}PSe_2$: 507.0263, found: 507.0264 (100) [M+H]. Note: ^{13}C signals denotated with * were observed in the 2D ^1H–^{13}C HMBC only.

4.3. Crystallographic Details

The crystallographic data for **1-O** were collected using a Rigaku XtaLAB P200 diffractometer using multi-layer mirror monochromated Mo Kα radiation at −180 °C (±1). The crystallographic data for **1-S** were collected using a Rigaku XtaLAB P100 diffractometer using multi-layer mirror monochromated Cu Kα radiation at −100 °C (±1). The crystallographic data for **1-O2** and **1-Se** were collected using a Rigaku SCX mini diffractometer using graphite monochromated Mo Kα radiation at −100 °C (±1) (Mo Kα = λ = 0.71073 Å; Cu Kα = λ = 1.54184 Å).

Intensity data were collected using ω steps accumulating area detector frames spanning at least a hemisphere of reciprocal space. All data were corrected for Lorentz, polarization, and long-term intensity fluctuations. Absorption effects were corrected on the basis of multiple equivalent reflections. The structures were solved by direct methods [26]. Non-hydrogen atoms were refined anisotropically, and hydrogen atoms were refined using the riding model.

The crystal structures were refined by full-matrix least squares against F2 (SHELXL) [27,28] using the CrystalStructure GUI [29]. Searches of the Cambridge Structural Database (CSD) were performed using the webCSD [30]. Images and manipulations of crystal structures and computed rotamers were obtained using OLEX-2 [31].

4.4. Computational Details

Geometries were fully optimized at the B3LYP level [32,33] (using a fine integration grid, i.e., 75 radial shells with 302 angular points per shell) with Curtis and Binning's 962(d) basis [34] on Se and 6-31+G(d,p) elsewhere. The solid-state structure was used as starting point for the optimizations of conformer **1-Se(A)**. The nature of the stationary points was verified by computation of the harmonic frequencies at the same level of theory, which were also used to compute thermodynamic corrections to obtain enthalpies and free energies (standard pressure and temperature). The structures were then re-optimized at the dispersion-corrected B3LYP-D3 [35] level using the same basis set and Becke–Johnson damping [36,37]. Single-point energies were refined for the B3LYP-D3 structures at the B3LYP-D3 level using 962+(d,f) basis on Se, i.e., including the recommended [38] diffuse s and p set and the f-function, and 6-311+G(d,p) elsewhere; an implicit solvent model was used in these single-point calculations, namely the Conductor-Like Polarizable Continuum Model (CPCM) [39,40], using the default settings in Gaussian09 and the parameters of bromobenzene. Wiberg bond indices (WBIs) were computed at that level from natural bond orbital (NBO) analysis. The WBI is a measure for the covalent character of a bond and adopts values close to 1 and 2 for true single and double bonds, respectively [41]. This and similar levels have performed well in previous studies of related acenaphthene chalcogen and pnictogen compounds [2–7,13,15,17]. Magnetic shieldings and spin–spin coupling constants (SSCCs) were computed at the GIAO-B3LYP level using IGLO DZ basis on H atoms and IGLO-basis II everywhere else (denoted ILGO-II), which was designed for computation of magnetic properties [42] and the CPCM model with the parameters of chloroform. The relative ^{77}Se shifts were referenced relative to Me_2Se (computed σ = 1652.1 ppm at the same level). Because the experimental standard for ^{31}P NMR, concentrated phosphoric acid, is difficult to model computationally, chemical shifts were first referenced to Ph_3PSe (computed σ = 235.1 ppm) and converted to the usual δ scale using the experimental chemical shift of that compound in $CDCl_3$, 43.2 ppm [43]. In the computations of SSCCs, the basis set was uncontracted for evaluating the Fermi contact contribution (keyword NMR = (Spin–Spin, Mixed) in Gaussian). All computations were performed using the Gaussian09 suite of programs [44].

5. Conclusions

A series of phosphorus and selenium *peri*-substituted acenaphthenes with the phosphorus atom oxidized by oxygen, sulfur, and selenium was synthesized and characterized by single-crystal X-ray diffraction and multinuclear NMR spectroscopy. For the Se(II) species, there were two major rotational conformers in solution, as identified by Variable-Temperature NMR experiments and supported with DFT calculations. Only one of these conformations was present in the solid state, as verified by X-ray crystallography and solid-state NMR spectroscopy.

Supplementary Materials: The following supporting information can be downloaded at: https://www.mdpi.com/article/10.3390/molecules28217297/s1, Figures S1–S5: Additional NMR spectra; Figures S6–S9: IR spectra of compounds; Table S1: Crystal and structure refinement data; computational detail: Cartesian coordinates in Å, B3LYP/6-31+G(d,p) optimized for rotamers of **1-Se**.

Author Contributions: A.E.T. carried out the required synthetic steps, collected all data (except X-ray data), and analyzed the data. A.M.Z.S. and C.L.C.-W. collected the X-ray data and solved the structures. P.K. provided supervision and research facilities. M.B. performed all computational analysis. B.A.C. designed the study, provided supervision, and wrote the manuscript. All authors have contributed to the proof-reading and editing of the manuscript. All authors have read and agreed to the published version of the manuscript.

Funding: This research received no external funding. We are grateful to the University of St Andrews School of Chemistry Undergraduate Project grants. Calculations were performed at a local high-performance computing facility maintained by H. Fruchtl.

Institutional Review Board Statement: Not applicable.

Informed Consent Statement: Not applicable.

Data Availability Statement: Accession codes CCDC 2298921-2298924 contain the supplementary crystallographic data for this paper. These data can be obtained free of charge via www.ccdc.cam.ac.uk/data_request/cif or by emailing data_request@ccdc.cam.ac.uk. The research data underpinning this publication can be accessed at https://doi.org/10.17630/8fe507af-e08a-4ce8-8edb-426149d527e6.

Acknowledgments: The authors thank Sharon Ashbrook and Daniel Dawson for acquisition of the SS MAS NMR data and Siobhan Smith and Tomáš Lébl for exclusive use of the spectrometers to acquire the variable-temperature solution-state NMR spectra of compounds **1-O**, **1-S**, and **1-Se**.

Conflicts of Interest: The authors declare no conflict of interest.

References

1. Chalmers, B.A.; Athukorala Arachcige, K.S.; Prentis, J.K.D.; Knight, F.R.; Kilian, P.; Slawin, A.M.Z.; Woollins, J.D. Sterically Encumbered Tin and Phosphorus *peri*-Substituted Acenaphthenes. *Inorg. Chem.* **2014**, *53*, 8795–8808. [CrossRef]
2. Chalmers, B.A.; Nejman, P.S.; Llewellyn, A.V.; Felaar, A.M.; Griffiths, B.L.; Portman, E.I.; Gordon, E.-J.L.; Fan, K.J.H.; Woollins, J.D.; Bühl, M.; et al. A Study of Through-Space and Through-Bond J$_{PP}$ Coupling in a Rigid Nonsymmetrical Bis(phosphine) and Its Metal Complexes. *Inorg. Chem.* **2018**, *57*, 3387–3398. [CrossRef]
3. Knight, F.R.; Randall, R.A.M.; Roemmele, T.L.; Boeré, R.T.; Bode, B.E.; Crawford, L.; Bühl, M.; Slawin, A.M.Z.; Woollins, J.D. Electrochemically Informed Synthesis: Oxidation versus Coordination of 5,6-Bis(phenylchalcogeno)acenaphthenes. *ChemPhysChem* **2013**, *14*, 2199–3203. [CrossRef] [PubMed]
4. Chalmers, B.A.; Bühl, M.; Athukorala Arachcige, K.S.; Slawin, A.M.Z.; Kilian, P. A Strutural, Spectroscopic, and Computational Examination of the Dative Interaction in Constrained Phosphine-Stibines and Phosphine-Stiboranes. *Chem. Eur. J.* **2015**, *21*, 7520–7531. [CrossRef] [PubMed]
5. Chalmers, B.A.; Bühl, M.; Athukorala Arachcige, K.S.; Slawin, A.M.Z.; Kilian, P. Geometrically Enforced Donor-Facilitated De-hydrocoupling Leading to an Isolable Arsanylidine-Phosphorane. *J. Am. Chem. Soc.* **2014**, *136*, 6247–6250. [CrossRef] [PubMed]
6. Nejman, P.S.; Curzon, T.E.; Bühl, M.; McKay, D.; Woollins, J.D.; Ashbrook, S.E.; Cordes, D.B.; Slawin, A.M.Z.; Kilian, P. Phosphorus-Bismuth *peri*-Substituted Acenaphthenes: A Synthetic, Structural, and Computational Study. *Inorg. Chem.* **2020**, *59*, 5616–5625. [CrossRef]
7. Nordheider, A.; Hupf, E.; Chalmers, B.A.; Knight, F.R.; Bühl, M.; Mebs, S.; Checinska, L.; Lork, E.; Camacho, P.S.; Ashbrook, S.E.; et al. *Peri*-Substituted Phosphorus–Tellurium Systems—An Experimental and Theoretical Investigation of the P···Te through-Space Interaction. *Inorg. Chem.* **2015**, *54*, 2435–2446. [CrossRef]

8. Hupf, E.; Lork, E.; Mebs, S.; Checinska, L.; Beckmann, J. Probing Donor−Acceptor Interactions in *peri*-Substituted Diphenylphosphinoacenaphthyl−Element Dichlorides of Group 13 and 15 elements. *Organometallics* **2014**, *33*, 7247–7259. [CrossRef]
9. Hupf, E.; Lork, E.; Mebs, S.; Beckmann, J. 6-Diphenylphosphinoacenapth-5-yl-mercurials as Ligands for d10 Metals. Observation of Closed-Shell Interactions of the Type Hg(II)···M; M = Hg(II), Ag(I), Au(I). *Inorg. Chem.* **2015**, *54*, 1847–1859. [CrossRef]
10. Furan, S.; Vogt, M.; Winkels, K.; Lork, E.; Mebs, S.; Hupf, E.; Beckmann, J. (6-Diphenylphosphinoacenapth-5-yl)indium and -nickel Compounds: Synthesis, Structure, Transmetalation, and Cross-Coupling Reactions. *Organometallics* **2021**, *40*, 1284–1295. [CrossRef]
11. Kordts, N.; Künzler, S.; Rathjen, S.; Sieling, T.; Großekappenberg, H.; Schmidtmann, M.; Müller, T. Silyl Chalconium Ions: Synthesis, Structure and Application in Hydrodefluorination Reactions. *Chem. Eur. J.* **2017**, *23*, 10068–10079. [CrossRef] [PubMed]
12. Hierso, J.C. Indirect Nonbonded Nuclear Spin−Spin Coupling: A Guide for the Recognition and Understanding of "Through-Space" NMR J Constants in Small Organic, Organometallics, and Coordination Compounds. *Chem. Rev.* **2014**, *114*, 4838–4867. [CrossRef] [PubMed]
13. Athukorala Arachcige, K.S.; Camacho, P.S.; Ray, M.J.; Chalmers, B.A.; Knight, F.R.; Ashbrook, S.E.; Bühl, M.; Kilian, P.; Slawin, A.M.Z.; Woollins, J.D. Sterically Restricted Tin Phosphines, Stabilised by Weak Intramolecular Donor-Acceptor Interactions. *Organometallics* **2014**, *33*, 2121–2430. [CrossRef]
14. Hupf, E.; Lork, E.; Mebs, S.; Beckmann, J. Intramolecularly Coordinated (6-(Diphenylphosphino)acenapth-5-yl)stannanes. Repulsion vs Attraction of P- and Sn-Containing Substituents in the *peri* Positions. *Organometallics* **2014**, *33*, 2409–2423. [CrossRef]
15. Knight, F.R.; Fuller, A.L.; Bühl, M.; Slawin, A.M.Z.; Woollins, J.D. Sterically Crowded *peri*-Substituted Naphthalene Phosphines and their P^V Derivatives. *Chem. Eur. J.* **2010**, *16*, 7617–7634. [CrossRef] [PubMed]
16. Malkina, O.L.; Hierso, J.-C.; Malkin, V.G. Distinguishing "Through-Space" from "Through-Bonds" Contribution in Indirect Nuclear Spin−Spin Coupling; General Approaches Applied to Complex J_{PP} and J_{PSe} Scalar Couplings. *J. Am. Chem. Soc.* **2022**, *144*, 10768–10784. [CrossRef]
17. Zhang, L.; Christie, F.A.; Tarcza, A.E.; Lancaster, H.G.; Taylor, L.J.; Bühl, M.; Malkinia, O.L.; Woollins, J.D.; Carpenter-Warren, C.L.; Cordes, D.B.; et al. Phosphine and Selenoether *peri*-Substituted Acenaphthenes and Their Transition-Metal Complexes: Structural and NMR Investigations. *Inorg. Chem.* **2023**, *62*, 16084–16100. [CrossRef]
18. Chernick, C.L.; Skinner, H.A. 285. Thermochemistry of organophosphorus compounds. Part II. Triethyl phosphate, tripropylphosphine oxide, and tributylphosphine oxide. *J. Chem. Soc.* **1956**, 1401–1405. [CrossRef]
19. Bartlett, P.D.; Meguerian, G. Reactions of Elemental Sulfur. I. The Uncatalysed Teaction of Sulfur with Triarylphosphines. *J. Am. Chem. Soc.* **1956**, *78*, 3701–3715. [CrossRef]
20. Capps, K.B.; Wixmerten, B.; Bauer, A.; Hoff, C.D. Thermochemistry of Sulfur Atom Transfer. Enthalpies of Reaction of Phosphines with Sulfur, Selenium, and Tellurium, and of Desulfurization of Triphenylarsenic Sulfide, Triphenylantimony Sulfide, and Benzyl Trisulfide. *Inorg. Chem.* **1998**, *37*, 2861–2864. [CrossRef]
21. Cinderalla, A.P.; Vulovic, B.; Watson, D.A. Palladium-Catalysed Cross-Coupling of Silyl Electrophiles with Akylzinc Halides: A Silyl-Negishi Reaction. *J. Am. Chem. Soc.* **2017**, *139*, 7741–7744. [CrossRef] [PubMed]
22. Kilian, P.; Milton, H.L.; Slawin, A.M.Z.; Woollins, J.D. Chlorides, Oxochlorides, and Oxoacids of 1,8-Diphosphanaphthalene: A System with Imposed Close P···P Interaction. *Inorg. Chem.* **2004**, *43*, 2252–2260. [CrossRef] [PubMed]
23. Eventova, V.A.; Belov, K.V.; Efimov, S.V.; Khodov, I.A. Conformational Screening of Arbidol Solvates: Investigation via 2D NOESY. *Pharmaceutics* **2023**, *15*, 226. [CrossRef]
24. Pyykko, P.; Atsumi, M. Molecular Single-Bond Covalent Radii for Elements 1–118. *Chem. Eur. J.* **2009**, *15*, 186–197. [CrossRef] [PubMed]
25. Mantina, M.; Chamberlin, A.C.; Valero, R.; Cramer, C.J.; Truhlar, D.G. Consistent ver der Waals Radii for the Whole Main Group. *J. Phys. Chem. A* **2009**, *113*, 5806–5812. [CrossRef] [PubMed]
26. Burla, M.C.; Caliandro, R.; Camalli, M.; Carrozzini, B.; Cascarano, G.L.; Giacovazzo, C.; Mallamo, M.; Mazzone, A.; Polidori, G.; Spagna, R. SIR2011: A new package for crystal structure determination and refinement. *J. Appl. Crystallogr.* **2012**, *45*, 357–361. [CrossRef]
27. Sheldrick, G.M. Crystal structure refinement with *SHELXL*. *Acta Crystallogr. C* **2015**, *71*, 3–8. [CrossRef]
28. Sheldrick, G.M. SHELXT—Integrated space-group and crystal-structure determination. *Acta Crystallogr. A* **2015**, *71*, 3–8. [CrossRef]
29. *CrystalStructure 4.3.0*; Rigaku Americas: The Woodlands, TX, USA; Rigaku Corporation: Tokyo, Japan, 2018.
30. Groom, C.R.; Bruno, I.J.; Lightfoot, M.P.; Ward, S.C. The Cambridge Structural Database. *Acta Crystallogr. B* **2016**, *B72*, 171–179. [CrossRef]
31. Dolomanov, O.V.; Bourhis, L.J.; Gildea, R.J.; Howard, J.A.K.; Puschmann, H. OLEX2: A complete structure solution, refinement and analysis program. *J. Appl. Crystallogr.* **2009**, *42*, 339–341. [CrossRef]
32. Becke, A.D. Density-functional thermochemistry. III. The role of exact exchange. *J. Chem. Phys.* **1993**, *98*, 5642–5648. [CrossRef]
33. Lee, C.; Yang, W.; Parr, R.G. Development of the Colle-Salvetti correlation-energy formula into a function of the electron density. *Phys. Rev. B* **1988**, *37*, 785–789. [CrossRef] [PubMed]
34. Binning, R.C.; Curtiss, L.A. Compact contracted basis sets for third-row atoms: Ga–Kr. *J. Comput. Chem.* **1990**, *11*, 1206–1216. [CrossRef]

35. Grimme, S.; Antony, J.; Ehrlich, S.; Kreig, H. A consistent and accurate ab initio parametrization of density function dispersion correction (DFT-D) for the 94 elements H-Pu. *J. Chem. Phys.* **2010**, *132*, 154104. [CrossRef] [PubMed]
36. Becke, A.D.; Johnson, E.R. Exchange-hole dipole moment and the dispersion interaction. *J. Chem. Phys.* **2005**, *122*, 154104. [CrossRef]
37. Johnson, E.R.; Becke, A.D. A post-Hartree-Fock model of intermolecular interactions: Inclusion of higher-order corrections. *J. Chem. Phys.* **2006**, *124*, 174104. [CrossRef] [PubMed]
38. Reed, A.E.; Curtiss, L.A.; Weinhold, F. Intermolecular Interactions from a Natural Bond Orbital, Donor-Acceptor Viewpoint. *Chem. Rev.* **1988**, *88*, 899–926. [CrossRef]
39. Barone, V.; Cossi, M. Quantum Calculation of Molecular Energies and Energy Gradients in Solution by a Conductor Solvent Model. *J. Phys. Chem. A* **1998**, *102*, 1995–2001. [CrossRef]
40. Cossi, M.; Rega, N.; Scalmani, G.; Barone, V. Energies, structures, and electronic properties of molecules in solution with the C-PCM solvation model. *J. Comput. Chem.* **2003**, *24*, 669–681. [CrossRef]
41. Wiberg, K.B. Application of the pople-santry-segal CNDO method to the cyclopropylcarbinyl and cyclobutyl cation and to bicyclobutane. *Tetrahedron* **1968**, *24*, 1083–1096. [CrossRef]
42. Kutzelnigg, W.; Fleischer, U.; Schindler, M. The IGLO-Method: Ab Initio Calculation and Interpretation of NMR Chemical Shifts and Magnetic Susceptibilities. In *NMR Basic Principles and Progress*; Springer: Berlin/Heidelberg, Germany, 1990; Volume 23. [CrossRef]
43. Albright, T.A.; Freeman, W.J.; Schweizer, E.E. Nuclear Magnetic Resonance Studies. IV. The Carbon and Phosphorus Nuclear Magnetic Resonance of Phosphine Oxides and Related Compounds. *J. Org. Chem.* **1975**, *40*, 3437–3441. [CrossRef]
44. Frisch, M.J.; Trucks, G.W.; Schlegel, H.B.; Scuseria, G.E.; Robb, M.A.; Cheeseman, J.R.; Scalmani, G.; Barone, V.; Mennucci, B.; Petersson, G.A.; et al. *Gaussian 09*; Rev. A.02; Gaussian Inc.: Wallingford, CT, USA, 2009.

Disclaimer/Publisher's Note: The statements, opinions and data contained in all publications are solely those of the individual author(s) and contributor(s) and not of MDPI and/or the editor(s). MDPI and/or the editor(s) disclaim responsibility for any injury to people or property resulting from any ideas, methods, instructions or products referred to in the content.

Article

Hydrogen Bonds Stabilize Chloroselenite Anions: Crystal Structure of a New Salt and Donor-Acceptor Bonding to SeO$_2$

René T. Boeré [1,2]

[1] Department of Chemistry and Biochemistry, University of Lethbridge, Lethbridge, AB T1K 3M4, Canada; boere@uleth.ca

[2] Canadian Centre for Research in Applied Fluorine Technologies (C-CRAFT), University of Lethbridge, Lethbridge, AB T1K 3M4, Canada

Abstract: The single-crystal X-ray diffraction structure characterizing a new 4-methylbenzamidinium salt of chloroselenite [C$_8$H$_{11}$N$_2$][ClSeO$_2$] is reported. This is only the second crystal structure report on a ClSeO$_2^-$ salt. The structure contains an extended planar hydrogen bond net, including a double interaction with both O atoms of the anion (an $R_2^2(8)$ ring in Etter notation). The anion has the shortest Se–Cl distances on record for any chloroselenite ion, 2.3202(9) Å. However, the two Se–O distances are distinct at 1.629(2) and 1.645(2) Å, attributed to weak anion–anion bridging involving the oxygen with the longer bond. DFT computations at the RB3PW91-D3/aug-CC-pVTZ level of theory reproduce the short Se–Cl distance in a gas-phase optimized ion pair, but free optimization of ClSeO$_2^-$ leads to an elongation of this bond. A good match to a known value for [Me$_4$N][ClSeO$_2$] is found, which fits to the Raman spectroscopic evidence for this long-known salt and to data measured on solutions of the anion in CH$_3$CN. The assignment of the experimental Raman spectrum was corrected by means of the DFT-computed vibrational spectrum, confirming the strong mixing of the symmetry coordinate of the Se–Cl stretch with both ν_2 and ν_4 modes.

Keywords: halochalcogenite(IV) ion; crystallography; H-bonding; chalcogen bonding; π-holes; DFT-computed vibrational spectra

Citation: Boeré, R.T. Hydrogen Bonds Stabilize Chloroselenite Anions: Crystal Structure of a New Salt and Donor-Acceptor Bonding to SeO$_2$. *Molecules* **2023**, *28*, 7489. https://doi.org/10.3390/molecules28227489

Academic Editor: Petr Kilián

Received: 17 October 2023
Revised: 3 November 2023
Accepted: 6 November 2023
Published: 8 November 2023

Copyright: © 2023 by the author. Licensee MDPI, Basel, Switzerland. This article is an open access article distributed under the terms and conditions of the Creative Commons Attribution (CC BY) license (https://creativecommons.org/licenses/by/4.0/).

1. Introduction

The chloroselenite ion, ClSeO$_2^-$, is found in several salts crystallized from the addition of chloride ions to selenium(IV)oxide in non-aqueous solutions [1]. Conceptually, chloroselenite ions are donor–acceptor adducts between the halogen and the chalcogen dioxide (Scheme 1) and they were first identified by direct addition reactions in aprotic solvents. The Cambridge Structural Database (release 2023.2.0) [2] currently lists just one crystal structure [Me$_4$N][ClSeO$_2$], assigned the CSD Refcode BIRHOZ, but to our knowledge, no atom coordinates are available for this structure in publications or databases. This structure is most consistent with type **A** in Scheme 1, described as having 'monomeric pyramidal anions' with *d*(Se–Cl) = 2.453(1); *d*(Se–O) = 1.632(2) Å [3–6]. The Se–Cl bond length, almost 14% longer than the sums of the covalent radii, is consistent with a weak donor–acceptor bond (for further details on this structure, see the Supplementary Materials). A 2,2′-bipyridium salt [C$_{10}$H$_9$N$_2$][ClSeO$_2$] is indexed in Chemical Abstracts (Registry number [27380-14-9]), but its crystal structure is apparently only available in an unpublished thesis [7]. The anion has been identified in various solvents via liquid-phase Raman spectroscopy as well as in isolated Me$_4$N$^+$ and Ph$_4$As$^+$ salts, largely through the systematic work of the Canadian chemist John Milne (1934–2022) [1]. The ^{77}Se NMR spectrum of solid [Me$_4$N][ClSeO$_2$] was noteworthy for having anisotropic shielding of >1000 ppm [8]. The dominance of 1:1 adducts between Cl$^-$ ions and SeO$_2$ in several aprotic solvents was originally established using UV-vis absorption spectroscopy [9], but the non-existence of the parent acid ClSe(O)OH in aqueous solutions of SeO$_2$ in either dilute or concentrated

HCl has been attested [10,11]. Similarly, there is no evidence for stable salts of simple metal cations M[ClSeO$_2$] despite reported attempts to obtain these [12].

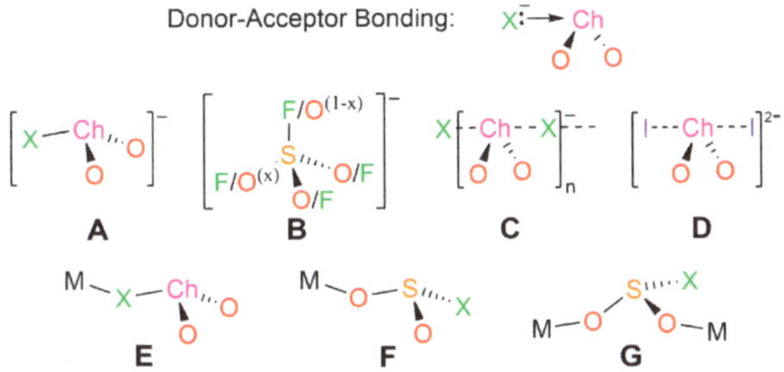

Scheme 1. Donor–acceptor bonding in halochalcogenite(IV) ions and schematics of the structure types **A–G** known from SC-XRD data; Ch = chalcogen (S, Se, Te); X = halide (F, Cl, Br, I).

The nature and scope of weak chemical bonds has become a major focus of research in recent years [13]. Chloroselenites are currently of interest in relation to the speciation and extraction of Se(IV) and Se(VI) for environmental concerns [14]. The electron affinities of SeO$_n$ clusters have been evaluated for similar motivations [15] and fundamental spectroscopy on SeO$_2$ rotational lines remain of interest for astronomic detection [16–18]. Selenium compounds are central to the surge in interest in chalcogen bonding [19–23]. Importantly, SeO$_2$ itself has also been identified as having electrostatic 'π-holes' [24], which may be of direct relevance to the formation of the halogen adducts [XChO$_2$]$^-$ (X = halogen; Ch = chalcogen). New attention to the higher oxochloroselenates, after a long hiatus, is bearing fruit with the report of an inclusion compound of Cl$_2$ in a tetramethylamino salt of [Se$_2$O$_2$Cl$_7$]$^{3-}$ [25], which harkens back to a much older structure [26]. The thermochemical properties of the fluoroselenite ion have been assessed in a large prospective study [27]. Less is known about the heavier halide adducts of SeO$_2$, and halotellurites remain rare. However, there is active research on all the halosulfites due to the recognition of the importance of SO$_2$ as a Lewis acid relative to is capability as a solvent medium and environmental hazard, with important structural [28–30] and computational studies [31]. The application of modern speciation, structural, and theoretical techniques makes the study of weakly bonded adducts, such as those encountered amongst the halochalcogenites [XChO$_2$]$^-$, more feasible than ever before.

Despite their simple constitution, there is a dearth of confirmed structural evidence on [XChO$_2$]$^-$ salts, all of which so far are from single-crystal X-ray diffraction (SC-XRD) data. Since the literature is very scattered, the current state of knowledge is briefly reviewed. FSO$_2$$^-$ salts are the best represented, with nine known structures, consistent with the accepted wisdom that this is the most stable member of its class (the order of X–Ch bond strength is believed to be F > Cl < Br < I, but this may be Ch-dependent) [32,33]. A phosphonium ylid salt of FSO$_2$$^-$ was the first reported structure (CSD refcode: LIHWAA) but it suffers from serious F/O positional disorder [34]. Another, aprotic, imidazolium salt (CSD refcode: TOSXEE), also displays a disordered anion [35]. This theme continues for the metal salts K[FSO$_2$], Rb[FSO$_2$] [36] and the α- and β- polymorphs of Cs[SO$_2$F] [30]. The recognition of O/F positional disorder led to a determined but only partially successful attempt to overcome the phenomenon with the preparation and structures of [(Me$_2$N)$_3$S]$^+$, [(Me$_2$N)$_3$SO]$^+$ and [(Me$_4$N)$_4$N]$^+$ salts (CSD refcodes: ADEJOI, ADEJUO and ADEKAN, respectively) [29]. The structure type that best describes all the FSO$_2$$^-$ salts is **B** in Scheme 1, due to the positional exchange of the very similar-sized O and F, and further positional disorder that often lends a pseudo-tetrahedral appearance to the anion in these structures (the value of x can range from 0 to 0.5).

Two chlorosulfite, ClSO$_2$$^-$, ion structures are in the CSD. The oldest structure (refcodes: POMBEY [37]) contains an isolated ion of type **A** with an S–Cl bond that is 23% longer than the sums of the covalent radi. Much more recently, structure XEGCAQ [38] was reported, wherein one oxygen coordinates to Li⁺, lending it structure type **F**. Fascinatingly, there are also two structures in the database, KIGZEF [39] and LAQYOR [40], which contain infinite chains in which the chloride ions bridge the SO_2 molecules more or less equally, i.e., structure type **C**. Similar, though far more symmetrical, chain structures of type **C** are displayed by [Et₄N][BrSO₂], LAYTUC [28] and by [Me₄N][BrSeO₂], BIRHUF [3–5].

The remainder are ISO_2^- salts, and these are the most structurally diverse of all. The [Ph₃PBz]⁺ salt BZTPPI [41] and WUKQUR [42] are both of type **A** (with S–I lengths 38% and 24% longer than sums of the covalent radii), although WUKQUR has a positional disorder that reduces the accuracy of the derived parameters. Structures MPICSO [43] and MPTPIS [44] are of type **E** in which SO_2 forms an adduct to a metal-coordinated iodide ion. The iodide adduct in WUQMED is another with oxygen coordinated to a metal, type **F**, whereas WUQLUS is a variant on this theme with both oxygen atoms of the SO_2 attached to separate metals, type **G** [42]. Finally, in DOTXOA, there is a discrete $I_2SO_2^{2-}$ ion of structure type **D** [45].

In summary, there are still relatively few known structures for this class, many of which are problematic, and there is a very wide structural diversity, particularly with regards to the X–Ch bond lengths. This situation is consistent with what may be anticipated from weak donor–acceptor bonding. Thus, when we happened on a good quality structure containing a chloroselenite ion, quite by accident, we immediately recognized its importance. Our structure is the first of its kind where paradigmatic hydrogen bonding to the $ClSeO_2^-$ ion has been established, as well as having the shortest Cl–Se bond for chloroselenites. Herein, we provide a full report on this interesting structure and analyze the anion geometry and bonding through extensive B3PW91/aug-cc-pVTZ density functional theory (DFT) computations.

2. Results

2.1. Formation of the Salt from a Hydrolysis Reaction

In our work on heterocyclic thiazyls and selenazyls, we have explored the synthesis of 1,2,4,6-thiatriazinyl radicals via the reduction of 1-chloro-1,2,4,6-thiatriazines [46,47] and are now extending this work to the selenium analogues. Attempts to recrystallize extremely insoluble selenatriazine **1** for purification used boiling acetonitrile (Scheme 2). **1** is analogous to 1,1-dichloro-3,5-diphenyl-4H-1,2,4,6-selenatriazine, which displays H-bonding in its crystal structure (CSD refcode: DUVDUT), likely the origin of the insolubility [48]. Colorless crystals of **2** were the only identifiable product of the reaction mixture, which has been unambiguously characterized by SC-XRD. This interesting reaction, producing at once the rare chloroselenite ion and the 4-methylbenzamidinium cation, may be contrasted with our earlier observation that hydrolysis of 1-chloro-3-phenyl-5-trifluoromethyl-1λ⁴,2,4,6-thiatriazine **3** [46] in the presence of air forms the covalent (imino(phenyl)methyl)sulfamyl(VI) chloride **4**. Thus, in the formation of **2**, selenium demonstrates its well-known resistance to adopting the highest group oxidation state, a characteristic that is usually attributed to the Scandide contraction [49]. The two products provide an appealing contrast, yet both are remarkable for retaining a Ch–Cl bond and are evidently stabilized by similar H-bonding networks (see below). The full structural and computational characterization of salt **2** follows. For a depiction of the interesting molecular structure of **4**, including its H-bonds, see Appendix B.

2.2. Crystallographic Characterization

From the SC-XRD data, a structure model for **2** with restrained full refinement of the hydrogen atom positions and displacements was developed by applying Hirsfeld atom refinement (HAR). This method employs custom aspherical atomic scattering factors, computed on the fly by density functional theory (DFT) methods, under the control of the NoSpherA2 package [50] within Olex2 release 1.5 [51]. This approach is particularly useful when H-bonding is present, as it avoids having to normalize E–H bond lengths as otherwise required with XRD structures [52]. Full details of the refinement strategy are provided in the Experimental section. The ion pair structure in **2** is shown in Figure 1, the extended H-bond network and inter-anion contacts are shown later. The derived interatomic parameters have been placed in Table 1, while the H-bond and short contact data are presented in Table 2.

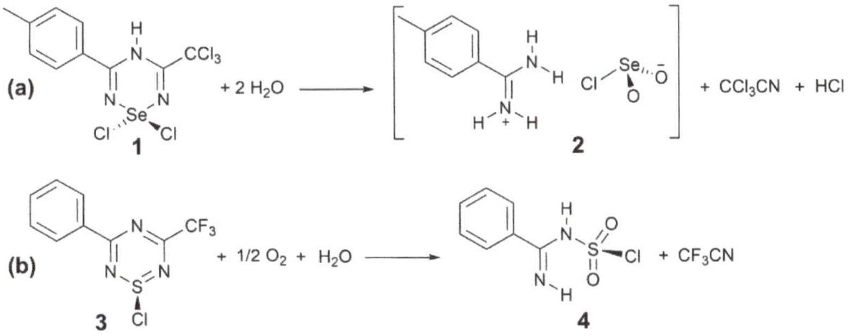

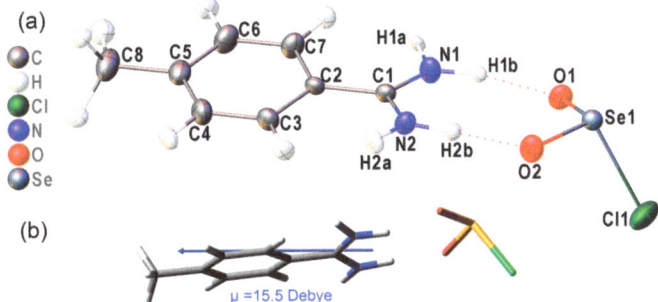

Scheme 2. Plausible reactions leading to (**a**) chloroselenite salt **2** from the dichloroselenatriazine **1** or (**b**) sulfamyl chloride **4** from chlorthiatriazine **3**.

Figure 1. (**a**) Displacement ellipsoids plots (50% probability) for the molecular structures of **2** as found in the crystal lattice. The second component of the CH_3 rotational disorder model is omitted. H-bonds are shown with red dotted lines. (**b**) Tubes plot, showing the DFT-computed dipole moment (blue vector; IUPAC convention) of an isolated ion pair in **2** from an RB3PW91-D3/aug-CC-pVTZ calculation.

The ion pair structure obtained for **2** (Figure 1a) consists of $ClSeO_2^-$ ions that are doubly hydrogen-bonded to the toluamidinium H atoms H1b and H2b. There is, to date, only one set of comparison data in the literature, the aforementioned BIRHOZ structure of $[Me_4N][ClSeO_2]$ from an apparently very accurate 143 K crystal structure [3]. The Se–Cl bond length of 2.453(1) Å in BIRHOZ is significantly longer than that found in **2**, but the apparently equal Se–O distances of 1.632(2) Å agree well with the average of Se1–O1 and Se1–O2 (1.637(1) Å). The accuracy of this report has been confirmed through a personal communication of the structure details (see the Supplementary Materials), so that the divergence between the two geometries will be considered in detail [6]. Krebs et al. further describe this structure in a review article, mentioning that an X-X deformation density analysis has been undertaken on the same salt at 120 K, but these data also remain unpublished: "deformation density maps clearly reveal the presence of lone-pair (E) density (maximum of $0.40 \pm 0.04\ e^-/\text{Å}^{-1}$ at a distance of ca. 0.75 Å from Se) consistent with model predictions for a pseudo-tetrahedral SO_2ClE arrangement with additional π density in the Se–O bonds and with a rather polar Se–Cl bond" [4]. From the HAR/NoSpherA2, we were able to extract a deformation density map (Figure 2) that corroborates this verbal description.

Table 1. Interatomic distances (Å) and angles (°) in the crystal structure of **2** and for $ClSO_2^-$ by DFT [1].

Atoms	$d_{Experiment}$	$d_{Computed}$	Atoms	$\angle_{Experiment}$	$\angle_{Computed}$
Se1–Cl1	2.3202(9)	2.491	O1–Se1–Cl1	100.86(9)	102.75
Se1–O1	1.645(2)	1.630	O2–Se1–Cl1	101.55(9)	102.75
Se1–O2	1.629(2)	1.630	O2–Se1–O1	104.90(11)	110.58
N1–C1	1.323(4)		N2–C1–N1	119.1(3)	
N2–C1	1.317(4)		C2–C1–N1	120.8(3)	
C1–C2	1.472(4)		C2–C1–N2	120.1(3)	
C2–C3	1.404(4)		C3–C2–C1	120.5(3)	
C2–C7	1.395(4)		C7–C2–C3	118.1(3)	
C3–C4	1.382(5)		C4–C3–C2	120.4(3)	
C4–C5	1.395(5)		C5–C4–C3	121.3(3)	
C5–C6	1.392(5)		C6–C5–C4	118.1(3)	
C5–C8	1.497(5)		C8–C5–C4	120.7(3)	
C6–C7	1.384(5)		C8–C5–C6	121.2(3)	
			C7–C2–C1	121.4(3)	
			C7–C6–C5	120.9(3)	

[1] Full atomic positional and derived data for SC-XRD experiments in the Supplementary Materials. DFT geometry optimized in the gas phase at the RB3PW91-D3/aug-CC-pVTZ level of theory.

Table 2. Hydrogen bonds and inter-anion contacts in the crystal structure of **2**.

D(–H)	:A	d(D–H)/Å	d(H···A)/Å	$d-\sum r_{vdW}$	d(D···A)/Å	Angle/°
Hydrogen-bonds						
N1–H1a	O2 [1]	1.045(17)	1.90(3)	−0.82	2.852(3)	150(3)
N1–H1b	O1	1.051(18)	1.843(19)	−0.877	2.880(4)	168(3)
N2–H2a	O1 [2]	1.030(17)	1.83(3)	−0.89	2.792(3)	153(4)
N2–H2b	O2	1.039(18)	1.95(2)	−0.77	2.948(3)	160(3)
Inter-anion Contacts						
O1	Se1 [3]	(angle is Se1–O1···Se1 [3])		−0.289	3.131(2)	108.8(1)
Cl1	Se1 [3]	(angle is Se1 [3]–Cl1 [3]···Se1)		−0.252	3.3980(9)	85.96(3)

Symmetry codes: [1] 1 + x,1.5-y,1/2 + z; [2] x,1.5-y,-1/2 + z; [3] 1 + x,y,z.

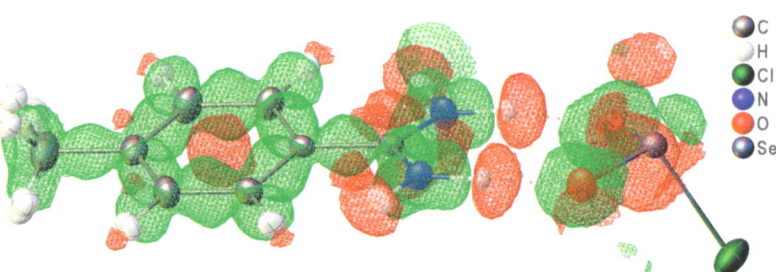

Figure 2. Deformation density map from the HAR/NoSpherA2 refinement of **2**. T If you are certain that the sharing Convention to which you subscribe permits this alteration, you are free to add the reference to the end of the reference list. The lone-pair density above and right of the Se atom is small but distinguishable.

2.3. Hydrogen Bonding in the Solid Lattice of **2**

There is an extensive H-bond network (Figure 3, Table 2) in **2** and all the H-bond parameters fit for standard electrostatic-covalent H-bonding according to the criteria of Jeffrey (summarized in Table A1 in Appendix B). The H-bonds form layered nets, wherein every second ion pair is reversed in a head-to-tail fashion that, as viewed in Figure 3, form a 'Vee' or roof shape, the horizontal of which aligns with the bifurcator of $\angle ac$. For extended views of these nets, see Figure S3 in the Supplementary Materials. There are both short

H-bonds between the amidinium and chloroselenite ions [d(D···A) 2.852(3) and 2.948(3) Å] and even shorter links to the next amidinium ions on both sides in the net [d(D···A) 2.852(3) and 2.880(4) Å]. The H-bond acceptor sites at O correspond to negative charge maxima on the computed electrostatic potential surface (Figure S4), which is otherwise unexceptional. Bonds of this length can be worth as much as 50 kJ/mol each (Table A1) so could add up to as much as 200 kJ/mol per formula unit. This is significant stabilization. The primary H-bonds (in Etter notation) are the discrete $D_1^1(2)$ links b and d of the cation to the facing anion, which thereby form an $R_2^2(8) > b < d$ ring, a standard motif for amidinium ions [53]. There are also $D_1^1(2)$ links a and c to two adjacent anions that are of comparable strength to the ring bonds. Many other infinite chain paths and much larger rings can also be identified in the network. Figure 3 also evidences classic π-π stacking, wherein ring carbon atom C3 is closely aligned with the centroid of the phenyl ring below, at a distance of 3.56 Å, with repeats of this interaction throughout the lattice.

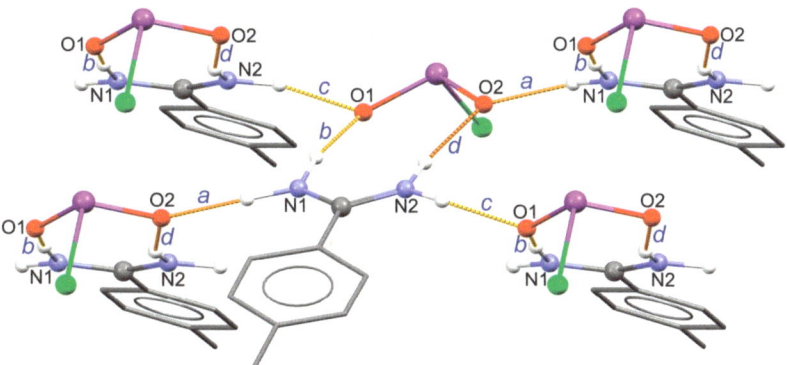

Figure 3. Hydrogen bond network in the crystal lattice of **2**. The Etter notation for the lower-level nets are indicated in the blue lettering: $D_1^1(2)$ a; $D_1^1(2)$ b; $D_1^1(2)$ c; $D_1^1(2)$ d; $C_2^2(6) > a < b$; $C_2^2(8) > a < c$; $C_2^1(6) > a < d$; $C_2^1(6) > b < c$; $R_2^2(8) > b < d$; $C_2^2(6) > c < d$. The relative H-bond strengths are color-coded, with yellow being stronger than orange.

2.4. Intra-Ionic Short Contacts in the Lattice of 2

We now consider how well the $ClSeO_2^-$ ions in **2** are isolated from each other. Metric parameters for contacts shorter than the sums of the Van der Waals' radii are included in Table 2. Figure 4a emphasizes the major interactions between $ClSeO_2^-$ ions by including only the ring-forming ion pairs for clarity, whereas Figure 4b shows the overall packing and deliberately includes all atoms to show the intermolecular environment. As is clear from the literature on general oxochlorochalcogenates(IV) [3], there are no truly isolated halochalcogenite ions in crystal lattices, although large organic cations such as PPh_4^+ or $AsPh_4^+$ sometimes come close, as in [$AsPh_4$][$OSeCl_3$] (CSD refcode: BIRGUE10 [54]), which at the very least tend to displace inter-anion contact with more benign donor–acceptor interactions with aromatic ring electron density. With smaller organic cations, inter-anion interactions are commonly observed, both in the forms of discrete dimers and infinite chain polymers. The strongest interactions occur with small monometallic cations, such as in alkali and alkaline earth metal salts, such as the type **C** direct anion chain structure in K[$FSeO_2$] [55]. However, many halochalcogenite anions of heavier halogens and chalcogens cannot exist with these small, focused charge, cations.

Table 3. Experimental (Raman) and computed vibrational spectra for ClSO$_2^-$ ions (cm^{-1}).

Band	Assignment	Symmetry	Experiment [1] CH$_3$CN Solution	Experiment [1] [Me$_4$N][ClSeO$_2$]	DFT Optimized [2]	DFT X-ray Geom. [3]
ν_1	ν_{sym}(SO$_2$)	A'	890 (p)	903 (s)	896 (vs,p)	886 (vs,p)
ν_2	ν(S–Cl)	A'	273 (p) [4]	267 (s) [4]	267 (w,p)	321 (m,p)
ν_3	δ_{sciss}(SO$_2$)	A'	380 (p)	396 (w)	367 (w,p)	382 (w,p)
ν_4	δ_{sym}(ClSO$_2^-$)	A'	200 (p) [4]	193 (vs) [4]	178 (m,p)	242 (m,p)
ν_5	ν_{asym}(SO$_2$)	A''	840 (dp)	841 (w)	912 (m,dp)	887 (s,dp)
ν_6	δ_{asym}(FSO$_2^-$)	A''	not obsv.	not obsv.	165 (w,dp)	189 (w,dp)

[1] As reported in Ref. [1]. [2] Frequency calculation of the Raman spectrum with full RB3PW91-D3/aug-CC-pVTZ geometry optimization of the anion geometry. [3] RB3PW91-D3/aug-CC-pVTZ-computed anion at the X-ray geometry in the crystal lattice of **2**, symmetrized to C_s. [4] Reversals of the 1978 assignments, based on the vibrational symmetries obtained by DFT; the colored bands draw attention to this switch. Notably, the ν_2 and ν_4 bands are strongly coupled, and hence both will reflect variations in S–Cl bond strength.

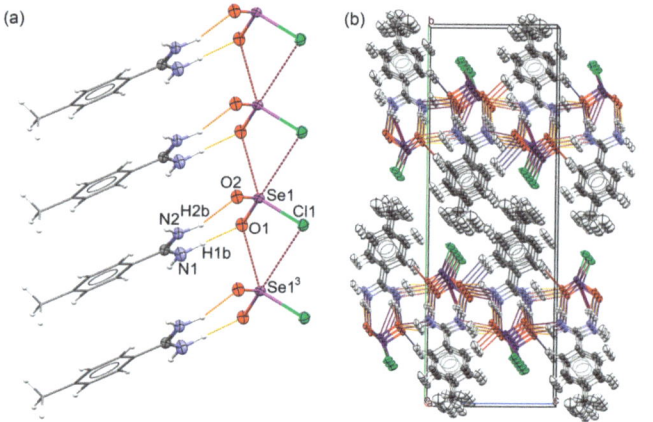

Figure 4. (a) The major inter-anion interactions in the crystal lattice of **2**. The chain of ClSeO$_2^-$ ions is aligned with the crystallographic *a* axis. Symmetry code: [3] 1 + x,y,z. See Table 3 for metric data. (b) Stacked packing diagram viewed down the *a*-axis direction.

Specific inter-anion contacts are found in the lattice of **2** (Figure 4a). The anions are doubly bridged by Se1–O1···Se1' and Se1–Cl1···Se1' contacts, forming discrete chains parallel to the crystallographic *a* axis in which the anions are stabilized in three directions by the H-bonds and the large chlorine atoms are surrounded by stacks of tolyl ring methyl groups (Figure 4b). Importantly, the direction of approach of Cl1 to Se1^3 on the next anion is close to linear with the opposing O1^3, i.e., the direction consistent with chalcogen bonding from a σ-hole at Se [23]. A consideration of the metrical data in Table 3, specifically the $d-\sum r_{vdW}$ values, indicates that these inter-anion contacts are relatively weak compared with the anion-cation H-bonding contacts. They are thus comparable (7–8.5% < $\sum r_{vdW}$) to the bridging Cl···Se contacts in the chain structure of [Me$_4$N][ClSeO$_2$] (see Supplementary Materials). This is further borne out by the comparison with the chain-forming Se–Cl···Se' contacts in 8-hydroxyquinolinium trichloro-oxyselenate, [C$_9$H$_8$NO][OSeCl$_3$], which displays $d-\sum r_{vdW} = -0.77$ Å (CSD refcode HQNLSE [56]). Other known structures of oxychlorselenium(IV) anion salts, which all show degrees of inter-anion contacts, are as follows: [PPh$_4$]$_2$[O$_2$Se$_2$Cl$_6$] (CSD refcode: BIRHAL10 [57]); [NEt$_4$]$_2$[O$_2$Se$_2$Cl$_6$] (CSD refcode: BORCAM [54]); [C$_{10}$H$_{10}$N$_2$][OSeCl$_4$] (CSD refcode: DPRYSE [58]); [NMe$_4$]$_3$[O$_2$Se$_2$Cl$_7$][Cl$_2$] (CSD refcode: EWILOO [25]); [N^{n}Pr$_4$]$_2$[O$_2$Se$_2$Cl$_6$] (CSD refcode: JUCDIU [59]) (CSD refcode: RAFYOM [60]).

2.5. DFT Computational Investigation of Structure

Surprisingly, no prior computational study of $ClSeO_2^-$ ions could be found in the literature, so it was considered essential to undertake a reliable DFT investigation. Based on precedents in the literature, the B3PW91 functional was selected for its proven efficacy in selenium chemistry [61], but we also chose to enhance it with Grimme's original D3 dispersion correction to improve its capability for also modelling the full H-bonded salt. For a basis set, aug-CC-pVTZ was selected because of its prior accuracy for selenium compounds [62]. The chosen RB3PW91-D3/aug-CC-pVTZ method was first validated by computing the structure of $SeO_2(g)$, which provides excellent agreement on geometry and molecular vibrations (see Appendix A).

In a first calculation, the ion pair at the crystal coordinates was computed, which confirms the primary directionality of the H-bonded geometry. A large dipole moment of 15.5 Debye orients almost parallel to the toluamidinium molecular plane and bifurcating the SeO_2 moiety (Figure 1b). Although far from a complete network, this primary H-bonding (i.e., the $R_2^2(8)$ net) does seem to be significantly structure directing (Table 2). Next, the full gas-phase geometry optimization of the ion pair was attempted, which did converge, albeit with a rather more curved overall structure ($\angle C2 \cdots Se1-Cl1 = 68.0°$) than that found in the lattice geometry ($\angle C2 \cdots Se1-Cl1 = 117.0°$).

Thereafter, the free $ClSeO_2^-$ ion was geometrically optimized, with full frequency calculations for comparison to the vibrational spectra in early literature reports, when IR, and especially Raman, spectroscopy were used as the chief characterization tools [1,33]. The geometry optimizes to effective C_s point symmetry, as expected, resulting in Se–O lengths of 1.630 Å, close to the average 1.637(1) Å of the two experimental values. Most surprisingly, however, the Se–Cl length *increases* to 2.491 Å, more than 7% longer than the experimental value of 2.3202(9) Å. This is far larger than the expected elongation from just using DFT at this level of theory. Moreover, these values are within the typical DFT accuracy (just 1.5% longer) reported for the BIRHOZ structure on $[Me_4N][ClSeO_2]$ [3].

Further support that the shorter Se–Cl length is a real effect from the H-bonding of the toluamidinium to the oxygen atoms of the $ClSO_2^-$ ion in **2** is provided by the above-mentioned gas-phase optimization of the cation-anion pair, where the Se–Cl length remains in the range 2.32 to 2.33 Å through all the optimization steps. Presumably, this is a primarily electrostatic effect, whereby withdrawing ED from the SeO_2 moiety enhances the presumed dative bond of the Cl^- nucleophile in its interaction with Se (see next section). We note that in oxytetrachloroselenate(IV) structures, H-bonding is known to induce *longer* Se–Cl bonds, e.g., in the structure $[C_4H_{10}NO][OSeCl_4]$ (CSD refcode: RAFYOM [60]) where the Se–Cl elongates to 2.776(2) Å from H-bonding to the morpholinium nitrogen (15% longer than the average of the three other equatorial bonds), or $[C_{10}H_{10}N_2][OSeCl_4]$ (CSD refcode: DPRYSE [58]) where the Se–Cl that is H-bonded to the bipyridinium NH elongates to 2.990(4) Å (24% longer than the equatorial average). Thus, H-bonding to the halogens causes longer Se–Cl bonds, whereas to the oxygen it causes shorter Se–Cl bonds. The observed behavior may also be related to variations in valence sharing.

Seeking further experimental confirmation, we noted that the BIRHOZ structure [3] on $[Me_4N][ClSeO_2]$ was one of the very salts originally investigated by Milne using Raman spectroscopy [1]. At that time, the vibrational data were only assigned using symmetry criteria, including experimental depolarization ratios, and by analogy to $ClSO_2^-$ [32]. The results of the RB3PW91-D3/aug-CC-pVTZ-computed spectra, conducted on (i) the optimized gas phase geometry and (ii) on the isolated anion at the X-ray geometry found in **2**, with assignments, and a comparison to the experimental Raman spectroscopy data are compiled in Table 3, and a comparison of one computed and experimental spectrum is shown in Figure 5.

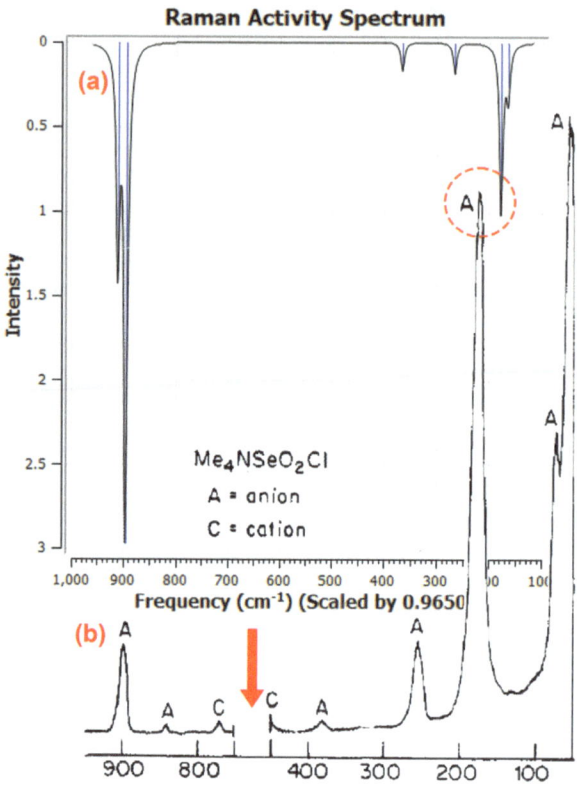

Figure 5. Overlay of (**a**) the RB3PW91-D3/aug-CC-pVTZ DFT Raman spectrum computed on a gas phase isolated ClSeO$_2^-$ anion with (**b**) the experimental spectrum reported on a crystalline powder sample of [Me$_4$N][ClSeO$_2$]; note that the bottom scale is interrupted at the red arrow, so the band positions must be interpolated. Bands below 100 cm^{-1} are lattice modes. Adapted with permission from LaHaie, P.; Milne, *J. Inorg. Chem.* **1979**, *18*, 632–637 [1]. Copyright (1979) American Chemical Society. Red text is used to distinguish from the original annotations of the underlying graphics.

First, the fit of the numerical data between the experimental and the computed anion values is remarkably good given the combination of experimental uncertainty and comparing gas-phase structures with solids and solutions (there is also very good agreement between the latter two). Importantly, the normal coordinate analysis in the DFT calculations contradict the assignment of ν_2 to the 200/193 cm^{-1} experimental bands, leading to a switch in the ν_2 and ν_4 assignments. A similar discrepancy has already been noted for the vibrational spectra of FSO$_2^-$, which is far and away the most thoroughly studied halochalcogenite anion to date [29]. These workers reported that a normal coordinate analysis with potential energy distribution indicates that the symmetry coordinate of the S–F stretch in FSO$_2^-$ contributes only 20% to ν_2, and instead has 40% ν_3 and 39% ν_4 character. For ClSeO$_2^-$, we find that ν_3 is not much involved, but there is this formal reversal of ν_2 and ν_4, as well as strong coupling that distributes a large amount of S–Cl stretch character to both modes. Viewed through this lens, it is apparent that these two bands are, respectively, 54 and 64 cm^{-1} higher in frequency when computed at the X-ray geometry of **2** than at the DFT-optimized geometry (Table 3 and Figure 6). This, and the good overall fit of the data, is a strong corroboration that the Se–Cl length in [Me$_4$N][ClSeO$_2$] corresponds to that reported for the BIRHOZ structure [3], and very probably is close to that adopted in CH$_3$CN solutions.

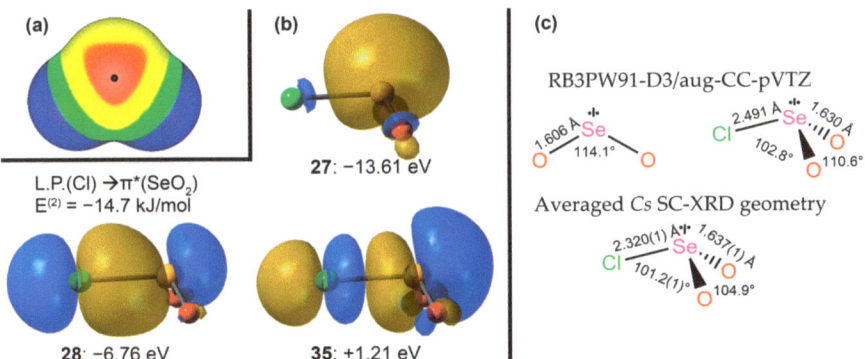

Figure 6. (**a**) Computed electrostatic potential surface at the 0.001 AU level of SeO$_2$; the position of the most positive electrostatic potential associated with the π-hole above (and below) Se is indicated by the black hemisphere with $V_{S,max}$ = 148 kJ/mol; adapted with permission from Murray, J. S. et al., *J. Mol. Model.* **2012**, *18*, 541–548 [24]. Copyright (2011) Springer Nature. (**b**) NBO isosurfaces from RB3PW91-D3/aug-CC-pVTZ DFT calculations on gas-phase optimized structure of ClSeO$_2^−$ for (**27**) the Se L.P, (**28**) the Cl L.P. interacting with the empty π*(SeO$_2$) orbital; (**35**) the corresponding out-of-phase interaction. Surfaces constructed at the 0.04 AU level. (**c**) Computed geometries (see text).

2.6. Donor–Acceptor Bonding in Hypervalent ClSeO$_2^−$ Using the NBO Formalism

After almost a century of debate, a combination of experimental and computational evidence has settled that the bonding in molecular SO$_2$ is not hypervalent and corresponds very closely to the description provided by the classical Lewis octet structure with nominal S–O bond orders of 1.5 [63]. SeO$_{2(g)}$ must certainly be described similarly, although under ambient conditions, it forms a solid polymerized via OSeO→Se dative bonding [49]. Indeed, our RB3PW91-D3/aug-CC-pVTZ DFT-computed structure for it has Wiberg bond indices of 1.45 for the Se–O bonds, and 0.31 between the two O atoms. The electrostatic π-holes detected in SeO$_2$ (Figure 6a) are oriented above and below the central Se atom and are perpendicular to the molecule plane [24]. Classical nucleophiles such as HCN or NH$_3$ are indeed computed to bind to the Se atom close to this perpendicular with interaction distances of 70–89% of the Σr_{vdW} [24].

Whether or not these π-holes are operative in ClSeO$_2^−$, it is clear from both the experimental and computed geometries that a chloride ion donates electron density to form a kind of dative or charge-transfer bonding, which is stronger than a mere intermolecular interaction, and that is definitely hypervalent [64]. We have applied a natural bond order (NBO) analysis using the NBO 3.1 component within Gaussian W16 on the geometry optimized in the gas phase for ClSeO$_2^−$ (Figure 6b,c). This clearly shows the Se L.P. orbital (#27) very close to co-planar with the SeO$_2$ atoms (and hence very similar to that of the educt—see Appendix A). The bond-forming NBO is the Cl L.P.→π*(SeO$_2$) orbital (#28), whilst the lowest-energy Rydberg NBO is its out-of-phase companion (#35). The Wiberg bond indices for the optimized geometry are reduced to 1.26 for the Se–O bonds and 0.19 between the O atoms, whilst the bond that forms between Cl and Se has an index of 0.44. This provides an excellent model for weak hyper-valent bonding in the chloroselenite ion and is reminiscent of the bonding models developed for the very well-known trihalide anions [65]. Since Cl→Se bonding has the net effect of occupying the SeO$_2$ π* molecular orbital, the π-bond order is expected to be reduced, which rationalizes the lower Wiberg indices for these bonds. And the low bond order for the X–ChO$_2$ bond is consistent with the very long X–Ch bond distances in most crystal structures of halochalcogenites and the observation of a wide range of bonding modes, ranging from well-defined XChO$_2^−$

molecular ions to 'solvated halide' geometries with almost equal X···Ch(O$_2$)···X determined in some SC-XRD structures (see the Introduction).

Returning now to the chloroselenite structure obtained in our salt **2** with the amidinium ion, it becomes understandable how H-bonding to the two anion oxygen atoms in **2** can have such a strong influence on the Se–Cl bond length, causing it to be more than 5% shorter than observed in the structure of [Me$_4$N][ClSeO$_2$] (BIRHOZ [3]). In the H-bonded adduct, computed in the gas phase, the NPA charge on Se increases from +1.66 in the optimized ion structure to +1.71, while the charge on Cl decreases from −0.62 to only −0.46. There is thus a clear rationale in the NBO analysis for the shorter Se–Cl bond observed in the structure of **2**, supporting the notion that H-bonding significantly stabilizes the chloroselenite in this amidinium salt.

3. Experimental

General synthetic methods for thiazyl and selenazyl chemistry are as previously described [46,66]. The isolation of **4** has been described previously [46].

3.1. Chemical Synthesis

Initially, 0.26 g of selenium(IV)-4-NH-dichloroselenatriazine, C$_{10}$H$_8$Cl$_5$N$_3$Se, was exposed to 10 mL of CH$_2$Cl$_2$ and 10 mL of CH$_3$CN and heated in an attempted purification by recrystallization. Upon removing all volatiles, 30 mL of CH$_3$CN was added and the mixture was heated to reflux, ensuring all solids were dissolved, and then filtered hot under nitrogen. After cooling, the filtrate was placed in a −10 °C freezer overnight, producing small amounts of solids which were removed by a second filtration. Subsequent cooling of the filtrate in a −30 °C freezer overnight produced well-formed colorless crystals, which were quickly isolated, dried under high vacuum, and submitted for crystallographic study.

3.2. Single-Crystal X-ray Crystallography

Suitable crystals were selected under a microscope, mounted on fine glass capillaries in Paratone™ oil, and cooled using the diffractometer cooling wand. Crystal and refinement data are summarized in Table 2. Data for **2** were collected on a Bruker Platform/SMART 1000 CCD diffractometer at the University of Alberta. Image collection, peak identification, cell and space group determination were controlled using SAINT. Multi-scan absorption correction was undertaken using SADABS. Initial structure solution was performed with SHELXS [67]. In view of the extensive H-bonding observed in this structure, refinement was completed in the independent atom model (IAM) using olex2.refine [68] within the Olex2 release 1.5 suite of programs [51]. After a detailed analysis and structure verification in the IAM, Hirsfeld atom refinement was continued using aspherical scattering factors with the aid of NoSpherA2 [50]. A detailed description of our workflow for HAR with aspherical form factors has been published [69].

In the case of **2**, HAR/NoSpherA2 quickly proved to be very successful. The electron density (ED) of each atom was computed using ORCA 5.0 [70] at the R2SCAN/def2-TZVP level of theory, whereafter the custom scattering factors for all atoms were computed using NoSpherA2, and refinement was completed with olex2.refine. When the H-atoms were refined anisotropically, it immediately became apparent that the tolyl methyl group is rotationally disordered, an effect that is normally ignored in IAM where the riding atom approach for H on C usually masks such subtleties. A two-part disorder model was adopted, and HAR resumed (which inter alia requires that two full sets of DFT calculations are required, one for each disorder component). This too proved successful. However, although the overall data quality is quite good (R_{int} = 3.28%), the data were obtained on an older sealed-tube diffractometer where the resolution was capped at 0.80 Å. This is probably the origin of the two ghost peaks for the Se atom (Fourier ripples), hindering the overall refinement quality. Hence, in the final refinement cycles the aromatic C-H, distances were restrained to 1.085 Å and N-H distances to 1.040 Å, which are the current best values from neutron refinement in this temperature range [71]. Similar 2- and 3-atom distance restraints were applied to the disordered H atoms in the CH$_3$ group, and the occupancies were frozen in a 50:50 ratio. This structure is not intended for validating the performance of the HAR/NoSpherA2 method (which we have previously done thoroughly [69,72–74]); instead, it aims for an accurate description of the extensive H-bonding network, and obtaining a deformation density map to show the non-bonded ED.

Crystal Data for C$_8$H$_{11}$ClN$_2$O$_2$Se (M = 281.601 g/mol): monoclinic, space group $P2_1/c$ (no. 14), a = 3.9773(3) Å, b = 28.477(2) Å, c = 9.5781(9) Å, β = 91.242(1)°, V = 1084.56(16) Å^3, Z = 4, T = 193.15 K, μ(Mo Kα) = 3.685 mm^{-1}, D_{calc} = 1.725 g/cm^3, 8245 reflections measured (4.48° $\leq 2\theta \leq$ 52.84°), 2194 unique (R_{int} = 0.0328, R_{sigma} = 0.0282) which were used in all calculations. The final R_1 was 0.0339 (I $\geq 2\sigma$(I)) and wR_2 was 0.0827 (all data). CCDC 2301348.

3.3. Computational Methods

Initial geometries for the DFT calculations were obtained from the X-ray coordinates (ignoring the second component of the toluene methyl group disorder). All calculations were performed with Gaussian W16 under GUI control from GaussView 6.0 and were run on an AMD Ryzen Threadripper 16-core 3.75 GHz PC under Windows 10 [75]. Minimum energy geometries were verified using harmonic vibrational analysis. The use of Grimme's D3 correction for dispersion, an attractive effect that is not readily accounted for by the bare B3PW31 functional, was applied in all cases. The suitability of the RB3PW91-D3/aug-CC-pVTZ level of theory was thoroughly investigated by computing the known structural and vibrational properties of SeO$_2$ (gas-phase structure), for which, see Appendix A. Atomic charges were computed using normal bond order analysis, and bond orders using the Wiberg definition, all with internal functions. Results computed with Gaussian W16 were visualized and, where required, plotted using GaussView.

4. Conclusions

The evidence from this paper is that significant Cl–Se bond shortening is induced in ClSeO$_2^-$ ions by the characteristic H-bonding from benzamidinium cations. This discovery could have major benefits for isolating stable salts of other halochalcogenite(IV) salts. For example, selectivity of the H-bonding for O is considered likely, which may help prevent O/F positional disorder in FChO$_2^-$ salts. These cations may also lead to success in obtaining crystalline adducts for the many X/Ch combinations for which there are no SC-XRD data. This work also emphasizes a point, made previously by others, that work in this area should use multiple, mutually integrated, techniques. For example, vibrational spectroscopy, when fully interpreted by adequate levels of computation, should be feasible on the proposed benzamidinium salts, enhancing the reliability of salt characterization and enabling direct comparisons, especially with Raman spectroscopy, to the structures adopted by these ions in solutions. Additionally, ^{77}Se NMR may become invaluable for characterizing the formation of haloselenite ions in solution.

Supplementary Materials: The following supporting information can be downloaded at: https://www.mdpi.com/article/10.3390/molecules28227489/s1, Supplementary report, with structural details on BIRHOZ, further information on **2** and a full crystal structure report for **2**. References [6,50,51,67,68] are cited in the Supplementary Materials.

Funding: Financial support for this work was provided by Discovery Grants from the Natural Sciences and Engineering Research Council of Canada and by the University of Lethbridge.

Data Availability Statement: CCDC 2301348 contains the supplementary crystallographic data for this paper. These data can be obtained free of charge via www.ccdc.cam.ac.uk/data_request/cif (accessed on 14 October 2023), or by emailing data_request@ccdc.cam.ac.uk, or by contacting The Cambridge Crystallographic Data Centre, 12 Union Road, Cambridge CB2 1EZ, UK; fax: +44-1223-336033 or from the Inorganic Crystal Structure Database, https://icsd.products.fiz-karlsruhe.de/ accessed on 14 October 2023.

Acknowledgments: The author thanks Jiamin Zhou for the synthesis of the sample and the X-ray Crystallography Laboratory at the University of Alberta for obtaining the X-ray intensity data. Daniel

Stuart is thanked for some early work on the structure refinement. The author is very grateful to Bernt Krebs for supplying the atom coordinate and displacement data for the BIRHOZ structure, via a personal communication.

Conflicts of Interest: The author declares no conflict of interest.

Appendix A

Computational method calibration: SeO_2 at the RB3PW91-D3/aug-CC-pVTZ level of theory computes d(Se–O) = 1.6026 Å; ∠(OSeO) 114.1°(cf. 1.6076(6); 113.83(8)° from microwave spectroscopy [76]). Vibrational spectrum: ν_2'' 362 cm^{-1}; ν_3'' 989 cm^{-1} (cf. 364; 968 cm^{-1} in the gas phase above solid SeO_2, >350 °C by infra-red spectroscopy [77]). This gives high confidence in the method so that agreement with experimental data for $ClSeO_2^-$ will reflect accurately the structural data for this weakly bound ion.

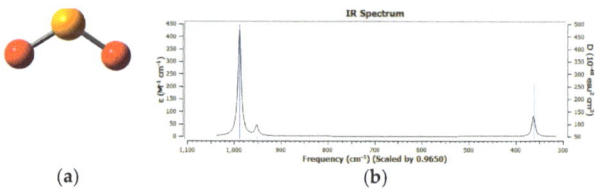

Figure A1. (a) Structure of SeO_2 (in the gas phase; it is a polymeric solid at RT). (b) Computed IR spectrum at the RB3PW91-D3/aug-CC-pVTZ level of theory.

The MO sequence $(1a_1)^2(1b_2)^2(2a_1)^2(2b_2)^2(3a_1)^2(1b_1)^2(1a_2)^2(3b_2)^2(4a_1)^2(2b_1)^0$ computed at our level is in remarkable agreement with Walsh's original theoretical predictions for AB_2 molecules at a bond angle of 114° [78], being $(1a_1)^2(1b_2)^2(2a_1)^2(1b_1)^2(3a_1)^2(2b_2)^2(1a_2)^2(3b_2)^2(4a_1)^2(2b_1)^0$, so that only $1b_1$ and $2b_2$ levels are exchanged (the π_1 MO is lower-lying in the generic Walsh interaction diagram, probably reflecting stronger π-overlap in the model system compared to that of the 4th period Se). Moreover, there is good agreement in the relative energy levels for the two highest filled and lowest virtual levels, such that the HOMO is the in-plane sulfur L.P. orbital by both approaches. Importantly for anion coordination, as in $[ClSO_2]^-$, the lowest unoccupied MO of SeO_2 is unambiguously the $2b_1$ π3* orbital dominated by the empty Se $4p_z$ AO.

Appendix B

The molecular and crystal structure of **4** has been reported previously in the supplementary data of Ref. [46] and is available from the CSD via refcode EZOWUM (or via its acquisition code: CCDC 851053). Allowing for the differing covalent radii of S and Se (1.02; 1.16 Å), the bond lengths at the chalcogen seem quite comparable to those in the $ClSeO_2^-$ ion in **2**, another indication of the relatively strong Se–Cl bond in that structure. However, the angles at S in sulfamyl chloride **4** are closer to tetrahedral values, as expected for a four-coordinate structure. Almost all the bonds and angles around the S atom in **4** fit well with the averages from 20 neutral comparator structures of other sulfamyl chlorides from a CSD search. The exception is d(S1–N1) which at 1.695(3) Å is 7% longer than the average of 1.59(3) Å. However, the average when restricted to just those ten comparators in which the attachment atom is not double bonded (to C or especially P) is closer, at 1.629(3) Å.

The H-bonds found in the lattice of **4** (Figure A2) are of only three types, but they are found to be slightly shorter even than those in **2**, and hence are expected to total about 150 kJ/mol in stabilization energy. The Etter notation for the lower-level nets are shown in the figure using blue lettering. Evidently, H-bonding is well able to stabilize this very reactive sulfamyl chloride.

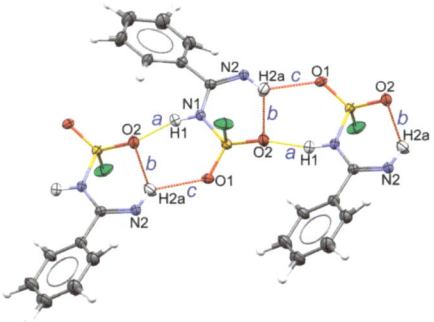

Figure A2. Hydrogen bond network in the crystal lattice of **4** (structure taken from CSD refcode EZOWUM). The Etter notation for the lower-level nets are indicated in the blue lettering: $C_1^1(4)$ *a*; $S_1^1(6)$ *b*; $C_1^1(6)$ *c*; $C_2^2(8) > a < c$; $R_2^2(10) > a > c$. The relative H-bond strengths are color-coded with yellow stronger than red.

Table A1 is a very useful compilation of H-bond properties, which enables a meaningful interpretation of the data found in the structures of **2** and **4**, as well as many other structures. These data have been compiled by the author from various literature sources and modified to be suitable for acceptors from both the second and the third periods of the main group elements.

Table A1. Ranking of Hydrogen Bond Strengths and Properties Adapted from Jeffrey *.

Parameter	Strong	Moderate	Weak
Interaction type:	strongly covalent	mostly electrostatic	electrostatic/dispersion
$d(\text{H}\cdots\text{A})$, Å #	1.2–1.5	1.5–2.2	>2.2
$d(\text{D}\cdots\text{A})$, Å #	2.2–2.5	2.5–3.2	>3.2
lengthening of D–H	0.08–0.25	0.02–0.08	<0.02
D–H *versus* H···A	D–H ≈ H···A	D–H < H···A	D–H << H···A
Directionality:	Strong	moderate	weak
∠D–H···A, °	170–180	>130	>90
Bond energy, kJ mol^{-1}	60–160	16–60	<16
IR shift $\Delta\bar{\nu}_{DH}$, cm^{-1}	25%	10–25%	<10%
^{1}H downfield shift, ppm	14–22	<14	

* Jeffrey, G. A. *An introduction to hydrogen bonding.* Oxford University Press: New York, 1997, Volume 12, pp. 330 [79]. # For a given donor type, the hydrogen-bond distance typically increases by over 0.5 Å from 2nd to 3rd period, 0.15 Å from 4th to 5th period, and 0.25 Å from 5th to 6th period acceptors [80].

References

1. LaHaie, P.; Milne, J. Chloro and oxochloro anions of selenium(IV). *Inorg. Chem.* **1979**, *18*, 632–637. [CrossRef]
2. Groom, C.R.; Bruno, I.J.; Lightfoot, M.P.; Ward, S.C. The Cambridge Structural Database. *Acta Crystallogr.* **2016**, *72*, 171–179. [CrossRef] [PubMed]
3. Krebs, B.; Hucke, M.; Schäffer, A. Strukturchemie von Selen(IV)-Sauerstoff-Halogenverindinungen. *Z. Kristallogr. Kristallgeom. Kristallphys. Kristallchem.* **1982**, *159*, 84–85.
4. Krebs, B.; Ahlers, F.-P. Developments in chalcogen-halide chemistry. *Adv. Inorg. Chem.* **1990**, *35*, 235–317.
5. Atwood, D.A. Selenium: Inorganic chemistry. In *Encyclopedia of Inorganic Chemistry*, 2nd ed.; King, R.B., Ed.; Wiley: Chichester, UK, 2005; Volume 3, pp. 4931–4955.
6. Schäffer, A. Cavity-Enhanced Optical Clocks. Ph.D. Thesis, University of Münster, Münster, Germany, 1984.
7. Wang, B.-C. Ab Initio and Density Functional Theory Studies of Nuclear Magnetic Resonance and Electron Spin Resonance Parameters of Biomolecules. Ph.D. Thesis, University of Arkansas, Fayetteville, AR, USA, 1968.
8. Collins, M.J.; Ratcliffe, C.I.; Ripmeester, J.A. CP/MAS ^{77}Se NMR in Solids. Chemical Shift Tensors and Isotropic Shifts. *J. Magn. Res.* **1986**, *68*, 172–179. [CrossRef]
9. Wasif, S.; Salama, S.B. Weak complexes of sulfur and selenium. IV. Complex of selenium dioxide and seleninyl dichloride with halide ions. *J. Chem. Soc. Dalton Trans.* **1975**, *21*, 2239–2241. [CrossRef]

10. Milne, J.; LaHaie, P. Chloroselenate(IV) Equilibria in Aqueous Hydrochloric Acid. *Inor. Chem.* **1979**, *18*, 3180–3183. [CrossRef]
11. Milne, J. Haloselenate(IV) formation and selenous acid dissociation equilibria in hydrochloric and hydrofluoric acids. *Can. J. Chem.* **1987**, *65*, 316–321. [CrossRef]
12. Paetzold, R.; Aurich, K. Investigation of selenium-oxygen compounds. XL. Alkali haloselenites and alkali halodiselenites. *Z. Chem.* **1966**, *6*, 265. [CrossRef]
13. Jiao, Y.; Chen, X.-Y.; Stoddart, J.F. Weak bonding strategies for achieving regio- and site-selective transformations. *Chem* **2022**, *8*, 414–438. [CrossRef]
14. Narita, H.; Maeda, M.; Tokoro, C.; Suzuki, T.; Tanaka, M.; Shiwaku, H.; Yaita, T. Extraction of Se(IV) and Se(VI) from aqueous HCl solution by using a diamide-containing tertiary amine. *RSC Adv.* **2023**, *13*, 17001–17007. [CrossRef]
15. Xu, W.; Bai, W. The selenium oxygen clusters SeO_n (n = 1–5) and their anions: Structures and electron affinities. *J. Mol. Struct. THEOCHEM* **2008**, *863*, 1–8. [CrossRef]
16. Crowther, S.A.; Brown, J.M. The 313 nm band system of SeO_2. Part 1: Vibrational structure. *J. Mol. Spect.* **2004**, *225*, 196–205. [CrossRef]
17. Crowther, S.A.; Brown, J.M. The 313 nm band system of SeO_2. Part 2: Rotational structure. *J. Mol. Spect.* **2004**, *225*, 206–221. [CrossRef]
18. Grein, F. Theoretical studies on the electronic spectrum of selenium dioxide. Comparison with ozone and sulfur dioxide. *Chem. Phys.* **2009**, *360*, 1–6. [CrossRef]
19. Chulanova, E.A.; Radiush, E.A.; Balmohammadi, Y.; Beckmann, J.; Grabowsky, S.; Zibarev, A.V. New charge-transfer complexes of 1,2,5-chalcogenadiazoles with tetrathiafulvalenes. *CrystEngComm* **2023**, *25*, 391–402. [CrossRef]
20. Rozhkov, A.V.; Zhmykhova, M.V.; Torubaev, Y.V.; Katlenok, E.A.; Kryukov, D.M.; Kukushkin, V.Y. Bis(perfluoroaryl)chalcolanes Ar^F_2Ch (Ch = S, Se, Te) as σ/π-Hole Donors for Supramolecular Applications Based on Noncovalent Bonding. *Cryst. Growth Des.* **2023**, *23*, 2593–2601. [CrossRef]
21. Biswal, H.S.; Sahu, A.K.; Galmés, B.; Frontera, A. Se···O/S and S···O Chalcogen Bonds in Small Molecules and Proteins: A Combined CSD and PDB Study. *ChemBioChem* **2022**, *23*, e202100498. [CrossRef]
22. Burguera, S.; Gomila, R.M.; Bauzá, A.; Frontera, A. Selenoxides as Excellent Chalcogen Bond Donors: Effect of Metal Coordination. *Molecules* **2022**, *27*, 8837. [CrossRef]
23. Scilabra, P.; Terraneo, G.; Resnati, G. The Chalcogen Bond in Crystalline Solids: A World Parallel to Halogen Bond. *Acc. Chem. Res.* **2019**, *52*, 1313–1324. [CrossRef]
24. Murray, J.S.; Lane, P.; Clark, T.; Riley, K.E.; Politzer, P. σ-Holes, π-holes and electrostatically-driven interactions. *J. Mol. Model* **2012**, *18*, 541–545. [CrossRef] [PubMed]
25. Usoltsev, A.N.; Korobeynikov, N.A.; Kolesov, B.A.; Novikov, A.S.; Abramov, P.A.; Sokolov, M.N.; Adonin, S.A. Oxochloroselenate(IV) with Incorporated $\{Cl_2\}$: The Case of Strong Cl···Cl Halogen Bonding. *Chem. Eur. J.* **2021**, *27*, 9292–9294. [CrossRef] [PubMed]
26. Hermodsson, Y. The crystal structure of $(CH_3)_4NCl·5SeOCl_2$. *Acta Chim. Scand.* **1967**, *21*, 1328–1342. [CrossRef]
27. Jackson, V.E.; Dixon, D.A.; Christe, K.O. Thermochemical Properties of Selenium Fluorides, Oxides, and Oxofluorides. *Inorg. Chem.* **2012**, *51*, 2472–2485. [CrossRef] [PubMed]
28. Kumar, A.; McGrady, G.S.; Passmore, J.; Grein, F.; Decken, A. Reversible SO_2 Uptake by Tetraalkylammonium Halides: Energetics and Structural Aspects of Adduct Formation Between SO_2 and Halide Ions. *Z. Anorg. Allg. Chem.* **2012**, *638*, 744–753. [CrossRef]
29. Lork, E.; Mews, R.; Viets, D.; Watson, P.G.; Borrmann, T.; Vij, A.; Boatz, J.A.; Christe, K.O. Structure of the SO_2F^- Anion, a Problem Case. *Inorg. Chem.* **2001**, *40*, 1303–1311. [CrossRef] [PubMed]
30. Kessler, U.; van Wullen, L.; Jansen, M. Structure of the Fluorosulfite Anion: Rotational Disorder of SO_2F^- in the Alkali Metal Fluorosulfites and Crystal Structures of α- and β-$CsSO_2F$. *Inorg. Chem.* **2001**, *40*, 7040–7046. [CrossRef]
31. Arnold, S.T.; Miller, T.M.; Viggiano, A.A. A Combined Experimental and Theoretical Study of Sulfur Oxyfluoride Anion and Neutral Thermochemistry and Reactivity. *J. Phys. Chem. A* **2002**, *106*, 9900–9909. [CrossRef]
32. Burow, D.F. Spectroscopic Studies of Halosulfinate Ions. *Inorg. Chem.* **1972**, *11*, 573–583. [CrossRef]
33. Milne, J.; LaHaie, P. Raman spectra of haloselenate(IV) ions—The SeO_2Br^- anion. *Spectrochim. Acta* **1983**, *39*, 555–557. [CrossRef]
34. Zhu, S.-Z.; Huang, Q.-C.; Wu, K. Synthesis and Structure of (Difluoromethyl)triphenylphosphonium Fluorosulfite. Evidence for Formation of Difluorosulfene as an Intermediate. *Inorg. Chem.* **1994**, *33*, 4584–4585. [CrossRef]
35. Kuhn, N.; Bohnen, H.; Fahl, J.; Bläser, D.; Boese, R. Imidazole Derivatives, XIX. Coordination or Reduction? On the Reaction of 1,3-Diisopropyl-4,5-dimethylimidazolylidene with Sulfur Halides and Sulfur Oxygen Halides. *Chem. Ber.* **1996**, *129*, 1579–1586. [CrossRef]
36. Kessler, U.; Jansen, M. Crystal Structures of Monofluorosulfites MSO_2F (M = K, Rb). *Z. Anorg. Allg. Chem.* **1999**, *625*, 385–388. [CrossRef]
37. Kuhn, N.; Bohnen, H.; Bläser, D.; Boese, R.; Maulitz, A.H. Selective Reduction of Sulfuric Chloride: The Structure of the Chlorosulfite Ion. *J. Chem. Soc. Chem. Commun.* **1994**, *19*, 2283–2284. [CrossRef]
38. Reuter, K.; Rudel, S.S.; Buchner, M.R.; Kraus, F.; von Hänisch, C. Crown Ether Complexes of Alkali-Metal Chlorides from SO_2. *Chem. Eur. J.* **2017**, *23*, 9607–9617. [CrossRef] [PubMed]
39. Awere, E.G.; Burford, N.; Haddon, R.C.; Parsons, S.; Passmore, J.; Waszczak, J.V.; White, P.S. X-ray Crystal Structures of the 1,3,2-Benzodithiazolyl Dimer and 1,3,2-Benzodithiazolium Chloride Sulfur Dioxide Solvate: Comparison of the Molecular and

Electronic Structures of the 10-π-Electron $C_6S_2N^+$ Cation and the $C_6H_4S2N^\bullet$ Radical and Dimer and a Study of the Variable-Temperature Magnetic Behavior of the Radical. *Inorg. Chem.* **1990**, *29*, 4821–4830.

40. Boyle, P.D.; Godfrey, S.M.; Pritchard, R.G. The reaction of N-methylbenzothiazole-2-selone and 1,1-dimethylselenoureawith sulfuryl chloride and dichlorine. *J. Chem. Soc. Dalton Trans.* **1999**, *23*, 4245–4250. [CrossRef]
41. Eller, P.G.; Kubas, G.J. Synthesis, Properties, and Structure of Iodosulfinate Salts. *Inorg. Chem.* **1978**, *17*, 894–897. [CrossRef]
42. Dankert, F.; Feyh, A.; von Hänisch, C. Chalcogen Bonding of SO_2 and s-Block Metal Iodides Near Room Temperature: A Remarkable Structural Diversity. *Eur. J. Inorg. Chem.* **2020**, *2020*, 2744–2756. [CrossRef]
43. Eller, P.G.; Kubas, G.J.; Ryan, R.R. Synthesis and Properties of Sulfur Dioxide Adducts of Organophosphinecopper(I) Iodides. Structures of the Dinuclear Compounds Tetrakis(methyldiphenylphosphine)di-μ-iodo-dicopper(I)-Sulfur Dioxide and Tris(triphenylphosphine)di-μ-Li-iodo-dicopper(I). *Inorg. Chem.* **1977**, *16*, 2454–2462. [CrossRef]
44. Snow, M.R.; Ibers, J.A. The Halogen to Sulfur Dioxide Bond. Structure of Iodo(sulfur dioxide)methylbis(triphenyl phosphine)platinum, $Pt(CH_3)(PPh_3)_2I\text{-}SO_2$. *Inorg. Chem.* **1973**, *12*, 224–229. [CrossRef]
45. Nagasawa, I.; Amitaa, H.; Kitagawa, H. A new type of iodosulfite ion formulated as $I_2SO_2^{2-}$. *Chem. Commun.* **2009**, *8*, 204–205. [CrossRef] [PubMed]
46. Boeré, R.T.; Roemmele, T.L.; Yu, X. Unsymmetrical $1\lambda^3$-1,2,4,6-Thiatriazinyls with Aryl and Trifluoromethyl Substituents: Synthesis, Crystal Structures, EPR Spectroscopy, and Voltammetry. *Inorg. Chem.* **2011**, *50*, 5123–5136. [CrossRef] [PubMed]
47. Suduweli Kondage, S.; Roemmele, T.L.; Boeré, R.T. Dispersion in Crystal Structures of 1-Chloro-3-aryl-5-trihalomethyl-$1\lambda^4$,2,4,6-thiatriazines: Towards an Understanding of the Supramolecular Organization of Covalent Thiazyl Chlorides. *Synlett* **2023**, *34*, 1113–1121.
48. Cordes, A.W.; Oakley, R.T.; Reed, R.W. Structure of 1,1-Dichloro-3,5-diphenyl-4H-1,2,4,6-selenatriazine. *Acta Cryst.* **1986**, *42*, 1889–1890. [CrossRef]
49. Greenwood, N.N.; Earnshaw, A. *Chemistry of the Elements*, 2nd ed.; Butterworth/Heinemann: Oxford, UK, 1997; pp. 775–779.
50. Kleemiss, F.; Dolomanov, O.V.; Bodensteiner, M.; Peyerimhoff, N.; Midgley, L.; Bourhis, L.J.; Genoni, A.; Malaspina, L.A.; Jayatilaka, D.; Spencer, J.L.; et al. Accurate crystal structures and chemical properties from NoSpherA2. *Chem. Sci.* **2021**, *12*, 1675–1692. [CrossRef] [PubMed]
51. Dolomanov, O.V.; Bourhis, L.J.; Gildea, R.J.; Howard, J.A.K.; Puschmann, H. OLEX2: A complete structure solution, refinement and analysis program. *J. Appl. Crystallogr.* **2009**, *42*, 339–341. [CrossRef]
52. Steiner, T. The Hydrogen Bond in the Solid State. *Angew. Chem. Int. Ed.* **2002**, *41*, 48–76. [CrossRef]
53. Etter, M.C. Encoding and decoding hydrogen-bond patterns of organic compounds. *Acc. Chem. Res.* **1990**, *23*, 120–126. [CrossRef]
54. Krebs, B.; Hucke, M.; Hein, M.; Schäffer, A. Monomeric and Dimeric Oxotrihalogenoselenates(IV): Preparation, Structure and Properties of $[As(C_6H_5)_4]SeOCl_3$ and $[N(C_2H_5)_4] SeOCl_3$. *Z. Naturforsch. B Chem. Sci.* **1983**, *38*, 20–29. [CrossRef]
55. Feldmann, C.; Jansen, M. On the Crystal Structures of the Monofluoroselenites $MSeO_2$,F (M = K, Rb, Cs). *Chem. Ber.* **1994**, *127*, 2173–2176. [CrossRef]
56. Cordes, A.W. The crystal structure of 8-hydroxyquinolinium trichlorooxyselenate. *Inorg. Chem.* **1967**, *6*, 1204–1208. [CrossRef]
57. Krebs, B.; Schäffer, A.; Hucke, M. Oxotrihalogenoselenates(IV): Preparation, Structure and Properties of $P(C_6H_5)_4SeOCl_3$, $P(C_6H_5)_4SeOBr_3$ and $As(C_6H_5)_4SeOBr_3$. *Z. Naturforsch. B Chem. Sci.* **1982**, *37*, 1410–1417. [CrossRef]
58. Wang, B.-C.; Cordes, A.W. The Crystal Structure of Dipyridinium(II) Oxytetrachloroselenate(IV), $C_{10}H_8N_2H_2^{2+}$ $SeOCl_4^{2-}$. A Highly Coordinated Selenium Compound. *Inorg. Chem.* **1970**, *9*, 1643–1650. [CrossRef]
59. James, M.A.; Knop, O.; Cameron, T.S. Crystal structures of $(n\text{-}Pr_4N)_2SnCl_6$, $(n\text{-}Pr_4N)[TeCl_4(OH)]$, $(n\text{-}Pr_4N)_2[Te_2Cl_{10}]$ (nominal), and $(n\text{-}Pr_4N)_2[Se_2O_2Cl_6]$ with observations on $Z_2L_{10}^{2n-}$ and $Z_2L_8^{2-}$ dimers in general. *Can. J. Chem.* **1992**, *70*, 1795–1821. [CrossRef]
60. Hasche, S.; Reich, O.; Beckmann, I.; Krebs, B. Stabilization of Oxohalogeno and Halogenochalcogenates(IV), by Proton Acceptors—Synthesis, Structures and Properties of $[C_4H_{10}NO]_2[SeOCl_4]$, $[C_4H_{10}NO]_2[Se_2Br_{10}]$ and $[(CH_3)_2CH(NH_2)(OH)][Te_3Cl_{13}]$ $(CH_3)_2CHCN$. *Z. Anorg. Allg. Chem.* **1997**, *623*, 724–734. [CrossRef]
61. Tuononen, H.M.; Suontamo, R.; Valkonen, J.U.; Laitinen, R.S.; Chivers, T. Conformations and Energetics of Sulfur and Selenium Diimides. *Inorg. Chem.* **2003**, *42*, 2447–2454. [CrossRef] [PubMed]
62. Lobring, K.C.; Hao, C.; Forbes, J.K.; Ivanov, M.R.J.; Bachrach, S.M.; Sunderlin, L.S. Bond Strengths in $ChCl_3^-$ and $ChOCl_3^-$ (Ch = S, Se, Te): Experiment and Theory. *J. Phys. Chem. A* **2003**, *107*, 11153–11160. [CrossRef]
63. Grabowsky, S.; Luger, P.; Buschmann, J.; Schneider, T.; Schirmeister, T.; Sobolev, A.N.; Jayatilaka, D. The Significance of Ionic Bonding in Sulfur Dioxide: Bond Orders from X-ray Diffraction Data. *Angew. Chem. Int. Ed.* **2012**, *51*, 6776–6779. [CrossRef]
64. Durrant, M.C. A quantitative definition of hypervalency. *Chem. Sci.* **2015**, *6*, 6614–6623. [CrossRef]
65. Landrum, G.A.; Goldberg, N.; Hoffmann, R. Bonding in the trihalides (X_3^-), mixed trihalides (X_2Y^-) and hydrogen bihalides (X_2H^-). The connection between hypervalent, electron-rich three-center, donor–acceptor and strong hydrogen bonding. *J. Chem. Soc. Dalton Trans.* **1997**, *19*, 3605–3613. [CrossRef]
66. Oakley, R.T.; Reed, R.W.; Cordes, A.W.; Craig, S.L.; Graham, J.B. 1,2,4,6-Selenatriazinyl Radicals and Dimers. Preparation and Structural Characterization of l-Chloro-3,5-diphenyl-1,2,4,6-selenatriazine ($Ph_2C_2N_3SeCl$) and Bis(3,5-diphenyl-1,2,4,6-selenatriazine) (($Ph_2C_2N_3Se)_2$). *J. Am. Chem. Soc.* **1987**, *109*, 7745–7749. [CrossRef]
67. Sheldrick, G.M. A short history of SHELX. *Acta Cryst.* **2008**, *64*, 112–122. [CrossRef] [PubMed]

68. Bourhis, L.J.; Dolomanov, O.V.; Gildea, R.J.; Howard, J.A.; Puschmann, H. The anatomy of a comprehensive constrained, restrained refinement program for the modern computing environment—Olex2 dissected. *Acta Crystallogr. Sect. A Found. Adv.* **2015**, *71*, 59–75. [CrossRef]
69. Boeré, R.T. Crystal structures of $CuCl_2 \cdot 2H_2O$ (Eriochalcite) and $NiCl_2 \cdot 6H_2O$ (Nickelbischofite) at low temperature: Full refinement of hydrogen atoms using non-spherical atomic scattering factors. *Crystals* **2023**, *13*, 293. [CrossRef]
70. Neese, F. Software update: The ORCA program system Version 5.0. *WIREs Comput. Mol. Sci.* **2022**, *12*, e1606. [CrossRef]
71. Allen, F.H.; Bruno, I.J. Bond lengths in organic and metalorganic compounds revisited: X-H bond lengths from neutron diffraction data. *Acta Crystallogr. Sect. B Struct. Sci. Cryst. Eng. Mater.* **2010**, *66*, 380–386. [CrossRef] [PubMed]
72. Hill, N.D.D.; Lilienthal, E.; Bender, C.O.; Boeré, R.T. Accurate Crystal Structures of $C_{12}H_9CN$, $C_{12}H_8(CN)_2$, and $C_{16}H_{11}CN$ Valence Isomers Using Nonspherical Atomic Scattering Factors. *J. Org. Chem.* **2022**, *87*, 16213–16229. [CrossRef]
73. Marszaukowski, F.; Boeré, R.T.; Wohnrath, K. Frustrated and Realized Hydrogen Bonding in 4-Hydroxy-3,5-ditertbutylphenylphosphine Derivatives. *Cryst. Growth Des.* **2022**, *22*, 2512–2533. [CrossRef]
74. Ibrahim, M.A.; Boeré, R.T. The copper sulfate hydration cycle. Crystal structures of $CuSO_4$ (Chalcocyanite), $CuSO_4 \cdot H_2O$ (Poitevinite), $CuSO_4 \cdot 3H_2O$ (Bonattite) and $CuSO_4 \cdot O\ 5H_2O$ (Chalcanthite) at low temperature using non-spherical atomic scattering factors. *New J. Chem.* **2022**, *46*, 5479–5488. [CrossRef]
75. Frisch, M.J.; Trucks, G.W.; Schlegel, H.B.; Scuseria, G.E.; Robb, M.A.; Cheeseman, J.R.; Scalmani, G.; Barone, V.; Petersson, G.A.; Nakatsuji, H.; et al. *Gaussian 16*; Revision C.01; Gaussian, Inc.: Wallingford, CT, USA, 2016.
76. Takeo, H.; Hirota, E.; Morino, Y. Equilibrium structure and potential function of selenium dioxide by microwave spectroscopy. *J. Mol. Spectr.* **1970**, *34*, 370–382. [CrossRef]
77. Konings, R.J.M.; Booij, A.S.; Kovcás, A. The infrared spectra of SeO_2 and TeO_2 in the gas phase. *Chem. Phys. Lett.* **1998**, *292*, 447–453. [CrossRef]
78. Walsh, A.D. The Electronic Orbitals, Shapes, and Spectra of Polyatomic Molecules. Part II. Non-hydride AB_2, and BAC Molecules. *J. Chem. Soc.* **1955**, 2266–2287. [CrossRef]
79. Jeffrey, G.A. *An Introduction to Hydrogen Bonding*; Oxford University Press: New York, NY, USA, 1997; Volume 12, p. 330.
80. Steiner, T. Hydrogen-Bond Distances to Halide Ions in Organic and Organometallic Crystal Structures: Up-to-date Database Study. *Acta Cryst.* **1998**, *54*, 456–463. [CrossRef]

Disclaimer/Publisher's Note: The statements, opinions and data contained in all publications are solely those of the individual author(s) and contributor(s) and not of MDPI and/or the editor(s). MDPI and/or the editor(s) disclaim responsibility for any injury to people or property resulting from any ideas, methods, instructions or products referred to in the content.

Article

Formation, Characterization, and Bonding of cis- and trans-[PtCl$_2${Te(CH$_2$)$_6$}$_2$], cis-trans-[Pt$_3$Cl$_6${Te(CH$_2$)$_6$}$_4$], and cis-trans-[Pt$_4$Cl$_8${Te(CH$_2$)$_6$}$_4$]: Experimental and DFT Study [†]

Marko Rodewald [1,‡], J. Mikko Rautiainen [2], Helmar Görls [1], Raija Oilunkaniemi [3], Wolfgang Weigand [1,*] and Risto S. Laitinen [3,*]

[1] Institute for Inorganic and Analytical Chemistry, Friedrich Schiller University of Jena, Humboldt Str. 8, 07743 Jena, Germany; marko.rodewald@uni-jena.de (M.R.); helmar.goerls@uni-jena.de (H.G.)
[2] Department of Chemistry and Nanoscience Center, University of Jyväskylä, P.O. Box 35, 40014 Jyväskylä, Finland; j.mikko.rautiainen@jyu.fi
[3] Laboratory of Inorganic Chemistry, Environmental and Chemical Engineering, University of Oulu, P.O. Box 3000, 90014 Oulu, Finland; raija.oilunkaniemi@oulu.fi
* Correspondence: wolfgang.weigand@uni-jena.de (W.W.); risto.laitinen@oulu.fi (R.S.L.)
[†] Dedicated to Professor J. Derek Woollins on the occasion of his retirement.
[‡] Current address: Leibniz Institute of Photonic Technology, Member of Leibniz Health Technologies, Member of the Leibniz Centre for Photonics in Infection Research (LPI), P.O. Box 100239, 07702 Jena, Germany.

Citation: Rodewald, M.; Rautiainen, J.M.; Görls, H.; Oilunkaniemi, R.; Weigand, W.; Laitinen, R.S. Formation, Characterization, and Bonding of cis- and trans-[PtCl$_2${Te(CH$_2$)$_6$}$_2$], cis-trans-[Pt$_3$Cl$_6${Te(CH$_2$)$_6$}$_4$], and cis-trans-[Pt$_4$Cl$_8${Te(CH$_2$)$_6$}$_4$]: Experimental and DFT Study. *Molecules* 2023, 28, 7551. https://doi.org/10.3390/molecules28227551

Academic Editor: Petr Kilián

Received: 16 October 2023
Revised: 4 November 2023
Accepted: 9 November 2023
Published: 12 November 2023

Copyright: © 2023 by the authors. Licensee MDPI, Basel, Switzerland. This article is an open access article distributed under the terms and conditions of the Creative Commons Attribution (CC BY) license (https://creativecommons.org/licenses/by/4.0/).

Abstract: [PtCl$_2${Te(CH$_2$)$_6$}$_2$] (**1**) was synthesized from the cyclic telluroether Te(CH$_2$)$_6$ and cis-[PtCl$_2$(NCPh)$_2$] in dichloromethane at room temperature under the exclusion of light. The crystal structure determination showed that in the solid state, **1** crystallizes as yellow plate-like crystals of the cis-isomer **1**$_{cis}$ and the orange-red interwoven needles of **1**$_{trans}$. The crystals could be separated under the microscope. NMR experiments showed that upon dissolution of the crystals of **1**$_{cis}$ in CDCl$_3$, it isomerizes and forms a dynamic equilibrium with the trans-isomer **1**$_{trans}$ that becomes the predominant species. Small amounts of cis-trans-[Pt$_3$Cl$_6${Te(CH$_2$)$_6$}$_4$] (**2**) and cis-trans-[Pt$_4$Cl$_8${Te(CH$_2$)$_6$}$_4$] (**3**) were also formed and structurally characterized. Both compounds show rare bridging telluroether ligands and two different platinum coordination environments, one exhibiting a cis-Cl/cis-Te(CH$_2$)$_6$ arrangement and the other a trans-Cl/trans-Te(CH$_2$)$_6$ arrangement. Complex **2** has an open structure with two terminal and two bridging telluroether ligands, whereas complex **3** has a cyclic structure with four Te(CH$_2$)$_6$ bridging ligands. The bonding and formation of the complexes have been discussed through the use of DFT calculations combined with QTAIM analysis. The recrystallization of the mixture of the 1:1 reaction from d_6-DMSO afforded [PtCl$_2${S(O)(CD$_3$)$_2$}{Te(CH$_2$)$_6$}] (**4**) that could also be characterized both structurally and spectroscopically.

Keywords: density functional calculations; NMR spectroscopy; platinum; tellurium; X-ray diffraction

1. Introduction

The advent of the versatile chemistry of crown-ethers and their complexes in the 1980s [1,2] has generated interest in the related macrocycles of heavier chalcogen elements (for some recent reviews, see refs. [3–18]). The information on the syntheses, structures, and coordination chemistry of thioethers is particularly extensive, but the related seleno- and telluroethers have also shown growing research activity in recent decades. The study of macrocyclic chalcogen heterocycles helps to gain insight into the chalcogen bonding and its applications in supramolecular chemistry and crystal engineering [19–25]. The chalcogen bonding interactions are most significant in the case of tellurium.

The information on cyclic saturated telluroethers is sparse compared to related thioethers and selenoethers. While the preparation of cyclic Te(CH$_2$)$_4$ [26], Te(CH$_2$)$_5$ [27], and 1,5,9-Te$_3$(CH$_2$)$_9$ [28] has been known for a long time, it is only recently that the crystal structures and bonding in some [Te(CH$_2$)$_m$]$_n$ (n = 1–4, m = 3–7) [29] have been discussed. The coordination chemistry of the cyclic telluroethers is also little studied with only [MCl$_2${Te(CH$_2$)$_4$}$_2$] (M = Pd, Pt) [30], [PtCl$_2${Te(CH$_2$)$_4$O}$_2$] [31], [MCl$_2${Te$_2$O$_4$(CH$_2$)$_{12}$}] (M = Pd. Pt) [32], [Rh$_2$(η^5-C$_5$H$_5$)$_2$(CO)(μ-η^1: η^1- CF$_3$C$_2$CF$_3$){Te(CH$_2$)$_4$}] [33], and [Ag{TeS$_2$(CH$_2$)$_8$}]$_n$[BF$_4$]$_n$ [34] being either structurally or spectroscopically characterized.

The current contribution is the continuation of our systematic investigation of the synthesis of the series of telluroether heterocycles [Te(CH$_2$)$_m$]$_n$ (n = 1–4; m = 3–7) [29] and the coordination complexes of telluroethers (see refs. [35–39] for some recent publications). Monotelluroethers Te(CH$_2$)$_m$ are liquid in ambient conditions, but species with higher Te contents are solid. The molecular structures and the packing of seven macrocyclic aliphatic telluroethers have been explored [29]. Te(CH$_2$)$_6$ was one of the monotelluroethers that was synthesized and characterized through NMR spectroscopy, but since it is a liquid at room temperature with a low melting point, its crystal structure determination could not be carried out. It was thought that the reaction with [PtCl$_2$(NCPh)$_2$] should enable it to coordinate with the platinum center, and the resulting complex would be crystalline. Its crystal structure determination would establish the molecular structure of Te(CH$_2$)$_6$. The monodentate also ligand avoids the formation of coordination polymers. It turned out that the reaction produced not only the expected cis- and trans-[PtCl$_2${Te(CH$_2$)$_6$}$_2$] (1$_{cis}$ and 1$_{trans}$, respectively), but small amounts of cis-trans-[Pt$_3$Cl$_6${Te(CH$_2$)$_6$}$_4$]·1¼CH$_2$Cl$_2$) (2·1¼CH$_2$Cl$_2$) and, depending on the molar ratio of the reagents, also cis-trans-[Pt$_4$Cl$_8${Te(CH$_2$)$_6$}$_4$]·4CDCl$_3$ (3·4CDCl$_3$). The attempts to produce the polynuclear complexes in better yields involved the use of d_6-dimethyl sulfoxide as the crystallization solvent, in which case crystals of the mononuclear [PtCl$_2${S(O)(CD$_3$)$_2$}{Te(CH$_2$)$_6$}] (4) were obtained. The crystal structures of 1$_{cis}$ and 2-4, the isomerization of 1$_{cis}$ to 1$_{trans}$, and the bonding features and formation of 2 and 3 are discussed in this paper.

2. Results and Discussion

2.1. General

The reaction of two equivalents of Te(CH$_2$)$_6$ with one equivalent of cis-[PtCl$_2$(NCPh)$_2$] in dichloromethane produces cis- and trans-[PtCl$_2${Te(CH$_2$)$_6$}$_2$] (1$_{cis}$ and 1$_{trans}$) (see Scheme 1). The ^{1}H, ^{125}Te{^{1}H}, and ^{195}Pt{^{1}H} NMR spectra recorded in CDCl$_3$ indicate that the trans-isomer is the main isomer in CDCl$_3$, as discussed in Section 2.2. Small amounts of insoluble, likely polymeric products were also formed.

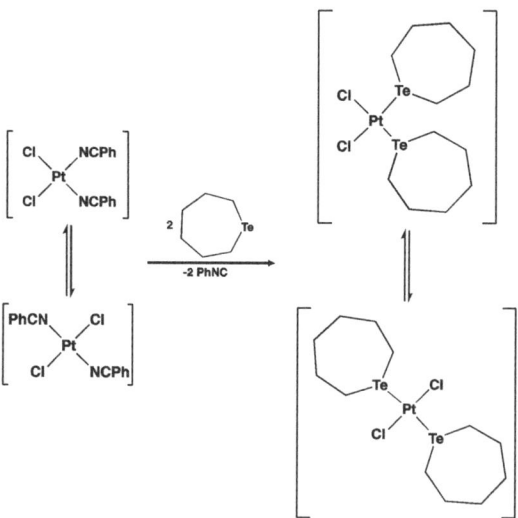

Scheme 1. Formation and isomerization of [PtCl$_2${Te(CH$_2$)$_6$}$_2$].

The syntheses of platinum and palladium complexes using related telluracyclopentane Te(CH$_2$)$_4$ have been reported to give [MX$_2$¦Te(CH$_2$)$_4$¦$_2$] (M = Pt, Pd; X = Cl, Br, I) [30]. The [PtCl$_2$¦Te(CH$_2$)$_4$¦$_2$] was deduced to be the *cis* isomer in the solid state on the basis of IR spectroscopy, but the complex was reported to exist as a mixture of *cis*- and *trans*-isomers in solution in the respective concentration ratio of, ca. 1:2. The palladium complex [PdCl$_2$¦Te(CH$_2$)$_4$¦$_2$] exists only as a *trans* isomer [30].

The crystallization of the product of the reaction of Te(CH$_2$)$_6$ and *cis*-[PtCl$_2$(NCPh)$_2$] from dichloromethane/pentane gave a mixture of yellow plates and orange-red needles. The yellow plate-shaped crystals were suitable for the determination of the crystal structure, which showed them to be *cis*-[PtCl$_2$¦Te(CH$_2$)$_6$¦$_2$] (**1**$_{cis}$) (see Section 2.3). The orange-red needles were intergrown and proved to be unsuitable for X-ray structure analysis. However, the NMR and mass spectroscopic information indicated them to be *trans*-[PtCl$_2$¦Te(CH$_2$)$_6$¦$_2$] (**1**$_{trans}$).

Small amounts of both yellow and orange-red crystals were manually separated under the microscope. They were dissolved in CDCl$_3$ for the recording of the NMR spectra (see Section 2.2). An additional small crop of red plate-shaped crystals was identified under the microscope and could be manually isolated based on their different crystal habits. The crystal structure was determined as *cis-trans*-[Pt$_3$Cl$_6$¦Te(CH$_2$)$_6$¦$_4$]·1¼CH$_2$Cl$_2$ (**2**·1¼CH$_2$Cl$_2$) (see Section 2.3). Because of the very small amount of **2**, no bulk analysis could be carried out.

The synthesis was repeated by using equimolar amounts of *cis*-[PtCl$_2$(NCPh)$_2$] and Te(CH$_2$)$_6$. Thin-layer chromatography indicated mostly the formation of *cis*- and *trans*-[PtCl$_2$¦Te(CH$_2$)$_6$¦$_2$] (**1**$_{cis}$ and **1**$_{trans}$) together with the presence of the starting materials. After recording the NMR spectra in CDCl$_3$, red well-shaped crystals grew in the NMR tube. The crystal structure determination showed these crystals to be *cis-trans*-[Pt$_4$Cl$_8$¦Te(CH$_2$)$_6$¦$_4$]·4CDCl$_3$ (**3**·4CDCl$_3$) (see Section 2.3).

During the attempts to crystallize **3** in larger amounts, the recrystallization in dimethyl sulfoxide was attempted. It yielded almost colorless crystals of [PtCl$_2$¦S(O)(CD$_3$)$_2$¦$_2$¦Te(CH$_2$)$_6$¦] (**4**) (see Section 2.3).

2.2. NMR Spectroscopy

The crystals of *cis*-[PtCl$_2$¦Te(CH$_2$)$_6$¦$_2$] were dissolved in deuterochloroform for recording the ^{125}Te{^{1}H} NMR and ^{195}Pt{^{1}H} NMR spectra (see Figure 1). All spectra can be interpreted in terms of a mixture of **1**$_{cis}$ and **1**$_{trans}$. The NMR spectroscopic information of **1**$_{cis}$ and **1**$_{trans}$ is compared with those of related species in Table 1.

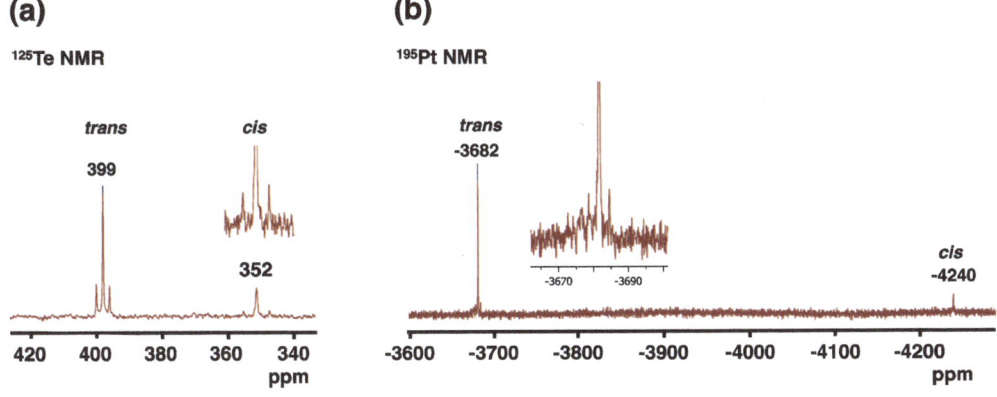

Figure 1. (a) The ^{125}Te{^{1}H} and (b) the ^{195}Pt{^{1}H} NMR spectra of the mixture of the *cis*- and *trans*-isomers of [PtCl$_2$¦Te(CH$_2$)$_6$¦$_2$].

Table 1. NMR spectroscopic information of some telluroether complexes of platinum [a].

Compound	δ(^{125}Te), ppm		$^1J_{PtTe}$, Hz		δ(^{195}Pt), ppm		Ref.
	cis	trans	cis	trans	cis	trans	
[PtCl$_2${Te(CH$_2$)$_6$}$_2$]	352	399	979	506	−4240	−3682	This work
[PtCl$_2${Te(CH$_2$)$_4$}$_2$]	457	459	665	321	−4251	−3715	[30]
[PtCl$_2$(TeMe$_2$)$_2$]	224	234	824	489	−4351	−3765	[40]
[PtCl$_2${Te(CH$_2$SiMe$_3$)$_2$}$_2$]	266	292	804	469	−4236	−3707	[41]
[PtCl$_2${S(O)(CD$_3$)$_2$}{Te(CH$_2$)$_6$}]	490		1052			−3830	This work

[a] The cis-designation refers to the relative positions of the two chloride ligands.

Based on the trends in the ^{125}Te and ^{195}Pt chemical shifts that have been reported previously [30,40,41], the tellurium resonance at 352 ppm has been assigned to the cis isomer **1**$_{cis}$ and that at 399 ppm to the trans isomer **1**$_{trans}$. It can also be seen in Table 1 that the magnitudes of the $^1J_{TePt}$ coupling constants in the cis-isomers are almost double compared to those of the corresponding trans-isomers. These relative values, as well as the trends in both the ^{125}Te and ^{195}Pt chemical shifts, reflect the stronger trans-influence of the tellurium donors compared to that of the chlorido ligand and support the assignments.

The cis -> trans isomerization of [PtCl$_2${Te(CH$_2$)$_6$}$_2$] was monitored using ^{1}H NMR spectroscopy (see Figure S1 in Supplementary Materials). The crop of crystals of **1**$_{cis}$ from which the crystal structure determination was carried out (Section 2.3) was dissolved in CDCl$_3$, and the spectra were recorded at room temperature for 30 s after the dissolution and again after 1.2 h. In solution, the trans-isomer **1**$_{trans}$ quickly became the dominant species. Already in the first spectrum (Figure S1a) recorded almost immediately after the dissolution of the crystals, the trans-isomer can be observed, and in the second spectrum (Figure S1b), it is clearly the predominant species. The assignment of the ^{1}H chemical shifts was verified by recording the ^{1}H NMR spectrum of the redissolved orange needles. Immediately after the dissolution, the resonances marked as trans in Figure S1 were the major signals in the spectrum and remained so throughout prolonged monitoring of the solution.

After 1.2 h, the cis:trans ratio was estimated from the integrated intensities of the resonances to be 1:1.8, which is very close to the ratio of 1:2 that Kemmitt et al. [30] estimated for the related [PtCl$_2${Te(CH$_2$)$_4$}$_2$] in dichloromethane. The intensity distribution in the ^{125}Te{^{1}H} and ^{195}Pt{^{1}H} NMR spectra (see Figure 1) bears semiquantitative agreement with the inferences from the ^{1}H spectrum in Figure S1b.

The ^{125}Te{^{1}H} NMR spectrum of the solidified and redissolved reaction mixture prior to separation of the products is shown in Figure S3 of Supplementary Materials. In addition to the resonances of **1**$_{cis}$ and **1**$_{trans}$, two weak signals at 489 ppm and 598 ppm were observed. The qualitative comparison of their relative signal intensities and the chemical shifts might indicate the presence of the trinuclear complex cis-trans-[Pt$_3$Cl$_6${Te(CH$_2$)$_6$}$_4$] (**2**), but this assignment remains tentative at best. The former resonance could be due to the terminal Te(CH$_2$)$_6$ ligand, and the latter due to the bridging ligand (see Section 2.3).

2.3. Crystal and Molecular Structures

Upon slow recrystallization from CH$_2$Cl$_2$/pentane (1:1) solution, intense yellow plate-shaped crystals and orange needles were obtained in addition to small crops of other products. The yellow plate-shaped crystals were suitable for the determination of the crystal structure and were shown to be cis-[PtCl$_2${Te(CH$_2$)$_6$}$_2$] (**1**$_{cis}$). The details of the data collection and structure determination are presented in Table S1 in Supplementary Materials. The molecular structure and the numbering of the atoms in **1**$_{cis}$ together with some selected bond parameters are shown in Figure 2.

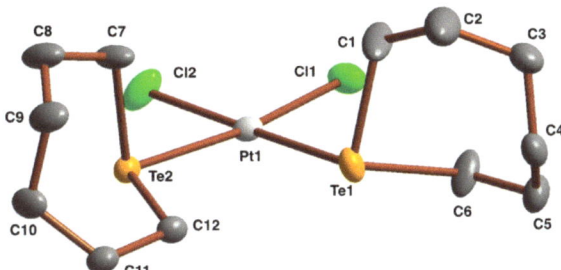

Figure 2. The molecular structure of *cis*-[PtCl$_2${Te(CH$_2$)$_6$}$_2$] (**1**$_{cis}$) indicating the numbering of the atoms. The thermal ellipsoids have been drawn at the 50% probability level. Hydrogen atoms have been omitted for clarity. Selected bond lengths (Å) and angles (°): Pt1-Te1 2.5145(7), Pt1-Te2 2.5266(6), Pt1-Cl1 2.340(2), Pt1-Cl2 2.359(2), Te1-C1 2.164(10), Te1-C6 2.129(9), Te2-C7 2.166(9), Te2-C12 2.138(8); Cl1-Pt1-Cl2 90.98(9), Cl1-Pt1-Te1 92.67(7), Cl1-Pt1-Te2 172.18(6), Cl2-Pt1-Te1 174.75(6), Cl2-Pt1-Te2 172.18(6), Te1-Pt1-Te2 94.01(2).

The platinum atom exhibits a slightly distorted square planar coordination geometry with all bond parameters showing their expected values (c.f., Pt-Te 2.4971(14)–2.541(14) Å; Pt-Cl 2.311(6)–2.356(6) Å [31,41–43]). Some *trans*-[PtCl$_2$(TeR$_2$)$_2$] show longer Pt-Te bond lengths but slightly shorter Pt-Cl bond distances (Pt-Te 2.5589(12)–2.5945(3) Å; Pt-Cl 2.275(5)–2.320(5) Å [31,41,43–45]). This is due to the relative strengths of the *trans*-influence of the telluroether and chlorido ligands.

In addition to **1**$_{cis}$ and **1**$_{trans}$, a few red, plate-shaped crystals of the trinuclear *cis-trans*-[Pt$_3$Cl$_6${Te(CH$_2$)$_6$}$_4$]·1¼CH$_2$Cl$_2$ (**2**·1¼CH$_2$Cl$_2$) could be manually separated from the reaction mixture after the crystallization from CH$_2$Cl$_2$/pentane. The crystals of tetranuclear *cis-trans*-[Pt$_4$Cl$_8${Te(CH$_2$)$_6$}$_4$]·4CDCl$_3$ (**3**·4CDCl$_3$) could be obtained upon recrystallization from CDCl$_3$ in the NMR tube after the recording of the NMR spectra of the equimolar reaction of the reagents. The molecular structures of **2**·1¼CH$_2$Cl$_2$ and **3**·4CDCl$_3$ are shown in Figure 3 (for details of the structure determination and the list of selected bond parameters, see Tables S1 and S2 in Supplementary Materials).

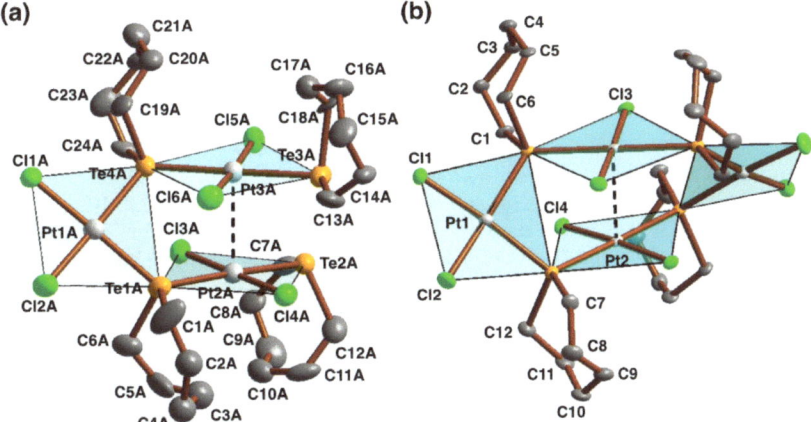

Figure 3. (**a**) Molecular structure of *cis-trans*-[Pt$_3$Cl$_6${Te(CH$_2$)$_6$}$_4$]·1¼CH$_2$Cl$_2$ (**2**·1¼CH$_2$Cl$_2$) and (**b**) *cis-trans*-[Pt$_4$Cl$_8${Te(CH$_2$)$_6$}$_4$]·4CDCl$_3$ (**3**·4CDCl$_3$) indicating the numbering of the atoms. The thermal ellipsoids have been drawn at the 50% probability level. Some Te(CH$_2$)$_6$ rings and the solvent molecules in **2**·1¼CH$_2$Cl$_2$ and **3**·4CDCl$_3$ are disordered. For clarity, only one disordered ligand-component is shown in the figure. The hydrogen atoms have also been omitted from both complexes for clarity.

The asymmetric unit of **2**·1¼CH$_2$Cl$_2$ contains two independent complexes (denoted **A** and **B**) with virtually the same conformations and bond parameters. There are also 2½ solvent molecules in the asymmetric unit. Some Te(CH$_2$)$_6$ ligands exhibit orientational disorder. Since the structures of both discrete complexes in the asymmetric unit are closely similar, only complex **A** is shown in Figure 3a. In the case of **3**·4CDCl$_3$, the asymmetric unit contains only half of the tetranuclear complex, with the other half being completed through symmetry. The geometry of **3** is closely related to that of **2**, as shown in Figure 3b.

Both complexes **2** and **3** show the simultaneous occurrence of very slightly distorted square-planar *cis-* and *trans-*PtCl$_2$Te$_2$ coordination environments. In the *cis*-moieties, the Pt-Te bonds range from 2.5045(7) to 2.5226(6) Å and the Pt-Cl bonds range from 2.309(3) to 2.331(5) Å. In the *trans-*PtTe$_2$Cl$_2$ moieties, the Pt-Te and Pt-Cl range from 2.5546(6) to 2.5774(15) Å and from 2.302(2) to 2.326(6) Å, respectively. The relative magnitudes of these metrical values demonstrate the stronger *trans*-influence of tellurium compared to chlorine.

The molecular structure of [PtCl$_2${S(O)(CD$_3$)$_2$}{Te(CH$_2$)$_6$}] (**4**) is shown in Figure 4 together with the labeling of atoms and selected bond parameters.

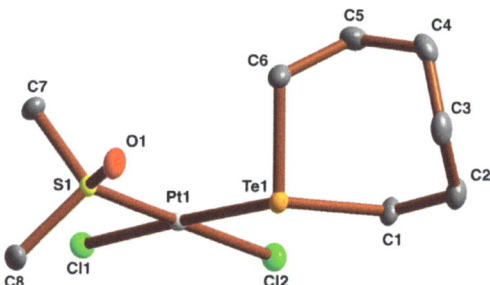

Figure 4. The molecular structure of [PtCl$_2${S(O)(CD$_3$)$_2$}{Te(CH$_2$)$_6$}] (**4**) indicating the numbering of the atoms. The thermal ellipsoids have been drawn at the 50% probability level. Hydrogen and deuterium atoms have been omitted for clarity. Selected bond lengths (Å) and angles (°): Pt1-Te1 2.5436(3), Pt1-S1 2.2049(7), Pt1-Cl1 2.3564(7), Pt1-Cl2 2.3203(7), S1-O1 1.479(2), Te1-C1 2.146(3), Te1-C6 2.176(3), Te1-Pt1-Cl1 176.33(2), Te1-Pt1-Cl2 92.32(2), Te1-Pt1-S1 87.68(2), Cl1-Pt1-Cl2 89.78(2), Cl1-Pt1-S1 90.27(3), Cl2-Pt1-S1 179.11(3).

The Pt-Te bond length of 2.5436(3) Å in **4** is in agreement with that of **1**$_{cis}$ and is consistent with the weak *trans*-influence of the chlorido ligand. The two Pt-Cl bond lengths are 2.3564(7) Å (*trans* to Te) and 2.3203(7) Å (*trans* to S) and reflect also the relative *trans*-influences of tellurium and sulfur. The Pt-S bond length is 2.2049(7) Å. The closest complex related to [PtCl$_2${S(O)(CD$_3$)$_2$}{Te(CH$_2$)$_6$}] (**4**) is [PtCl$_2${S(O)(CH$_3$)$_2$}{S(CH$_3$)$_2$}] [46] that shows a Pt-S(O)(CH$_3$)$_2$ distance of 2.211(4) Å. The Pt-Cl bond lengths are 2.299(3) and 2.320(4) Å, and the Pt-S(CH$_3$)$_2$ distance is 2.294(3) Å. In the Cambridge database, there are nine crystal structure determinations for [PtCl$_2${S(O){CH$_3$}$_2$}$_2$] containing the *trans-*Cl-Pt-S(O)CH$_3$ arrangement [47]. The Pt-S bonds span a range of 2.2302(18)–2.2537(12) Å (average 2.242(10) Å), and the Pt-Cl bonds vary in the range of 2.3068(8)–2.3251(5) Å (average 2.317(5) Å).

In the solid state, all complexes **1**$_{cis}$, **2**, and **3** show short Te···Cl and Pt···Pt contacts. These secondary bonding interactions comprise chalcogen bonds (Te···Cl) and metallophilic interactions (Pt···Pt) that, in the case of **1**$_{cis}$, link the mononuclear *cis*-[PtCl$_2${Te(CH$_2$)$_6$}$_2$] complexes into dimers exhibiting respective contacts of 3.659(3)–3.676(2) Å and 3.5747(5) Å. While the chalcogen bonds and metallophilic interactions in **2** and **3** are structurally similar to those in **1**$_{cis}$ (see Table S2), in the case of the latter polynuclear complexes, they are *intra*-molecular (see Figure 5). Similar linking of the square-planar MX$_2$E$_2$ (M = Pt, Pd; E = Se, Te) coordination spheres into dimers has also been observed in other chalcogenoether complexes provided that the steric bulk of the organic group bonded to the chalcogen atom does not prevent the dimer formation. Typical examples are *cis*-[PdCl$_2${SeMe(C$_4$H$_3$S)}$_2$] [48], *trans*-[PtI$_2$(TeMePh)$_2$] [49], *cis*-[PdCl$_2${(Te(C$_6$H$_4$)OMe-1,4)$_2$CH$_2$}] [50], and *cis*-[PdBr$_2${(TePh)$_2$(CH$_2$)$_3$}] [51]. The QTAIM analysis of these weak secondary bonding interactions is discussed in Section 2.4.4.

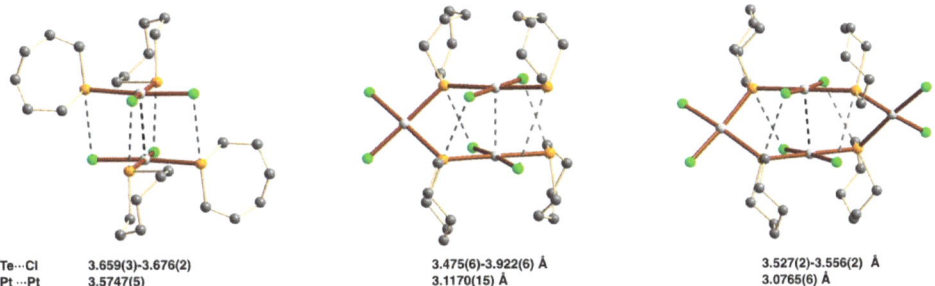

Te···Cl	3.659(3)-3.676(2)	3.475(6)-3.922(6) Å	3.527(2)-3.556(2) Å
Pt···Pt	3.5747(5)	3.1170(15) Å	3.0765(6) Å

Figure 5. Chalcogen bonds and metallophilic interactions in **1**$_{cis}$, **2**, and **3**.

Interestingly, [PtCl$_2${S(O)(CD$_3$)$_2$}{Te(CH$_2$)$_6$}] (**4**) does not show Pt···Pt close contacts (the closest distance is 4.5696(3) Å). The Te···Cl contacts of 3.7214(8)–3.8645(8) Å, on the other hand, link the complexes into *quasi*-single-strand polymers (see Figure 6).

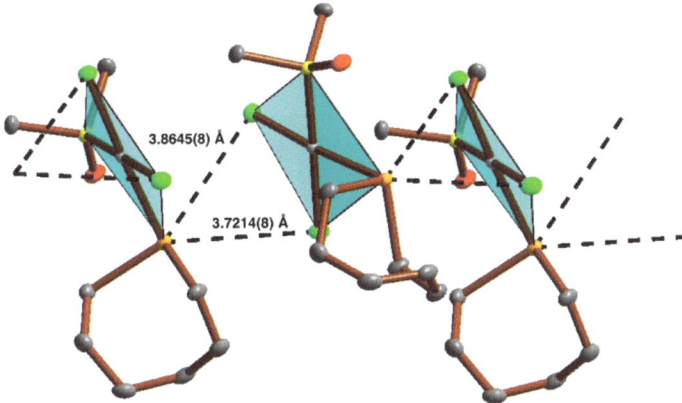

Figure 6. The Te···Cl chalcogen bonds in **4**.

2.4. Density Functional Theory (DFT) Computations

2.4.1. General

We carried out PBE0-D3/def2-TZVP calculations to study the bonding, secondary bonding interactions, as well as energetics of the *cis-trans* isomerization of [PtCl$_2${Te(CH$_2$)$_6$}$_2$] and the formation of tri- and tetranuclear complexes.

2.4.2. Optimized Geometries

The PBE0-D3/def2-TZVP optimized geometries of [PtCl$_2${Te(CH$_2$)$_6$}$_2$] (**1**$_{cis}$ and **1**$_{trans}$), *cis-trans*-[Pt$_3$Cl$_6${Te(CH$_2$)$_6$}$_4$] (**2**), and *cis-trans*-[Pt$_4$Cl$_8${Te(CH$_2$)$_6$}$_4$] (**3**) agree well with the experimental information despite the fact that the experimental information is from the crystalline state and the computational data are calculated in vacuum. The PBE0-D3/def2-TZVP-optimized coordinates are shown in Table S3 and the computed geometries are shown in Table S4 in Supplementary Materials. The total energies in vacuum are presented in Table S5 and those in dichloromethane in Table S6.

2.4.3. Cis-trans Isomerization

In vacuum at the PBE0-D3/def2-TZVP level of theory, trans-[PtCl$_2${Te(CH$_2$)$_6$}$_2$] (1_{trans}) is 33 kJ mol^{-1} more stable than the cis-isomer 1_{cis} (see Table S5). By contrast, in dichloromethane, the cis-isomer is 2 kJ mol^{-1} more stable than the trans-isomer (see Table S6). An earlier MP4(SDQ)//MP2 comparison of the Gibbs energies in vacuum and dichloromethane involving cis- and trans-[PtCl(SnCl$_3$)(PH$_3$)$_2$] (ΔG_{298}(cis-trans) = −32 and 9 kJ mol^{-1}, respectively) [52] exhibited the same trend. The computed equilibrium constant for the cis -> trans isomerization of [PtCl$_2${Te(CH$_2$)$_6$}$_2$] in dichloromethane utilizing the Gibbs energy change is 0.4. This is consistent with the variable-temperature ^{1}H NMR spectroscopic study of the cis -> trans isomerization of [PdCl$_2$(PMePh$_2$)$_2$] that has been concluded to show a value of 0.30 [53] in 1,2-dichloroethane. In chloroform, trans-[PtCl$_2${Te(CH$_2$)$_6$}$_2$] is interestingly 6 kJ mol^{-1} (equilibrium constant 11.2) more stable than the cis-form. Since all NMR spectra were recorded in deuterochloroform, this finding explains the observed cis -> trans isomerization of [PtCl$_2${Te(CH$_2$)$_6$}$_2$] by NMR spectroscopy (see Section 2.2). The route for the cis -> trans isomerization is generally considered to involve the formation of a five-coordinate intermediate. This intermediate can undergo Berry pseudorotation resulting in the isomerization. Other mechanisms have also been suggested (for early discussion of the possible isomerization mechanisms, see ref. [54]).

2.4.4. Chalcogen Bonding and Metallophilic Interactions

The QTAIM analysis of the secondary bonding Te···Cl and Pt···Pt interactions are shown in Table 2. The Pt-Te bond lengths and experimental close contacts in the crystals have been included for comparison. The use of QTAIM in the characterization of metal–metal bonding has recently been reviewed [55]. In addition to the classical descriptors of electron density (ρ) and the Laplacian ($\nabla^2\rho$) at the bond critical point that are expected to be large and negative, respectively, for covalent bonds, the relative magnitudes of the kinetic G_b and the potential V_b energy densities have been used to classify bonds. The relative magnitudes have been examined either via the electronic energy density ($H_b = G_b + V_b$) [56,57] or the $|V_b|/G_b$ ratio [58]. The Laplacian for the bonds between heavy atoms tends to give positive values regardless of the bond type [59]. Therefore, the energy densities have been good additional descriptors for defining metal bonding. The negative H_b values have been taken as a sign of covalent bonding and the positive values as indicators of closed-shell interactions [60]. The $|V_b|/G_b$ ratio further distinguishes the regions of shared-shell interactions with $|V_b|/G_b > 2$, intermediate regions corresponding to $1 < |V_b|/G_b < 2$, and closed-shell interactions with $|V_b|/G_b < 1$ [55]. The intermediate region includes metal–metal and donor–acceptor interactions [59].

In the case of complexes 1_{cis}, 2, and 3, the calculated $\nabla^2\rho$ for Pt···Pt, Pt–Te, and Te···Cl are all positive, as can be expected for interactions between the heavy atoms. For the Pt-Te bonds, the values of H_b are negative and $|V_b|/G_b$ ratios are in the intermediate region between 1.64 and 1.75, consistent with the donor–acceptor bonds. The electron densities at bond critical points (ρ) and the delocalization indices (DIs) that are close to the single bond values [61] corroborate this classification of the coordinative bonds. Because of the trans-influence, the Pt-Te bonds are slightly longer than single bonds and cause the delocalization indices of the Pt-Te(trans) bonds to be smaller than those of the Pt-Te(cis) bonds. By comparison, H_b of the Pt···Pt interactions are still negative but $|V_b|/G_b$ ratios that fall between 1.06 and 1.15 (see Table 2) are much closer to the limit of closed-shell interactions. The low level of electron sharing in the Pt···Pt interactions is reflected by the ρ values 0.122, 0.172, and 0.202 e Å^{-3} and the DI values 0.18, 0.26, and 0.30 for 1_{cis}, 2, and 3, respectively. By comparison, the QTAIM analysis in a recent study on platinum complexes of phenylpyridine, triazolyl-phenylpyridine, and imidazolyl-phenylpyridine that form head-to-tail dimers via the Pt···Pt interactions in the solid state show the ρ values of 0.132–0.150 e Å^{-3} that are between those of 1_{cis} and 2 [62]. The $|V_b|/G_b$ ratios of Pt···Pt interactions of 1.08–1.10 are also very similar to those found in this contribution. In both cases, the metallophilic Pt···Pt interactions show weak covalence.

Table 2. QTAIM analysis results (electron density at bond critical point, ρ; delocalization index, DI; Laplacian of electron density, $\nabla^2\rho$; kinetic energy density, G_b; and potential energy density, V_b) of the Pt$\cdots$Pt and Te$\cdots$Cl secondary bonding interactions in cis-[PtCl$_2${Te(CH$_2$)$_6$}$_2$] (1_{cis}), cis-trans-[Pt$_3$Cl$_6${Te(CH$_2$)$_6$}$_4$] (**2**), and cis-trans-[Pt$_4$Cl$_8${Te(CH$_2$)$_6$}$_4$] (**3**).

Parameter	Expl.	PBE0-D3/def2-TZVP	QTAIM					
			ρ (e Å^{-3})	DI	$\nabla^2\rho$ (e Å^{-5})	G_b (kJ mol^{-1} bohr^{-3}) [a]	V_b (kJ mol^{-1} bohr^{-3}) [a]	E_{int} (kJ mol^{-1} bohr^{-3}) [a,b]
1_{cis}								
Pt-Te(cis)	2.5145(7)–2.5266(6)	2.537–2.548	0.619/0.628	0.98/1.03	1.15/1.18	130/133	−228/−234	−114/−117
Pt$\cdots$Pt	3.5747(5)	3.403	0.122	0.18	1.0	29	−31	−16
Te$\cdots$Cl	3.659(3)–3.676(2)	3.516–3.531	0.074–0.068	0.10	0.74–0.67	14	−13–(−14)	−7
2								
Pt-Te(cis)	2.5140(15)–2.5219(16)	2.515–2.515	0.630–0.634	0.94	1.83–1.85	152–153	−254–(−256)	−127–(−128)
Pt-Te(trans)	2.5560(15)–2.5774(15)	2.572–2.581	0.578–0.588	0.81–0.92	1.13–1.77	118–132	−206–(−217)	−103–(−109)
Pt$\cdots$Pt	3.1170(11)–3.1499(13)	3.167	0.172	0.26	1.42	44	−50	−25
Te$\cdots$Cl	3.481(7)–3.964(6)	3.508–3.628	0.074–0.061	0.10–0.07	0.71–0.59	13–17	−11–(−14)	−6–(−7)
3								
Pt-Te(cis)	2.5043(6)–2.5226(6)	2.510	0.638	0.95	1.88	155	−259	−130
Pt-Te(trans)	2.5547(6)–2.5577(6)	2.561	0.595	0.84	1.55	132	−222	−111
Pt$\cdots$Pt	3.0764(6)	3.075	0.202	0.30	1.71	54	−62	−31
Te$\cdots$Cl	3.527(2)–3.556(2)	3.613	0.061	0.06	0.61	14	−11	−6

[a] 1 bohr = 0.52918 Å. [b] E_{int} is defined as $V_b/2$ [63].

The weak Te$\cdots$Cl contacts are classified as closed-shell interactions by the positive H_b values of 0–3, as shown in Table 2. This is also reflected by the small values of ρ (0.061–0.074 e Å^{-3}) and DI (0.06–0.10). They can be compared to the intermolecular Te$\cdots$Te chalcogen bonds in solid macrocyclic telluroethers and are of the same order of magnitude [29]. The relative strengths of the Pt$\cdots$Pt and Te$\cdots$Cl interactions can be qualitatively estimated using E_{int} calculated from V_b [63], although some caution should be exercised when drawing conclusions, as the reliability of the relationship has been questioned [64]. Comparison of the E_{int} values shows that in all complexes 1_{cis}, **2**, and **3**, the metallophilic Pt$\cdots$Pt interactions (E_{int} = −16–(−31) kJ mol^{-1} bohr^{-3}) are stronger than single Te$\cdots$Cl interactions (E_{int} = −6–(−7) kJ mol^{-1} bohr^{-3}). The stronger attraction between the Pt centers compared to that between Te and Cl could explain the observation that in all three complexes, the square-planar coordination plane is slightly concave with Pt$\cdots$Pt showing the closest distance between these distorted planes (see Figure S4). However, there are four Te$\cdots$Cl interactions in each complex structure compared to one Pt$\cdots$Pt interaction, suggesting that the total stabilization of the complexes due to Te$\cdots$Cl interactions is on par with the Pt$\cdots$Pt interaction.

2.4.5. Formation of cis-trans-[Pt$_3$Cl$_6${Te(CH$_2$)$_6$}$_4$] and cis-trans-[Pt$_4$Cl$_8${Te(CH$_2$)$_6$}$_4$]

The reaction of cis-[PtCl$_2$(NCPh)$_2$] and Te(CH$_2$)$_6$ afforded small amounts of cis-trans-[Pt$_3$Cl$_6${Te(CH$_2$)$_6$}$_4$] (**2**) and cis-trans-[Pt$_4$Cl$_8${Te(CH$_2$)$_6$}$_4$] (**3**) that could be identified and structurally characterized through X-ray diffraction. The total PBE0-D3/def2-TZVP energies of all species in dichloromethane (see Table S6) can be used to calculate Gibbs energy changes in the formation of complexes 1_{cis}, **2**, and **3** from cis-[PtCl$_2$(NCPh)$_2$] and Te(CH$_2$)$_6$ that are shown in Table S7. One possible route for the formation of the polynuclear complexes is shown in Scheme 2 together with the Gibbs energy changes in the individual reaction steps. The relative Gibbs energies of the reaction products and intermediates with respect to cis-[PtCl$_2$(NCPh)$_2$] are also shown in Scheme 2.

Whereas the energetics in Scheme 2 is favorable for the formation of **2** and **3** (see also Table S6), it seems that only small amounts of these complexes are formed during the syntheses. Though the solid starting material is cis-[PtCl$_2$(NCPh)$_2$], trans-[PtCl$_2$(NCPh)$_2$] is 7 kJ mol^{-1} more stable and therefore, in dichloromethane solution, virtually all cis-[PtCl$_2$(NCPh)$_2$] is converted into the trans-isomer. Because, in dichloromethane, cis-[PtCl$_2${Te(CH$_2$)$_6$}$_2$] is 2 kJ mol^{-1} more stable than the trans-isomer (Table S6), the isomeric composition in equilibrium is, ca., 70% cis-[PtCl$_2${Te(CH$_2$)$_6$}$_2$] and 30% trans-[PtCl$_2${Te(CH$_2$)$_6$}$_2$]. The trans-isomer is therefore the limiting reactant in the formation of **2** and **3**. In fact, the formation of **3** was only observed in the 1:1 reaction of cis-[PtCl$_2$(NCPh)$_2$] and Te(CH$_2$)$_6$.

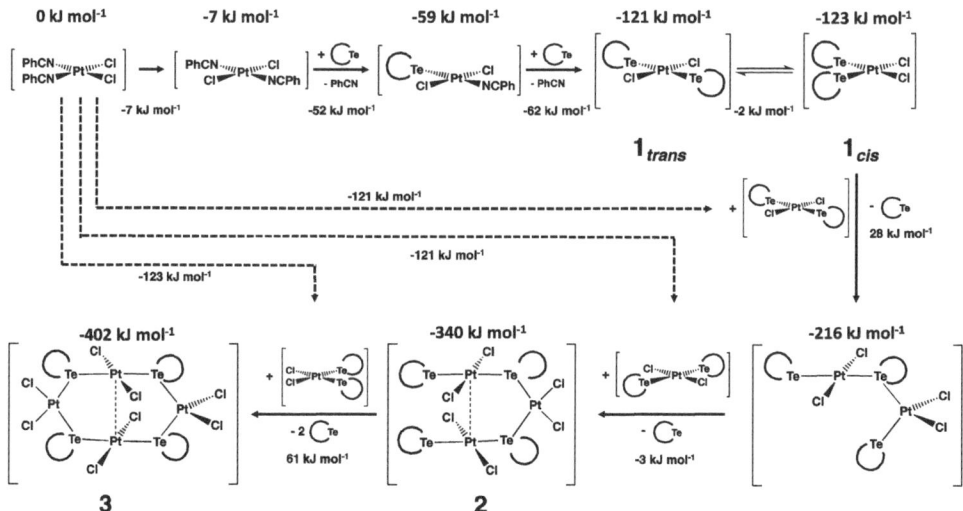

Scheme 2. The PBE0-D3/def2-TZVP energetics for the possible route of the formation of cis- and trans-[PtCl$_2${Te(CH$_2$)$_6$}] (**1**$_{cis}$ and **1**$_{trans}$), cis-trans-[Pt$_3$Cl$_6${Te(CH$_2$)$_6$}$_4$] (**2**), and cis-trans-[Pt$_4$Cl$_8${Te(CH$_2$)$_6$}$_4$] (**3**) in dichloromethane. The relative Gibbs energies of the reaction products and intermediates with respect to the starting complex cis-[PtCl$_2$)(NCPh)$_2$] are shown above each complex, and the Gibbs energy changes in individual transformation steps in connection are shown with the reaction arrows.

While we have not observed the formation of cis-trans-[Pt$_2$Cl$_4${Te(CH$_2$)$_6$}$_3$] in this reaction, we have previously reported the formation of small amounts of the structurally related cis-trans-[Pt$_2$Cl$_4${Te(CH$_2$SiMe$_3$)$_2$}$_3$] [40].

3. Materials and Methods

3.1. General Procedures

All manipulations involving air- and moisture-sensitive materials were conducted under a nitrogen atmosphere using Schlenk techniques. Dichloromethane and chloroform were distilled over CaH$_2$ and hexane over Na/benzophenone under a nitrogen atmosphere prior to use. Ethanol was degassed by bubbling nitrogen through the solvent for at least 15 min. Semiconductor-grade tellurium was freshly ground. All other reagents were used as purchased without further purification. The preparation of Te(CH$_2$)$_6$ has been reported earlier [29].

3.2. Spectroscopy

3.2.1. NMR Spectroscopy

^{1}H, ^{13}C{^{1}H}, ^{125}Te{^{1}H}, and ^{195}Pt{^{1}H} NMR spectra of **1**$_{cis}$ and **1**$_{trans}$ were recorded in CDCl$_3$, and those of **4** in d$_6$-DMSO were recorded on a Bruker Avance III 400 spectrometer operating at 400.13, 100.61, 126.24, and 86.02 MHz, respectively. The respective pulse widths were 13.0, 9.70, 6.0, and 10.0 μs, and the corresponding relaxation delay was 2.0 s for each nucleus. The deuterated solvent was used as the ^{2}H lock. All resonances were indirectly referenced by using the deuterium signal of the solvent for the lock to the frequency that relates to the resonance frequency of the TMS protons at exactly 400.130000 MHz. Chemical shifts for the ^{125}Te resonances are given relative to dimethyl telluride through indirect referencing (the tellurium resonance ν_0(Te) was calculated by using the ratio $\Xi = \nu_{0,H}$(Te)/ν_0(TMS) = 31.549769% [61]). Chemical shifts for ^{195}Pt are given relative to Na$_2$[PtCl$_6$] (1.2 M in D$_2$O) also through indirect referencing

($\Xi = \nu_{0,H}(Pt)/\nu_0(TMS) = 21.96784\%$) [65]. The ^{1}H and ^{13}C are reported relative to tetramethyl silane TMS [66].

3.2.2. Mass Spectrometry

Electron ionization mass spectra were recorded on Finnigan MAT SSQ 710 and Finnigan MAZ95XL spectrometers. The energy of the electrons was 70 eV.

3.3. X-ray Diffraction

The crystals of **1–4** were coated with Paratone oil and mounted in a nylon CryoLoop, and the intensity data were collected on a Bruker Nonius Kappa CCD diffractometer at 133 K using graphite monochromated Mo Kα radiation (λ = 0.71073 Å; 55 kV, 25 mA) [67,68]. Crystal data and the details of structure determinations are given in Table S1. The data were corrected for Lorenz and polarization effects, after which semi-empirical absorption correction was applied to net intensities using SADABS [69]. The structures were solved through direct methods using SHELXT [70] and refined using SHELXL-2018 [71]. After the full-matrix least-squares refinement of the non-hydrogen atoms with anisotropic thermal parameters, the hydrogen atoms were placed in calculated positions in the CH$_2$ groups (C-H = 0.99 Å). In the final refinement, the hydrogen atoms were riding with the carbon atoms they were bonded to. The isotropic thermal parameters of the hydrogen atoms were fixed at 1.5 times that of the corresponding carbon atoms. The scattering factors for the neutral atoms were those incorporated with the program.

Some Te(CH$_2$)$_6$ ligands and solvent molecules in **2**·1¼CH$_2$Cl$_2$ and **3**·4CDCl$_3$ were disordered. The disorder was resolved through appropriate restraining of the anisotropic displacement parameters and some bond lengths. Some parts of the disorder model were introduced by utilizing the program *DSR* [72].

3.4. Computational Details

All calculations were performed using the Gaussian 16 program [73] by employing the PBE0 hybrid functional [74–76] together with the def2-TZVP basis sets [77,78]. The combination of the PBE0 functional and the def2-TZVP basis set has been shown to be suitable for computational characterization of compounds of heavy p-block elements (see the discussion in ref. [79]). Implicit C-PCM solvent model was applied to treat the solvation effects [80,81], and dispersion forces were treated by using the D3BJ version of Grimme's empirical correction with Becke–Johnson damping parameterized for the PBE0 functional [82–84]. Full structure optimization was carried out for each species considered in this work and the frequencies were calculated for the optimum geometries to ascertain the nature of the stationary points. The quantum theory of atoms in molecules (QTAIM) was used to study *inter*-molecular interactions in the [PtCl$_2$\{Te(CH$_2$)$_6$\}] (**1**$_{cis}$, **1**$_{trans}$), as well as *intra*-molecular interactions in the *cis-trans*-[Pt$_3$Cl$_6$\{Te(CH$_2$)$_6$\}$_4$] (**2**) and *cis-trans*-[Pt$_4$Cl$_8$\{Te(CH$_2$)$_6$\}$_4$] (**3**) structures [85]. AIMAll software was used for the QTAIM calculations [86].

3.5. Reaction of Te(CH$_2$)$_6$ with cis-[PtCl$_2$(NCPh)$_2$]

Te(CH$_2$)$_6$ (46.2 mg, 0.218 mmol) was dissolved in 50 mL CH$_2$Cl$_2$ and crystalline *cis*-[PtCl$_2$(NCPh)$_2$] (50.0 mg, 0.106 mmol) was added. The mixture was stirred under exclusion of light for 20 h. The solvent was removed under reduced pressure and the crude product was extensively dried at 40 °C and 1 mbar to remove any benzonitrile, yielding 71.7 mg (theoretical: 74.4 mg) of an odorless yellow solid with few orange crystals in it. TLC analysis (silica, chloroform) showed two spots at Rf = 0.89 and 0.25 corresponding to *trans*-[PtCl$_2$\{Te(CH$_2$)$_6$\}$_2$] (**1**$_{trans}$) and *cis*-[PtCl$_2$\{Te(CH$_2$)$_6$\}$_2$] (**1**$_{cis}$) beside a very faint spot at Rf = 0.58 and few minor spots at Rf < 0.25. The substances corresponding to the latter were mostly removed through column chromatography (silica, chloroform), yielding 54.5 mg of substance. The product was dissolved in CH$_2$Cl$_2$/pentane and, upon slow evaporation, both **1**$_{trans}$ and **1**$_{cis}$ crystallized, with the former giving orange to red, heavily intergrown needles and the latter giving intensely yellow plates suitable for XRD analysis. Most of the substance crystallized in the *cis* form. Small quantities of both isomers could be separated manually. The mixture obtained after the column chromatography still contained a minor side product that could be observed by means of NMR spectroscopy. Therefore, while it is not possible to give exact yields of 1_{trans} and 1_{cis}, the combined overall yield of about 70% can be estimated based on NMR spectroscopy. Few crystals of *cis-trans*-[Pt$_3$Cl$_6${Te(CH$_2$)$_6$}$_4$]·1¼CH$_2$Cl$_2$ (**2**·1¼CH$_2$Cl$_2$) could also be isolated from the crystalline product mixture due to their different crystal habit. The compound was identified using X-ray diffraction.

cis-[PtCl$_2${Te(CH$_2$)$_6$}$_2$] (**1**$_{cis}$): yellow plates, Rf = 0.25, m.p.: beginning brown coloration at ~140 °C, melting at ~180 °C accompanied by quick decomposition, MS: 690 u/e (M$^+$), EA: calcd. C 35.61% H 2.13% N 5.93% Cl 15.02% found C 35.88% H 2.13%, N 6.14% Cl 15.34%. NMR (CDCl$_3$): δ_{Te} = 352 ppm, $^1J_{TePt}$ = 979 Hz; δ_{Pt} = −4240 ppm.

trans-[PtCl$_2${Te(CH$_2$)$_6$}$_2$] (**1**$_{trans}$): orange to red interwoven needles, Rf = 0.89, m.p.: decomposition at ~150 °C under black coloration, MS: 690 u/e (M$^+$). NMR (CDCl$_3$): δ_{Te} = 399 ppm, $^1J_{TePt}$ = 506 Hz; δ_{Pt} = −3682 ppm.

The reaction was also repeated by using 33.9 mg (0.160 mmol) of Te(CH$_2$)$_6$ and 75.5 mg (0.160 mmol) of *cis*-[PtCl$_2$(NCPh)$_2$] in 20 mL of dichloromethane. The reaction progressed in an analogous manner to a 2:1 reaction with 1_{cis} and 1_{trans} as main products. In addition, the presence of starting materials was observed in the reaction mixture. A few well-shaped red crystals were formed in the NMR tube. They were identified through single-crystal X-ray diffraction as *cis-trans*-[Pt$_4$Cl$_8${Te(CH$_2$)$_6$}$_4$]·4CDCl$_3$ (**3**·4CDCl$_3$) (**3**). In order to increase the solubility of the solid material, the recrystallization from dimethyl sulfoxide was attempted. After the NMR measurement in d$_6$-dimethyl sulfoxide, pale yellow, almost colorless crystals were formed that, upon the crystal structure determination, proved to be [PtCl$_2${S(O)(CD$_3$)$_2$}{Te(CH$_2$)$_6$}] (**4**). NMR (d$_6$-DMSO): δ_{Te} = 490 ppm, $^1J_{TePt}$ = 1052 Hz; δ_{Pt} = −3830 ppm.

4. Conclusions

The coordination of Te(CH$_2$)$_6$ to the Pt(II) center was examined through the reaction of Te(CH$_2$)$_6$ with [PtCl$_2$(NCPh)$_2$] (2:1) in dichloromethane. The initial objective was to obtain information about the molecular structure of the Te(CH$_2$)$_6$ as a ligand in the complex, because the free compound is a thermally unstable and light-sensitive liquid with a low melting point. The main products in the reaction were *cis*- and *trans*-[PtCl$_2${Te(CH$_2$)$_6$}$_2$]. In dichloromethane, the isomers exist as a mixture, but upon crystallization, the *cis* isomer seems to be the dominant species in the solid state. It is likely that in non-polar solvents, the more polar *cis*-isomer is less soluble than the *trans*-isomer. The molecular structures of *cis*-[PtCl$_2${Te(CH$_2$)$_6$}] and the trinuclear and tetranuclear by-products *cis-trans*-[Pt$_3$Cl$_6${Te(CH$_2$)$_6$}$_4$] and *cis-trans*-[Pt$_4$Cl$_8${Te(CH$_2$)$_6$}$_4$] were determined using X-ray diffraction. These polynuclear complexes show the simultaneous presence of both *cis*-Cl and *trans*-Cl isomers. The secondary bonding interactions involving the Te···Cl chalcogen bonds and Pt···Pt metallophilic interactions were explored through the use of PBE0-D3/def2-TZVP calculations and discussed using the QTAIM analysis. It turned out that in all complexes, the discrete Pt···Pt interaction is stronger than any single Te···Cl contact. The total strength of the four Te···Cl interactions in each solid lattice is, however, comparable to that of the single Pt···Pt interaction.

Cis-trans-[Pt$_3$Cl$_6${Te(CH$_2$)$_6$}$_4$] is formed in the 1:2 reaction of *cis*-[PtCl$_2$(NCPh)$_2$] with Te(CH$_2$)$_6$, and *cis-trans*-[Pt$_4$Cl$_8${Te(CH$_2$)$_6$}$_4$] is obtained from the 1:1 reaction. While the overall energetics is favorable to the formation of both complexes, only a few crystals of **2**·1¼CH$_2$Cl$_2$ were obtained upon crystallization from dichloromethane/pentane. **3**·4CDCl$_3$ crystallized on the walls of the NMR tube after the recording of the spectra in both cases. Since the equilibrium composition in dichloromethane contains, ca., 70% of *cis*-[PtCl$_2${Te(CH$_2$)$_6$}$_2$] and 30% of *trans*-[PtCl$_2${Te(CH$_2$)$_6$}$_2$], the latter is the limiting reactant to the formation of tri- and tetranuclear complexes. A significantly larger excess of the cis-[PtCl$_2$(NCPh)$_2$] reagent might improve their yields, but that is the subject for a future study.

In an attempt to improve the yields of cis-trans-[Pt$_3$Cl$_6${Te(CH$_2$)$_6$}$_4$] and cis-trans-[Pt$_4$Cl$_8${Te(CH$_2$)$_6$}$_4$], crystallization experiments were performed using dimethyl sulfoxide. This led to the formation of [PtCl$_2${S(O)(CD$_3$)$_2$}{Te(CH$_2$)}] that could be characterized using X-ray diffraction and NMR spectroscopy.

Supplementary Materials: The following supporting information can be downloaded at: https://www.mdpi.com/article/10.3390/molecules28227551/s1, Figure S1. ^{1}H NMR spectra of the mixture of cis- and trans-[PtCl$_2${Te(CH$_2$)$_6$}$_2$] (a) 30 s and (b) 1.2 h after the dissolution of the mixture of cis-[PtCl$_2${Te(CH$_2$)$_6$}$_2$]. Figure S2. Fluxionality of the Te(CH$_2$)$_6$ ligand in cis- and trans-[PtCl$_2${Te(CH$_2$)$_6$}$_2$]. Figure S3. (a) The ^{125}Te{^{1}H} and (b) the ^{195}Pt{^{1}H} NMR spectra of the reaction mixture of cis-[PtCl$_2$(NCPh)$_2$] and Te(CH$_2$)$_6$ after solidification but prior to manual separation of the crystals. Figure S4. The Pt···Pt interactions lead the square-planar coordination planes to become concave in 1_{cis}, **2**, and **3**. Table S1. Crystal data and refinement details for the X-ray structure determinations of cis-[PtCl$_2${Te(CH$_2$)$_6$}$_2$] (1_{cis}), cis-trans-[Pt$_3$Cl$_6${Te(CH$_2$)$_6$}$_4$]·1¼CH$_2$Cl$_2$ (**2**·1¼CH$_2$Cl$_2$), cis-trans-[Pt$_4$Cl$_8${Te(CH$_2$)$_6$}$_4$]·4CDCl$_3$ (**3**·4CDCl$_3$), and [PtCl$_2${S(O)(CD$_3$)$_2$}{Te(CH$_2$)$_6$}] (**4**). Table S2. Selected bond lengths (Å) and angles (°) of cis-trans-[Pt$_3$Cl$_6${Te(CH$_2$)$_6$}$_4$]·1¼CH$_2$Cl$_2$ (**2**·1¼CH$_2$Cl$_2$) and cis-trans-[Pt$_4$Cl$_8${Te(CH$_2$)$_6$}$_4$]·4CDCl$_3$ (**3**·4CDCl$_3$). Table S3. Atomic coordinates (Å) of the PBE0-D3/def2-TZVP optimized species discussed in this contribution. Table S4. PBE0-D3/def2-TZVP optimized geometries of the [Pt$_n$Cl$_{2n}${Te(CH$_2$)$_6$}$_m$] ($n = 1–4$; $m = 2–4$). Table S5. Total energies of optimized species at PBE0-D3/def2-TZVP level of theory in vacuum (Hartree). Table S6. Total energies of optimized species at PBE0-D3/def2-TZVP level of theory in dichloromethane (Hartree). Table S7. Gibbs PBE0-D3/def2-TZVP formation energies of 1_{cis}, 1_{trans}, **2**, and **3** from cis-[PtCl$_2$(NCPh)$_2$] and Te(CH$_2$)$_6$ in dichloromethane (kJ mol^{-1}). CCDC Deposition Numbers 2298160–2298162 and 2301073 contain the supplementary crystallographic data for this paper. These data are provided free of charge by the joint Cambridge Crystallographic Data Centre and Fachinformationszentrum Karlsruhe Access Structure service www.ccdc.cam.ac.uk/structures.

Author Contributions: Conceptualization, R.S.L., R.O. and W.W.; synthesis and spectroscopic characterization, M.R.; quantum chemical calculations, J.M.R.; X-ray structural determination, H.G.; writing—original draft preparation, R.O. and R.S.L.; writing—review, all authors. All authors have read and agreed to the published version of the manuscript.

Funding: The research received no external funding.

Institutional Review Board Statement: Not applicable.

Informed Consent Statement: Not applicable.

Data Availability Statement: The data presented in this study are openly available in the article.

Acknowledgments: Provision of computational resources by Heikki Tuononen (University of Jyväskylä) (J.M.R.) is gratefully acknowledged.

Conflicts of Interest: The authors declare no conflict of interest.

References

1. Pedersen, C.J. Cyclic polyethers and their complexes with metal salts. *J. Am. Chem. Soc.* **1967**, *89*, 7017–7036. [CrossRef]
2. Gokel, G.W.; Negin, S.; Catwell, R. Crown Ethers. In *Comprehensive Supramolecular Chemistry II*; Atwood, J.L., Gokel, G.W., Barbour, L.J., Eds.; Elsevier: Amsterdam, The Netherlands, 2017; Volume 3, pp. 3–48.
3. Levason, W.; Orchard, S.D.; Reid, G. Recent developments in the chemistry of selenoethers and telluroethers. *Coord. Chem. Rev.* **2002**, *225*, 159–199. [CrossRef]
4. Segi, M. Acyclic dialkyl tellurides. *Sci. Synth.* **2007**, *39*, 1069–1082.
5. Barton, A.J.; Genge, A.R.J.; Hill, N.J.; Levason, W.; Orchard, S.D.; Patel, B.; Reid, G.; Ward, A.J. Recent developments in thio-, seleno-, and telluroether ligand chemistry. *Heteroat. Chem.* **2002**, *13*, 550–560. [CrossRef]
6. Panda, A. Chemistry of selena macrocycles. *Coord. Chem. Rev.* **2009**, *253*, 1056–1098. [CrossRef]
7. Levason, W.; Reid, G. Recent developments in the chemistry of thio-, seleno-, and telluroethers. In *Handbook of Chalcogen Chemistry*; Devillanova, F.A., Ed.; RSC Publishing: Cambridge, UK, 2007; pp. 81–106.
8. Levason, W.; Reid, G. Macrocyclic thio-, seleno- and telluroether ligands. In *Comprehensive Coordination Chemistry II*; McCleverty, J.A., Meyer, T.J., Eds.; Elsevier: Amsterdam, The Netherlands, 2004; Volume 1, pp. 399–410.

9. Sommen, G.L. Rings containing selenium and tellurium. In *Comprehensive Heterocyclic Chemistry III*; Katritzky, A.R., Ramsdenm, C.A., Scriven, E.F.V., Taylor, R.J.K., Eds.; Elsevier: Amsterdam, The Netherlands, 2008; Volume 14, pp. 863–900.
10. Levason, W.; Reid, G.; Zhang, W.-J. The chemistry of the p-block elements with thioether, selenoether and telluroether ligands. *Dalton Trans.* 2011, *40*, 8491–8506. [CrossRef]
11. Moorefield, C.N.; Newcombe, G.R. Eight-membered and larger rings. *Progr. Heterocycl. Chem.* 2020, *31*, 649–669.
12. Singh, A.K.; Sharma, S. Recent developments in the ligand chemistry of tellurium. *Coord. Chem. Rev.* 2000, *209*, 49–98. [CrossRef]
13. Levason, W.; Reid, G. Macrocyclic and polydentate thio- and seleno-ether ligand complexes of the p-block elements. *J. Chem. Soc. Dalton Trans.* 2001, 2953–2960. [CrossRef]
14. Levason, W.; Reid, G. Early transition metal complexes of polydentate and macrocyclic thio- and seleno-ethers. *J. Chem. Res. Synop.* 2002, *467–472*, 1001–1022. [CrossRef]
15. Jain, V.K.; Chauhan, R.S. New vistas in the chemistry of platinum group metals with tellurium ligands. *Coord. Chem. Rev.* 2016, *306*, 270–301. [CrossRef]
16. Gahan, L.R. Cyclononanes: The extensive chemistry of fundamentally simple ligands. *Coord. Chem. Rev.* 2016, *311*, 168–223. [CrossRef]
17. Oilunkaniemi, R.; Laitinen, R.S. Synthesis and molecular structures of cyclic selenoethers and their derivatives. In *Comprehensive Inorganic Chemistry III*; Reedijk, J., Poeppelmeier, K.R., Laitinen, R.S., Eds.; Elsevier: Amsterdam, The Netherlands, 2023; Volume 1, pp. 527–555.
18. Laitinen, R.S.; Oilunkaniemi, R.; Weigand, W. Structure, bonding, and ligand chemistry of macrocyclic seleno- and telluroethers. In *Chalcogen Chemistry: Fundamentals and Applications*; Lippolis, V., Santi, C., Lenardão, E.J., Braga, A.L., Eds.; Royal Society of Chemistery: Cambridge, UK, 2023; pp. 530–566.
19. Boyle, P.D.; Godfrey, S.M. The reactions of sulfur and selenium donor molecules with dihalogens and interhalogens. *Coord. Chem. Rev.* 2001, *223*, 265–299. [CrossRef]
20. Cozzolino, A.F.; Elder, P.J.W.; Vargas-Baca, I. A survey of tellurium-centered secondary-bonding supramolecular synthons. *Coord. Chem. Rev.* 2011, *255*, 1426–1438. [CrossRef]
21. du Mont, W.-W.; Hrib, C.G. Halogen-chalcogen X-E (X = F, Cl, Br, I.; E = S, Se, Te) chemistry. In *Handbook of Chalcogen Chemistry: New Perspectives in Sulfur, Selenium and Tellurium*, 2nd ed.; Devillanova, F.A., du Month, W.-W., Eds.; RSC Publishing: Cambridge, UK, 2013; Volume 2, pp. 273–316.
22. Chivers, T.; Laitinen, R.S. Tellurium: A maverick among the chalcogens. *Chem. Soc. Rev.* 2015, *44*, 1725–1739. [CrossRef] [PubMed]
23. Gleiter, R.; Haberhauer, G.; Werz, D.B.; Rominger, F.; Bleiholder, C. From noncovalent chalcogen-chalcogen interactions to supramolecular aggregates: Experiments and calculations. *Chem. Rev.* 2018, *118*, 2010–2041. [CrossRef]
24. Kolb, S.; Oliver, G.A.; Werz, D.B. Chemistry evolves, terms evolve, but phenomena do not evolve: From chalcogen-chalcogen interactions to chalcogen bonding. *Angew. Chem. Int. Ed.* 2020, *59*, 22306–22310. [CrossRef]
25. Kolb, S.; Oliver, G.A.; Werz, D.B. Chalcogen bonding in supramolecular structures, anion recognition, and catalysis. In *Comprehensive Inorganic Chemistry III*; Reedijk, J., Poeppelmeier, K.R., Laitinen, R.S., Eds.; Elsevier: Amsterdam, The Netherlands, 2023; Volume 1, pp. 602–650.
26. Morgan, G.T.; Burgess, H. XLIV. -cycloTelluropantane. *J. Chem. Soc.* 1928, 321–329. [CrossRef]
27. Morgan, G.T.; Burstall, F.H. XXIV.-cycloTellurobutane (Tetrahydrotellurophen). *J. Chem. Soc.* 1931, 180–184. [CrossRef]
28. Takaguchi, Y.; Horn, E.; Furukawa, N. Preparation and X-ray Structure Analysis of 1,1,5,5,9,9-Hexachloro-1,5,9-tritelluracyclododecane (Cl_6([12]aneTe_3)) and Its Redox Behavior. *Organometallics* 1996, *15*, 5112–5115. [CrossRef]
29. Rodewald, M.; Rautiainen, J.M.; Niksch, T.; Görls, H.; Oilunkaniemi, R.; Weigand, W.; Laitinen, R.S. Chalcogen-bonding interactions in telluroether heterocycles [Te$(CH_2)_m]_n$ (n = 1-4; m = 3-7). *Chem. Eur. J.* 2020, *26*, 13806–13818. [CrossRef]
30. Kemmitt, T.; Levason, W.; Oldroyd, R.D.; Webster, M. Palladium and platinum complexes of telluracyclopentane. Structure of trans-[Pd{Te$(CH2)_4$}$2Cl2$]. *Polyhedron* 1992, *11*, 2165–2169.
31. Singh, A.K.; Kadarkaraisamy, M.; Husebye, S. The first metal complexes of 1,4-oxatellurane: Synthesis and crystal structure of its platinum(II) complex. *J. Chem. Res.* 2000, 64–65. [CrossRef]
32. Hesford, M.J.; Levason, W.; Matthews, M.L.; Reid, G. Synthesis and complexation ofm the mixed tellurium-oxygen macrocycles 1-tellura-4,7,13,16-tetraoxacyclooctadecane, [18]aneO_4Te_2 and their selenium analogues. *Dalton Trans.* 2003, 2852–2858. [CrossRef]
33. Devery, M.P.; Dickson, R.S.; Skelton, B.W.; White, A.H. Coordination and rearrangement of organic chalcogenides on a rhodium-rhodium bond: Reactions with strained-ring cyclic thioethers and with selenium and tellurium ligands. *Organometallics* 1999, *18*, 5292–5298. [CrossRef]
34. Levason, W.; Orchard, S.D.; Reid, G. Synthesis and properties of the first series of mixed thioether/telluroether macrocycles. *Chem. Commun.* 2001, 427–428. [CrossRef]
35. Vigo, L.; Poropudas, M.J.; Salin, P.; Oilunkaniemi, R.; Laitinen, R.S. Versatile coordination chemistry of rhodium complexes containing the bis(trimethylsilylmethyl)tellane ligand. *J. Organomet. Chem.* 2009, *694*, 2053–2060. [CrossRef]
36. Vigo, L.; Salin, P.; Oilunkaniemi, R.; Laitinen, R.S. Formation and structural characterization of mercury complexes from Te{(R)(CH_2SiMe_3)} (R = Ph, CH_2SiMe_3) and $HgCl_2$. *J. Organomet. Chem.* 2009, *694*, 3134–3141. [CrossRef]
37. Poropudas, M.J.; Vigo, L.; Oilunkaniemi, R.; Laitinen, R.S. Versatile Solid-State Coordination Chemistry of Telluroether Complexes of Silver(I) and Copper(I). *Dalton Trans.* 2013, *42*, 16868–16877. [CrossRef]

38. Taimisto, M.; Bajorek, T.; Rautiainen, J.M.; Pakkanen, T.A.; Oilunkaniemi, R.; Laitinen, R.S. Experimental and computational investigation on the formation pathway of [RuCl$_2$(CO)$_2$(ERR')$_2$] (E = S, Se, Te; R, R' = Me, Ph) from [RuCl$_2$(CO)$_3$]$_2$ and ERR'. *Dalton Trans.* **2022**, *51*, 11747–11757. [CrossRef]
39. Taimisto, M.; Poropudas, M.J.; Rautiainen, J.M.; Oilunkaniemi, R.; Laitinen, R.S. Ruthenium-assisted tellurium abstraction in bis(thiophen-2-yl) ditelluride. *Eur. J. Inorg. Chem.* **2023**, *26*, e202200772. [CrossRef]
40. Kemmitt, T.; Levason, W. Synthesis and multinuclear NMR studies of [M{o-C$_6$H$_4$(TeMe)$_2$}X$_2$] (M = Pd, Pt; X = Cl, Br, I). The presence of a characteristic ring contribution to ^{125}Te chemical shifts. *Inorg. Chem.* **1990**, *29*, 731–735. [CrossRef]
41. Vigo, L.; Oilunkaniemi, R.; Laitinen, R.S. Formation and characterization of platinum and palladium complexes of bis(trimethylsilylmethyl) tellane. *Eur. J. Inorg. Chem.* **2008**, 284–290. [CrossRef]
42. Knorr, M.; Guyon, F.; Jourdain, I.; Kneifel, S.; Frenzel, J.; Strohmann, C. (Phenylthiomethyl)silanes and (butyltelluromethyl)-silanes as novel bifunctional ligands for the construction of dithioether-, ditelluroether- and transition metal-silicon complexes. *Inorg. Chim. Acta* **2003**, *350*, 455–466. [CrossRef]
43. Prasad, R.R.; Singh, H.B.; Butcher, R.J. Synthesis, structure and reactivity of b-chalcocyclohexanals: Dichalcogenides and chalcogenides. *J. Organomet. Chem.* **2016**, *814*, 42–56. [CrossRef]
44. Singh, A.K.; Kumar, J.S.; Butcher, R.J. N-{2-(4-methoxyphenyltelluro)ethyl}morpholine (L^1) and its platinum(II) and ruthenium(II) complexes. *Synthesis and crystal structure of L1 and trans-[PtCl2(L1)2]. Inorg. Chim. Acta* **2001**, *312*, 163–169.
45. Kolay, S.; Kumar, M.; Wadawale, A.; Das, D.; Jain, V.K. Role of anagostic interactions in cycloplatination of telluroethers: Synthesis and structural characterization. *J. Organomet. Chem.* **2015**, *794*, 40–47. [CrossRef]
46. Kapoor, P.; Lövqvist, K.; Oskarsson, A. Cis/trans influences in platinum(II) complexes. X-ray crystal structures of cis-dichloro(dimethyl sulfide)(dimethyl sulfoxide)platinum(II) and cis-dichloro(dimethyl sulfide)(dimethyl phenyl phosphine)platinum(II). *J. Mol. Struct.* **1998**, *470*, 39–47. [CrossRef]
47. *ConQuest*; Version 2022.3.0; Cambridge Crystallographic Data Center: Cambridge, UK, 2022.
48. Oilunkaniemi, R.; Komulainen, J.; Laitinen, R.S.; Ahlgrén, M.; Pursiainen, J. Trends in the structure and bonding of [MCl$_2${(C$_4$H$_3$S)ECH$_3$}$_2$] (M = Pd, Pt; E = Te, Se). *J. Organomet. Chem.* **1998**, *571*, 129–138. [CrossRef]
49. Levason, W.; Webster, M.; Mitchell, C.J. Structure of trans-diiodobis(methyl phenyl telluride)platinum(II). *Acta Crystallogr. Sect. C* **1992**, *48*, 1931–1933. [CrossRef]
50. Drake, J.E.; Yang, J.; Khalid, A.; Srivastava, V.; Singh, A.K. Palladium(II) and platinum(II) complexes of bis (4-methoxyphenyltelluro) methane. Crystal structure of [{meso-(4-MeOC$_6$H$_4$Te)$_2$CH$_2$}(Ph$_2$PCH$_2$CH$_2$PPh$_2$)Pd(II)](ClO$_4$)·H$_2$O and [meso-(4-MeOC$_6$H$_4$Te)$_2$CH$_2$]Pd(II)Cl$_2$. *Inorg. Chim. Acta* **1997**, *254*, 57–62. [CrossRef]
51. Kemmitt, T.; Levason, W.; Webster, M. Chelating ditelluroether complexes of palladium and platinum: Synthesis and multinuclear NMR Studies. Structure of dibromo(meso-1,3-bis(phenyltelluro)propane)palladium(II), [Pd{meso -PhTe(CH$_2$)$_3$TePh}Br$_2$]. *Inorg. Chem.* **1989**, *28*, 692–696. [CrossRef]
52. Rocha, R.W.; de Almeida, W.B. On the cis->trans isomerization of the square-planar [PtCl(SnCl$_3$)(PH$_3$)$_2$] compound: Ab initio gas phase reaction mechanism and solvent effects using continuum models. *J. Braz. Chem. Soc.* **2000**, *11*, 112–120. [CrossRef]
53. Redfield, D.A.; Nelson, J.H. Equilibrium energetics of cis-trans isomerization for two square-planar palladium(II)-Phosphine Complexes. *Inorg. Chem.* **1973**, *12*, 15–19. [CrossRef]
54. Anderson, G.K.; Cross, R.J. Isomerisation mechanisms of square-planar complexes. *J. Chem. Soc. Rev.* **1980**, *9*, 185–215. [CrossRef]
55. Lepetit, C.; Fau, P.; Fajerwerg, K.; Kahn, M.L.; Silvi, B. Topological analysis of the metal-metal bond: A tutorial review. *Coord. Chem. Rev.* **2017**, *345*, 150–181. [CrossRef]
56. Cremer, D.; Kraka, E. Chemical bonds without bonding electron density—Does the difference electron-density analysis suffice for a description of the chemical bond? *Angew. Chem. Int. Ed. Engl.* **1984**, *23*, 627–628. [CrossRef]
57. Cremer, D.; Kraka, E. A description of the chemical bond in terms of local properties of electron density and energy. *Croat. Chem. Acta* **1984**, *57*, 1259–1281.
58. Espinosa, E.; Alkorta, I.; Elguero, J.; Molins, E. From weak to strong interactions: A comprehensive analysis of the topological and energetic properties of the electron density distribution involving X–H···F–Y systems. *J. Chem. Phys.* **2002**, *117*, 5529–5542. [CrossRef]
59. Macchi, P.; Sironi, A. Chemical bonding in transition metal carbonyl clusters: Complementary analysis of theoretical and experimental electron densities. *Coord. Chem. Rev.* **2003**, *238–239*, 383–412. [CrossRef]
60. Macchi, P.; Proserpio, D.M.; Sironi, A. Experimental Electron Density in a Transition Metal Dimer: Metal-Metal and Metal-Ligand Bonds. *J. Am. Chem. Soc.* **1998**, *120*, 13429–13435. [CrossRef]
61. Firme, C.L.; Antunes, O.A.C.; Esteves, P.M. Relation between bond order and delocalization index of QTAIM. *Chem. Phys. Lett.* **2009**, *468*, 129–133. [CrossRef]
62. Sivchik, V.; Kochetov, A.; Eskelinen, T.; Kisel, K.S.; Solomatina, A.I.; Grachova, E.V.; Tunik, S.P.; Hirva, P.; Koshevoy, I.O. Modulation of Metallophilic and π–π Interactions in Platinum Cyclometalated Luminophores with Halogen Bonding. *Chem. Eur. J.* **2021**, *27*, 1787–1794. [CrossRef]
63. Espinosa, E.; Molins, E.; Lecomte, C. Hydrogen bond strengths revealed by topological analyses of experimentally observed electron densities. *Chem. Phys. Lett.* **1998**, *285*, 170–173. [CrossRef]
64. Spackman, M.A. How reliable are intermolecular interaction energies estimated from topological analysis of experimental electron densities? *Cryst. Growth. Des.* **2015**, *15*, 5624–5628. [CrossRef]

65. Harris, R.K.; Becker, E.D.; Cabral de Menezes, S.M.; Goodfellow, R.; Granger, P. NMR nomenclature: Nuclear spin properties and conventions for chemical shifts. IUPAC recommendations 2001. *Magn. Reson. Chem.* **2002**, *40*, 489–505. [CrossRef]
66. Fulmer, G.R.; Miller, A.J.M.; Sherden, N.H.; Gottlieb, H.E.; Nudelman, A.; Stoltz, B.M.; Bercaw, J.E.; Goldberg, K.I. NMR chemical shifts of trace impurities: Common laboratory solvents, organics, and gases in deuterated solvents relevant to the organometallic chemist. *Organometallics* **2010**, *29*, 2176–2179. [CrossRef]
67. Nonius, B.V. *COLLECT*; Data Collection Software; Nonius BV: Delft, The Netherlands, 1998.
68. Otwinowski, Z.; Minor, W. *Macromolecular Crystallography, Part A, Methods in Enzymology*; Carter, C.W., Sweet, R.M., Eds.; Academic Press: Cambridge, UK, 1997; Volume 276, pp. 307–326.
69. Krause, L.; Herbst-Irmer, R.; Sheldrick, G.M.; Stalke, D. Comparison of silver and molybdenum microfocus X-ray sources for single-crystal structure determination. *J. Appl. Crystallogr.* **2015**, *48*, 3–10. [CrossRef]
70. Sheldrick, G.M. SHELXT—Integrated space-group and crystal-structure determination. *Acta Crystallogr.* **2015**, *A71*, 3–8. [CrossRef]
71. Sheldrick, G.M. A short history of SHELX. *Acta Crystallogr.* **2008**, *A64*, 112–122. [CrossRef]
72. Kratzert, D.; Krossing, I. Recent improvements in DSR. *J. Appl. Cryst.* **2018**, *51*, 928–934. [CrossRef]
73. Frisch, M.J.; Trucks, G.W.; Schlegel, H.B.; Scuseria, G.E.; Robb, M.A.; Cheeseman, J.R.; Scalmani, G.; Barone, V.; Petersson, G.A.; Nakatsuji, H.; et al. *Gaussian 16, Revision C.01*; Gaussian, Inc.: Wallingford, CT, USA, 2019.
74. Perdew, J.P.; Burke, K.; Ernzerhof, M. Generalized gradient approximation made simple. *Phys. Rev. Lett.* **1996**, *77*, 3865–3868, Erratum in *Phys. Rev. Lett.* **1997**, *78*, 1396. [CrossRef]
75. Perdew, J.P.; Ernzerhof, M.; Burke, K. Rationale for mixing exact exchange with density functional approximations. *J. Chem. Phys.* **1996**, *105*, 9982–9985. [CrossRef]
76. Adamo, C.; Barone, V. Toward reliable density functional methods without adjustable parameters: The PBE0 model. *J. Chem. Phys.* **1999**, *110*, 6158–6170. [CrossRef]
77. Weigend, F.; Ahlrichs, R. Balanced basis sets of split valence, triple zeta valence and quadruple zeta valence quality for H to Rn: Design and assessment of accuracy. *Phys. Chem. Chem. Phys.* **2005**, *7*, 3297–3305. [CrossRef]
78. Weigend, F.; Häser, M.; Patzelt, H.; Ahlrichs, R. RI-MP2: Optimized auxiliary basis sets and demonstration of efficiency. *Chem. Phys. Lett.* **1998**, *294*, 143–152. [CrossRef]
79. Maaninen, T.; Tuononen, H.M.; Kosunen, K.; Oilunkaniemi, R.; Hiitola, J.; Laitinen, R.; Chivers, T. Formation, structural characterization and calculated NMR chemical shifts of selenium-nitrogen compounds from $SeCl_4$ and ArNHLi (Ar supermesityl, mesityl). *Z. Anorg. Allg-. Chem.* **2004**, *630*, 1947–1954. [CrossRef]
80. Barone, V.; Cossi, M. Quantum calculation of molecular energies and energy gradients in solution by a conductor solvent model. *J. Phys. Chem. A* **1998**, *102*, 1995–2001. [CrossRef]
81. Cossi, M.; Rega, N.; Scalmani, G.; Barone, V. Energies, structures, and electronic properties of molecules in solution with the C-PCM solvation model. *J. Comp. Chem.* **2003**, *24*, 669–681. [CrossRef]
82. Grimme, S.; Antony, J.; Ehrlich, S.; Krieg, H. A consistent and accurate *ab initio* parametrization of density functional dispersion correction (DFT-D) for the 94 elements H-Pu. *J. Chem. Phys.* **2010**, *132*, 154104/1–154104/19. [CrossRef]
83. Burns, L.A.; Vazquez-Mayagoitia, A.; Sumpter, B.G.; Sherrill, C.D. Density-functional approaches to noncovalent interactions: A comparison of dispersion corrections (DFT-D), exchange-hole dipole moment (XDM) theory, and specialized functions. *J. Chem. Phys.* **2011**, *134*, 084107/1–084107/25. [CrossRef]
84. Grimme, S.; Ehrlich, S.; Goerigk, L. Effect of the damping function in dispersion corrected density functional theory. *J. Comp. Chem.* **2011**, *32*, 1456–1465. [CrossRef]
85. Bader, R.F.W. *Atoms in Molecules: A Quantum Theory*; Oxford University Press: Oxford, UK, 1990.
86. Keith, T.A. *TK Gristmill Software*; Version 19.10.12; AIMAll: Overland Park, KS, USA, 2019; Available online: aim.tkgristmill.com (accessed on 9 February 2022).

Disclaimer/Publisher's Note: The statements, opinions and data contained in all publications are solely those of the individual author(s) and contributor(s) and not of MDPI and/or the editor(s). MDPI and/or the editor(s) disclaim responsibility for any injury to people or property resulting from any ideas, methods, instructions or products referred to in the content.

Article

A Convenient One-Pot Synthesis of a Sterically Demanding Aniline from Aryllithium Using Trimethylsilyl Azide, Conversion to β-Diketimines and Synthesis of a β-Diketiminate Magnesium Hydride Complex

Nikita Demidov, Mateus Grebogi, Connor Bourne, Aidan P. McKay, David B. Cordes and Andreas Stasch *

EaStCHEM School of Chemistry, University of St Andrews, North Haugh, St Andrews KY16 9ST, UK; nd49@st-andrews.ac.uk (N.D.); mateus0303@icloud.com (M.G.); cb376@st-andrews.ac.uk (C.B.); apm31@st-andrews.ac.uk (A.P.M.); dbc21@st-andrews.ac.uk (D.B.C)
* Correspondence: as411@st-andrews.ac.uk; Tel.: +44-(0)-1334-463-382

Citation: Demidov, N.; Grebogi, M.; Bourne, C.; McKay, A.P.; Cordes, D.B.; Stasch, A. A Convenient One-Pot Synthesis of a Sterically Demanding Aniline from Aryllithium Using Trimethylsilyl Azide, Conversion to β-Diketimines and Synthesis of a β-Diketiminate Magnesium Hydride Complex. *Molecules* **2023**, *28*, 7569. https://doi.org/10.3390/molecules28227569

Academic Editor: Yves Canac

Received: 19 October 2023
Revised: 8 November 2023
Accepted: 9 November 2023
Published: 13 November 2023

Copyright: © 2023 by the authors. Licensee MDPI, Basel, Switzerland. This article is an open access article distributed under the terms and conditions of the Creative Commons Attribution (CC BY) license (https://creativecommons.org/licenses/by/4.0/).

Abstract: This work reports the one-pot synthesis of sterically demanding aniline derivatives from aryllithium species utilising trimethylsilyl azide to introduce amine functionalities and conversions to new examples of a common N,N'-chelating ligand system. The reaction of TripLi (Trip = 2,4,6-iPr$_3$-C$_6$H$_2$) with trimethylsilyl azide afforded the silyltriazene TripN$_2$N(SiMe$_3$)$_2$ in situ, which readily reacts with methanol under dinitrogen elimination to the aniline TripNH$_2$ in good yield. The reaction pathways and by-products of the system have been studied. The extension of this reaction to a much more sterically demanding terphenyl system suggested that TerLi (Ter = 2,6-Trip$_2$-C$_6$H$_3$) slowly reacted with trimethylsilyl azide to form a silyl(terphenyl)triazenide lithium complex in situ, predominantly underwent nitrogen loss to TerN(SiMe$_3$)Li in parallel, which afforded TerN(SiMe$_3$)H after workup, and can be deprotected under acidic conditions to form the aniline TerNH$_2$. TripNH$_2$ was furthermore converted to the sterically demanding β-diketimines $^{R\text{Trip}}$nacnacH (=HC{RCN(Trip)}$_2$H), with R = Me, Et and iPr, in one-pot procedures from the corresponding 1,3-diketones. The bulkiest proligand was employed to synthesise the magnesium hydride complex [{($^{i\text{Pr}\text{Trip}}$nacnac)MgH}$_2$], which shows a distorted dimeric structure caused by the substituents of the sterically demanding ligand moieties.

Keywords: aniline synthesis; azides; β-diketiminates; magnesium hydride; metal-halogen exchange; organolithium reagents; sterically demanding N-ligands; terphenyl ligands; triazenes

1. Introduction

Sterically demanding N-ligands [1–6] have been driving advances in numerous areas of chemical research. In main group chemistry, for example, the introduction of sterically demanding N-ligands and related species has led to the discovery of compound classes with low coordination numbers in a variety of oxidation states, which stabilised molecular entities in unusual bonding modes that were found to show unique properties and novel reactivities [7–9]. In addition to effects on compound properties and reactivity instilled by the ligand class attached to central elements, steric effects have a strong influence on compound properties, including on coordination numbers and allowing or preventing certain reaction pathways. The interplay of steric demand, including considerations of repulsion from sheer size and ligand shape [10,11], and attractive effects between ligands, substituents, and central elements from London dispersion forces [12–14], paints a more complex picture of the effects bulky ligands have on various compound classes. As such, surprising effects on the chemistry of unusual compound classes stabilised by sterically demanding ligands have often been discovered by exploratory investigation of the sterics and electronics of their ligands. For example, in magnesium β-diketiminate chemistry, which is of relevance here, relatively small changes in ligand sterics have led to significantly different product outcomes for magnesium hydride [15,16] and low oxidation state magnesium [17–19] chemistry.

The introduction of new sterically demanding ligands bearing different substituents to the chemistry of the elements led to novel classes of compounds [7] that, for the bulkiest systems, show coordination numbers down to one. In selected cases, these have displayed unique structures, bonding modes, and reactivities, for example in Al [20], Ge [21], P [22], Sb [23] and Bi [24] species, but the steric bulk can also pose significant new challenges for the syntheses of these new (pro)ligand entities. Due to the increased steric demand in these ligands, common synthetic techniques may prove less applicable to the task, or suitable, convenient starting materials are not readily commercially available. Thus, facile synthetic techniques are required to expand our toolset to access new ligands. Here, we explore the conversion of bulky aryl halides to reactive organolithium compounds, their further transformation to aniline derivatives, the synthesis to a common proligand class, and an example metal complex.

2. Results and Discussion

The aim of the first part of the study was to find a convenient one-pot procedure that would allow the synthesis of substituted anilines [25–28] from aryl lithium compounds with techniques and methods suitable for, and familiar to, the synthetic inorganic and organometallic communities. Many methods for electrophilic amination reactions have been developed [29–34], including some to selectively prepare primary amines, but many also show some drawbacks or limitations, and an important consideration for the work described herein was that it could be applied to highly sterically demanding systems. Other synthetic routes to sterically demanding anilines have been successfully developed, but these require either multi-step protocols and/or high pressure set-ups [35–38] or modify the substituents via palladium-catalysed cross-coupling reactions [39]. For this study we decided on using the 2,4,6-triisopropylphenyl substituent, Trip, due to the commercial availability of starting materials, rarer use of the respective aniline compared with the ubiquitous Dip (2,6-diisopropylphenyl) congener, and the use of isopropyl substituents as suitable groups for ^{1}H NMR spectroscopic investigations.

2.1. Lithiation of TripBr

In order to devise a convenient aniline synthesis via an organometallic route, the first consideration was to revisit the generation of TripLi **1** from commercially available TripBr **2**. Some crystallographically characterised TripLi species [40,41] are known, and are prepared using metal-halogen exchange [42,43], with n-butyllithium as the lithiating agent. The metal-halogen exchange reaction is an equilibrium system [44] that forms a stabilised mixture, and the thermodynamics and kinetics are influenced by solvent effects [45–47], and the steric and electronic nature of the substituents [48,49]. Furthermore, side reactions such as C–C coupling [50] and ether cleavage [51] can lead to consumption of lithium reagent and substrate, potentially hampering efficient conversion to the desired product. In addition, we envisaged that further conversions of in situ generated aryllithium species for the synthesis of sterically demanding systems might require harsher conditions for onward reactivity. As such, we were interested in a quite robust lithiation protocol that can tolerate non-cryogenic conditions, e.g., room temperature, at least for onward reactivity, and would also work for electron-rich aryl substituents. To briefly test and ensure sufficient in situ lithiation for further conversions, TripBr **2** was treated with 1.05 equivalents of nBuLi under varying conditions and the mixture was hydrolysed and the product ratio was analysed by ^{1}H NMR spectroscopy, see Scheme 1 and Table 1. The relative percentages of the main products TripBr **2**, i.e., unreacted starting material, TripH **3** as a proxy for hydrolysed TripLi **1**, and the coupled product TripnBu **4** were added to 100% and represent the main products. Traces of TripOH **5**, likely from the reaction of TripLi **1** with trace amounts of air, for example, formed during the quenching process, were also present in some samples.

Inspecting the lithiation results summarised in Table 1, diethyl ether (entry 1) as a donor led to insufficient conversion. Using only one or a half equivalent of the more powerful donor solvent THF per Li centre (entries 2 and 3) afforded poor conversion of TripBr **2**. More THF per Li as a donor (2–5 equivalents, entries 4–7) provided good conversion of **2** to **1**, by implication, but also saw increasing quantities of Trip*n*Bu **4** formed from direct C–C coupling [50]. The latter issue could be suppressed by cooling the reaction solution to −20 °C (entry 8), which decreased the coupling to Trip*n*Bu **4** and was beneficial for the lithiation of TripBr **2**, possibly due to forming more reactive lower aggregates for entropic reasons [52] and/or less ether cleavage [51]. To test how competitive ether cleavage is under the conditions, the experiment for entry 7 (Table 1) was repeated, but the *n*BuLi was added to the solvent mixture and left for 30 min at room temperature before TripBr **2** was added and reacted for 30 min before hydrolysis and analysis (entry 9). This experiment provided significant unreacted **2** likely because some *n*BuLi degraded via ether cleavage under these conditions. This highlights that ether cleavage is a significant issue even for relatively low THF concentrations and suggests that these issues are remedied at the lower temperature used in the conditions of entry 8. Going forward, the conditions for entry 8 were used in subsequent sections.

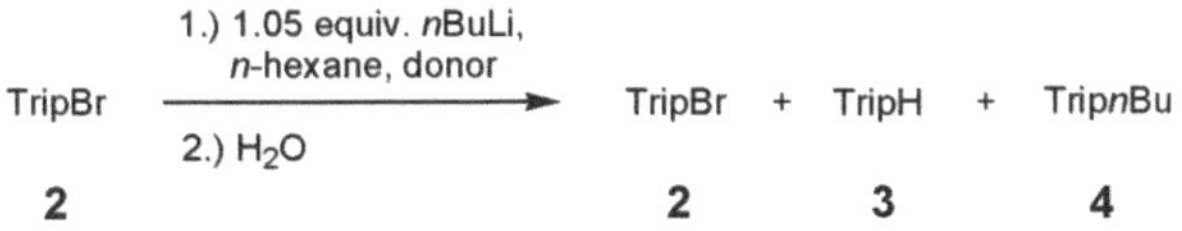

Scheme 1. Lithiation of TripBr **2**.

Table 1. Lithiation study of TripBr **2**.

Entry	Donor, Equiv. per Li [a]	Temperature, T	TripBr **2**	TripH **3** (c.f. TripLi **1**)	Trip*n*Bu **4**
1	Et$_2$O, 17:1	r.t.	41.5	58.5	~0
2	THF, 0.5:1	r.t.	92.2	7.8	~0
3	THF, 1:1	r.t.	68.0	32.0	~0
4	THF, 2:1	r.t.	23.2	74.8	2.0
5	THF, 3:1	r.t.	12.9	83.3	3.8
6	THF, 4:1	r.t.	9.8	84.7	5.4
7	THF, 5:1	r.t.	6.6	85.7	7.7
8	THF, 5:1	−20 °C	trace	>97	trace
9	THF, 5:1 (+*n*BuLi first) [b]	r.t.	36.1	61.8	2.1

[a] 1.05 equivalents of *n*BuLi solution were added dropwise to TripBr in *n*-hexane plus donor solvent as given to afford a 0.395 M reaction solution, stirred for 30 min, then hydrolysed with water and the organic residues were analysed by ^{1}H NMR spectroscopy (CDCl$_3$), reporting the percentages of the products **2**, **3** and **4** from their relative ratios. The experiments were conducted with the same reagent concentrations as single experiments only to allow for a brief study to find favourable conditions. Estimated error +/− 1–2%. In addition, trace amount of TripOH, presumably from traces of oxygen were present in some samples. [b] *n*BuLi was first reacted with the solvent mixture for 30 min before TripBr was added and the experiment continued.

An alternative to using *n*BuLi is the direct reaction of TripBr **2** with lithium metal in diethyl ether for one hour under reflux to TripLi **1**, see Scheme 2, which is straightforward and afforded a good overall yield after in situ conversion to the desired aniline product, vide infra.

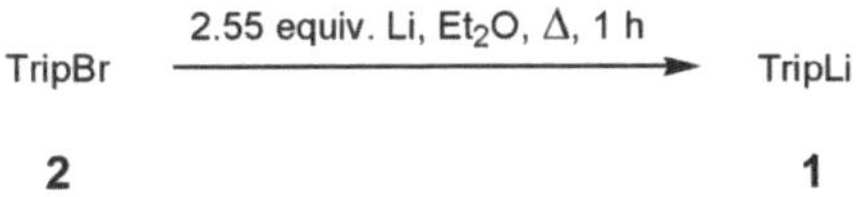

Scheme 2. Lithiation of TripBr **2**.

2.2. Reaction of TripLi 1 with TMSN$_3$

Reactions of organometallic compounds and organic azides [29,53–56] can form triazenes [57], including sterically demanding triazenides [58,59]. In the past, the group of N. Wiberg has studied reactions of s-block organometallics with silylazides and the synthesis and properties of silyltriazenes [60–67]. Subsequently, reactions of specialised organic azides with Grignard or organolithium reagents have been utilised as NH$_2^+$ synthons to successfully afford amines or anilines that show various advantages and limitations. Trimethylsilylmethyl azide [68–70] and azidomethyl phenyl sulfide [71–73] have been shown to be effective in reactions with Grignard reagents, which can be challenging to form for electron-rich substrates [74], but these azide systems struggle when reacted with organolithium reagents. Diphenylphosphoryl azide affords satisfactory to good yields with either of these classes of organometallic reagents but is more expensive and requires harsh hydrolysis conditions [75]. Vinyl or allyl azides have shown high efficiency [76,77], but are not commercially available and are likely more hazardous to handle [53]. To introduce an alternative that uses a commercially available and relatively stable azide, we proposed that the reaction between an organolithium species, and by implication other electropositive organometallics, and a silylazide would lead to a silyl(organo)triazenide lithium complex, which could be easily worked up to a primary amine in a one-pot procedure. Nucleophilic attack of an organometallic substituent on the (outer) azide nitrogen atom also seemed like a procedure that could be facile even for highly sterically demanding systems, as suggested by the syntheses of bulky triazenides [58,59] and terphenyl azide compounds [78–80]. Although syntheses for TripNH$_2$ **6** are known [81–85], most involve the synthesis and reduction of TripNO$_2$, which has some drawbacks such as expensive and/or hazardous reagents, multi-step protocols, and relies on the accessibility of the nitro derivative.

Initially, the reaction of TripLi **1**, prepared according to either Scheme 1 or Scheme 2, with one equivalent of trimethylsilyl azide (azidotrimethylsilane, Me$_3$SiN$_3$) at room temperature proceeded rapidly. We found that simple quenching with (wet) methanol provided rapid gas formation, and crude TripNH$_2$ **6** was obtained after workup—an observation that was in line with our expectation of a pathway via an intermediate silyl(organo)triazenide lithium complex, complex **7** in Scheme 3 (grey arrow). The yield of TripNH$_2$ **6**, however, did not significantly exceed 40%, and large quantities of TripH **3** were obtained alongside **6**, independent of reaction times. Furthermore, an insoluble precipitate formed during the reaction. Adding further Me$_3$SiN$_3$, however, increased the yield of TripNH$_2$ **6**, and the consistently formed insoluble by-product precipitated from the reaction mixture early on. The latter was identified as LiN$_3$ via its properties, NMR, and IR spectroscopic studies, and suggests that further silylation of **7** with Me$_3$SiN$_3$ occurs, i.e., the azide is acting as a pseudohalide, resulting in the formation of disilyl(organo)triazene TripN$_2$N(SiMe$_3$)$_2$ **8**. Evidence for the formation of **8** comes from an NMR spectroscopic study that shows an intermediate with two chemically identical SiMe$_3$ groups by integration and a ^{15}N NMR resonance [86,87] detected via a 2D ^{1}H-^{15}N HMBC NMR experiment of a silyl-bound nitrogen atom at δ 187 ppm in deuterated benzene. Derivatives of **8**, such as PhN$_2$N(SiMe$_3$)$_2$, have been obtained by Wiberg previously, and some of these products decomposed with dinitrogen elimination [64]. Treating **8** with methanol gave rapid gas evolution and afforded TripNH$_2$ **6**. Reactions performed on a larger scale also afforded small and varying quantities of TripN(SiMe$_3$)$_2$ **9**, which could be structurally characterised (Figure 1), alongside the main product TripNH$_2$ **6**. Compound **9** is typically present in low percentages as a by-product from these reactions, and it is likely that during the synthesis, some TripN$_2$N(SiMe$_3$)$_2$ **8** decomposes and loses dinitrogen to form TripN(SiMe$_3$)$_2$ **9** (Scheme 3). In contrast to **8**, a ^{15}N NMR resonance of δ 43 ppm (in CDCl$_3$, via a 2D ^{1}H/^{15}N HMBC NMR experiment) was found for TripN(SiMe$_3$)$_2$ **9**. As expected, the nitrogen centre in **9** is planar (sum of angles: ca. 360°) in its molecular structure (Figure 1), c.f. the structure of N(SiMe$_3$)$_3$ [88], and the metrical features are as expected.

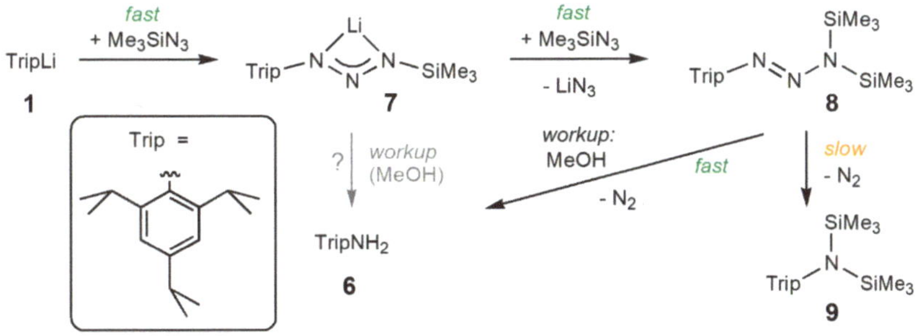

Scheme 3. Reactions of TripLi **1** with Me$_3$SiN$_3$.

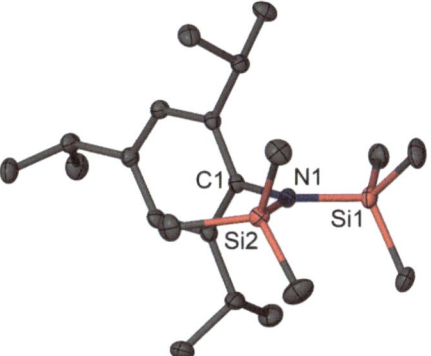

Figure 1. Molecular structure of TripN(SiMe$_3$)$_2$ **9**, 30% thermal ellipsoids. Hydrogen atoms omitted. Selected bond lengths (Å) and angles (°): Si1–N1 1.7497(19), Si2–N1 1.7554(19), N1–C1 1.450(3); Si1–N1–Si2 123.82(10), C1–N1–Si1 120.42(14), C1–N1–Si2 115.75(14).

A simple aqueous workup of the reaction mixtures afforded crude TripNH$_2$ **6** in good, isolated yields. The crude product can be dried under vacuum, removing some volatile by-products, which would include small molecules from ether cleavage and could include smaller amines (e.g., nBuNH$_2$, b.p. ca. 78 °C) from side reactions. This leaves predominantly TripNH$_2$ **6** as the main product, but also the possible high-boiling by-products TripBr **2** from insufficient lithiation, TripH **3** from protonolysis, Trip$n$Bu **4** from C–C coupling, TripOH **5** from oxidation and TripN(SiMe$_3$)$_2$ **9** from loss of dinitrogen of the intermediate **8**. Gratifyingly, none of the Trip-containing by-products readily reacted with acids and the crude product could be treated with aqueous HCl and petroleum ether or hexane as a two-phase system to afford solid TripNH$_3^+$Cl$^-$ **10** (as a hydrate) for purification that could be further washed with petroleum ether or hexane. Treatment with bases regenerated purer TripNH$_2$ **6**. Isolated yields of TripNH$_2$ **6** after this purification from TripBr were 86% (typically ca. 75–86%) via the nBuLi route, and 81% via the Li metal route. In addition, the compound can also be purified by column chromatography on alumina with petroleum ether/dichloromethane.

2.3. Extension to a Sterically Demanding Terphenyl System

To study if the above method can be conveniently transferred to another, more sterically demanding ligand system, we investigated the terphenyl substituent 2,6-bis(2,4,6-triisopropylphenyl)phenyl, Ter, in this reaction [78,80,89]. Initially, TerLi(OEt$_2$) **11**(OEt$_2$) was prepared from TerI **12** and nBuLi [89] to study its reaction with Me$_3$SiN$_3$ on a small scale by NMR spectroscopy. These experiments showed that the reaction is very slow and that significant quantities of TerH **13** are formed alongside some N-containing main product, later identified as a TerN(SiMe$_3$)Li **14** derivative. A significant change in the reaction kinetics is not surprising when the steric demand of the system is changed dramatically. It is proposed that the hydrogen on TerH **13** originated from Me$_3$SiN$_3$ during the reaction and a brief study was undertaken to test the influence of donor solvent addition. When TerLi(OEt$_2$) **11**(OEt$_2$) was reacted with Me$_3$SiN$_3$ in deuterated benzene, after four days at room temperature a ratio of TerH **13** to TerN(SiMe$_3$)Li **14** of approximately 2:1 was obtained. A reaction with additional five equivalents of the donor solvent diethyl ether under the same conditions was slower (18 days) and the main product ratio changed to approximately 2:3 (**13**:**14**). Changing the donor solvent to THF (5 equivalents) provided an approximate ratio of 1:4 after 7 days at room temperature and 8 h at 60 °C, showing that TerN(SiMe$_3$)Li **14** can be afforded as the dominant product (also see Table S1). Other observations that were made showed that, in contrast to the reaction with TripLi **1**, only one equivalent of Me$_3$SiN$_3$ was required in this bulkier system, and that before **14** forms, the main intermediate, presumed to be TerN$_2$N(SiMe$_3$)Li **15**, the Ter-equivalent of compound **7** (Scheme 3), was produced, see Scheme 4. Due to the very slow formation of TerN$_2$N(SiMe$_3$)Li **15** from the starting materials, nitrogen loss from **15**, yielding TerN(SiMe$_3$)Li **14**, becomes competitive in parallel. Workup with (wet) methanol afforded TerN(SiMe$_3$)H **16** from TerN(SiMe$_3$)Li **14**, whereas reaction mixtures with incomplete conversion and larger quantities of the intermediate TerN$_2$N(SiMe$_3$)Li **15** afforded TerNH$_2$ **17** after workup, see Scheme 4 for a summary of these pathways. Some evidence for this was obtained by 2D ^{1}H-^{15}N HMBC NMR experiments in deuterated benzene where the ^{15}N NMR signal of the Me$_3$Si-bound nitrogen centre could be measured and compared. These were found at δ 317 ppm for the intermediate TerN$_2$N(SiMe$_3$)Li **15**, δ 115 ppm for TerN(SiMe$_3$)Li **14** and δ 67 ppm for TerN(SiMe$_3$)H **16** [76,77]. Furthermore, the reaction of TerN(SiMe$_3$)Li **14** to TerN(SiMe$_3$)H **16** was found to be reversible; treatment of **14** with an excess of methanol in an NMR tube afforded **16**, and this mixture could then treated be with an excess of solid MeLi to afford **14** again. So far, we found no evidence for further reaction of TerN$_2$N(SiMe$_3$)Li **15**, with Me$_3$SiN$_3$ to TerN$_2$N(SiMe$_3$)$_2$ and LiN$_3$ (Scheme 4), as was analogously observed for the Trip system (Scheme 3). The molecular structures of TerN(SiMe$_3$)H **16**, shown in Figure 2, and the solvate TerN(SiMe$_3$)H·C$_6$H$_6$ **16**·C$_6$H$_6$, were determined and highlight the steric demand around the N atom with a C–N–Si angle of ca. 130°. The molecular structure infers that in solution, the silyl methyl groups of **16** can reside above the flanking Trip-aryls which is likely the reason for the upfield-shifted resonance for the protons of the SiMe$_3$ group (δ −0.34 ppm).

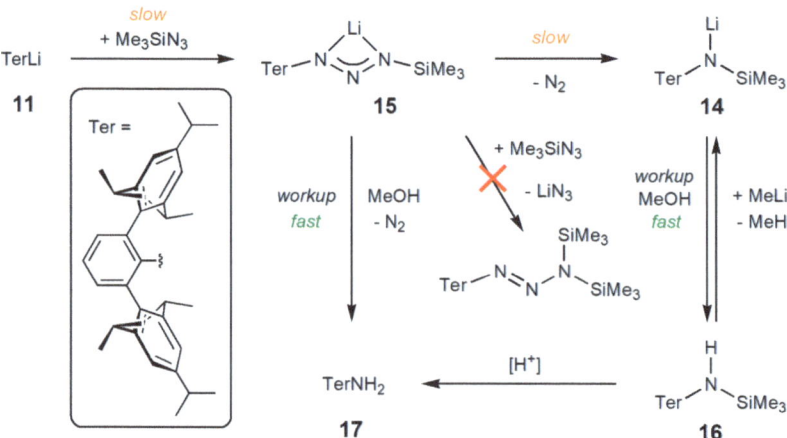

Scheme 4. Reactions of TerLi **11** with Me$_3$SiN$_3$ at room temperature.

C1
N1
Si1
H1

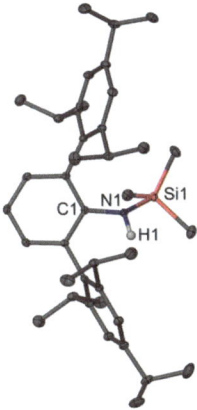

Figure 2. Molecular structure of TerN($SiMe_3$)H **16**, 30% thermal ellipsoids. Only the NH hydrogen atom is shown. Selected bond lengths (Å) and angles (°) for **16**: Si1–N1 1.7439(10), N1–C1 1.4080(14); C1–N1–Si1 129.12(7); metrical data for TerN($SiMe_3$)H·C_6H_6, **16**·C_6H_6: Si1–N1 1.7423(12), N1–C1 1.4061(17); C1–N1–Si1 133.11(10).

On a preparative scale, TerI **12** was converted with *n*BuLi in an *n*-hexane and diethyl ether mixture to crude TerLi(OEt_2) **11**(OEt_2) and then reacted in situ with Me_3SiN_3 in toluene and 10 equivalents of THF at 60 °C for 20 h, followed by workup with (wet) methanol at 0 °C to afford TerN($SiMe_3$)H **16** in 60% isolated yield after recrystallisation from diethyl ether. TerN($SiMe_3$)H **16** has been found to desilylate to $TerNH_2$ **17** when treated with aqueous HCl, or slowly in wet chloroform or with silica gel.

2.4. Synthesis of ^{RTrip}nacnacH Compounds

With the aniline $TripNH_2$ **6** in hand, we studied its conversion to β-diketimine proligands [2] to progress towards β-diketiminate complexes. The three β-diketimines ^{RTrip}nacnacH, =HC{RCN(Trip)}$_2$H, with R = Me (**18**), Et (**19**), and *i*Pr (**20**), were prepared by one-pot condensation reactions between $TripNH_2$ **6** and appropriate 1,3-diketones under acidic conditions, followed by aqueous workup steps under basic conditions, see Scheme 5. Previously, the ^{tBuTrip}nacnacH ligand [83] was prepared via a multi-step route, and a selection of other sterically demanding ^{MeAr}nacnacH proligands, where Ar represents a robust substituent larger than Dip, have been reported in recent years [90–97].

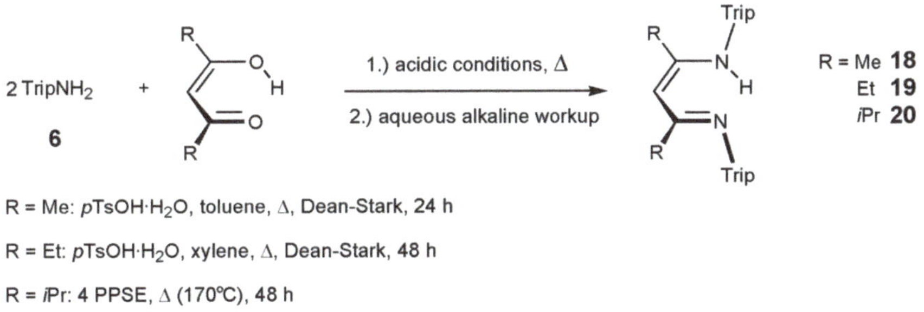

2 TripNH2
6
R
O
H
O
R
1.) acidic conditions, Δ
2.) aqueous alkaline workup
Trip
R
N
H
N
R
Trip
R = Me 18
Et 19
iPr 20
R = Me: pTsOH·H2O, toluene, Δ, Dean-Stark, 24 h
R = Et: pTsOH·H2O, xylene, Δ, Dean-Stark, 48 h
R = iPr: 4 PPSE, Δ (170°C), 48 h

Scheme 5. Synthesis of ^{RTrip}nacnacH compounds **18**–**20**.

The synthesis of ^{MeTrip}nacnacH **18**, generally followed an established protocol [98] and afforded an isolated yield of 76%. *β*-Diketimines with an ethyl backbone, ^{EtTrip}nacnacH **19** (66% yield), are not common, and we have modified an established procedure [98] used for related ligands. The route to the isopropyl backbone-substituted ^{iPrTrip}nacnacH **20** (67%

yield) uses a protocol we have recently introduced preparing the related [iPrDip]nacnacH [99], employing the powerful acidic dehydrating agent polyphosphoric acid trimethylsilylester, PPSE [100,101].

The three proligands **18–20** were structurally characterised, see Figure 3, and show the expected overall structure for this ligand class. The molecular structures of **18** and **19** each crystallised with a full molecule in the asymmetric unit and show some preference for alternating long and short N–C/C–C bonds in the ligand backbone, and, accordingly, some localisation of the NH hydrogen atom.

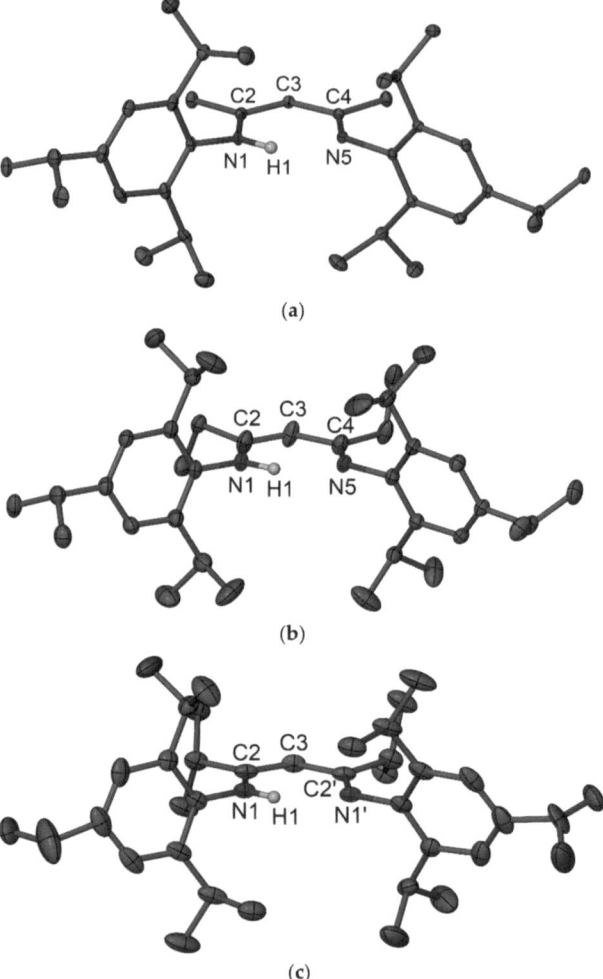

Figure 3. Molecular structures of [MeTrip]nacnacH·C_7H_8 **18**·C_7H_8 (**a**), [EtTrip]nacnacH **19** (**b**) and [iPrTrip]nacnacH **20** (**c**), 30% thermal ellipsoids. The toluene molecule in **18**·C_7H_8 is omitted. Only NH nitrogen atoms are shown including only one NH position for **20**. Selected bond lengths (Å) and angles (°): **18**·C_7H_8: N1–C2 1.341(3), C2–C3 1.383(3), C3–C4 1.424(3), N5–C4 1.315(3); C2–N1–C6 123.8(2), C4–N5–C23 121.0(2), N1–C2–C3 121.3(2), C2–C3–C4 125.4(2), N5–C4–C3 121.0(2); **19**: N1–C2 1.346(3), C2–C3 1.372(3), C4–C3 1.428(3), N5–C4 1.305(3); C2–N1–C6 124.87(18), C4–N5–C25 121.83(17), N1–C2–C3 121.5(2), C2–C3–C4 126.8(2), N5–C4–C3 120.06(19); **20**: N1–C2–1.338(3), C2–C3–1.392(3); N1–C2–C3 121.0(2), C2–N1–C4 122.7(2), C2–C3–C2′ 127.2(3).

2.5. Synthesis and Characterization of [{(^{iPrTrip}nacnac)MgH}$_2$] 21

To study the impact of the introduction of the Trip substituent to a β-diketiminate system, we used the bulkiest proligand reported herein, ^{iPrTrip}nacnacH 20, to prepare a magnesium hydride complex [16,93,102] for comparison to other related molecules of the type ^{RAr}nacnacH [{(^{RAr}nacnac)MgH}$_2$], where ^{RAr}nacnac = {HC{RCN(Ar)} and R is typically an alkyl group and Ar is a sterically demanding aryl substituent. Using an established protocol [103], ^{iPrTrip}nacnacH 20 was treated with MgnBu$_2$ in toluene to form the expected intermediate complex [(^{iPrTrip}nacnac)MgnBu]. After removal of all volatiles, the oily residue was taken up in n-hexane and reacted for two days at 60 °C with phenyl silane which precipitated clean [{(^{iPrTrip}nacnac)MgH}$_2$] 21 in 41% isolated yield (Scheme 6). Complex 21 could be recrystallised from hot benzene to form large colourless crystals that were structurally characterised, see Figure 4. The complex crystallised as a dimeric system, as is found for most other related complexes of the type [{(nacnac)MgH}$_2$], with a comparable steric bulk [15,93,103,104], although a few examples of the type [(nacnac)MgH] are known with a monomeric solid state structure with a terminal hydride species and three-coordinate Mg centre [18,91]. The least-square-planes of the two essentially planar β-diketiminate magnesium chelate rings are rotated by approximately 47.6° relative to each other which must be due to the to the alternating "interlocking" contact of the isopropyl groups of the two ligand units which is visualised in the space-filling model in Figure 5, showing the hydrocarbyl units of both β-diketiminates in different colours. For comparison, β-diketiminate magnesium chelate rings are approximately co-planar in [{(^{MeDip}nacnac)MgH}$_2$] [103], approximately orthogonal to each other in [{(^{tBuDip}nacnac)MgH}$_2$] [104] but show a similar rotation in [{(^{MeDIPeP}nacnac)MgH}$_2$] (DIPeP = 2,6-di(3-pentyl)phenyl) with ca. 42° [93] between metal-ligand planes to accommodate the various bulky substituents on the ligands. An analysis of the buried volume [11] for the Mg centre in 21 (V_{buried} = 53.5%) provided a similar value compared to those for known structurally characterised [{(^{RDip}nacnac)MgH}$_2$] complexes (V_{buried} = 51.1–56.3%), but hints at a more even distribution of the bulk around the metal centre when compared to those of [{(^{RDip}nacnac)MgH}$_2$] as judged from inspection the distribution in the four quadrants (see Table S2 in the Supporting Information).

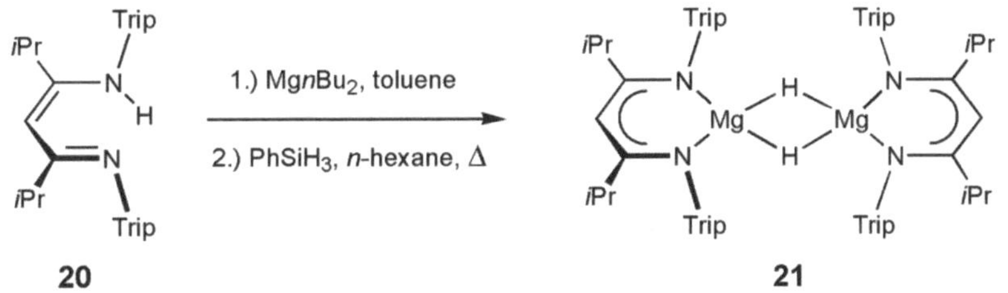

Scheme 6. Synthesis of [{(^{iPrTrip}nacnac)MgH}$_2$] 21.

In solution, [{(^{iPrTrip}nacnac)MgH}$_2$] 21 shows ^{1}H NMR resonances for a symmetric compound with a sharp singlet at δ 3.96 ppm for the magnesium hydride resonance in the expected region. The room temperature ^{1}H NMR spectrum shows one broad and two sharp septets, and one broad and three sharp doublets for the isopropyl hydrogen atoms. The broad septet and one broad doublet resonance are associated with one ortho isopropyl group including one methyl group that likely experiences the steric influence from the dimeric interlocked "geometry." The broad plus one sharp doublet merge above 60 °C and at 80 °C, three resolved septets and three doublets are observed.

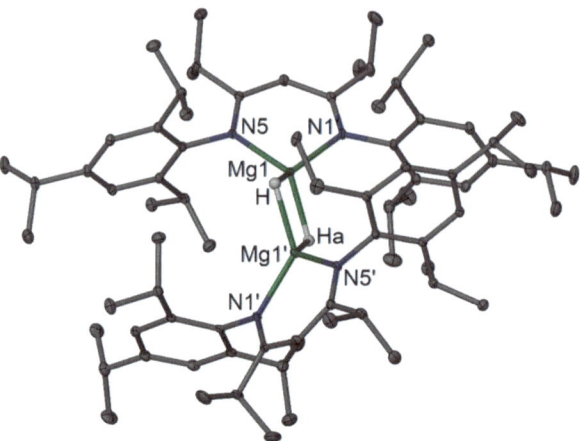

Figure 4. Molecular structures of [{(iPrTripnacnac)MgH}$_2$] **21**, 30% thermal ellipsoids. Only MgH hydrogen atom are shown. Selected bond lengths (Å) and angles (°): Mg1–N1 2.0796(12), Mg1–N5 2.0829(12), Mg1–H 2.029(13), Mg1′–H 2.029(13), Mg1–HA 1.979(14), Mg1···Mg1 2.9502(9), N1–C2 1.3341(18), C2–C3 1.3981(19), C3–C4 1.3987(19), N5–C4 1.3338(18); N1–Mg1–N5 94.40(5), C2–N1–C6 116.64, C4–N5–C27 116.73(11), N1–C2–C3 124.64(13), N5–C4–C3 124.40(13), C2–C3–C4 132.32(14).

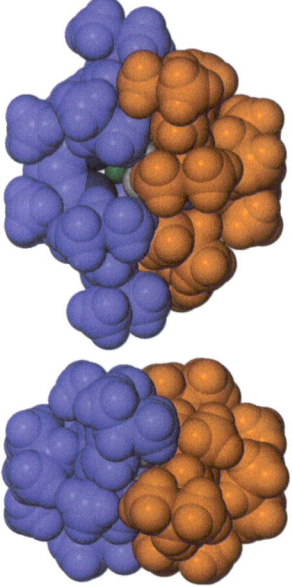

Figure 5. Space-filling model (two views) showing some central atoms (Mg green, N blue, H grey) and the hydrocarbyl groups from the two ligand units in two colours (lavender-purple, rusty orange) showing the "interlocking" isopropyl substituents.

3. Conclusions

Sterically demanding aryllithium compounds can be converted with trimethylsilyl azide to triazene-class intermediates in a one-pot reaction, installing a nitrogen atom at the aryl group and forming aniline derivatives. For two different systems, one with a bulky Trip (Scheme 3) substituent, and one with an extremely bulky terphenyl substituent

(Scheme 4) slightly different (long-lived) intermediates were observed, but the general reactivity is comparable but also influenced by the reaction kinetics resulting from the steric demand of the substituents. An aryllithium can react with trimethylsilyl azide to form silyl(aryl)triazenide lithium, which, for R = Trip, reacted with further trimethylsilyl azide and converted to TripN$_2$N(SiMe$_3$)$_2$ **8**. Upon workup with methanol and resulting in dinitrogen formation, this afforded the aniline TripNH$_2$ **6** in good yield. For the bulkier terphenyl system, the in situ generated silyl(aryl)triazenide lithium species did not react with further trimethylsilyl azide. Instead, this lithiated species formed only slowly enough that it (slowly) lost dinitrogen in parallel and predominantly formed TerN(SiMe$_3$)H **16** in a one-pot synthesis after treatment with methanol. This silylated aniline derivative is a promising sterically demanding amide proligand in its own right but can also be readily desilylated with acids to afford TerNH$_2$ **17**. This work shows the main types of intermediates for two systems with highly different steric demands, and gives practical considerations around their synthesis, optimisation, and workup, with the view that the protocol can be easily extended to other substituents and organometallic species for the synthesis of anilines, primary amines, and silylated derivatives. This includes optimised conditions for metal-halogen exchange with n-butyllithium that supress side reactions but allow conversions at relatively high temperatures. Furthermore, TripNH$_2$ **6** has been further converted in acid-promoted one-pot condensation reactions to the sterically demanding proligands ^{MeTrip}nacnacH **18**, ^{EtTrip}nacnacH **19**, and ^{iPrTrip}nacnacH **20**, the latter of which was converted to the magnesium hydride complex [{(^{iPrTrip}nacnac)MgH}$_2$] **21**. The significant effects of the steric influence and dispersion forces of the 16 isopropyl groups in **21** will likely impact compound properties and future reactivity.

4. Materials and Methods

4.1. Experimental Details

All manipulations were carried out using standard Schlenk and glove box techniques under an atmosphere of high purity argon or dinitrogen. Tetrahydrofuran, diethyl ether, toluene and n-hexane were either dried and distilled under inert gas over LiAlH$_4$ or taken from an MBraun solvent purification system and degassed prior to use. ^{1}H, ^{7}Li, ^{13}C{^{1}H} and ^{1}H-^{15}N HMBC NMR spectra were recorded on a Bruker AVII 400, Bruker AVIII 500, Bruker AVIII-HD 500 or Bruker AVIII-HD 700 spectrometer (Bruker, Billerica, MA, USA) in deuterated benzene or chloroform and were referenced to the residual ^{1}H or ^{13}C{^{1}H} resonances of the solvent used, external LiCl in D$_2$O, or external liquid NH$_3$, respectively. Abbreviations: s = singlet, d = doublet, t = triplet, q = quartet, quint = quintet, sext = sextet, sept = septet, br = broad, m = multiplet, e.g., brs, broad singlet; ad denotes an apparent doublet. The IR spectrum was recorded on a neat solid using a Shimadzu IRAfinity 1S IR spectrometer. Melting points were determined in air and are uncorrected. 1-bromo-2,4,6-triisopropylbenzene (TripBr **2**) was degassed and stored over molecular sieves under inert atmosphere. Trimethylsilyl azide (azidotrimethylsilane) and phenylsilane were degassed and stored under inert atmosphere. TerI **12** and TerLi(OEt$_2$) **11**(OEt$_2$) were prepared according to the literature [89]. All other compounds were used as received from chemical suppliers.

CAUTION! Azides are potentially explosive and highly toxic substances, and all manipulations must be carried out by trained workers. Trimethylsilyl azide is considered relatively stable but must not be mixed with acids or water and certain other substances. The aqueous layer during the workup steps below will, or can, contain lithium azide, should be treated accordingly (e.g., collect as a dilute alkaline aqueous solution), and should not be mixed with other waste. A reaction between ionic azides and halogenated reagents and solvents, such as dichloromethane, must be avoided to prevent the formation of explosive azides. The hazards and risks of procedures are dependent on scale. Use suitable gloves when working with azides. Even though no issues were encountered during the work, the use of a blast shield is strongly suggested [53,105,106].

4.2. Syntheses and Formation of Trip Compounds (1–6, 9, 10)

4.2.1. General Procedure for the Optimisation of the Lithium-Bromide Exchange

A Schlenk flask was charged with TripBr **2** (1-bromo-2,4,6-triisopropylbenzene; 0.50 mL, 1.97 mmol), *n*-hexane and the amount of THF or diethyl ether to afford the desired solvent ratio as shown in Table 1. The amount of solvent used for each reaction was determined by calculating the total amount of donor solvent and *n*-hexane (accounting for hexanes from *n*BuLi solution) needed to make a 0.395 M solution of TripBr **2** with the desired donor solvent to Li centre ratio. For an example involving 5 eq. of THF per Li see Section 4.2.2. The reaction mixture was brought to the required temperature and *n*BuLi (1.30 mL, 1.6 M solution in hexanes, 1.05 eq.) was added dropwise. The reaction mixture was stirred for 30 min and quenched with H_2O (ca. 5 mL). All volatiles were removed in vacuo, dichloromethane (ca. 10 mL) and water (ca. 10 mL) were added, and the organic layer was separated. All volatiles were removed in vacuo affording a crude oil that was analysed with ^1H NMR spectroscopy in $CDCl_3$. Since TripBr **2**, TripH **3** and Trip*n*Bu **4** (plus an occasional minor quantity of TripOH **5** from O_2 capture) are the only Trip-containing products observed, the conversion was calculated directly from the ratio of their aromatic protons, and their sum was assumed as 100% for a qualitative analysis of the lithiation (Table 1).

4.2.2. Synthesis of TripNH$_2$ 6

Method 1: A Schlenk flask was charged with TripBr **2** (4.00 mL, 15.79 mmol), *n*-hexane (22.2 mL) and THF (7 mL, ca. 5 THF per Li). The reaction mixture was cooled to $-20\ ^\circ C$ using a cold bath and *n*BuLi (10.8 mL, 1.6 M solution in hexanes, 1.09 eq.) was added dropwise over approximately two minutes. The reaction mixture was stirred for 30 min at $-20\ ^\circ C$ and was then placed into a room temperature water bath. After a period of 1–2 min, Me_3SiN_3 (4.80 mL, 36.5 mmol, 2.31 eq.) was added dropwise and the reaction mixture was stirred for a further 30 min. A white precipitate of LiN_3 was formed during that period. The reaction mixture was placed in an ice-water bath ($0\ ^\circ C$), the stopper of the flask was removed under a gentle flow of inert gas, and subsequently, methanol (10 mL) was added slowly added over a period of 5 min under vigorous gas evolution. Stirring of the mixture was continued until all gas evolution ceased, after which all volatiles were removed in vacuo and the crude product was redissolved in methanol (10 mL) to ensure complete conversion to aniline **6**. After standing for 10 min, all volatiles were removed in vacuo, and DCM (ca. 30 mL) and water (ca. 30 mL) were added, and organic layer was separated. [*Note:* Be aware that the aqueous layer contains dissolved LiN_3. Ensure that Me_3SiN_3 has reacted and is largely consumed in the procedure; if unsure, use an alkaline solution for carefully quenching of the reaction mixture to ensure no significant quantities of HN_3, are produced.] Removal of all volatiles in vacuo afforded crude TripNH$_2$ **6** as a yellow-orange oil which may be sufficiently pure for some applications.

Purification: Various methods of purification could be used (e.g., column chromatography using alumina (90), eluent: petroleum ether (40–60 $^\circ C$) followed by dichloromethane), however, a batch scale purification method via conversion to an anilinium chloride, TripNH$_3^+$Cl$^-$ **10** was found to work best. Conc. aq. HCl (ca. 37%, 12 mL) and petroleum ether (40–60 $^\circ C$ fraction, 10 mL) were added, which produced an off-white precipitate of TripNH$_3^+$Cl$^-$ **10** as a hydrate. The resulting suspension was stirred for 10 min, after which the solid (powder) was isolated by filtration and washed with petroleum ether (10 mL). [*Note:* A small and varying quantity of TripN(SiMe$_3$)$_2$ **9** was obtained in crystalline form by evaporation of the petroleum ether filtrate.] The TripNH$_3^+$Cl$^-$ **10** hydrate solid was redissolved in dichloromethane (40 mL) and (saturated) aqueous Na_2CO_3 (50 mL) was added, and the resulting mixture was stirred for 30 min. The organic layer was separated, the solvent removed in vacuo, and the product was dried under vacuum and afforded TripNH$_2$ **6** of sufficient purity. Yield: 2.98 g (86%).

Method 2: A Schlenk flask was charged with lithium granules (280 mg, 0.5% sodium, 4–10 mesh, 40.3 mmol, 2.55 eq.), CBr_4 for activation (ca. 2–3 mg), diethyl ether (ca. 30 mL) and equipped with a reflux condenser and a nitrogen inlet. A solution of TripBr **2** (4.00 mL, 15.8 mmol) in diethyl ether (ca. 8 mL) was added in portions over 30 min to a stirring suspension of lithium granules which initiated an exothermic reaction. After the addition has been completed, the resulting mixture was refluxed for 60 min, cooled to room temperature, allowed to settle, and filtered. To the filtrate, Me$_3$SiN$_3$ (4.80 mL, 36.5 mmol, 2.31 eq.) was added dropwise at room temperature and the reaction mixture was stirred for a further 30 min. Workup with methanol and purification were carried out as described above in *Method 1*. Yield: 2.82 g (81%). ^{1}H NMR (499.9 MHz, CDCl$_3$, 298 K) δ = 1.29 (d, J_{HH} = 7.0 Hz, 6H, Ar-*p*-CH(C*H*$_3$)$_2$), 1.33 (d, J_{HH} = 6.9 Hz, 12H, Ar-*o*-CH(C*H*$_3$)$_2$), 2.88 (sept, J_{HH} = 6.9 Hz, 1H, Ar-*p*-C*H*(CH$_3$)$_2$), 2.99 (sept, J_{HH} = 6.8 Hz, 2H, Ar-*o*-C*H*(CH$_3$)$_2$), 3.66 (brs, 2H, N*H*$_2$), 6.96 (s, 2H, Ar-*H*).

4.2.3. Data for LiN$_3$

The compound is formed during the synthesis of TripNH$_2$ **6** and could be isolated directly from the reaction mixture by filtration. The solid also showed a positive flame test for Br suggesting minor LiBr contamination from the reaction mixture. ^{7}Li NMR (155.5 MHz, D$_2$O, 294 K) δ = −0.10 (s, *Li*N$_3$). IR (ATR), $v\sim$/cm^{-1}: 2127 (s).

4.2.4. Data for TripN$_2$N(SiMe$_3$)$_2$ **8**

Solid TripLi **1** was obtained by performing a lithiation of TripBr as descried for the synthesis of **6** above, via *Method 2*, and storing the concentrated diethyl ether solution at −40 °C. After isolation and briefly drying under vacuum, the obtained crystalline material showed approximately 45% TripLi by weight (with the rest assumed to be ^{1}H NMR-silent LiBr as judged by integration against an internal standard). TripLi (ca. 3.3 mg, 15.7 µmol) was dissolved in C$_6$D$_6$ (0.5 mL) in a J. Young's NMR tube and Me$_3$SiN$_3$ (4.6 µL, 35 µmol, ca. 2.23 eq.) was added at 20 °C. Analysis by ^{1}H NMR spectroscopy showed immediate consumption of TripLi **1** and formation of **8**. ^{1}H NMR (700.1 MHz, C$_6$D$_6$, 295 K) δ = 0.34 (s, 18H, Si(C*H*$_3$)$_3$), 1.26 (d, J_{HH} = 6.9 Hz, 6H, Ar-*p*-CH(C*H*$_3$)$_2$), 1.28 (d, J_{HH} = 7.0 Hz, 12H, Ar-*o*-CH(C*H*$_3$)$_2$), 2.83 (sept, J_{HH} = 7.0 Hz, 1H, Ar-*p*-C*H*(CH$_3$)$_2$), 3.21 (sept, J_{HH} = 6.9 Hz, 2H, Ar-*o*-C*H*(CH$_3$)$_2$), 7.15 (s, 2H, Ar-*H*). ^{13}C{^{1}H} NMR (176.0 MHz, C$_6$D$_6$, 295 K) δ = 2.0 (Si(*C*H$_3$)$_3$), 24.2 (Ar-*o*-CH(*C*H$_3$)$_2$), 24.5 (Ar-*p*-CH(*C*H$_3$)$_2$), 28.3 (Ar-*o*-*C*H(CH$_3$)$_2$), 34.8 (Ar-*p*-*C*H(CH$_3$)$_2$), 121.4 (Ar-*C*), 140.9 (Ar-*C*), 145.7 (Ar-*C*), 146.6 (Ar-*C*). ^{1}H-^{15}N HMBS NMR (700.1/70.9 MHz, C$_6$D$_6$, 295 K) δ ≈ 187 (TripN$_2$*N*(SiMe$_3$)$_2$).

4.2.5. Data for TripN(SiMe$_3$)$_2$ **9**

The compound was occasionally isolated during the purification of TripNH$_2$ **6** as described above, especially when the reaction was carried out on a multigram scale (>30 mmol). It remains unclear what factors favour its formation, as preliminary experiments with altered stoichiometry or prolonged reaction time have shown no clear pattern, although heat appears to favour dinitrogen elimination. The compound sublimes at ca. 50 °C (ca. 0.05 mbar) affording colourless crystals that were suitable for X-ray crystallographic analysis. ^{1}H NMR (700.1 MHz, CDCl$_3$, 295 K) δ = 0.06 (s, 18H, Si(C*H*$_3$)$_3$), 1.17 (d, J_{HH} = 6.9 Hz, 12H, Ar-*o*-CH(C*H*$_3$)$_2$), 1.21 (d, J_{HH} = 6.9 Hz, 6H, Ar-*p*-CH(C*H*$_3$)$_2$), 2.82 (sept, J_{HH} = 6.9 Hz, 1H, Ar-*p*-C*H*(CH$_3$)$_2$), 3.41 (sept, J_{HH} = 6.9 Hz, 2H, Ar-*o*-C*H*(CH$_3$)$_2$), 6.84 (s, 2H, Ar-*H*). ^{13}C{^{1}H} NMR (176.0 MHz, CDCl$_3$, 295 K) δ = 2.7 (Si(*C*H$_3$)$_3$), 24.3 (Ar-*p*-CH(*C*H$_3$)$_2$), 25.3 (Ar-*o*-CH(*C*H$_3$)$_2$), 27.6 (Ar-*o*-*C*H(CH$_3$)$_2$), 33.8 (Ar-*p*-*C*H(CH$_3$)$_2$), 121.4 (Ar-*C*), 140.5 (Ar-*C*), 144.1 (Ar-*C*), 146.2 (Ar-*C*). ^{1}H NMR (700.1 MHz, C$_6$D$_6$, 295 K) δ = 0.16 (s, 18H, Si(C*H*$_3$)$_3$), 1.22 (d, J_{HH} = 6.9 Hz, 6H, Ar-*p*-CH(C*H*$_3$)$_2$), 1.29 (d, J_{HH} = 6.9 Hz, 12H, Ar-*o*-CH(C*H*$_3$)$_2$), 2.79 (sept, J_{HH} = 6.9 Hz, 1H, Ar-*p*-C*H*(CH$_3$)$_2$), 3.58 (sept, J_{HH} = 6.9 Hz, 2H, Ar-*o*-C*H*(CH$_3$)$_2$), 7.05 (s, 2H, Ar-*H*). ^{1}H-^{15}N HMBS NMR (700.1/70.9 MHz, C$_6$D$_6$, 295 K) δ ≈ 43 (TripN(SiMe$_3$)$_2$).

4.2.6. Data for TripNH$_3^+$Cl$^-$·1.75 H$_2$O, 10·1.75 H$_2$O

The compound was formed during the purification of TripNH$_2$ **6** as described above, and the water content was estimated by integration of the ^{1}H NMR spectrum. ^{1}H NMR (499.9 MHz, CDCl$_3$, 298 K) δ = 1.24 (d, J_{HH} = 6.9 Hz, 6H, Ar-*p*-CH(C*H*$_3$)$_2$), 1.30 (d, J_{HH} = 6.5 Hz, 12H, Ar-*o*-CH(C*H*$_3$)$_2$), 1.68 (brs, ca. 3.5H, *H*$_2$O), 2.89 (sept, J_{HH} = 6.9 Hz, 1H, Ar-*p*-C*H*(CH$_3$)$_2$), 3.70 (sept, J_{HH} = 5.9 Hz, 2H, Ar-*o*-C*H*(CH$_3$)$_2$), 7.06 (s, 2H, Ar-*H*), 10.30 (brs, 3H, N*H*$_3$). ^{13}C{^{1}H} NMR (125.7 MHz, CDCl$_3$, 298 K) δ = 24.1 (Ar-*p*-CH(*C*H$_3$)$_2$), 24.4 (Ar-*o*-CH(*C*H$_3$)$_2$), 28.6 (Ar-*o*-*C*H(CH$_3$)$_2$), 34.3 (Ar-*p*-*C*H(CH$_3$)$_2$), 122.4 (Ar-*C*), 123.5 (Ar-*C*), 142.7 (Ar-*C*), 149.5 (Ar-*C*).

4.3. Syntheses and Formation of Ter Compounds (14–17)

4.3.1. TerN(SiMe$_3$)Li 14

TerLi(OEt$_2$) **11**(OEt$_2$) (9.5 mg, 16.9 μmol) was dissolved in C$_6$D$_6$ (0.5 mL) in a J. Young's NMR tube, after which THF (7.0 μL, 5.1 eq.) and Me$_3$SiN$_3$ (2.4 μL, 18 μmol, 1.07 eq.) were added. The sample was heated for 24 h at 60 °C, after which analysis by ^{1}H NMR spectroscopy showed complete consumption of TerLi·Et$_2$O **11**(OEt$_2$) and formation of TerN(SiMe$_3$)Li **14**. Further context is described in the main text. ^{1}H NMR (700.1 MHz, C$_6$D$_6$, 295 K) δ = −0.18 (s, 9H, Si(C*H*$_3$)$_3$), 1.24 (d, J_{HH} = 6.8 Hz, 12H, Trip-*o*-CH(C*H*$_3$)$_2$), 1.28 (d, J_{HH} = 7.0 Hz, 12H, Trip-*p*-CH(C*H*$_3$)$_2$), 1.40 (d, J_{HH} = 6.9 Hz, 12H, Trip-*o*-CH(C*H*$_3$)$_2$), 2.84 (sept, 2H, Trip-*p*-C*H*(CH$_3$)$_2$), 3.47 (m, 4H, Trip-*o*-C*H*(CH$_3$)$_2$), 6.87 (t, J_{HH} = 7.3 Hz, 1H, *p*-C$_6$*H*$_3$), 7.23 (d, J_{HH} = 7.3 Hz, 2H, *m*-C$_6$*H*$_3$), 7.24 (s, 4H, *m*-Trip). ^{7}Li NMR (155.5 MHz, C$_6$D$_6$) δ = −0.81 (N(SiMe$_3$)*Li*). ^{13}C{^{1}H} NMR (176.0 MHz, C$_6$D$_6$, 295 K) δ = 3.3 (Si(*C*H$_3$)$_3$), 23.8 (Trip-*o*-CH(*C*H$_3$)$_2$), 24.4 (Trip-*p*-CH(*C*H$_3$)$_2$), 26.4 (Trip-*o*-CH(*C*H$_3$)$_2$), 30.5 (Trip-*o*-*C*H(CH$_3$)$_2$), 34.7 (Trip-*p*-*C*H(CH$_3$)$_2$), 112.9 (Ar-*C*), 121.7 (Ar-*C*), 131.4 (Ar-*C*), 132.9 (Ar-*C*), 140.9 (Ar-*C*), 147.6 (Ar-*C*), 148.4 (Ar-*C*), 159.0 (Ar-*C*). ^{1}H-^{15}N HMBS NMR (700.1/70.9 MHz, C$_6$D$_6$, 295 K) δ ≈ 115 (TerN(SiMe$_3$)Li).

4.3.2. Partial NMR Data for TerN$_2$N(SiMe$_3$)Li 15

Compound TerN$_2$N(SiMe$_3$)Li **15** is formed as an intermediate during the reaction between TerLi(OEt$_2$) **11**(OEt$_2$) and Me$_3$SiN$_3$ in C$_6$D$_6$ with or without additional donor solvents. It was observed using ^{1}H NMR spectroscopy by tracking the apparent triplet at 6.98 ppm (*p*-C$_6$*H*$_3$), doublet at 7.12 ppm (*m*-C$_6$*H*$_3$) and a singlet at 0.03 ppm (Si(C*H*$_3$)$_3$). ^{1}H-^{15}N HMBS NMR (700.1/70.9 MHz, C$_6$D$_6$, 295 K) δ ≈ 317 (TerN$_2$*N*(SiMe$_3$)Li).

4.3.3. Partial NMR Data for TerH

TerH is formed as a by-product during the reaction between TerLi(OEt$_2$) **11**(OEt$_2$) and Me$_3$SiN$_3$ in C$_6$D$_6$ with or without the addition of donor solvents. It was observed using ^{1}H NMR spectroscopy by tracking the singlet at 7.21 ppm (*m*-Trip).

4.3.4. Synthesis of TerN(SiMe$_3$)H 16

A Schlenk flask was charged with TerI **12** (1.00 g, 1.64 mmol), *n*-hexane (ca. 25 mL) and Et$_2$O (ca. 10 mL). The reaction mixture was cooled to −50 °C using a cold bath and *n*BuLi (1.03 mL, 1.6 M solution in hexanes, ca. 1 eq.) was added dropwise. The reaction mixture was allowed to slowly warm to room temperature and stirred for an additional 2 h. All volatiles were removed in vacuo affording a white powder which was extensively dried for 1h, after which toluene (ca. 20 mL), THF (1.34 mL, 10 eq.) and Me$_3$SiN$_3$ (0.30 mL, 2.28 mmol, 1.4 eq.) were added. The resulting mixture was heated for 20 h to 60 °C, after which it was placed in an ice bath and methanol (10 mL) was added slowly over a period of 5 min, after which all volatiles were removed in vacuo. Dichloromethane (ca. 20 mL) and water (ca. 20 mL) were added, and organic layer was separated. Removal of volatiles in vacuo afforded crude product as a white solid, which was purified by recrystallisation from cold (room temperature to −40 °C) diethyl ether and afforded as two crops. Crystals suitable for X-ray crystallographic analysis were obtained upon storage of concentrated diethyl ether solution at 6 °C. Yield = 0.56 g (60%). ^{1}H NMR (700.1 MHz, C$_6$D$_6$, 295 K) δ = −0.34 (s, 9H,

Si(CH$_3$)$_3$), 1.16 (d, J_{HH} = 6.8 Hz, 12H, Trip-o-CH(CH$_3$)$_2$), 1.29 (d, J_{HH} = 7.0 Hz, 12H, Trip-p-CH(CH$_3$)$_2$), 1.38 (d, J_{HH} = 6.9 Hz, 12H, Trip-o-CH(CH$_3$)$_2$), 2.87 (sept, J_{HH} = 6.9 Hz, 2H, Trip-p-CH(CH$_3$)$_2$), 3.06 (sept, J_{HH} = 6.8 Hz, 4H, Trip-o-CH(CH$_3$)$_2$), 3.24 (s, 1H, N(SiMe$_3$)H), 6.92 (t, J_{HH} = 7.5 Hz, 1H, p-C$_6$H$_3$), 7.19 (d, J_{HH} = 7.5 Hz, 2H, m-C$_6$H$_3$), 7.23 (s, 4H, m-Trip). ^{13}C{^{1}H} NMR (176.0 MHz, C$_6$D$_6$, 295 K) δ = 0.4 (Si(CH$_3$)$_3$), 23.8 (Trip-o-CH(CH$_3$)$_2$), 24.4 (Trip-p-CH(CH$_3$)$_2$), 26.2 (Trip-o-CH(CH$_3$)$_2$), 30.9 (Trip-o-CH(CH$_3$)$_2$), 34.9 (Trip-p-CH(CH$_3$)$_2$), 119.3 (Ar-C), 121.6 (Ar-C), 131.3 (Ar-C), 131.4 (Ar-C), 135.7 (Ar-C), 144.9 (Ar-C), 148.0 (Ar-C), 149.0 (Ar-C). ^{1}H-^{15}N HMBS NMR (700.1/70.9 MHz, C$_6$D$_6$, 295 K) $\delta \approx$ 67 (TerN(SiMe$_3$)H).

4.3.5. Deprotection of TerN(SiMe$_3$)H 16: Synthesis of TerNH$_2$ 17

NMR scale experiments have shown that TerN(SiMe$_3$)H 16 could be deprotected with silica or by prolonged standing in wet CDCl$_3$ to obtain the corresponding aniline TerNH$_2$ 17, a known compound [78]. On a large scale it is more convenient to deprotect 16 using aqueous HCl. TerN(SiMe$_3$)H 16 (120 mg, 0.20 mmol) was dissolved in diethyl ether (12 mL), and conc. aq. HCl (37%, 0.5 mL) was added dropwise, and the resulting solution was stirred at room temperature for 1 h, after which aqueous (saturated) Na$_2$CO$_3$ solution (20 mL) was added, and the resulting mixture was stirred for 30 min. The organic layer was separated, and the aqueous layer was extracted with dichloromethane (ca. 15 mL). The organic fractions were combined, and the solvent was removed in vacuo, affording solid TerNH$_2$ 17 which was isolated and dried under vacuum. (Note: an additional extraction step with dichloromethane may be required if insoluble material is present, e.g., NaHCO$_3$). Yield = 90 mg (86%). ^{1}H NMR (500.1 MHz, CDCl$_3$, 295 K) δ = 1.10 (d, J_{HH} = 6.9 Hz, 12H, Trip-o-CH(CH$_3$)$_2$), 1.12 (d, J_{HH} = 6.8 Hz, 12H, Trip-o-CH(CH$_3$)$_2$), 1.30 (d, J_{HH} = 6.9 Hz, 12H, Trip-p-CH(CH$_3$)$_2$), 2.75 (sept, J_{HH} = 6.8 Hz, 4H, Trip-o-CH(CH$_3$)$_2$), 2.94 (sept, J_{HH} = 6.9 Hz, 2H, Trip-p-CH(CH$_3$)$_2$), 3.14 (brs, 2H, NH$_2$), 6.81 (t, J_{HH} = 7.4 Hz, 1H, p-C$_6$H$_3$), 6.96 (d, J_{HH} = 7.4 Hz, 2H, m-C$_6$H$_3$), 7.08 (s, 4H, m-Trip).

4.4. Syntheses of ^{RTrip}nacnacH Compounds 18–20

4.4.1. ^{MeTrip}nacnacH 18

A round bottom flask was charged with pentane-2,4-dione (1.07 mL, 10.42 mmol, 1.0 equiv), para-toluenesulfonic acid monohydrate, pTsOH·H$_2$O (2.18 g, 11.46 mmol, 1.10 eq.), TripNH$_2$ 6 (4.81 g, 21.9 mmol, 2.10 eq.) and toluene (90 mL), and equipped with a Dean-Stark trap and a reflux condenser. The flask was placed in an oil bath and the reaction mixture was heated under reflux for 24 h to remove the water. Subsequently, most of the solvent (ca. 85 mL) was distilled off via the Dean-Stark trap leaving a dark brown oily residue. The oil was taken up in saturated aqueous Na$_2$CO$_3$ solution (ca. 100 mL) and dichloromethane (ca. 100 mL) and stirred until two clear phases formed. The organic layer was separated, and all volatiles were reduced in vacuo giving a dark viscous oil. Addition of methanol (20 mL) results in almost immediate precipitation of ^{MeTrip}nacnacH 18 as an off-white solid which is subsequently isolated by filtration and washed with cold methanol (ca. 20 mL). Concentrating the supernatant solution to ca. 10 mL and storing at −40 °C afforded additional colourless crystals of 6 that were suitable for X-ray crystallographic analysis. Yield = 4.00 g (76%). M.p. 158–161 °C. ^{1}H NMR (500.1 MHz, CDCl$_3$, 295 K) δ = 1.12 (d, J_{HH} = 6.9 Hz, 12H, Ar-o-CH(CH$_3$)$_2$), 1.21 (d, J_{HH} = 6.9 Hz, 12H, Ar-o-CH(CH$_3$)$_2$), 1.25 (d, J_{HH} = 6.9 Hz, 12H, Ar-p-CH(CH$_3$)$_2$), 1.73 (s, 6H, NCCH$_3$), 2.87 (sept, J_{HH} = 6.9 Hz, 2H, Ar-p-CH(CH$_3$)$_2$), 3.09 (sept, J_{HH} = 6.9 Hz, 4H, Ar-o-CH(CH$_3$)$_2$), 4.85 (s, 1H, NC(CH$_3$)CH), 6.95 (s, 4H, Ar-H), 12.15 (s, 1H, NH). ^{1}H NMR (400.1 MHz, C$_6$D$_6$, 294 K) δ = 1.21 (d, J_{HH} = 6.9 Hz, 12H, Ar-o-CH(CH$_3$)$_2$), 1.29 (d, J_{HH} = 7.0 Hz, 12H, Ar-o-CH(CH$_3$)$_2$), 1.30 (d, J_{HH} = 6.9 Hz, 12H, Ar-p-CH(CH$_3$)$_2$), 1.71 (s, 6H, NCCH$_3$), 2.88 (sept, J_{HH} = 6.9 Hz, 2H, Ar-p-CH(CH$_3$)$_2$), 3.35 (sept, J_{HH} = 6.9 Hz, 4H, Ar-o-CH(CH$_3$)$_2$), 4.90 (s, 1H, NC(CH$_3$)CH), 7.17 (s, 4H, Ar-H), 12.61 (s, 1H, NH). ^{13}C{^{1}H} NMR (100.6 MHz, C$_6$D$_6$, 295 K): δ = 20.8 (NCCH$_3$), 23.7 (Ar-o-CH(CH$_3$)$_2$), 24.6 (Ar-o-CH(CH$_3$)$_2$ or Ar-p-CH(CH$_3$)$_2$), 24.6 (Ar-o-CH(CH$_3$)$_2$ or Ar-p-CH(CH$_3$)$_2$), 28.8 (Ar-o-CH(CH$_3$)$_2$), 34.8 (Ar-p-CH(CH$_3$)$_2$), 94.1 (NC(CH$_3$)CH), 121.4 (Ar-C), 139.2 (Ar-C), 142.7 (Ar-C), 145.9 (Ar-C), 161.7 (NCCH$_3$).

4.4.2. ^{EtTrip}nacnacH 19

A round bottom flask was charged with heptane-3,5-dione (1.45 mL, 1.37 g, 10.42 mmol), para-toluenesulfonic acid monohydrate, pTsOH·H$_2$O (2.24 g, 11.46 mmol, 1.10 eq.), TripNH$_2$ **6** (4.94 g, 21.93 mmol, 2.10 eq.) and xylene (mixture of isomers, 90 mL), and equipped with a Dean-Stark trap, a reflux condenser, and a nitrogen inlet with oil bubbler. The flask was placed in an oil bath and nitrogen gas was blown through the apparatus for a few minutes to displace most of the air. The reaction mixture was refluxed under a very slow nitrogen flow for 48 h to remove the water. Subsequently, the majority of the solvent (ca. 80 mL) was distilled off via the Dean-Stark trap leaving a dark brown oily residue. The oil was taken up in a saturated aqueous Na$_2$CO$_3$ solution (ca. 100 mL) and dichloromethane (ca. 100 mL) and stirred until two clear phases formed. The organic layer was separated, and all volatiles were reduced in vacuo giving a dark viscous oil that was dried under vacuum. Addition of methanol (30 mL) with prolonged sonication resulted in the precipitation of ^{EtTrip}nacnacH **19** as an off-white solid which was subsequently isolated by filtration and washed with cold methanol (ca. 20 mL). Colourless crystals suitable for X-ray crystallographic analysis were obtained by redissolving **19** in boiling methanol and subsequent cooling to room temperature. Yield = 3.77 g (66%). M.p. 142–145 °C. ^{1}H NMR (500.1 MHz, CDCl$_3$, 295 K) δ = 1.08 (t, J_{HH} = 7.6 Hz, 6H, NCCH$_2$CH$_3$), 1.10 (d, J_{HH} = 6.8 Hz, 12H, Ar-o-CH(CH$_3$)$_2$), 1.19 (d, J_{HH} = 6.9 Hz, 12H, Ar-o-CH(CH$_3$)$_2$), 1.25 (d, J_{HH} = 6.9 Hz, 12H, Ar-p-CH(CH$_3$)$_2$), 2.05 (q, J_{HH} = 7.6 Hz, 4H, NCCH$_2$CH$_3$), 2.87 (sept, J_{HH} = 6.9 Hz, 2H, Ar-p-CH(CH$_3$)$_2$), 3.07 (sept, J_{HH} = 6.9 Hz, 4H, Ar-o-CH(CH$_3$)$_2$), 4.91 (s, 1H, NC(CH$_2$CH$_3$)CH), 6.94 (s, 4H, Ar-H), 12.14 (s, 1H, NH). ^{1}H NMR (499.9 MHz, C$_6$D$_6$, 298 K) 0.98 (t, J_{HH} = 7.6 Hz, 6H, NCCH$_2$CH$_3$), 1.22 (d, J_{HH} = 6.8 Hz, 12H, Ar-o-CH(CH$_3$)$_2$), 1.30 (d, J_{HH} = 6.9 Hz, 12H, Ar-p-CH(CH$_3$)$_2$), 1.30 (d, J_{HH} = 6.9 Hz, 12H, Ar-o-CH(CH$_3$)$_2$), 2.13 (q, J_{HH} = 7.6 Hz, 4H, NCCH$_2$CH$_3$), 2.88 (sept, J_{HH} = 6.9 Hz, 2H, Ar-p-CH(CH$_3$)$_2$), 3.38 (sept, J_{HH} = 6.9 Hz, 4H, Ar-o-CH(CH$_3$)$_2$), 5.09 (s, 1H, NC(CH$_2$CH$_3$)CH), 7.18 (s, 4H, Ar-H), 12.64 (s, 1H, NH). ^{13}C{^{1}H} NMR (125.7 MHz, C$_6$D$_6$, 298 K): δ = 12.1 (NCCH$_2$CH$_3$), 23.6 (Ar-o-CH(CH$_3$)$_2$), 24.6 (Ar-p-CH(CH$_3$)$_2$), 25.0 (Ar-o-CH(CH$_3$)$_2$), 26.8 (NCCH$_2$CH$_3$), 28.7 (Ar-o-CH(CH$_3$)$_2$), 34.8 (Ar-p-CH(CH$_3$)$_2$), 89.1 (NC(CH$_2$CH$_3$)CH), 121.4 (Ar-C), 138.8 (Ar-C), 142.8 (Ar-C), 145.8 (Ar-C), 166.9 (NCCH$_3$).

4.4.3. ^{iPrTrip}nacnacH 20

A Schlenk flask with reflux condenser and nitrogen inlet was charged with P$_4$O$_{10}$ (10.1 g, 35.6 mmol) and hexamethyldisiloxane (25.0 mL, 117.6 mmol), and the mixture was dissolved in dry dichloromethane (25 mL). The reaction mixture was heated to reflux for 2 h under a gentle flow of nitrogen before cooling to 20 °C. All volatiles were removed in vacuo, affording a colourless, viscous syrup of PPSE. 2,6-dimethylheptane-3,5-dione (1.80 mL, 1.64 g, 10.5 mmol) and TripNH$_2$ **6** (4.70 g, 21.42 mmol, 2.04 eq.) were added to the flask under a gentle flow of nitrogen. The reaction mixture was then slowly heated to 170 °C and stirred for 48 h at this temperature. The reaction mixture was then cooled to ca. 95 °C and an aqueous NaOH solution (8.0 g in 100 mL) was carefully (exothermic reaction!) and slowly added via the top of the reflux condenser with vigorous stirring. After cooling, the formed solid residue was extracted with dichloromethane (ca. 80 mL), the organic layer was separated off, and all volatiles were removed in vacuo. Addition of methanol (25 mL) resulted in almost immediate precipitation of ^{iPrTrip}nacnacH **20** as an off-white solid which was subsequently isolated by filtration and washed with cold methanol (ca. 20 mL). Colourless crystals suitable for X-ray crystallographic analysis were obtained by storing an n-hexane solution of **20** at 6 °C for two months. Yield = 3.90 g (67%). M.p. 204–207 °C. ^{1}H NMR (500.1 MHz, CDCl$_3$, 295 K) δ = 1.08 (ad, 24H, Ar-o-CH(CH$_3$)$_2$ and NC(CH(CH$_3$)$_2$), 1.22 (d, J_{HH} = 6.9 Hz, 12H, Ar-o-CH(CH$_3$)$_2$), 1.25 (d, J_{HH} = 6.9 Hz, 12H, Ar-p-CH(CH$_3$)$_2$), 2.43 (sept, J_{HH} = 6.8 Hz, 2H, NC(CH(CH$_3$)$_2$)), 2.87 (sept, J_{HH} = 6.9 Hz, 2H, Ar-p-CH(CH$_3$)$_2$), 3.07 (sept, J_{HH} = 6.8 Hz, 4H, Ar-o-CH(CH$_3$)$_2$), 4.89 (s, 1H, NC(CH(CH$_3$)$_2$)CH), 6.93 (s, 4H, Ar-H), 11.76 (s, 1H, NH). ^{1}H NMR (499.9 MHz, C$_6$D$_6$, 298 K) δ = 1.05 (d, J_{HH} = 6.8 Hz, 12H, NC(CH(CH$_3$)$_2$)), 1.23 (d, J_{HH} = 6.8 Hz, 12H, Ar-o-CH(CH$_3$)$_2$), 1.29 (d, J_{HH} = 6.9 Hz, 12H, Ar-p-CH(CH$_3$)$_2$), 1.34 (d, J_{HH} = 6.9 Hz, 12H, Ar-o-CH(CH$_3$)$_2$), 2.60 (sept, J_{HH} = 6.8 Hz, 2H,

NC(CH(CH$_3$)$_2$)), 2.86 (sept, J_{HH} = 7.0 Hz, 2H, Ar-p-CH(CH$_3$)$_2$), 3.39 (sept, J_{HH} = 6.9 Hz, 4H, Ar-o-CH(CH$_3$)$_2$), 5.12 (s, 1H, NC(CH(CH$_3$)$_2$)CH), 7.18 (s, 4H, Ar-H), 12.36 (s, 1H, NH). ^{13}C{^{1}H} NMR (125.7 MHz, C$_6$D$_6$, 298 K): δ = 22.2 (NC(CH(CH$_3$)$_2$)), 23.6 (Ar-o-CH(CH$_3$)$_2$), 24.6 (Ar-p-CH(CH$_3$)$_2$), 25.7 (Ar-o-CH(CH$_3$)$_2$), 28.4 (Ar-o-CH(CH$_3$)$_2$), 30.4 (NC(CH(CH$_3$)$_2$)), 34.7 (Ar-p-CH(CH$_3$)$_2$), 84.0 (NC(CH(CH$_3$)$_2$)CH), 121.5 (Ar-C), 138.4 (Ar-C), 142.9 (Ar-C), 145.6 (Ar-C), 171.8 (NC(CH(CH$_3$)$_2$)).

4.5. Synthesis of [{(^{iPrTrip}nacnac)MgH}$_2$] 21

A J Youngs flask was charged with ^{iPrTrip}nacnacH **20** (1.00 g, 1.79 mmol) and toluene (ca. 20 mL). The resulting solution was cooled using an ice bath and MgnBu$_2$ (2.33 mL, 1.0 M solution in heptane, 1.30 eq.) was added. The solution was then briefly heated to 50 °C for ca. 30 min and then stirred at room temperature for 16 h. The resulting reaction mixture was reduced in vacuo, redissolved in n-hexane (ca. 20 mL), a very small quantity of an insoluble precipitate was filtered off, and phenylsilane (0.29 mL, 2.35 mmol, 1.32 eq.) was added. The resulting mixture was heated to 60 °C for 48 h producing a fine white precipitate of **21**, that was isolated by hot filtration, washed with n-hexane (ca. 10 mL) and dried under vacuum. Colourless crystals of **21** suitable for X-ray crystallographic analysis were obtained by cooling a saturated solution in benzene from 60 °C to room temperature. Yield = 0.43 g (41%). ^{1}H NMR (499.9 MHz, C$_6$D$_6$, 298 K) δ = 0.94 (d, J_{HH} = 6.7 Hz, 24H, NCCH(CH$_3$)$_2$), 1.00 (brs, 24H, Ar-o-CH(CH$_3$)$_2$), 1.34 (d, J_{HH} = 7.1 Hz, 24H, Ar-o-CH(CH$_3$)$_2$), 1.34 (d, J_{HH} = 7.0 Hz, 24H, Ar-p-CH(CH$_3$)$_2$), 2.53 (sept, J_{HH} = 6.8 Hz, 4H, NCCH(CH$_3$)$_2$), 2.89 (sept, J_{HH} = 7.1 Hz, 4H, Ar-p-CH(CH$_3$)$_2$), 3.19 (brsept, 8H, Ar-o-CH(CH$_3$)$_2$), 3.96 (s, 2H, Mg-H), 4.88 (s, 2H, NC(CH(CH$_3$)$_2$)CH), 7.08 (s, 8H, Ar-H). ^{1}H NMR (499.9 MHz, C$_6$D$_6$, 353 K) δ = 0.99 (ad, 48H, Ar-o-CH(CH$_3$)$_2$ and NCCH(CH$_3$)$_2$), 1.30 (d, J_{HH} = 6.8 Hz, 24H, Ar-o-CH(CH$_3$)$_2$), 1.33 (d, J_{HH} = 6.9 Hz, 24H, Ar-p-CH(CH$_3$)$_2$), 2.54 (sept, J_{HH} = 6.6 Hz, 4H, NCCH(CH$_3$)$_2$), 2.88 (sept, J_{HH} = 7.0 Hz, 4H, Ar-p-CH(CH$_3$)$_2$), 3.17 (sept, J_{HH} = 6.9 Hz, 8H, Ar-o-CH(CH$_3$)$_2$), 3.93 (s, 2H, Mg-H), 4.91 (s, 2H, NC(CH(CH$_3$)$_2$)CH), 7.05 (s, 8H, Ar-H). ^{13}C{^{1}H} NMR (125.7 MHz, C$_6$D$_6$, 353 K): δ = 23.3 (NC(CH(CH$_3$)$_2$) or Ar-o-CH(CH$_3$)$_2$), 24.2 (Ar-o-CH(CH$_3$)$_2$ or Ar-p-CH(CH$_3$)$_2$), 24.4 (Ar-o-CH(CH$_3$)$_2$ or Ar-p-CH(CH$_3$)$_2$), 26.3 (NC(CH(CH$_3$)$_2$) or Ar-o-CH(CH$_3$)$_2$), 28.1 (Ar-o-CH(CH$_3$)$_2$), 32.0 (NC(CH(CH$_3$)$_2$)), 34.5 (Ar-p-CH(CH$_3$)$_2$), 85.5 (NC(CH(CH$_3$)$_2$)CH), 121.9 (Ar-C), 142.5 (Ar-C), 143.2 (Ar-C), 144.8 (Ar-C), 179.9 (NC(CH(CH$_3$)$_2$)).

4.6. X-Ray Crystallographic Details

X-ray diffraction data for compounds **16**, **16**·C$_6$H$_6$, **18**·C$_7$H$_8$, **20**, **21** were collected using a Rigaku FR-X Ultrahigh Brilliance Microfocus RA generator/confocal optics with XtaLAB P200 diffractometer [Mo Kα radiation (λ = 0.71073 Å)]. Diffraction data for compounds **9** and **19** were collected using a Rigaku MM-007HF High Brilliance RA generator/confocal optics with XtaLAB P100 or P200 diffractometers [Cu Kα radiation (λ = 1.54187 Å)]. Data for all compounds analysed were collected and processed (including correction for Lorentz, polarization, and absorption) using CrysAlisPro. [107] Structures were solved by dual space (SHELXT) [108] or direct (SIR2011) [109] methods. All structures were refined by full-matrix least-squares against F^2 (SHELXL-2019/3) [110]. Non-hydrogen atoms were refined anisotropically, and hydrogen atoms were refined using a riding model except for those on nitrogen atoms in **16**, **16**·C$_6$H$_6$, **18**·C$_7$H$_8$, and **19**, which were located from the difference Fourier map and refined isotropically subject to a distance restraint. The hydride hydrogen atoms in **21** were also located from the difference Fourier map and refined isotropically without distance restraints. All calculations were performed using the Olex2 [111] interface. Selected crystallographic data are presented below. CCDC 2301678–2301684 contains the supplementary crystallographic data for this paper. These data can be obtained free of charge from The Cambridge Crystallographic Data Centre via www.ccdc.cam.ac.uk/structures.

Crystal data for TripN(SiMe$_3$)$_2$ **9**: CCDC 2301680, C$_{21}$H$_{41}$NSi, M = 363.73, colourless prism, 0.06 × 0.05 × 0.04 mm^3, orthorhombic, space group Aea2 (No. 41), a = 37.6202(5),

$b = 12.11603(15)$, $c = 15.74019(18)$ Å, $V = 7174.49(16)$ Å^3, $Z = 12$, $D_c = 1.010$ g cm^{-3}, $F_{000} = 2424$, $\mu = 1.343$ mm^{-1}, $T = 173$ K, $2\theta_{max} = 151.4°$, 66,014 reflections collected, 7404 unique ($R_{int} = 0.0494$). Final $GoF = 1.020$, $R_1 = 0.0339$, $wR_2 = 0.0863$, R indices based on 7141 reflections with $I > 2\sigma(I)$ (refinement on F^2), 359 parameters, 22 restraints. The molecule crystallised with 1.5 molecules in the asymmetric unit. The 4-isopropyl group in the half molecule is disordered and was refined with two positions for each atom using geometry restraints.

Crystal data for TerN(SiMe$_3$)H **16**: CCDC 2301681, C$_{39}$H$_{59}$NSi, $M = 569.96$, colourless prism, $0.16 \times 0.11 \times 0.04$ mm^3, monoclinic, space group $P2_1/c$ (No. 14), $a = 19.3572(3)$, $b = 9.47187(15)$, $c = 19.6361(3)$ Å, $b = 95.6547(14)°$, $V = 3582.73(10)$ Å^3, $Z = 4$, $D_c = 1.057$ g cm^{-3}, $F_{000} = 1256$, $\mu = 0.091$ mm^{-1}, $T = 173$ K, $2\theta_{max} = 58.8°$, 75,801 reflections collected, 8820 unique ($R_{int} = 0.0310$). Final $GoF = 1.035$, $R_1 = 0.0413$, $wR_2 = 0.1035$, R indices based on 7244 reflections with $I > 2\sigma(I)$ (refinement on F^2), 410 parameters, 4 restraints. The molecule crystallised with a full molecule in the asymmetric unit. One 4-isopropyl group is disordered and was refined with two positions for the atoms of the outer methyl groups using geometry restraints.

Crystal data for TerN(SiMe$_3$)H·C$_6$H$_6$ **16**·C$_6$H$_6$: CCDC 2301683, C$_{45}$H$_{65}$NSi, $M = 648.07$, colourless plate, $0.07 \times 0.06 \times 0.02$ mm^3, monoclinic, space group $P2_1/n$ (No. 14), $a = 9.1204(2)$, $b = 19.8896(4)$, $c = 22.6572(5)$ Å, $b = 92.2103(19)°$, $V = 4106.97(15)$ Å^3, $Z = 4$, $D_c = 1.048$ g cm^{-3}, $F_{000} = 1424$, $\mu = 0.086$ mm^{-1}, $T = 125$ K, $2\theta_{max} = 58.3°$, 178,448 reflections collected, 10365 unique ($R_{int} = 0.0671$). Final $GoF = 1.017$, $R_1 = 0.0490$, $wR_2 = 0.1076$, R indices based on 7382 reflections with $I > 2\sigma(I)$ (refinement on F^2), 454 parameters, 7 restraints. The molecule crystallised with a full molecule and one benzene molecule in the asymmetric unit. One 2-isopropyl group is disordered and was refined with two positions for the atoms of one methyl group and the methine-H using geometry restraints.

Crystal data for ^{MeTrip}nacnacH·C$_7$H$_8$ **18**·C$_7$H$_8$: CCDC 2301684, C$_{42}$H$_{62}$N$_2$, $M = 594.93$, colourless prism, $0.12 \times 0.04 \times 0.03$ mm^3, orthorhombic, space group $Pbca$ (No. 61), $a = 17.3978(6)$, $b = 17.1598(6)$, $c = 24.9732(8)$ Å, $V = 7455.6(4)$ Å^3, $Z = 8$, $D_c = 1.060$ g cm^{-3}, $F_{000} = 2624$, $\mu = 0.060$ mm^{-1}, $T = 120$ K, $2\theta_{max} = 58.6°$, 160,281 reflections collected, 9438 unique ($R_{int} = 0.1475$). Final $GoF = 1.022$, $R_1 = 0.0952$, $wR_2 = 0.1760$, R indices based on 5281 reflections with $I > 2\sigma(I)$ (refinement on F^2), 416 parameters, 1 restraint. The compound crystallised with one full molecule plus one toluene molecule in the asymmetric unit.

Crystal data for ^{EtTrip}nacnacH **19**: CCDC 2301678, C$_{37}$H$_{58}$N$_2$, $M = 530.85$, colourless plate, $0.12 \times 0.11 \times 0.01$ mm^3, monoclinic, space group $P2_1/c$ (No. 14), $a = 9.2583(6)$, $b = 25.5248(15)$, $c = 15.1348(8)$ Å, $b = 99.624(6)°$, $V = 3526.3(4)$ Å^3, $Z = 4$, $D_c = 1.000$ g cm^{-3}, $F_{000} = 1176$, $\mu = 0.421$ mm^{-1}, $T = 173$ K, $2\theta_{max} = 141.3°$, 30,594 reflections collected, 6182 unique ($R_{int} = 0.0831$). Final $GoF = 1.060$, $R_1 = 0.0547$, $wR_2 = 0.1336$, R indices based on 3814 reflections with $I > 2\sigma(I)$ (refinement on F^2), 390 parameters, 9 restraints. The compound crystallised with a full molecule in the asymmetric unit. One backbone ethyl group is disordered and was modelled and refined with two positions for each atom.

Crystal data for ^{iPrTrip}nacnacH **20**: CCDC 2301679, C$_{39}$H$_{62}$N$_2$, $M = 558.90$, colourless prism, $0.12 \times 0.06 \times 0.04$ mm^3, orthorhombic, space group $Ibca$ (No. 73), $a = 16.1089(8)$, $b = 16.8884(10)$, $c = 26.3413(18)$ Å, $V = 7166.2(7)$ Å^3, $Z = 8$, $D_c = 1.036$ g cm^{-3}, $F_{000} = 2480$, $\mu = 0.059$ mm^{-1}, $\lambda = 0.71073$ Å, $T = 125$ K, $2\theta_{max} = 58.3°$, 38,250 reflections collected, 4425 unique ($R_{int} = 0.0410$). Final $GoF = 1.032$, $R_1 = 0.0850$, $wR_2 = 0.1897$, R indices based on 2675 reflections with $I > 2\sigma(I)$ (refinement on F^2), 224 parameters, 34 restraints. The compound crystallised with half a molecule in the asymmetric unit. The 4-isopropyl group is disordered and was modelled with two positions for each atom using geometry restraints.

Crystal data for [{(^{iPrTrip}nacnac)MgH}$_2$] **21**: CCDC 2301682, C$_{39}$H$_{62}$N$_2$Mg, $M = 1166.42$, colourless block, $0.09 \times 0.09 \times 0.03$ mm^3, orthorhombic, space group $Pbcn$ (No. 60), $a = 16.0213(5)$, $b = 17.4646(5)$, $c = 25.6440(7)$ Å, $V = 7175.3(4)$ Å^3, $Z = 4$, $D_c = 1.080$ g cm^{-3}, $F_{000} = 2576$, $\mu = 0.077$ mm^{-1}, $T = 125$ K, $2\theta_{max} = 58.2°$, 77,501 reflections collected, 8645 unique ($R_{int} = 0.0521$). Final $GoF = 1.031$, $R_1 = 0.0484$, $wR_2 = 0.1125$, R indices based on 6107 reflections with $I > 2\sigma(I)$ (refinement on F^2), 399 parameters, 0 restraints. This compound crystallised with half a molecule in the asymmetric unit.

Supplementary Materials: The following supporting information can be downloaded at: https://www.mdpi.com/article/10.3390/molecules28227569/s1; Table S1: In-situ NMR scale study of reaction between TerLi(OEt2) 11(OEt2) and Me3SiN3; IR spectrum (Figure S1), NMR spectroscopy (Figure S2–S38), Buried volume information, Table S2: Buried volume for [{(RArnacnac)MgH}2] complexes for the four quadrants and in total and Figures S39–S40.

Author Contributions: N.D. performed most experiments and compound characterisations, guided the main body of work, wrote the experimental section, and contributed to the results and discussion. M.G. and C.B. carried out experiments and characterisation. A.P.M. and D.B.C. conducted the X-ray crystallographic analyses. A.S. conceived and supervised the project and wrote the main section of the manuscript with input from all authors. All authors have read and agreed to the published version of the manuscript.

Funding: We are grateful to the EPSRC DTG (EP/T518062/1, EP/R513337/1), the School of Chemistry and the University of St Andrews for support.

Institutional Review Board Statement: Not applicable.

Informed Consent Statement: Not applicable.

Data Availability Statement: X-ray crystallographic data are available via the CCDC; please see the X-ray Section 4.5. The research data (NMR spectroscopy) supporting this publication can be accessed at https://doi.org/10.17630/419a7764-7072-412a-b008-115c696fd9cd.

Conflicts of Interest: The authors declare no conflict of interest.

References

1. Tsai, Y.-C. The chemistry of univalent metal β-diketiminates. *Coord. Chem. Rev.* **2012**, *256*, 722–758. [CrossRef]
2. Bourget-Merle, L.; Lappert, M.F.; Severn, J.R. The chemistry of β-diketiminatometal complexes. *Chem. Rev.* **2002**, *102*, 3031–3065. [CrossRef] [PubMed]
3. Jones, C. Bulky guanidinates for the stabilization of low oxidation state metallacycles. *Coord. Chem. Rev.* **2010**, *254*, 1273–1289. [CrossRef]
4. Kays, D.L. Extremely bulky amide ligands in main group chemistry. *Chem. Soc. Rev.* **2016**, *45*, 1004–1018. [CrossRef]
5. Cao, C.-S.; Shi, Y.; Xu, H.; Zhao, B. Metal–metal bonded compounds with uncommon low oxidation state. *Coord. Chem. Rev.* **2018**, *365*, 122–144. [CrossRef]
6. Noor, A. Coordination Chemistry of Bulky Aminopyridinates with Main Group and Transition Metals. *Top. Curr. Chem.* **2021**, *379*, 6. [CrossRef] [PubMed]
7. Power, P.P. Main-group elements as transition metals. *Nature* **2010**, *463*, 171–177. [CrossRef]
8. Weetman, C.; Inoue, S. The road travelled: After main-group elements as transition metals. *ChemCatChem* **2018**, *10*, 4213–4228. [CrossRef]
9. Weetman, C. Main Group Multiple Bonds for Bond Activations and Catalysis. *Chem. Eur. J.* **2021**, *27*, 1941–1954. [CrossRef]
10. Forster, H.; Vögtle, F. Steric Interactions in Organic Chemistry: Spatial Requirements of Substituents. *Angew. Chem. Int. Ed. Engl.* **1977**, *16*, 429–441. [CrossRef]
11. Poater, A.; Cosenza, B.; Correa, A.; Giudice, S.; Ragone, F.; Scarano, V.; Cavallo, L. SambVca: A Web Application for the Calculation of the Buried Volume of N-Heterocyclic Carbene Ligands. *Eur. J. Inorg. Chem.* **2009**, *2009*, 1759–1766. [CrossRef]
12. Wagner, J.P.; Schreiner, P.R. London Dispersion in Molecular Chemistry—Reconsidering Steric Effects. *Angew. Chem. Int. Ed.* **2015**, *54*, 12274–12296. [CrossRef] [PubMed]
13. Liptrot, D.J.; Power, P.P. London dispersion forces in sterically crowded inorganic and organometallic molecules. *Nat. Rev. Chem.* **2017**, *1*, 0004. [CrossRef]
14. Mears, K.L.; Power, P.P. Beyond Steric Crowding: Dispersion Energy Donor Effects in Large Hydrocarbon Ligands. *Acc. Chem. Res.* **2022**, *55*, 1337–1348. [CrossRef]
15. Lalrempuia, R.; Kefalidis, C.E.; Bonyhady, S.J.; Schwarze, B.; Maron, L.; Stasch, A.; Jones, C. Activation of CO by Hydrogenated Magnesium(I) Dimers: Sterically Controlled Formation of Ethenediolate and Cyclopropanetriolate Complexes. *J. Am. Chem. Soc.* **2015**, *137*, 8944–8947. [CrossRef]
16. Mukherjee, D.; Okuda, J. Molecular Magnesium Hydrides. *Angew. Chem. Int. Ed.* **2018**, *57*, 1458–1473. [CrossRef]
17. Green, S.P.; Jones, C.; Stasch, A. Stable Magnesium(I) Compounds with Mg-Mg Bonds. *Science* **2007**, *318*, 1754–1757. [CrossRef]
18. Rösch, B.; Gentner, T.X.; Eyselein, J.; Langer, J.; Elsen, H.; Harder, S. Strongly reducing magnesium(0) complexes. *Nature* **2021**, *592*, 717–721. [CrossRef]

19. Jones, C. Dimeric magnesium(I) β-diketiminates: A new class of quasi-universal reducing agent. *Nat. Rev. Chem.* **2017**, *1*, 0059. [CrossRef]
20. Queen, J.D.; Lehmann, A.; Fettinger, J.C.; Tuononen, H.M.; Power, P.P. The Monomeric Alanediyl: AlAr^{iPr8} (Ar^{iPr8} = C$_6$H-2,6-(C$_6$H$_2$-2,4,6Pr$^i_3$)$_2$-3,5-Pr$^i_2$): An Organoaluminum(I) Compound with a One-Coordinate Aluminum Atom. *J. Am. Chem. Soc.* **2020**, *142*, 20554–20559. [CrossRef]
21. Rit, A.; Campos, J.; Niu, H.; Aldridge, S. A stable heavier group 14 analogue of vinylidene. *Nat. Chem.* **2016**, *8*, 1022–1026. [CrossRef] [PubMed]
22. Liu, L.; Ruiz, D.A.; Munz, D.; Bertrand, G. A Singlet Phosphinidene Stable at Room Temperature. *Chem* **2016**, *1*, 147–153. [CrossRef]
23. Wu, M.; Li, H.; Chen, W.; Wang, D.; He, Y.; Xu, L.; Ye, S.; Tan, G. A triplet stibinidene. *Chem* **2023**, *9*, 1–12. [CrossRef]
24. Pang, Y.; Nöthling, N.; Leutzsch, M.; Kang, L.; Bill, E.; van Gastel, M.; Reijerse, E.; Goddard, R.; Wagner, L.; SantaLucia, D.; et al. Synthesis and isolation of a triplet bismuthinidene with a quenched magnetic response. *Science* **2023**, *380*, 1043–1048. [CrossRef] [PubMed]
25. Pratley, C.; Fenner, S.; Murphy, J.A. Nitrogen-Centered Radicals in Functionalization of sp^2 Systems: Generation, Reactivity, and Applications in Synthesis. *Chem. Rev.* **2022**, *122*, 8181–8260. [CrossRef]
26. Romanazzi, G.; Petrelli, V.; Fiore, A.M.; Mastrorilli, P.; Dell'Anna, M.M. Metal-based Heterogeneous Catalysts for One-Pot Synthesis of Secondary Anilines from Nitroarenes and Aldehydes. *Molecules* **2021**, *26*, 1120. [CrossRef]
27. Saha, B.; De, S.; Dutta, S. Recent Advancements of Replacing Existing Aniline Production Process with Environmentally Friendly One-Pot Process: An Overview. *Crit. Rev. Environ. Sci. Techn.* **2013**, *43*, 84–120. [CrossRef]
28. Kienle, M.; Dubbaka, S.R.; Brade, K.; Knochel, P. Modern Amination Reactions. *Eur. J. Org. Chem.* **2007**, *2007*, 4166–4176. [CrossRef]
29. Cenini, S.; Gallo, E.; Caselli, A.; Ragaini, F.; Fantauzzi, S.; Piangiolino, C. Coordination chemistry of organic azides and amination reactions catalyzed by transition metal complexes. *Coord. Chem. Rev.* **2006**, *250*, 1234–1253. [CrossRef]
30. Intrieri, D.; Zardi, P.; Caselli, A.; Gallo, E. Organic azides: "Energetic reagents" for the intermolecular amination of C–H bonds. *Chem. Commun.* **2014**, *50*, 11440–11453. [CrossRef]
31. Corpet, M.; Gosmini, C. Recent Advances in Electrophilic Amination Reactions Electrophilic Amination Reactions. *Synthesis* **2014**, *46*, 2258–2271. [CrossRef]
32. Starkov, P.; Jamison, T.F.; Marek, I. Electrophilic Amination: The Case of Nitrenoids. *Chem. Eur. J.* **2015**, *21*, 5278–5300. [CrossRef]
33. Zhou, Z.; Kürti, L. Electrophilic Amination: An Update. *Synlett* **2019**, *30*, 1525–1535. [CrossRef]
34. O'Neil, L.G.; Bower, J.F. Electrophilic Aminating Agents in Total Synthesis. *Angew. Chem. Int. Ed.* **2021**, *60*, 25640–25666. [CrossRef] [PubMed]
35. Meiries, S.; Le Duc, G.; Chartoire, A.; Collado, A.; Speck, K.; Arachchige, K.S.A.; Slawin, A.M.Z.; Nolan, S.P. Large yet Flexible N-Heterocyclic Carbene Ligands for Palladium Catalysis. *Chem. Eur. J.* **2013**, *19*, 17358–17368. [CrossRef] [PubMed]
36. Savka, R.; Plenio, H. Metal Complexes of Very Bulky N,N'-Diarylimidazolylidene N-Heterocyclic Carbene (NHC) Ligands with 2,4,6-Cycloalkyl Substituents. *Eur. J. Inorg. Chem.* **2014**, *2014*, 6246–6253. [CrossRef]
37. Steele, B.R.; Georgakopoulos, S.; Micha-Screttas, M.; Screttas, C.G. Synthesis of New Sterically Hindered Anilines. *Eur. J. Org. Chem.* **2007**, *2007*, 3091–3094. [CrossRef]
38. Oleinik, I.I.; Oleinik, I.V.; Abdrakhmanov, I.B.; Ivanchev, S.S.; Tolstikov, G.A. Design of arylimine postmetallocene catalytic systems for olefin polymerization: I. Synthesis of substituted 2-cycloalkyl- and 2,6-dicycloalkylanilines. *Russ. J. Gen. Chem.* **2004**, *74*, 1423–1427. [CrossRef]
39. Atwater, B.; Chandrasoma, N.; Mitchell, D.; Rodriguez, M.J.; Pompeo, M.; Froeze, R.D.J.; Organ, M.G. The Selective Cross-Coupling of Secondary Alkyl Zinc Reagents to Five-Membered-Ring Heterocycles Using Pd-PEPPSI-IHept^{Cl}. *Angew. Chem. Int. Ed.* **2015**, *54*, 9502–9506. [CrossRef]
40. Bartlett, R.A.; Dias, H.V.R.; Power, P.P. Isolation and X-ray crystal structures of the organolithium etherate complexes, [Li(Et$_2$O)$_2$(CPh$_3$)] and [{Li(Et$_2$O)(2,4,6-(CHMe$_2$)$_3$C$_6$H$_2$)}$_2$]. *J. Organomet. Chem.* **1988**, *341*, 1–9. [CrossRef]
41. Ruhlandt-Senge, K.; Ellison, J.J.; Wehmschulte, R.J.; Pauer, F.; Power, P.P. Isolation and Structural Characterization of Unsolvated Lithium Aryls. *J. Am. Chem. Soc.* **1993**, *115*, 11353–11357. [CrossRef]
42. Reich, H.J. Role of Organolithium Aggregates and Mixed Aggregates in Organolithium Mechanisms. *Chem. Rev.* **2013**, *113*, 7130–7178. [CrossRef] [PubMed]
43. Bailey, W.F.; Patricia, J.J. The Mechanism Of The Lithium–Halogen Interchange Reaction: A Review Of The Literature. *J. Organomet. Chem.* **1988**, *352*, 1–46. [CrossRef]
44. Applequist, D.E.; O'Brien, D.F. Equilibria in Halogen-Lithium Interconversions. *J. Am. Chem. Soc.* **1962**, *85*, 743–748. [CrossRef]
45. Jedlicka, B.; Crabtree, R.H.; Siegbahn, P.E.M. Origin of Solvent Acceleration in Organolithium Metal-Halogen Exchange Reactions. *Organometallics* **1997**, *16*, 6021–6023. [CrossRef]
46. Tai, O.; Hopson, R.; Williard, P.G. Aggregation and Solvation of n-Butyllithium. *Org. Lett.* **2017**, *19*, 3966–3969. [CrossRef]
47. Bailey, W.F.; Luderer, M.R.; Jordan, K.P. Effect of Solvent on the Lithium-Bromine Exchange of Aryl Bromides: Reactions of n-Butyllithium and tert-Butyllithium with 1-Bromo-4-tert-butylbenzene at 0 °C. *J. Org. Chem.* **2006**, *71*, 2825–2828. [CrossRef]
48. Gorecka-Kobylinska, J.; Schlosser, M. Relative Basicities of ortho-, meta-, and para-Substituted Aryllithiums. *J. Org. Chem.* **2009**, *74*, 222–229. [CrossRef]

49. Shi, L.; Chu, Y.; Knochel, P.; Mayr, H. Kinetics of Bromine-Magnesium Exchange Reactions in Substituted Bromobenzenes. *J. Org. Chem.* **2009**, *74*, 2760–2764. [CrossRef]
50. Merrill, R.E.; Negishi, E.-I. Tetrahydrofuran-Promoted Aryl-Alkyl Coupling Involving Organolithium Reagents. *J. Org. Chem.* **1974**, *39*, 3452–3453. [CrossRef]
51. Maercker, A. Ether Cleavage with Organo-Alkali-Metal Compounds and Alkali Metals. *Angew. Chem. Int. Ed. Engl.* **1987**, *26*, 972–989. [CrossRef]
52. McGarrity, J.F.; Ogle, C.A.; Brich, Z.; Loosli, H.-R. A Rapid-Injection NMR Study of the Reactivity of Butyllithium Aggregates in Tetrahydrofuran. *J. Am. Chem. Soc.* **1985**, *107*, 1810–1815. [CrossRef]
53. Bräse, S.; Gil, C.; Knepper, K.; Zimmermann, V. Organic Azides: An Exploding Diversity of a Unique Class of Compounds. *Angew. Chem. Int. Ed.* **2005**, *44*, 5188–5240. [CrossRef] [PubMed]
54. Chiba, S. Application of Organic Azides for the Synthesis of Nitrogen-Containing Molecules. *Synlett* **2012**, *1*, 21–44. [CrossRef]
55. Xie, S.; Sundhoro, M.; Houk, K.N.; Yan, M. Electrophilic Azides for Materials Synthesis and Chemical Biology. *Acc. Chem. Res.* **2020**, *53*, 937–948. [CrossRef] [PubMed]
56. Zhu, L.; Kinjo, R. Reactions of main group compounds with azides forming organic nitrogen-containing species. *Chem. Soc. Rev.* **2023**, *52*, 5563–5606. [CrossRef] [PubMed]
57. Kimball, D.B.; Haley, M.M. Triazenes: A Versatile Tool in Organic Synthesis. *Angew. Chem. Int. Ed.* **2002**, *41*, 3338–3351. [CrossRef]
58. Hauber, S.-O.; Lissner, F.; Deacon, G.B.; Niemeyer, M. Stabilization of Aryl—Calcium, —Strontium, and—Barium Compounds by Designed Steric and π-Bonding Encapsulation. *Angew. Chem. Int. Ed.* **2005**, *44*, 5871–5875. [CrossRef]
59. McKay, A.I.; Cole, M.L. Structural diversity in a homologous series of donor free alkali metal complexes bearing a sterically demanding triazenide. *Dalton Trans.* **2019**, *48*, 2948–2952. [CrossRef]
60. Wiberg, N.; Nerudu, B. Darstellung, Eigenschaften und Struktur einiger Silylazide. *Chem. Ber.* **1966**, *99*, 740–749. [CrossRef]
61. Wiberg, N.; Joo, W.-C. Zur Umsetzung von Silylaziden mit Grignard-Verbindungen. *J. Organomet. Chem.* **1970**, *22*, 333–340. [CrossRef]
62. Wiberg, N.; Joo, W.-C.; Olbert, P. Zum Mechanismus der Umsetzung von Silylaziden mit Grignard-Verbindungen (Zur Frage der Reaktivität von Grignard-Verbindungen). *J. Organomet. Chem.* **1970**, *22*, 341–348. [CrossRef]
63. Wiberg, N.; Joo, W.-C. Zum Mechanismus der Thermolyse von Silylazid-Grignard-Addukten (Zur Frage der Reaktivität Der N-Diazoniumgruppe). *J. Organomet. Chem.* **1970**, *22*, 349–356. [CrossRef]
64. Wiberg, N.; Pracht, H.J. Darstellung und Eigenschaften einiger Silyltriazene. *Chem. Ber.* **1972**, *105*, 1377–1387. [CrossRef]
65. Wiberg, N.; Pracht, H.J. Zur 1.3-Wandertendenz von Silylgruppen in Silyltriazenen. *Chem. Ber.* **1972**, *105*, 1388–1391. [CrossRef]
66. Wiberg, N.; Pracht, H.J. Zur *cis-trans*-Isomerie von Silyltriazenen. *Chem. Ber.* **1972**, *105*, 1392–1398. [CrossRef]
67. Wiberg, N.; Pracht, H.J. Zur gehinderten Rotation in Silyltriazenen. *Chem. Ber.* **1972**, *105*, 1399–1402. [CrossRef]
68. Nishiyama, K.; Tanaka, N. Synthesis and reactions of trimethylsilylmethyl azide. *J. Chem. Soc. Chem. Commun.* **1983**, 1322–1323. [CrossRef]
69. Sieburth, S.M.; Somers, J.J.; O'Hare, H.K. α-Alkyl-a-aminosilanes. 1. Metalation and alkylation between silicon and nitrogen. *Tetrahedron* **1996**, *52*, 5669–5682. [CrossRef]
70. Kulyabin, P.S.; Izmer, V.V.; Goryunov, G.P.; Sharikov, M.I.; Kononovich, D.S.; Uborsky, D.V.; Canich, J.-A.M.; Voskoboynikov, A.Z. Multisubstituted C_2-symmetric ansa-metallocenes bearing nitrogen heterocycles: Influence of substituents on catalytic properties in propylene polymerization at higher temperatures. *Dalton Trans.* **2021**, *50*, 6170–6180. [CrossRef]
71. Trost, B.M.; Pearson, W.H. Azidomethyl Phenyl Sulfide. A Synthon for NH_2^+. *J. Am. Chem. Soc.* **1981**, *103*, 2483–2485. [CrossRef]
72. Trost, B.M.; Pearson, W.H. Sulfur activation of azides toward addition of organometallics. Amination of aliphatic carbanions. *J. Am. Chem. Soc.* **1983**, *105*, 1054–1056. [CrossRef]
73. Trost, B.M.; Pearson, W.H. A synthesis of the naphthalene core of streptovaricin D via A synthon of NH_2^+. *Tetrahedron Lett.* **1983**, *24*, 269–272. [CrossRef]
74. Krasovskiy, A.; Straub, B.F.; Knochel, P. Highly efficient Reagents for Br/Mg exchange. *Angew. Chem. Int. Ed.* **2005**, *45*, 159–162. [CrossRef] [PubMed]
75. Mori, S.; Aoyama, T.; Shioiri, T. New Methods and Reagents in Organic Synthesis. 60. A New Synthesis of Aromatic and Heteroaromatic Amines Using Diphenyl Phosphorazidate (DPPA). *Chem. Pharm. Bull.* **1986**, *34*, 1524–1530. [CrossRef]
76. Hassner, A.; Munger, P.; Belinka, B.A., Jr. A novel amination of aromatic and heteroaromatic compounds. *Tetrahedron Lett.* **1982**, *23*, 699–702. [CrossRef]
77. Kabalka, G.W.; Li, G. Synthesis of aromatic amines using allyl azide. *Tetrahedron Lett.* **1997**, *38*, 5777–5778. [CrossRef]
78. Twamley, B.; Hwang, C.-S.; Hardman, N.J.; Power, P.P. Sterically encumbered terphenyl substituted primary pnictanes $ArEH_2$ and their metallated derivatives ArE(H)Li (Ar = -C_6H_3-2,6-$Trip_2$; Trip = 2,4,6-triisopropylphenyl; E = N, P, As, Sb). *J. Organomet. Chem.* **2000**, *609*, 152–160. [CrossRef]
79. Wright, R.J.; Steiner, J.; Beaini, S.; Power, P.P. Synthesis of the sterically encumbering terphenyl silyl and alkyl amines HN(R)Ar^{Mes2} (R = Me and $SiMe_3$), their lithium derivatives LiN(R)Ar^{Mes2}, and the tertiary amine Me_2NAr^{Mes2}. *Inorg. Chim. Acta* **2006**, *359*, 1939–1946. [CrossRef]
80. Gavenonis, J.; Schüwer, N.; Tilley, T.D.; Boynton, J.N.; Power, P.P. 2,6-Dimesitylaniline ($H_2NC_6H_3$-2,6-Mes_2) and 2,6-bis(2,4,6-triisopropylphenyl)aniline ($H_2NC_6H_3$-2,6-$Trip_2$). In *Inorganic Syntheses*; Power, P.P., Ed.; John Wiley & Sons: Hoboken, NJ, USA, 2018; Volume 37, pp. 98–105.

81. Grubert, L.; Jacobi, D.; Buck, K.; Abraham, W.; Mügge, C.; Krause, E. Pseudorotaxanes and Rotaxanes Incorporating Cycloheptatrienyl Stations—Synthesis and Co-Conformation. *Eur. J. Org. Chem.* **2001**, *2001*, 3921–3932. [CrossRef]
82. Liu, J.-Y.; Zheng, Y.; Li, Y.-G.; Pan, L.; Li, Y.-S.; Hu, N.-H. Fe(II) and Co(II) pyridinebisimine complexes bearing different substituents on ortho- and para-position of imines: Synthesis, characterization and behavior of ethylene polymerization. *J. Organomet. Chem.* **2005**, *690*, 1233–1239. [CrossRef]
83. Sadique, A.R.; Gregory, E.A.; Brennessel, W.W.; Holland, P.L. Mechanistic Insight into N=N Cleavage by a Low-Coordinate Iron(II) Hydride Complex. *J. Am. Chem. Soc.* **2007**, *129*, 8112–8121. [CrossRef] [PubMed]
84. Hering-Junghans, C.; Schulz, A.; Thomas, M.; Villinger, A. Synthesis of mono-, di-, and triaminobismuthanes and observation of C–C coupling of aromatic systems with bismuth(III) chloride. *Dalton Trans.* **2016**, *45*, 6053–6059. [CrossRef] [PubMed]
85. Wang, T.; Hoffmann, M.; Dreuw, A.; Hasagić, E.; Hu, C.; Stein, P.M.; Witzel, S.; Shi, H.; Yang, Y.; Rudolph, M.; et al. A metal-free direct arene C−H amination. *Adv. Synth. Cat.* **2021**, *363*, 2783–2795. [CrossRef]
86. Axenrod, T.; Mangiaracina, P.; Pregosin, P.S. A ^{13}C- and ^{15}N-NMR. Study of Some 1-Aryl-3, 3-dimethyl Triazene Derivatives. *Helv. Chim. Act.* **1976**, *59*, 1655–1660. [CrossRef]
87. Ide, S.; Iwasawa, K.; Yoshino, A.; Yoshida, T.; Takahashi, K. NMR study of the lithium salts of aniline derivatives. *Magn. Reson. Chem.* **1987**, *25*, 675–679. [CrossRef]
88. Merz, K.; Bieda, R. In situ Crystallization of N(SiMe$_3$)$_3$ and As(SiMe$_3$)$_3$: Trigonal planar or pyramidal coordination of the central atoms? *Z. Kristallogr.* **2014**, *229*, 635–638. [CrossRef]
89. Barnett, B.R.; Mokhtarzadeh, C.C.; Figueroa, J.S.; Lummis, P.; Wang, S.; Queen, J.D. m-Terphenyl IODO and Lithium Reagents Featuring 2,6-bis-(2,6-diisopropylphenyl) Substitution Patterns and an m-Terphenyl Lithium Etherate Featuring the 2,6-bis-(2,4,6TRIISOPROPYLPHENYL) Substitution Pattern. In *Inorganic Syntheses*; Power, P.P., Ed.; John Wiley & Sons: Hoboken, NJ, USA, 2018; Volume 37, pp. 89–98.
90. Moore, D.R.; Cheng, M.; Lobkovsky, E.B.; Coates, G.W. Mechanism of the Alternating Copolymerization of Epoxides and CO_2 Using β-Diiminate Zinc Catalysts: Evidence for a Bimetallic Epoxide Enchainment. *J. Am. Chem. Soc.* **2003**, *125*, 11911–11924. [CrossRef]
91. Arrowsmith, M.; Maitland, B.; Kociok-Köhn, G.; Stasch, A.; Jones, C.; Hill, M.S. Mononuclear Three-Coordinate Magnesium Complexes of a Highly Sterically Encumbered β-Diketiminate Ligand. *Inorg. Chem.* **2014**, *53*, 10543–10552. [CrossRef]
92. Ma, M.-T.; Shen, X.C.; Yu, Z.J.; Yao, W.W.; Du, L.T.; Xu, L.S.B. β-Diketiminate Magnesium Complexes: Syntheses, Crystal Structure and Catalytic Hydrosilylation. *Chin. J. Inorg. Chem.* **2016**, *32*, 1857–1866. Available online: http://www.ccspublishing.org.cn/article/doi/10.11862/CJIC.2016.232 (accessed on 7 November 2023).
93. Gentner, T.X.; Rösch, B.; Ballmann, G.; Langer, J.; Elsen, H.; Harder, S. Low Valent Magnesium Chemistry with a Super Bulky β-Diketiminate Ligand. *Angew. Chem. Int. Ed.* **2019**, *58*, 607–611. [CrossRef]
94. Bakhoda, A.G.; Jiang, Q.; Badiei, Y.M.; Bertke, J.A.; Cundari, T.R.; Warren, T.H. Copper-Catalyzed C(sp^3)-H Amidation: Sterically Driven Primary and Secondary C-H Site-Selectivity. *Angew. Chem. Int. Ed.* **2019**, *58*, 3421–3425. [CrossRef] [PubMed]
95. Yuvaraj, K.; Douair, I.; Jones, D.D.L.; Maron, L.; Jones, C. Sterically controlled reductive oligomerisations of CO by activated magnesium(I) compounds: Deltate vs. ethenediolate formation. *Chem. Sci.* **2020**, *11*, 3516–3522. [CrossRef] [PubMed]
96. Rösch, B.; Gentner, T.X.; Eyselein, J.; Friedrich, A.; Langer, J.; Harder, S. Mg–Mg bond polarization induced by a superbulky β-diketiminate ligand. *Chem. Commun.* **2020**, *56*, 11402–11405. [CrossRef]
97. Jones, D.D.L.; Watts, S.; Jones, C. Synthesis and Characterization of Super Bulky β-Diketiminato Group 1 Metal Complexes. *Inorganics* **2021**, *9*, 72. [CrossRef]
98. Mindiola, D.J.; Holland, P.L.; Warren, T.H. Complexes of Bulky β-Diketiminate Ligands. In *Inorganic Syntheses*; Rauchfuss, T.B., Ed.; John Wiley & Sons: Hoboken, NJ, USA, 2010; Volume 35, pp. 1–55.
99. Burnett, S.; Bourne, C.; Slawin, A.M.Z.; van Mourik, T.; Stasch, A. Umpolung of an Aliphatic Ketone to a Magnesium Ketone-1, 2-diide Complex with Vicinal Dianionic Charge. *Angew. Chem. Int. Ed.* **2022**, *61*, e202204472. [CrossRef]
100. López, S.E.; Restrepo, J.; Salazar, J. Polyphosphoric acid trimethylsilylester: A useful reagent for organic synthesis. *J. Chem. Res.* **2007**, *2007*, 497–502. [CrossRef]
101. Yamamoto, K.; Watanabe, H. Composition of "Polyphosphoric Acid Trimethylsilyl Ester (PPSE)"and Its Use as a Condensation Reagent. *Chem. Lett.* **1982**, *11*, 1225–1228. [CrossRef]
102. Roy, M.M.D.; Omaña, A.A.; Wilson, A.S.S.; Hill, M.S.; Aldridge, S.; Rivard, E. Molecular Main Group Metal Hydrides. *Chem. Rev.* **2021**, *121*, 12784–12965. Available online: https://pubs.acs.org/doi/epdf/10.1021/acs.chemrev.1c00278 (accessed on 7 November 2023). [CrossRef]
103. Green, S.P.; Jones, C.; Stasch, A. Stable Adducts of a Dimeric Magnesium(I) Compound. *Angew. Chem. Int. Ed.* **2008**, *47*, 9079–9083. [CrossRef]
104. Bonyhady, S.J.; Jones, C.; Nembenna, S.; Stasch, A.; Edwards, A.J.; McIntyre, G.J. β-Diketiminate-Stabilized Magnesium(I) Dimers and Magnesium(II) Hydride Complexes: Synthesis, Characterization, Adduct Formation, and Reactivity Studies. *Chem. Eur. J.* **2010**, *16*, 938–955. [CrossRef] [PubMed]
105. Treitler, D.S.; Leung, S. How Dangerous Is Too Dangerous? A Perspective on Azide Chemistry. *J. Org. Chem.* **2022**, *87*, 11293–11295. [CrossRef] [PubMed]
106. Conrow, R.E.; Dean, W.D. Diazidomethane Explosion. *Org. Proc. Res. Dev.* **2008**, *12*, 1285–1286. [CrossRef]
107. *CrysAlisPro v1.171.42.94a Rigaku Oxford Diffraction*; Rigaku Corporation: Tokyo, Japan, 2023.

108. Sheldrick, G.M. SHELXT—Integrated space-group and crystal structure determination. 2. *Acta Crystallogr. Sect. A Found. Adv.* **2015**, *71*, 3–8. [CrossRef] [PubMed]
109. Burla, M.C.; Caliandro, R.; Camalli, M.; Carrozzini, B.; Cascarano, G.L.; Giacovazzo, C.; Mallamo, M.; Mazzone, A.; Polidori, G.; Spagna, R. SIR2011: A new package for crystal structure determination and refinement. *J. Appl. Crystallogr.* **2012**, *45*, 357–361. [CrossRef]
110. Sheldrick, G.M. Crystal structure refinement with SHELXL. *Acta Crystallogr. Sect. C Struct. Chem.* **2015**, *71*, 3–8. [CrossRef]
111. Dolomanov, O.V.; Bourhis, L.J.; Gildea, R.J.; Howard, J.A.K.; Puschmann, H. OLEX2: A complete structure solution, refinement and analysis program. *J. Appl. Crystallogr.* **2009**, *42*, 339–341. [CrossRef]

Disclaimer/Publisher's Note: The statements, opinions and data contained in all publications are solely those of the individual author(s) and contributor(s) and not of MDPI and/or the editor(s). MDPI and/or the editor(s) disclaim responsibility for any injury to people or property resulting from any ideas, methods, instructions or products referred to in the content.

Poly(imidazolyliden-yl)borato Complexes of Tungsten: Mapping Steric vs. Electronic Features of *Facially* Coordinating Ligands

Callum M. Inglis, Richard A. Manzano, Ryan M. Kirk, Manab Sharma, Madeleine D. Stewart, Lachlan J. Watson and Anthony F. Hill *

Research School of Chemistry, Australian National University, Canberra, ACT 2601, Australia
* Correspondence: a.hill@anu.edu.au

Abstract: A convenient synthesis of [HB(HImMe)$_3$](PF$_6$)$_2$ (ImMe = N-methylimidazolyl) is decribed. This salt serves in situ as a precursor to the tris(imidazolylidenyl)borate Li[HB(ImMe)$_3$] pro-ligand upon deprotonation with ^{n}BuLi. Reaction with [W($\equiv$CC$_6$H$_4$Me-4)(CO)$_2$(pic)$_2$(Br)] (pic = 4-picoline) affords the carbyne complex [W($\equiv$CC$_6$H$_4$Me-4)(CO)$_2${HB(ImMe)$_3$}]. Interrogation of experimental and computational data for this compound allow a ranking of familiar tripodal and facially coordinating ligands according to steric (percentage buried volume) and electronic (ν_{CO}) properties. The reaction of [W($\equiv$CC$_6$H$_4$Me-4)(CO)$_2${HB(ImMe)$_3$}] with [AuCl(SMe$_2$)] affords the heterobimetallic semi-bridging carbyne complex [WAu(μ-CC$_6$H$_4$Me-4)(CO)$_2$(Cl){HB(ImMe)$_3$}].

Keywords: organometallic compounds; tungsten; carbyne; carbene

1. Introduction

The poly(pyrazolyl)borate class of chelates developed by Trofimenko, colloquially known as 'scorpionates' [1–3], have found broad application in diverse of areas of coordination and bioinorganic and organometallic chemistry. Key features that have contributed to their widespread deployment include (i) ease of synthesis; (ii) functionalization at both the bridgehead boron and pyrazolyl rings to provide a range of steric and electronic properties; (iii) kinetic stability of the chelated cage once coordinated to a metal centre; (iv) their so-called 'octahedral enforcer' nature, whereby the topology of the cage especially favours octahedral coordination geometries; and (v) the extension of the principle to the replacement of the pyrazol-1-yl arms with a range of other heterocycles that bridge boron and the metal to which they coordinate. Amongst these, the hydrotris(3,5-dimethylpyrazol-1-yl)borate ligand (HB(pzMe$_2$)$_3$, Scheme 1) has proven to be especially useful in presenting a moderate degree of steric protection to the remaining three ligands in an octahedral metal complex.

N-heterocyclic carbenes (NHC) have emerged over the last three decades, from being rather niche ligands of fundamental interest, to highly effective supporting co-ligands for the development of robust materials and, in particular, catalysts [4–6]. Fehlhammer first demonstrated the confluence of poly(azolyl)borate and NHC chemistries with reports of the first tris(N-alkylimidazolylidenyl)borates (HB(ImR^1)$_3$, R^1 = Me, Et, ^{i}Pr; Scheme 2) [7–10], and whilst the trimethyl derivative HB(ImMe)$_3^-$ most closely resembles the topology of the Tp* scorpionate, its chemistry has been scarcely developed beyond the original Fehlhammer work. Rather, the ligand class has been functionally elaborated to include (i) sterically imposing N-subtituents (R^1 = ^{t}Bu, Cy, adamantly, mesityl and 2,6-diisopropylphenyl) [11–14], (ii) macrocyclic variants [15–21], (iii) extension to bidentate examples [5,22–34], (iv) replacement of the bridgehead borohydride with phenyl or fluoro groups [35–37] and (v) substitution of the imidazolylidene bridges by triazolylidenes or benzoimidazolylidenes [35–38].

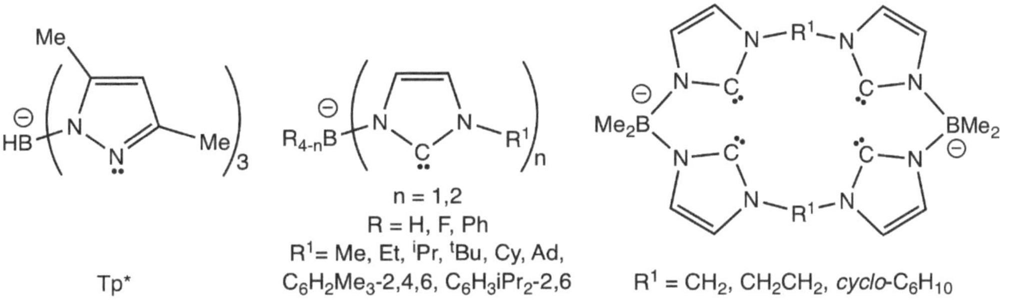

Scheme 1. Selected pyrazolyl and imidazolylidenyl borates. Tp* = hydrotris(dimethylpyrazolyl)borate.

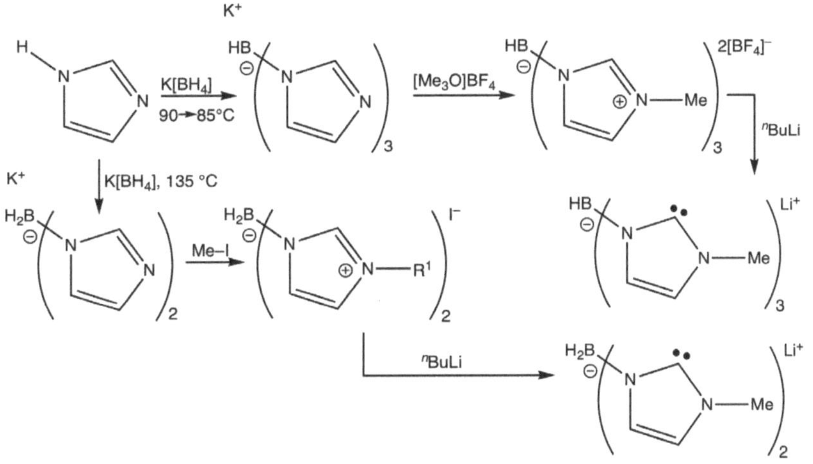

Scheme 2. Fehlhammer's syntheses of bis- and tris(N-methylimidazolylidenyl)borates [3].

Amongst the multitude of catalytic processes catalysed by NHC-supported mediators, the advent of Grubbs's second generation alkene metathesis catalyst and related analogues [39–42] has led to a plethora of complexes that feature both NHC and conventional alkylidene ligands. These serve to demonstrate the vastly different nature and reactivity of the metal–carbon 'multiple' bonds, whereby productive metathesis involves the alkylidene ligand exclusively, while the NHC ligand remains innocent. That said, an early report by Lappert described the metathesis of electron-rich alkenes by an NHC complex devoid of alkylidene ligands [43]. In contrast, alkylidyne complexes with metal–carbon triple bonds that are supported by NHC ligands are somewhat scarcer [44–57] with most examples having emerged from the groups of Esteruelas and Buchmeiser. The intersection of poly(imidazolylidenyl)borates with the chemistry of metal–carbon multiple bonds would appear limited to a single macrocyclic complex [Fe(=CPh$_2$)({Me$_2$B(C$_3$N$_2$H$_2$)$_2$C$_6$H$_{10}$}$_2$)] [21]. Given the important role that poly(pyrazolyl)borate ligands have played in the development of alkylidyne chemistry [58], herein we report the first carbyne complex ligated by a poly(imidazolylidenyl)borate, [W(≡CC$_6$H$_4$Me-4)(CO)$_2${HB(ImMe)$_3$}] (HB(ImMe)$_3$ = hydrotris(3-methylimidazoylyliden-1-yl)borate) which provides an opportunity to benchmark the donor properties of the HB(ImMe)$_3$ ligand against more familiar tripodal tridentate ligands. The complex also serves as a precursor to the first heterometallic complex of a poly(imidazolylidenyl)borate viz. [WAu(μ-CC$_6$H$_4$Me-4)Cl(CO)$_2${HB(ImMe)$_3$}].

2. Results

2.1. Pro-Ligand Synthesis

Fehlhammer's original synthetic approach (Scheme 2) [7] involved threefold alkylation of potassium hydrotris(imidazol-1-yl)borate with Meerwein's salt [Me$_3$O]BF$_4$, this latter reagent being the most expensive component. Apart from blazing the original trail, Fehlhammer's approach allows for the installation of various carbene alkyl N-substituents at a late stage on a common late synthetic intermediate.

We have developed an alternative synthesis that borrows from protocols developed for more sterically encumbered examples described by Smith [11–14]. Whilst demonstrating no new principles here, our approach does offer both convenience and economy, employing cheap commercially available reagents (Scheme 3).

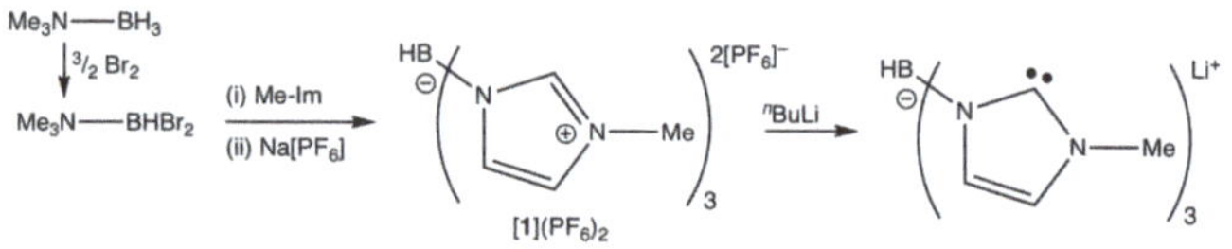

Scheme 3. Alternative syntheses of tris(N-methylimidazolylidenyl)borate salts.

The reaction of [Me$_3$N·BH$_3$] with bromine affords [Me$_3$N·BHBr$_2$] [59], which may be generated in situ without isolation. Subsequent treatment with N-methylimidazole affords the salt [HB(ImMeH)$_3$]Br$_2$ ([1]Br$_2$). This salt, whilst forming in high yields, is difficult to manipulate as it is exceedingly deliquescent and upon filtration under ambient air rapidly forms a sticky syrup. This behaviour is potentially problematic since the subsequent step calls for deprotonation via strong, moisture-sensitive bases, e.g., ^{n}BuLi or KN(SiMe$_3$)$_2$. Metathesis with aqueous Na[PF$_6$], however, results in ready recovery of the hexafluorophosphate salt [HB(ImMeH)$_3$](PF$_6$)$_2$ ([1](PF$_6$)$_2$), which is not hygroscopic and crystallizes free of water as confirmed via a crystallographic analysis (Figure 1).

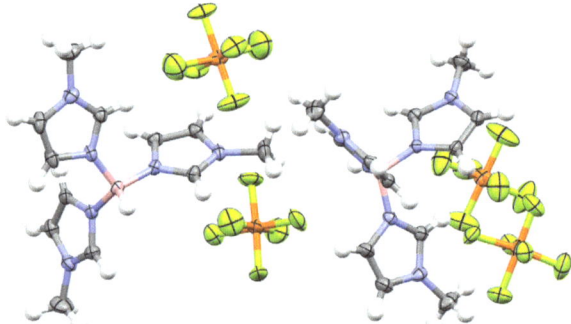

Figure 1. Structure of the hydrotris(N-methylimidazolyl)boronium salt [HB(ImMeH)$_3$](PF$_6$)$_2$ ([1](PF$_6$)$_2$ (two crystallographically independent molecules shown, 50% displacement ellipsoids, major occupancies of positionally disordered PF$_6$ anions shown).

2.2. Ligand Installation

For installation of the pro-ligand on a suitable alkylidyne precursor, the 4-toluidyne complex *trans,cis,cis*-[W(≡CC$_6$H$_4$Me-4)(CO)$_2$(pic)$_2$Br] (pic = 4-picoline) (**2a**) was chosen to exploit the lability of the bromide and 4-picoline ligands. Whilst this complex has not been previously reported on, its synthesis (Scheme 4) is unremarkable and mirrors that of the known xylyl or mesityl analogues [60–62]. Synthetic procedures are presented alongside those for the molybdenum analogue (**2b**) in the Experimental section in addition to a crystallographic analysis.

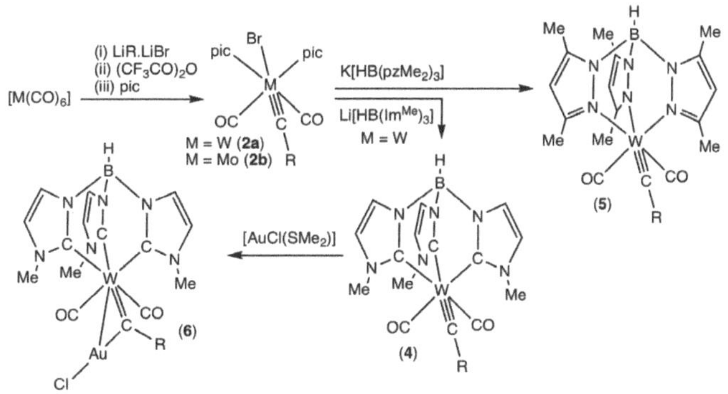

Scheme 4. Synthesis of mono- and bi-metallic toluidyne complexes ligated by the HB(Im^{Me})$_3$ ligand (R = C$_6$H$_4$Me-4, pic = 4-picoline = NC$_6$H$_4$Me-4).

The pro-ligand salt [1] (PF$_6$)$_2$ was dissolved in tetrahydrofuran and cooled (dry ice/propanone) before addition of 3 equivalents of ^{n}BuLi, followed by slow warming to room temperature to provide a yellow solution of Li[HB(ImMe)$_3$] (Li [3]) which was re-cooled and treated with **2a**. Re-warming to room temperature resulted in a colour change to dark brown as the infrared absorptions for the starting material (**2a**: ν$_{CO}$ = 1986, 1898) were replaced with those of the new product (**4**: ν$_{CO}$ = 1958, 1873 cm^{-1}). After stirring for 3 h, the product was isolated via column chromatography to yield a bright orange microcrystalline powder.

Spectroscopic data were consistent with the formulation of the desired product [W(≡CC$_6$H$_4$Me-4)(CO)$_2${HB(ImMe)$_3$}] (**4**). Amongst these, the most conspicuous datum is that for the carbyne resonance in the ^{13}C{^{1}H} NMR spectrum (CD$_2$Cl$_2$: δ$_C$ = 280.7, $^1J_{WC}$ = 171.4 Hz). Consistent with the inferred C_s symmetry of the molecule, the carbonyls gave rise to a single resonance (δ$_C$ = 223.3, $^1J_{WC}$ = 132.0 Hz) while the tungsten-bound carbon nuclei of the NHC donors gave rise to two resonances at a ratio of 2:1 with markedly different chemical shifts and $^1J_{WC}$ couplings (δ$_C$/$^1J_{WC}$ = 192.0/95.2), 181.3/44.7). With the exception of the complexes [Pt{H$_2$B(ImR^1)$_2$}$_2$] (R^1 = Me, Et) for which $^1J_{PtC}$ values were not reported [8], and [Rh(CO)(L){X$_2$B(ImR)$_2$] (L = CO, PPh$_3$, PCy$_3$; X = H, F; R = Ph, Cy) [31], poly(imidazolylidenyl)borates have not previously been coordinated to metal nuclei with usefully spin-active ($I = \frac{1}{2}$) isotopes.

As **4** is the first tungsten complex of such a ligand, it provides an opportunity to demonstrate the special feature of HB(ImR^1)$_3$ chelates cf. poly(pyrazolyl)borates; scalar couplings observed in the ^{13}C{^{1}H} NMR spectra may serve as reporters to interrogate metal–carbon bonding. Thus, whilst the chemical shift and associated coupling for the carbon nuclei trans to the carbonyl ligands are unremarkable (e.g., cf. the conventional NHC complex [W{=C(ND^{i}PP)$_2$C$_2$H$_2$}(CO)$_5$]: δ$_C$ = 187.9, $^1J_{WC}$ = 105.7 Hz, DiPP = C$_6$H$_3^i$Pr$_2$-2,6) [63], the resonance for the carbon trans to the carbyne is shifted some 11 ppm to higher field and displays a dramatically reduced coupling to tungsten-183 (44.7 Hz). These may be taken as indicating a weaker W–C interaction which in turn reflects the pronounced trans influence of the alkylidyne ligand, a feature well-documented in the structural chemistry of alkylidyne complexes ligated via poly(pyrazolyl)borate ligands [58]. As to the impact of the HB(ImMe)$_3$ ligand on the remaining co-ligands, comparison with the known complex [W(≡CC$_6$H$_4$Me-4)(CO)$_2$(Tp*)](**5**) [64] (Tp* = hydrotris(dimethylpyrazoyl)borate, prepared here from K[Tp*] and **2a**, see Experimental) is useful. The carbyne and carbonyl resonances for the Tp* derivative appeared at almost identical frequencies to those of the HB(ImMe)$_3$ complex [δ$_C$($^1J_{WC}$/Hz) = 279.2 (186.6), 224.0 (166.2)]; however, in both cases, the magnitudes of $^1J_{WC}$ values were significantly larger for **5** than for **4**. Insofar as these may be taken as being indicative of the strength of the metal–carbon interaction, it would appear that the NHC donors weaken both the carbyne and carbonyl binding. This is, however, difficult to reconcile with the ν_{CO}-associated infrared data which comprise A_1 and B_1 modes observed at 1958 and 1873 cm^{-1} in dichloromethane (ATR: 1949, 1867 cm^{-1}). These are amongst the lowest observed for neutral complexes of the form [W($\equiv$CC$_6$H$_4$Me-4)(CO)$_2$(L)] where L is one of a range of nominally tripodal facially capping ligands [58,64–67]. These values are even lower than for the π-donor ligand HB(mt)$_3$ (1967, 1875 cm^{-1}; mt = 2-mercapto-N-methyl-imidazol-1-yl) [67] and Kläui's (η^5-C$_5$H$_5$)Co(PO$_3$Me$_2$)$_3$ ligand [68]. It would therefore appear that the HB(ImMe)$_3$ ligand makes the tungsten centre especially electron rich and this may be verified using cyclic voltammetry (Figure 2). For both **4** and **5**, sweeping the voltage to ca +2 V reveals two oxidation processes, neither of which appear reversible. Limiting the sweep to ca 1.0 V indicates that the reversibility of first oxidation event increases with increasing sweep rate. For **5**, ΔE_p increases slightly with increased scan rate from 0.180 (0.1 Vs^{-1}) to 0.250 V (0.3 Vs^{-1}) suggesting the oxidation is essentially reversible with $E_\frac{1}{2}$ = 0.34 V ($E_{p,c}$ = 0.43 at 0.1 Vs^{-1}). For **4** the dependence of ΔE_p on sweep rate is more significant, increasing from 0.170 V at 0.1 Vs^{-1} ($E_{p,c}$ = 0.33 V) to 0.630 V at 5 Vs^{-1}($E_{p,c}$ = 0.64 V) is observed. Thus, fast sweep rates are required to observe a reasonable degree of reversibility, with, however, an almost identical half-wave potential ($E_\frac{1}{2}$ 0.345 V) to that of **5**. Chemical oxidation of tris(pyrazolyl)borate carbyne complexes of tungsten is typically accompanied by decarbonylation [65,69–71], which most likely accounts for the poor reversibility at slow sweep rates or higher voltages.

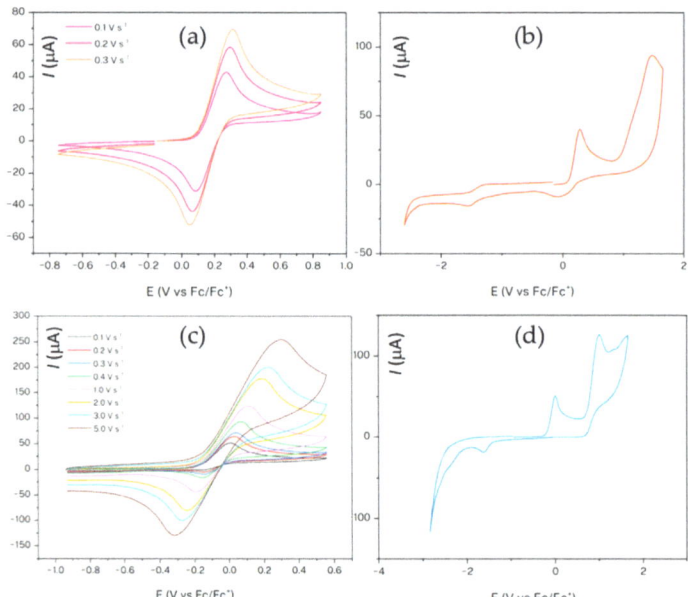

Figure 2. Cyclic voltammetry of [W($\equiv$CC$_6$H$_4$Me-4)(CO)$_2$(L)] (L = HB(pzMe$_2$)$_3$ **5**, HB(ImMe)$_3$ **4**) (Silver wire *pseudo*-reference electrode, anaerobic 1 mM in CH$_2$Cl$_2$, 0.1 M [NBu$_4$][PF$_6$] supporting electrolyte; ferrocene reference $E_{1/2}$ = 0.460 V cf. Ag/Ag$^+$ = 0). (**a**) Reversibility CV at varied scan rates of **5** (0 V → +1.0 V → –0.6 V). (**b**) Full window CV of **5** (0 V → +1.8 V → –2.5 V, υ = 0.1 Vs^{-1}). (**c**) Reversibility CV at varied scan rates of **4** (0 V → +0.9 V → –0.6 V). (**d**) Full window CV of **4** (0.6 V → +2.0 V → –2.5 V, υ = 0.1 Vs^{-1}).

2.3. Quantification of Steric and Electronic Features

A popular and time-honoured method for assessing the donor properties of ligands involves their impact on infrared frequencies of carbonyl co-ligands. This is traditionally assayed, in the case of phosphines, using the Tolman electronic parameter ν^T, viz. the frequency of the A_1 mode of CO vibrations in a host of complexes of the form [Ni(L)(CO)$_3$] [72]. Although similar scales may be developed for NHC ligands coordinated to the 'Ni(CO)$_3$' fragment [73–75], the toxicity of nickel carbonyl has led to the advent of alternative scales based on the RhCl(CO)$_2$ fragment (average of A_1 and B_1 modes) as the preferred platform, alongside metrics derived from NMR data for the NHC bound to selenium (=Se, δ_{Se}), phenylphosphinidine (=PPh, δ_P) or PdBr$_2$\{C(N^{i}Pr)$_2$C$_6$H$_4$\} (δ_C) fragments [76]. These methods are not directly applicable to H$_n$B(ImR^1)$_{4-n}$ complexes due to their negative charge and chelation. While it would be reasonable to presume that, as with conventional neutral NHC ligands, these will be potent net donors, it would be useful to be able to benchmark both the electronic and steric features of poly(imidazolylidenyl)borate ligands against those of more familiar facially capping nominally tridentate (κ^3, η^5 or η^6) ligands, of which there are many. Smith has already suggested such a ranking for a small number of facial/tripodal ligands based on the ν_{NO} stretching frequencies of complexes of the form [Ni(NO)(L)] [37]. Such ligands may be grouped according to their charge (neutral, mono- or di-anionic) which in turn impacts the charge of the derived complexes (cationic, neutral or anionic, respectively). In the case of complexes of the form [W($\equiv$CC$_6$H$_4$Me-4)(CO)$_2$(L)]$^{x+}$, a number of these have been compared in terms of the experimentally determined infrared data for the cis-dicarbonyl oscillator [67,77–89]. In addition to the frequencies of the observed symmetric and antisymmetric modes (A_1 $\nu_{s(CO)}$, B_1 $\nu_{as(CO)}$), the two numbers may be condensed into a singular Cotton–Kraihanzel force constant [90]. This is reasonable in the case of [W($\equiv$CR)(CO)$_2$(L)]$^{x+}$ because the two carbonyl ligands are chemically equivalent, i.e., both individual CO oscillators are identical. This is perhaps less appropriate in the 'RhCl(CO)$_2$' system, where in any event the simple arithmetic mean is usually employed.

Our previous collation was based on experimentally determined ν_{CO} values with the caveat that some were acquired from solid-state mesurements (Nujol mulls, KBr discs, ATR, etc.) while others were obtained from a variety of solvents. Infrared data for metal carbonyls are prone to significant perturbation in the solid state due to different crystal modifications or crystallographically independent molecules within the same crystal which in each case place the CO ligand(s) in different environments. The solvent-dependent nature of IR data for metal carbonyls, due to which both the frequency and broadening are significantly impacted by the choice of solvent, has long been recognized [91]. Thus, gas phase data, when measurement is viable, typically produce higher frequencies than are found in aliphatic hydrocarbons, and while such solvents provide the sharpest and therefore best-resolved peaks, comparatively few carbonyl complexes are sufficiently soluble. Dichloromethane has therefore become the solvent of choice offering the most accommodating solubility characteristics and reasonably narrow peaks.

To obviate these imponderables, we have collated infrared data for a range of complexes [W($\equiv$CC$_6$H$_4$Me-4)(CO)$_2$(L)]$^{x+}$ derived from computational interrogation (Table 1). Our intention is *not* to provide the most precise current state-of-the-art investigation of the intimate bonding and thermodynamic properties of such complexes but rather to derive a readily accessible and computationally economic comparative scale. A useful corollary of this approach is that the optimised geometries used for frequency calculations may be employed to directly calculate the percentage buried volume (%V_{bur}) [92,93] of each ligand L. The %V_{bur} approach to quantifying the steric impact of a ligand is especially suitable for ligands with irregular topologies, and for phosphines, such analysis reassuringly returns a correlation approximately linear with Tolman's cone angle (θ^T = 3.95x%V_{bur} + 31.5) [94]. Accordingly, a scatter plot of the Cotton–Kraihanzel force constant k_{CO} vs. %V_{bur} (Figure 3) may be presented for ligands L that is reminiscent of the familiar ν^T vs. θ^T plot used to map phosphine electronic and steric space [72]. For this purpose, with this combination of density functional, basis set and anharmonic scaling factor the value of the Cotton–Kraihanzel force constant reduces to the following equation:

$$k_{CO} \text{ [Ncm}^{-1}\text{]} = 1.7426 \times 10^{-6} \text{ Ncm} \times (\nu_s^2 + \nu_{as}^2)$$

Table 1. Experimental [a] and calculated [b] infra-red and steric [c] properties of [W($\equiv$CC$_6$H$_4$Me)(CO)$_2$(L)]$^{x+}$.

	L	x	$\nu_{(CO)}$/cm^{-1} Experimental [a]	k_{CK}/Ncm^{-1} [d]	$\nu_{(CO)}$/cm^{-1} Calculated [b]	k_{CK}/Ncm^{-1} $\lambda_1(\lambda_2)$ [i]	$\nu_{(WC)}$/cm^{-1}	%Vol$_{bur}$ [c]	Ref
1	κ^3-HB(ImMe)$_3$	0	1958, 1873	14.80	1969, 1907	15.15 (14.76)	1334	52.4	-
2	κ^3-HB(pzMe$_2$)$_3$ [g]	0	1971, 1889 [c]	15.07	1980, 1912	15.27 (14.86)	1350	50.7	[64]
3	η^5-C$_2$B$_9$H$_9$Me$_2$	1−	1956, 1874	14.82	1970, 1900	15.10 (14.71)	1354	49.6	[79]
4	κ^3-CpCo(PO$_3$Me$_2$)$_3$	0	1961, 1859	14.74	1980, 1906	15.23 (14.83)	1353	44.0	[68]
5	κ^3-HB(mt)$_3$	0	1967, 1875	14.91	1983, 1916	15.33 (14.93)	1352	48.7	[67]
6	η^5-C$_2$B$_9$H$_{11}$	1−	1965, 1880	14.93	1974, 1906	15.18 (14.77)	1356	44.3	[79]
7	κ^3-Me3[9]aneN$_3$ [e]	1+	1975, 1879 [f]	15.00	2003, 1940	15.68 (15.27)	1347	52.5	[80]
8	κ^3-HC(py)$_3$ [e]	1+	1988, 1894 [b,f]	15.22	2007, 1949	15.78 (15.37)	1346	46.2	[81]
9	κ^3-[9]aneS3 [e,h]	1+	2007, 1925 [f]	15.59	2029, 1980	16.20 (15.78)	1346	46.0	[81]
10	η^5-C$_5$H$_5$	0	1982, 1902	15.24	1997, 1941	15.64 (15.23)	1348	35.2	[82]
11	κ^3-HB(pz)$_3$ [k]	0	1986, 1903	15.28	1998, 1934	15.49 (15.11)	1347	43.3	[84]
12	η^5-C$_5$Me$_5$	0	1981, 1910 [c,j]	15.29	1989, 1933	15.51 (15.12)	1349	42.4	[86]
13	κ^3-HC(pz)$_3$	1+	1995, 1912	15.42	2016, 1959	15.93 (15.52)	1347	41.7	[87]
14	η^6-C$_2$B$_{10}$H$_{10}$Me$_2$	1−	1990, 1930	15.52	1981, 1932	15.44 (15.04)	1352	53.5	[89]
15	κ^3-P(py)$_3$ [e]	1+	2007, 1925 [f]	15.62	2008, 1951	15.80 (15.39)	1349	47.9	[81]
16	κ^3-MeC(CH$_2$Ph$_2$)$_3$ [e,g]	1+	1999, 1934 [b,f]	15.62	2095, 2037	17.01 (15.46)	n.r.	59.8	[81]
17	κ^3-HC(pzMe$_2$)$_3$	1+	–	–	2002, 1941	15.68 (15.27)	1349	49.1	–
18	κ^3-MeC(CH$_2$Pme$_2$)$_3$	1+	–	–	2021, 1974	16.09 (15.67)	1342	51.5	–
19	η^6-C$_6$H$_6$	1+	–	–	2051, 2017	16.68 (16.25)	1356	39.3	–
20	η^6-C$_6$Me$_6$	1+	–	–	2030, 1989	16.28 (15.85)	1351	45.9	–
21	η^6-C$_6$Et$_6$	1+	–	–	2019, 1975	16.08 (15.66)	1351	53.3	–
22	η^5-C$_9$H$_7$ (indenyl)	0	–	[h]	2002, 1949	15.74 (15.32)	1348	37.3	[61]
23	η^5-C$_{13}$H$_9$ (fluorenyl)	0	–	–	1999, 1941	15.65 (15.24)	1356	40.4	–
24	η^5-C$_5$Ph$_5$ [g]	0	–	–	2077, 2015	16.88 (15.34)	1532	48.3	–
25	η^5-C$_5$Cl$_5$	0	–	–	2012, 1962	15.92 (15.51)	1354	40.8	–
26	η^5-C$_5$H$_3$(SiMe$_3$)$_2$-1,3	0	–	–	1987, 1931	15.48 (15.08)	1307	54.5	–
27	η^5-C$_5$Me$_4$N	0	–	–	1992, 1937	15.56 (15.16)	1350	39.2	–
28	η^5-C$_5$Me$_4$P	0	–	–	1990, 1937	15.55 (15.15)	1348	41.7	–
29	η^5-C$_5$Me$_4$As	0	–	–	1989, 1936	15.53 (15.13)	1348	37.5	–
30	η^5-C$_5$H$_5$BH	0	–	–	2006, 1953	15.80 (15.39)	1351	40.8	–
31	κ^3-MeB(CH$_2$PPh$_2$)$_3$ [g]	0	–	–	2072, 2003	16.74 (15.22)	1519	59.8	–
32	κ^3-MeB(CH$_2$Pme$_2$)$_3$	0	–	–	1988, 1933	15.50 (15.10)	1339	51.2	–
33	κ^3-MeB(CH$_2$Sme)$_3$	0	–	–	1996, 1935	15.58 (15.17)	1348	49.7	–
34	κ^3-HB(mt^{Se})$_3$	0	–	–	1981, 1915	15.31 (14.91)	1357	49.7	–
35	κ^3-HB(ImEt)$_3$	0	–	–	1968, 1906	15.13 (14.74)	1322	54.4	–
36	κ^3-HB(Im^{i}Pr)$_3$	0	–	–	1970, 1907	15.16 (14.76)	1340	53.1	–
37	κ^3-HB(Im^{t}Bu)$_3$	0	–	–	1950, 1881	14.80 (14.41)	1334	59.5	–
38	κ^3-HB(ImPh)$_3$	0	–	–	1981, 1919	15.34 (14.94)	1329	54.3	–
39	κ^3-HB(ImCF$_3$)$_3$	0	–	–	1999, 1947	15.70 (15.29)	1337	57.1	–

[a] Unless otherwise indicated, data were determined from dichloromethane solutions. [b] DFT:ωB97X-D/6-31G*/LANL2Dζ(W)/Gas-phase, anharmonic scaling factor 0.9420. [c] Percentage buried volume calculated [92] for a 3.5 Å sphere centred on tungsten with H-atoms included. [d] Cotton–Kraihanzel force constant [90]. [e] Experimental data for benzylidyne. [f] KBr pellet. [g] Values in italics were determined at the reduced PM3tm level of theory. [h] [Mo($\equiv$CC$_6$H$_3$Me$_2$-2,6)(CO)$_2$(η^5-C$_9$H$_7$)] has ν_{CO} = 1998, 1925 cm^{-1} [61]. [i] λ_1 = 0.9420, λ_2 = 0.9297. [j] Measured in n-hexane. [k] The complex [W($\equiv$CC$_6$H$_4$Me-4)(CO)$_2$[B(pz)$_4$]] has identical ν_{CO} values to those for [W($\equiv$CC$_6$H$_4$Me-4)(CO)$_2$[HB(pz)$_3$]], i.e., replacing the remote B–H substituent with pz has negligible electronic impact. n.r. = not identified with confidence or heavily coupled with other oscillators.

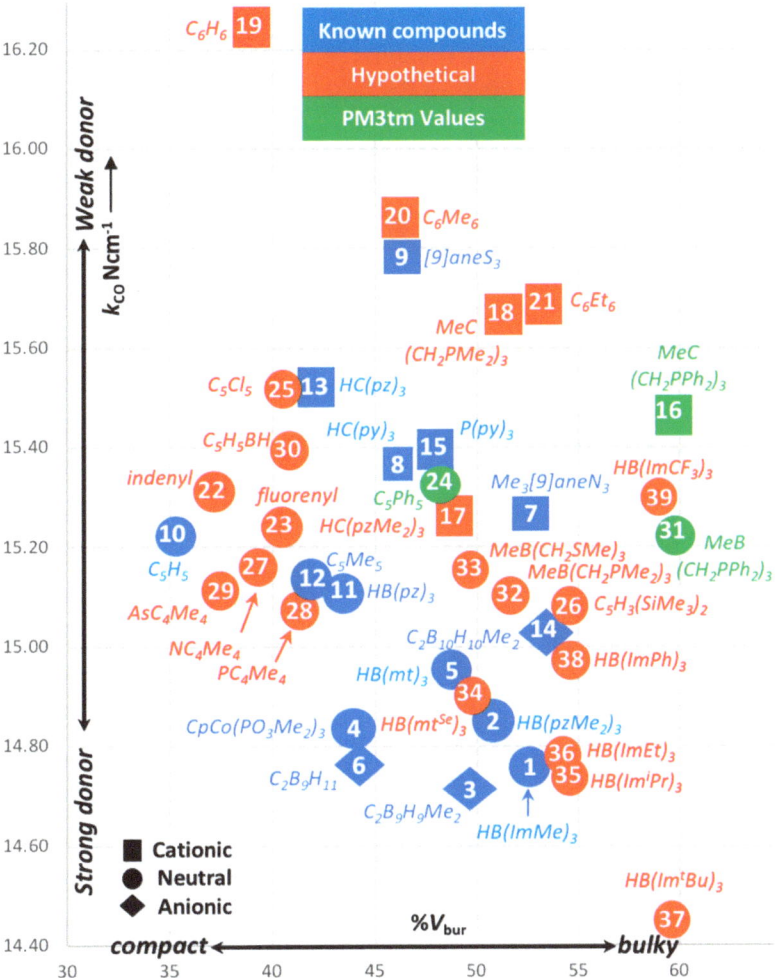

Figure 3. Electronic (k_{CO}) vs. steric (%V_{bur}) map for a range of facially coordinating ligands derived computationally (DFT:ωB97X-D/LANL2Dζ(W)). A small number (shown in green) were calculated at the semi-empirical PM3tm level of theory due to their large atom count, for which the ordinate positions should be treated with appropriate reservation.

The ωB97X-D [95,96] functional was employed with the 6-31G* basis set [97] in combination with the LANL2Dζ effective core potential for tungsten [98–100], and while much more sophisticated levels of theory are certainly available, this selection represents a balance between utility and computational economy for these medium-sized molecules. For larger ligands 'L', where steric bulk has or might be an intentional design feature, %V_{bur} values obtained at the simpler semi-empirical PM3tm level of theory are used, as we are here only concerned with molecular topologies (Figure 4). Taking complexes of the ligands HB(pzMe$_2$)$_3$, HB(ImMe)$_3$ and MeC(CH$_2$PMe$_2$)$_3$ as test cases, the variation in %V_{bur} calculated between ωB97X-D/6-31G*/LANL2Dζ and PM3tm methods was <3%, i.e., within the magnitude of molecular libration. Vibrational frequencies, whilst calculated to ensure local minima had been located, were imprecise at the PM3tm level and considered of little use. Accordingly, the ordinate location of such ligands in Figure 3 (shown in green) should be viewed with considerable caution. These were derived with little rigour by simply scaling the PM3tm k_{CO} values by 0.9089, this being the ratio of k_{CO} values calculated at the PM3tm and DFT levels of theory for **4** and **5**. That said, the peripheral inclusion of sterically obtrusive substituents in ligands often results in rather limited transmission of inductive electronic effects to the metal centre itself, as seen, for example, in experimental data for L = η^5-C_5H_5 (k_{CO} = 15.24 Nm^{-1}) and η^5-C_5Me_5 (k_{CO} = 15.29 Nm^{-1}). Similarly, experimental data are not available for toluidyne complexes of all ligands L, in which cases experimental data for the corresponding phenyl or xylyl carbynes are instead provided alongside those calculated for the toluidyne.

Table 2. Calculated (TT-DFT) [a] electronic absorptions of interest, natural atomic charges (Z) and Löwdin bond orders (LBO) for selected complexes [W(≡CC$_6$H$_4$Me)(CO)$_2$(L)]$^{x++}$.

	L	x	λ_{max}/nm $d_{xy} \to \pi^*_{W\equiv C}$	λ_{max}/nm $\pi_{W\equiv C} \to \pi^*_{W\equiv C}$	Z(W)	Z(C)	LBO (W≡C)	r(W≡C)/Å
1	κ^3-HB(ImMe)$_3$ (**4**)	0	433	332	+0.748	−0.316	2.37	1.833
2	κ^3-HB(pzMe$_2$)$_3$ (**5**)	0	406	316	+1.013	−0.268	2.40	1.811
3	η^5-C$_2$B$_9$H$_9$Me$_2$	1−	435	359	+0.831	−0.214	2.35	1.810
4	κ^3-CpCo(PO$_3$Me$_2$)$_3$	0	431	374	+1.177	−0.299	2.40	1.802
5	κ^3-HB(mt)$_3$	0	444	335	+0.685	−0.256	2.42	1.800
6	η^5-C$_2$B$_9$H$_{11}$	1−	428	358	+0.845	−0.230	2.36	1.810
7	κ^3-Me$_3$[9]aneN$_3$ [b]	1+	400	377	+0.858	−0.188	2.39	1.812
8	κ^3-HC(py)$_3^x$	1+	403 [b]	377 [b]	+0.909	−0.213	2.40	1.813
9	κ^3-[9]aneS$_3$	1+	377	330	+0.405	−0.131	2.35	1.818
10	η^5-C$_5$H$_5$	0	420	319	+0.851	−0.270	2.40	1.815
11	κ^3-HB(pz)$_3$	0	412	313	+0.979	−0.253	2.42	1.810
12	η^5-C$_5$Me$_5$	0	430	326	+0.870	−0.284	2.41	1.814
13	κ^3-HC(pz)$_3$ [b]	1+	405	337	+0.886	−0.190	2.41	1.811
14	η^6-C$_2$B$_{10}$H$_{10}$Me$_2$	1−	417	372	+0.732	−0.171	2.34	1.811
15	κ^3-P(py)$_3$ [b]	1+	384	332	+0.911	−0.214	2.40	1.809
17	κ^3-HC(pzMe$_2$)$_3$	1+	386	319	+0.920	−0.208	2.40	1.810
18	κ^3-MeC(cH$_2$PMe$_2$)$_3$	1+	390	335	+0.146	−0.130	2.33	1.830
19	η^6-C$_6$H$_6$	1+	356	381	+0.697	−0.105	2.32	1.820
20	η^6-C$_6$Me$_6$	1+	386	333	+0.754	−0.134	2.35	1.813
21	η^6-C$_6$Et$_6$	1+	379	336	+0.759	−0.130	2.33	1.814
22	η^5-C$_9$H$_7$ (indenyl)	0	415	354	+0.889	−0.269	2.45	1.802
23	η^5-C$_{13}$H$_9$ (fluorenyl)	0	436	357	+0.909	−0.237	2.45	1.798
25	η^5-C$_5$Cl$_5$	0	422	323	+0.851	−0.214	2.41	1.805
26	η^5-C$_5$H$_3$(SiMe$_3$)$_2$-1,3	0	418	318	+0.849	−0.257	2.40	1.812
27	η^5-C$_4$Me$_4$N	0	415	326	+0.939	−0.273	2.40	1.811
28	η^5-C$_4$Me$_4$P	0	391	365	+0.755	−0.247	2.37	1.816
29	η^5-C$_4$Me$_4$As	0	390	366	+0.740	−0.255	2.37	1.816
30	η^5-C$_5$H$_4$BH	0	383	323	+0.769	−0.164	2.37	1.811
32	κ^3-MeB(cH$_2$PMe$_2$)$_3$	0	412	329	+0.274	−0.213	2.37	1.825
33	κ^3-MeB(cH$_2$SMe)$_3$	0	422	323	+0.587	−0.224	2.40	1.809
34	κ^3-HB(mt^{Se})$_3$	0	443	338	+0.631	−0.263	2.42	1.800
35	κ^3-HB(ImEt)$_3$	0	424	329	+0.763	−0.324	2.35	1.835
36	κ^3-HB(Im^{i}Pr)$_3$	0	437	335	+0.749	−0.318	2.38	1.830
37	κ^3-HB(Im^{t}Bu)$_3$	0	423	322	+0.864	−0.294	2.33	1.820
38	κ^3-HB(ImPh)$_3$	0	449	335	+0.960	−0.300	2.37	1.828
39	κ^3-HB(ImCF$_3$)$_3$	0	426	329	+0.689	−0.238	2.37	1.824

[a] TD-DFT:ωB97X-D/6-31G*/LANL2Dζ(W)/gas-phase. [b] $\pi^*_{W\equiv C}$ does not correspond to the LUMO due to low-lying ligand(L)-centred virtual orbitals.

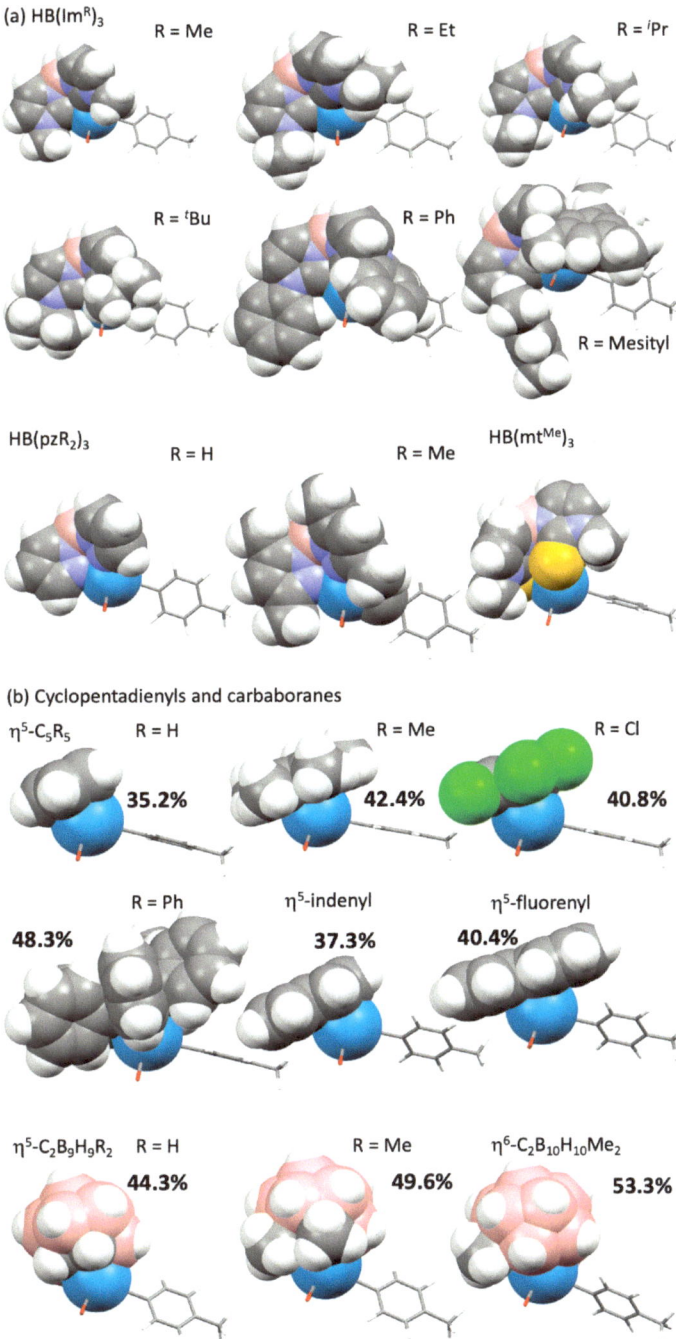

Figure 4. *Cont.*

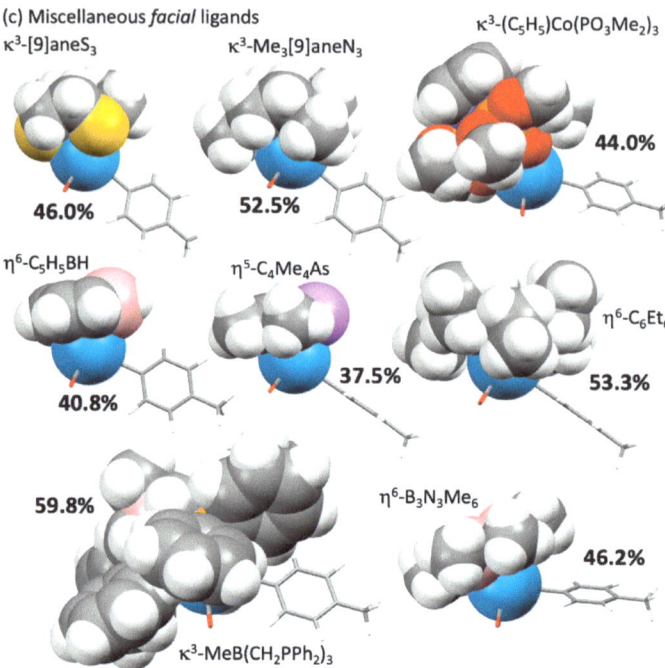

Figure 4. Corey–Pauling–Koltun representations of facial ligand from Tables 1 and 2 in the complexes [W(≡CC$_6$H$_4$Me-4)(CO)$_2$(L)] toluidyne and carbonyl ligands simplified). (**a**) Poly(azolyl)borates; (**b**) cyclopentadienyls and carbaboranes; (**c**) miscellaneous facial ligands.

A bonus of the requisite frequency calculations is that the vibrational mode for the W≡C bond may be readily identified, though in contrast to similar essentially 'pure' vibrations for terminal oxo (M≡O) and toluidyne (M≡N) ligands, this is by necessity coupled to the vibration of the C–C bond connecting it to the aryl substituent. This mode appears within a remarkably narrow frequency range (1345–1356 cm^{-1}), with the exception of **4** (1334 cm^{-1}), perhaps also reflecting the electron-releasing nature of the HB(ImMe)$_3$ ligand. The intensity of this mode, however, varies substantially, such that in some cases it is unlikely to be unambiguously identified in experimental IR spectra. This invariance in the value of ν_{WC} is also reflected in the derived Löwden bond orders (Table 2) for this bond, which fall within the very narrow range of 2.32–2.41. This is despite considerable variation in the calculated natural charge on tungsten (+0.405 to +1.177), while that for carbon is *comparatively* invariant (−0.105 to −0.299); i.e., electroneutrality would appear to balance charge distribution within the 'LW' unit so as to not significantly transmit this influence to the carbyne ligand.

Table 1 presents ν_{CO} frequencies corrected by an anharmonic scaling factor (λ_1) of 0.9740 as implemented in the *SPARTAN20*® software for the ωB97X-D/6-31G* combination [101,102], which, however, still overestimates these frequencies relative to those observed experimentally. Calculated vibrational frequencies generally exceed experimentally determined values due to incomplete incorporation of electron correlation, neglect of mechanical anharmonicity and the use of finite basis sets [103–105].

This overestimation is assumed to be relatively uniform, allowing for the development of generic scaling factors (λ) derived via least-squares analysis of calculated vs. experimental frequencies for various test sets of molecules. Such test sets typically involve small molecules comprising first and second row elements but rarely metals. Moreover, single scaling factors are not universally appropriate for the entire vibrational spectroscopy range (400–4000 cm^{-1}) [106], and the fundamental modes from which they are derived generally fall below the range of interest to organometallic chemists (1800–2200 cm^{-1}). For the present discussion, it therefore seems appropriate to consider an alternative scaling factor (λ_2 = 0.9297), which we have derived from consideration of 18 experimental and fundamental modes from Table 2, with the caveat that only data measured in dichloromethane solutions were used, discarding those from solid-state or alkane solution measurements. Gas phase data were calculated, since there seemed little benefit in introducing further artificial approximations such as conductor-like polarizable continuum, molecular electron density (SMD) or conductor-like screening models (COSMO) [107–111] when the aim was to construct an approximate but internally consistent steric–electronic map rather than to seek out absolute values.

The data points may be loosely grouped according to the charge on the complex, with the general observation that as this increased from anionic through neutral to cationic, so too did the k_{CO} value. It should, however, be noted that these groupings are not well separated. Rather, some cationic complexes are coordinated by strong net σ-donors, e.g., N,N',N''-trimethyltriazacyclononane (Me$_3$[9]aneN$_3$, Entry 7) and tris(dimethylpyrazolyl)methane (HC(pzMe$_2$)$_3$, Entry 17), such that comparatively low values are observed for v_{CO} and k_{CO}. Likewise, the icosohedral dicarbollide complexes [W(≡CC$_6$H$_4$Me-4)(CO)$_2$(η^5-C$_2$B$_9$H$_9$R$_2$)]$^-$ (R = H, Me), whilst anionic, have frequencies not dissimilar to those of neutral **4** (Entry 1) and **5** (Entry 2), while the anionic docosohedral example [W(≡CC$_6$H$_4$Me-4)(CO)$_2$(η^6-C$_2$B$_{10}$H$_{10}$Me$_2$)]$^-$ has a considerably higher k_{CO} value 15.04 Ncm^{-1}. There is no correlation obvious to us between the net charge on the complex and derived WC bond orders or W≡C bond lengths for the carbyne ligand.

2.4. Sub-Series of Ligands

Tables 1 and 2 along with Figures 3 and 4 contain a number of as yet hypothetical derivatives that have yet to be prepared but which would appear to be entirely plausible based on the demonstrated viability of the ligands L in other systems. Some comments on sub-classes now follow.

2.4.1. Hydrotris(N-R^1-imidazolylidenyl)borates

Central to this communication are the tris(imidazolylidene)borates HB(ImR^1)$_3$. From Figure 3, it is clear that the ligand HB(ImMe)$_3$ occupies a position in a somewhat sparsely populated area of the electronic–steric map, being both strongly basic and also imparting considerable steric prophylaxis upon the carbonyl and carbyne co-ligands akin to that provided by the popular HB(pzMe$_2$)$_3$ ligand. The experimental and calculated values for k_{CO} are comparable to those for Stone's dicarbollide complexes (L = η^5-C$_2$B$_9$H$_9$R$_2$ R = H, Me) [79,88] which, however, carry a net negative charge, and so it must be assumed much of the negative charge resides within the carbaborane cage.

As expected, the $\%V_{bur}$ value for **4** is close to that of **5**. Smith has developed synthetic routes to the pro-ligand salts that carry *N*-substituents of varying bulk (^{t}Bu, Cy, C$_6$H$_2$Me$_3$-2,4,6) [4] and accordingly entries **1** (R^1 = Me, **4**), **35** (R^1 = Et), **36** (R^1 = ^{i}Pr), **37** (R^1 = ^{t}Bu) and **38** (R^1 = Ph) survey the sequential inclusion of increasing steric bulk at the position β to the metal. All attempts to geometrically minimize, or indeed even reasonably construct, the derivative with R^1 = mesityl met with spectacular failure, perhaps indicating a step too far, though this ligand has been successfully installed on four-coordinate nickel [37]. The phenyl derivative **38**, however, is able to accommodate unsubstituted aryl groups by allowing them to interdigitate between the carbonyl and carbyne ligands such that the aryl planes are near colinear with the W⋯B vector. A very approximate value for the $\%V_{bur}$ of 56.6% is provided by the hypothetical and implausible (distorted) octahedral complex [WMe$_3${HB(ImMes)$_3$}] (PM3tm level of theory). While it is not dissimilar to the value (59.8%) estimated for L = neutral MeC(CH$_2$PPh$_2$)$_3$ (**16**) and anionic MeB(CH$_2$PPh$_2$)$_3$ (**31**), inclusion of this excessive steric bulk would seem problematic. It should, however, be noted that a rich organometallic chemistry has emerged for the dihydro*bis*(*N*-

mesitylimidazolylidenyl)borate ligand coordinated to tantalum [33,34], for which the bidentate variant presents a considerably reduced steric impact, e.g., V_{bur} = 39.8% in pseudo-octahedral [TaMe$_4${H$_2$B(ImMes)$_2$}]. The trifluoromethylimidazolylidenyl derivate (Entry 39) was also considered and found to be a rather modest net donor (ν_{CO} = 15.2 Ncm^{-1}) while presenting a comparatively occlusive encapsulating pocket (V_{bur} = 57.1%). The only currently available synthesis of N-trifluoromethylimidazole [111] is, however, not particularly amenable to the scales needed for an exploration of the HB(ImCF$_3$)$_3$ ligand. Figure 5 depicts the steric maps that arise from %V_{bur} calculations and shows the progression in steric encumbrance as the N-substituents are replaced along the alkyl series R^1 = Me, Et, ^{i}Pr, ^{t}Bu alongside those for R = Ph and CF$_3$.

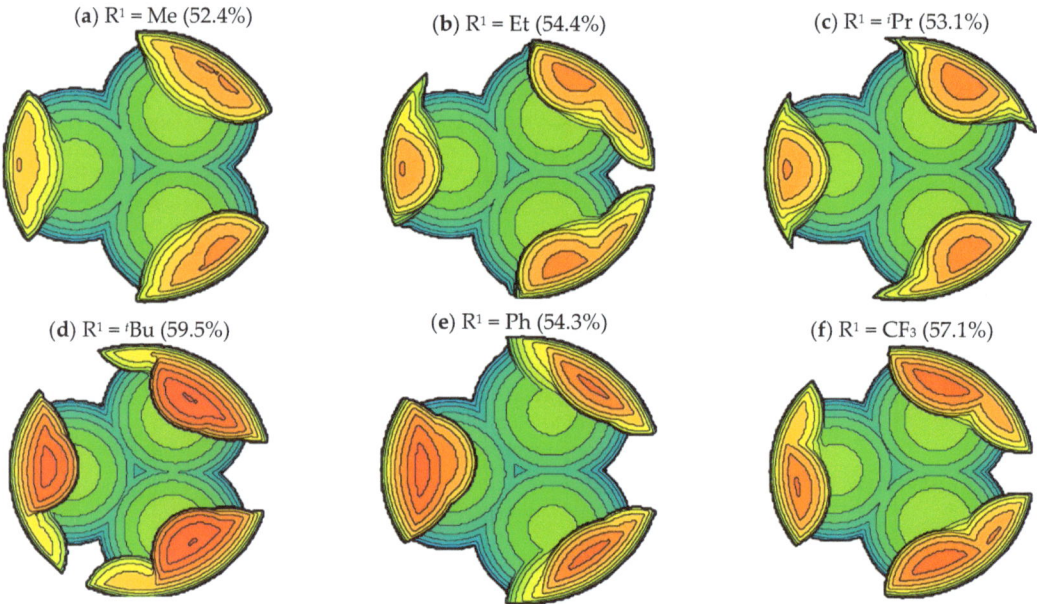

Figure 5. Steric maps [92,93] and %V_{bur} values in parentheses for a series of hydrotris(N-R^1-imidazolylidenyl)borates HB(ImR^1)$_3$ where (**a**) R^1 = Me; (**b**) R^1 = Et; (**c**) R^1 = ^{i}Pr; (**d**) R^1 = ^{t}Bu; (**e**) R^1 = Ph; (**f**) R^1 = CF$_3$.

What is immediately apparent from Figure 4 is that replacement of the 'parent' N-methylimidazole, which is both commercially available and cheap, with ethyl, iso-propyl or phenyl imidazoles actually results in very modest variation in the steric impact around the coordination sphere of the metal because the groups can direct their bulk away from the carbonyl and carbyne ligands. It is only with the ^{t}Bu (and to a lesser extent the CF$_3$) derivative that this bulk is unavoidably directed towards the metals centre. This is clear when the 3.5 Å value typically and arbitrarily employed in %V_{bur} calculations is replaced by 4.0, 5.0 and 6.0 Å (Figure 6), respectively. Thus, inclusion of phenyl, primary or secondary alkyl groups appears to have rather a modest steric influence directly on the metal coordination sphere but may contribute in a secondary manner to compound longevity by reducing the collisional cross section (Arrhenius pre-exponential factor) for proceeding reactions. It seems that only with tertiary alkyl (e.g., ^{t}Bu) or *ortho*-substituted aryl substituents (e.g., mesityl) that a significant impact on the steric profile is likely to manifest in the reactivity.

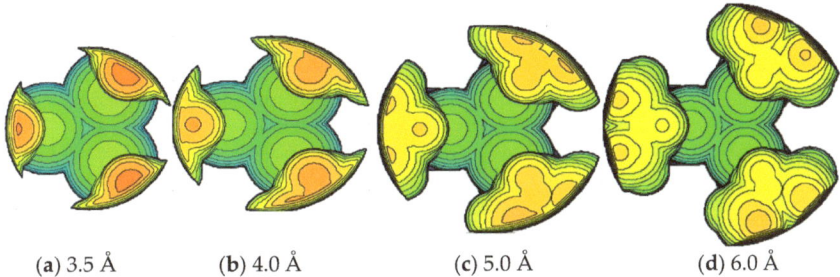

(a) 3.5 Å   (b) 4.0 Å   (c) 5.0 Å   (d) 6.0 Å

Figure 6. Steric map dependence on radius of coordination sphere employed.

An intriguing question does, however, arise when the steric bulk is exaggerated, in that whilst this might be expected to increase the donor strength of the NHC:→W interaction, the inter-ligand repulsion is such that there is a notable increase in the W–C bond lengths of both the NHC donors cis (mean value) and trans to the carbyne (Table 3).

Table 3. Steric Impact of *N*-substituents in the Complexes [W(≡CC$_6$H$_4$Me)(CO)$_2$\{HB(ImR)$_3$\}].

R	Mean W–C Å	Mean W–C$_{cis}$ Å	W–C$_{trans}$ Å	TR ^{a}
Me	2.262	2.226	2.335	1.049
CF$_3$	2.276	2.232	2.365	1.060
Et	2.268	2.233	2.339	1.047
^{i}Pr	2.268	2.232	2.341	1.049
Ph	2.277	2.237	2.357	1.054
^{t}Bu	2.349	2.312	2.424	1.048

^{a} (W–C$_{trans}$)/(Mean W–C$_{cis}$).

Thus, the simple σ-basicity vs. π-acidity of the free NHC is only part of the story if the metal–donor bond length increases (weakens?) significantly. This does not appear to be the case in the present system, in that while the ^{t}Bu derivative has especially long NHC–W bond lengths, it is nevertheless the most potent net donor (k_{CO} = 14.41 Ncm^{-1}) of all the ligands considered. In the case of the complexes [Ni(NO)(L)] where L represents a sub-set of ligands considered in Tables 1 and 2 (η^5-C$_5$Me$_5$, Tp*, Hb(mt^{tBu})$_3$ and PhB(CH$_2$PPh$_2$)$_3$ [112–116]) alongside those for selected tris(imidazolylidenyl)borates RB(ImR^1)$_3$ (R = H, Ph; R^1 = Me, ^{t}Bu, Mesilyl, CH$_2$Cy [37]), Smith employed nitrosyl stretching frequency as a measure of the relative donor ability of 'L'. Similar σ-donor/π-acceptor arguments apply as they do to CO with the caveat that depending on the electronic nature of the metal centre, the nitrosyl may bend; i.e., lower values for ν$_{NO}$ may indicate an electron rich metal centre *or* bending, which becomes more prevalent for late-transition metal centres with high *d*-occupancies [117]. In the case of four-coordinate nitrosyls of nickel, the situation is complicated by subtleties in the electronic nature of the nickel that remain moot [47,49]. While Smith was consistent in reporting data from the same essentially non-coordinating solvent toluene (or sometimes THF), data from other sources were acquired from a variety of media (not always stated) including the solid state (KBr, Nujol, Ar$_{(s)}$, etc.). The selenoimidazolylborate is a case in point for which the reported solid-state IR spectrum comprised two ν$_{NO}$ bands [114]. Since the crystal structure revealed a single crystallographically independent molecule, one might assume the second vibrational mode was due to an alternative crystal modification in the bulk sample. Given the two bands differ by 11 cm^{-1} and the entire Tolman ν^T scale only spans 45 cm^{-1}, the importance of using solution derived data, preferably from a common solvent, is demonstrated.

2.4.2. Cyclopentadienyl Derivatives

In terms of percentage buried volume, the cyclopentadienyl ligand is somewhat unassuming (V_{bur} = 35.2%), and this is most commonly 'bulked out' via permethylation (**12**: L = C$_5$Me$_5$ V_{bur} = 42.4%), inclusion of trimethylsilyl substituents (**26**: L = C$_5$H$_3$(SiMe$_3$)$_2$ V_{bur} = 54.5%) or benzannulation with either one (**22**: L = indenyl V_{bur} = 37.3%) or two (**23**: L = fluorenyl V_{bur} = 40.4%) benzo rings. This imbues variable electron-releasing nature in the series C$_5$H$_3$(SiMe$_3$)$_2$ > C$_5$Me$_5$ > C$_5$H$_5$ ≈ fluorenyl > indenyl. A subtlety emerges from the geometry minimization of the indenyl derivative, which reveals a structural basis for Basolo's 'indenyl effect' [61,118,119]. Incipient ring slippage ($\eta^5 \rightarrow \eta^3$) might be inferred, given that the angle between the tungsten, the cyclopentadienyl ring centroid and the unique carbon atom is slightly acute (84.8°), such that the unique carbon (2.319 Å) and adjacent *pseudo* η^3-carbons (2.355, 2.352 Å) are noticeably closer to tungsten than are the benzo-fused carbons (mean: 2.534 Å). The C$_6$H$_4$ unit makes an angle of *ca* 5.9° with the three non-ring-fused carbons of the cyclopentadienyl ring. This slippage places the benzenoid ring trans to the carbyne ligand, as might be expected based on the characteristic trans influence of carbyne ligands. Experimentally acquired structural data are not currently available for indenyl, fluorenyl or bis(trimethylsilyl)cyclopentadienyl ligated carbynes through which to further explore this question, though enhanced reactivity in associative ligand addition reactions has been noted for the indenyl carbyne [Mo(≡CC$_6$H$_3$Me$_2$-2,6)(CO)$_2$(η^5-C$_9$H$_7$)] [61].

Although no examples exist of carbyne complexes bearing by the perchlorocyclopentadienyl ligand (**25**: L = η^5-C$_5$Cl$_5$), the tricarbido bimetallic complex [ReMn(μ_2-C$_3$)(CO)$_2$(NO)(PPh$_3$)(η^5-C$_5$H$_5$)(η^5-C$_5$Cl$_5$)]$^+$ described by Gladysz [120,121] might be viewed as possessing a degree of manganese carbyne character. Perchlorination results in a modest increase in the steric bulk of the ligand (40.8 cf. 42.4% for η-C$_5$Me$_5$) but a quite substantial decrease in donor ability (k_{CO} = 15.51 Ncm^{-1}). Perphenylation, in contrast, has only a modest effect on the net basicity of the ligand (k_{CO} = 15.34 Ncm^{-1}), while the buried volume increases significantly (V_{bur} = 48.3%) due to the requisite orientation of the aryl groups to near orthogonal to the cyclopentadienyl plane. The tetraphenylcyclopentadienyl carbyne complex [W(≡CPh)(PPh$_2$C$_6$H$_4$CH=CHPh)(η^5-C$_5$HPh$_4$)] [122] and a single rather exotic pentaphenylcyclopentadienyl complex [W(≡CPh)(NCMe)(η^2-C$_{60}$)(η^5-C$_5$Ph$_5$)] [123] have been described.

2.4.3. Arene Derivatives

While hexahapto arene co-ligated carbyne complexes such as **18**, **20** and **21** appear unknown, a manifold of intriguing molybdenum carbyne complexes bearing the C$_6$H$_4$(C$_6$H$_4$P^{i}Pr$_2$-2)$_2$-1,4 trans-coordinating diphosphine have been shown by Agapie to enter into variable degrees of arene-molybdenum interaction during transformations that demonstrate the interplay of carbyne and carbido ligands [124–127]. It therefore seems reasonable to anticipate that compounds akin to Entries **18**, **20** and **21** will emerge. It is apparent that conclusions similar to those for cyclopentadienyl substituents will result, except that the overall complex bears a positive charge, providing a point of connection with group 7 carbynes [M(≡CR)(CO)$_2$(η^5-C$_5$H$_5$)]$^+$ (M = Mn, Re) [128–132]. The hexaethylbenzene derivative **21** would appear to present a sterically quite encapsulating environment (V_{bur} = 53.3%) cf. the hexamethyl analogue (V_{bur} = 45.9%) due to the 3-up/3-down mutual disposition of the ethyl substituents. This feature has been employed to favour unusual regiochemistry in selective alkane binding by the 'W(CO)$_2$(η^6-C$_6$Et$_6$)' fragment [133]. Finally, we note that the inorganic benzene B$_3$N$_3$Me$_3$ has, as expected, a steric profile similar (V_{bur} = 46.2%) to that of C$_6$Me$_6$ (V_{bur} = 45.9%), and the non-planar ring is a comparable net donor to the tungsten centre (k_{CO} = 15.84 Ncm^{-1} vs. 15.85 Ncm^{-1} for **20**). This is also implicit from infrared data for [Cr(CO)$_3$(η^6-B$_3$N$_3$Me$_6$)] (Cyclohexane: ν_{CO} = 1963, 1867 cm^{-1}) vs. [Cr(CO)$_3$(η^6-C$_6$Me$_6$)] (ν_{CO} = 1962, 1888 cm^{-1}) provided in a publication in which Werner indicated that [W(CO)$_3$(η^6-B$_3$N$_3$Me$_6$)] also appeared viable [134,135].

2.4.4. Pnictolyl Ligands

Schrock has explored the utility of high oxidation state carbene and carbyne complexes ligated by σ- and η^5-pyrollyl ligands [136], though low oxidation variants have yet to emerge. Carbynes ligated by the heavier pnictolyl ligands η^5-AC$_4$R$_4$ (A = P, As), however, remain unknown, though both ligands have been shown to serve as ersatz cyclopentadienyls [137–140]. With the ready availability of synthetic routes to anionic pnictolyl reagents, it may be presumed that complexes of the form [M(≡CR)(CO)$_2$(η^5-AC$_4$R$'_4$)] (A = P, As, Sb) will emerge in the future, given that, like carbynes, arsolyl ligands have been shown to support intermetallic bonding [141–143].

2.4.5. Toluidyne Orientation

Perusal of the structures, experimentally or computationally derived, reveals a broad range of orientations of the toluidine ring with respect to the nominal coordination axes. This is of secondary importance in that for all examples, the ^{1}H NMR spectra involve a simple AA'BB' pattern indicating free rotation on the ^{1}H NMR (and ^{13}C) NMR timescale(s). Arbitrarily adopting the cationic carbyne formalism ([CF]$^+$, [NO]$^+$ and CO being isoelectronic molecules), coordinated to a d^6-ML$_5$ fragment, the two carbyne acceptor orbitals vary in energy by only 0.2 eV, as do the two metal retrodative orbitals (HOMO-1, HOMO-2) of, e.g., the 'W(CO)$_2$(Tp)' fragment (Figure 7). The HOMO itself is invariably associated with metal–carbonyl π-bonding and is orthogonal (δ-symmetry) to the W–Carbyne vector. Accordingly, any conformational preference should be presumed to reflect interligand steric factors and/or intermolecular packing effects. For the majority of structurally characterized carbyne complexes of the M(CO)$_2$(Tp*) fragment; for example, the carbyne substituent typically nestles between two dimethylpyrazolyl groups. NB: The molecular orbitals of the actual carbyne complex are, as they must be, independent of the arbitrary electron allocation to hypothetical constituent fragments; i.e., similar interpretation based on [CC$_6$H$_4$Me]$^{3-}$ and d^2-ML$_5{}^{3+}$ or neutral CC$_6$H$_4$Me-4 and d^5-ML$_5$ deconstructions lead to the same conclusion.

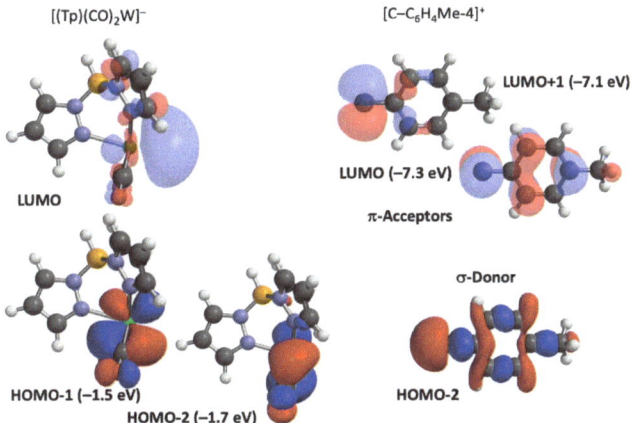

Figure 7. Valence orbitals of the hypothetical [CC$_6$H$_4$Me-4]$^+$ and d^6-[W(CO)$_2$(Tp)] fragments.

2.5. A Heterobimetallic Hydrotris(imidazolylidenyl)borate Complex

To date, the tris(imidazolylidenyl)borate class of ligands has only been employed in monometallic systems; however, terminal carbyne ligands have an extensively documented propensity to support metal–metal bond formation, as championed by Stone [144]. In particular, the addition of gold(I) reagents to monometallic carbyne complexes [145–155] is of interest due to the tendency of the carbyne to adopt a semi-bridging rather than the more common symmetrical bridging geometry. This is considered to arise when the carbyne bridges electronically disparate metals, and therefore, the late high d-occupancy metal (d^{10} gold(I) or platinum(0)) is considered to act as a σ-donor (Z-type metal–ligand bonding [156]) to the carbyne carbon. Accordingly, the reaction of **4** with [AuCl(SMe$_2$)] was investigated and found to readily provide the bimetallic complex [WAu(μ-CC$_6$H$_4$Me-4)Cl(CO)$_2${HB(ImMe)$_3$}] (**6**, Scheme 4). The complex is somewhat unstable in solution, slowly depositing elemental gold during unsuccessful attempts to slowly obtain crystallographically serviceable crystals. The formulation, however, rests reliably on spectroscopic data which may be compared with precedents for other carbyne and tungsten substituents. The reaction is accompanied by a shift in the ν_{CO} absorptions to a higher frequency (CH$_2$Cl$_2$: 1971, 1879 cm^{-1}) than those of the precursor in the same solvent (1958, 1873 cm^{-1}). The carbyne carbon resonance in the ^{13}C{^{1}H} NMR spectrum appears at δ_C = 277.7, and while this is only marginally shifted from that of the precursor (280.7 ppm), there is a dramatic decrease in the value of $^1J_{WC}$ (85 Hz cf. 171.3 Hz for **4**), which is consistent with the increase coordination number (reduced *s*-character) of both tungsten and carbon. The resonances due to the imidazolylidene donors appear at 187.7 [$^1J_{CW}$ = 90 Hz], 173.7 [$^1J_{CW}$ = 71 Hz] in a similar region to the precursor but with more similar values for $^1J_{WC}$ (90, 71 Hz) once the trans influence of the carbyne is alleviated upon gold adduct formation.

While the ^{1}H and ^{13}C{^{1}H} NMR spectra each confirm a locally C_s symmetric environment around the tungsten, at least on these timescales, they do not distinguish between the AuCl unit lying syn or anti to the imidazolylidene units; however, based on precedent from the sterically similar HB(pzMe$_2$)$_3$ ligand, it seems likely that the AuCl unit nestles between two imidazolylidene rings. This geometry was adequately modelled (Figure 8) at the ωB97X-D/6-31G/LANL2Dζ/gas-phase level of DFT, from which it would appear that the W–C bond clearly retains its considerable multiple-bond character (W–C = 1.913 Å). The W–C–C (148.9°) and Au–C–C (121.5°) angles indicate semi rather than symmetrical bridging such that the C–C and W–Au vectors form an obtuse angle of 101.4°. Despite numerous (>80) examples of structurally authenticated W–Au bonds, only two have bonds that are not supported by bridging ligands, viz. the compounds [WAu(CO)$_3$(PPh$_3$)(η^5-C$_5$H$_4$R)] (R = H 2.698 Å [157] and CH$_2$CH$_2$NHMe$_2$$^+Cl^-$ 2.712 Å [158]). The optimized Au–W bond length for **6** (2.812 Å) is therefore comparable to these, though towards the longer end of the range. The infrared ν_{CO} absorptions are noted at 1955 and 1899 cm^{-1} (λ_2), while TD-DFT analysis suggests that the colour of the complex may be attributed to absorptions calculated at 420 nm (W–C ≈ z-axis: HOMO-LUMO; d_{xy}-W=Cπ*), 357 (HOMO-LUMO+1; d_{xy}-WAuσ*) and 344 nm (HOMO-1-LUMO; W=Cπ-W=Cπ*), the first two of which involve considerable charge transfer.

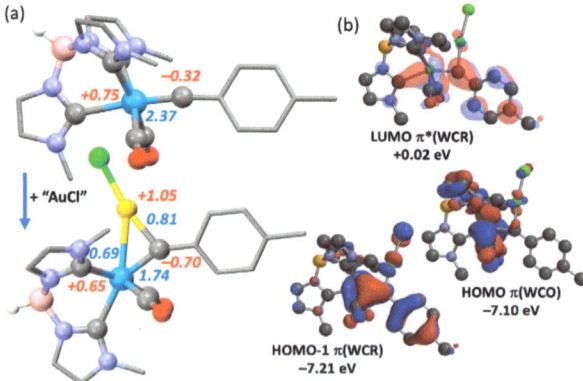

Figure 8. (**a**) Optimized geometries of [W(≡CC$_6$H$_4$Me-4)(CO)$_2${HB(ImMe)$_3$}] (**4**) and the heterobimetallic complex [WAu(μ-CC$_6$H$_4$Me-4)Cl(CO)$_2${HB(ImMe)$_3$}] (**6**) indicating changes in Löwdin bond order (blue) and natural atomic charge (red) upon 'AuCl' adduct formation (ωB97X-D/6-31G*/LANL2Dζ, hydrogen atoms omitted, tolyl and imidazolyl groups simplified). (**b**) Frontier molecular orbitals of interest for **6** at Isovalue = 0.032 $\sqrt{(e/au^2)}$.

3. Experimental

3.1. General Considerations

Experimental work was performed using standard Schlenk techniques with pure dry nitrogen or argon, or in an argon atmosphere glovebox, unless otherwise specified. All solvents used in the syntheses were dried and degassed. Unless otherwise indicated, reagents were used as received from commercial suppliers.

Infrared data were obtained using a Shimadzu FTIR-8400 for solutions and a Perkin Elmer FTIR Spectrum 2 for solid-state ATR measurements. NMR spectra were measured using Bruker Avance 400, Bruker Avance 600 or Bruker Avance 800 spectrometers at the temperatures indicated. Chemical shifts (δ) are reported in ppm with coupling constants in Hz, all referenced to the appropriate solvent resonance. Multiplicities indicated do not include the satellites for the ^{183}W isotopomers, the couplings for which are listed separately. Positive ion high-resolution electrospray ionisation mass spectroscopy (ESI) data were provided by the ANU Research School of Chemistry Joint Mass Spectrometry team; an acetonitrile matrix was used for all samples. Single-crystal X-ray diffraction (XRD) crystallographic data were acquired with a SuperNova CCD diffractometer using Mo-Kα radiation (λ = 0.71073 Å), employing CrysAlis PRO software [159] (https://www.agilent.com/ accessed on 20 November 2023), refined with Olex2 [160], and structural models were depicted using Mercury [161]. Elemental microanalytical data were not acquired [162].

Computational studies were performed using the *SPARTAN20* suite of programs [40]. Cyclic voltammetry (CV) was performed using a PalmSens 4 Potentiostat/Galvanostat/Impedance Analyser and carried out in a single-compartment 3-electrode glass cell, with a 3 mm glassy carbon working electrode, platinum wire counter electrode and silver wire pseudo-reference electrode. Analyte solutions were prepared at 1 mM in dichloromethane with 0.1 M [NBu$_4$][PF$_6$] supporting electrolyte. Solutions were sparged with N$_2$ bubbled through dichloromethane prior to measurements, then maintained under an atmosphere of N$_2$ during voltammetry. All measurements were referenced to ferrocene, which was added to the solution following each measurement.

Infrared and NMR spectra for all new compounds are provided in the accompanying Supplementary Materials.

3.2. Synthesis of [W($\equiv$CC$_6$H$_4$Me-4)(CO)$_2$(pic)$_2$(Br)] (2a)

Note: the following synthesis is a modified version of the synthesis of cis,cis,trans-[W($\equiv$CC$_6$H$_3$Me$_2$-2,6)(CO)$_2$(pic)$_2$Br] [15]. A solution of 4-bromotoluene (6.568 g, 38.40 mmol) in diethyl ether (60 mL) was cooled to 0 °C before lithium (0.618 g, 89.0 mmol, hammered and cut wire) was added. This was stirred vigorously at 0 °C for 30 min before being allowed to slowly warm to room temperature and being stirred for a further 3 h. The lithium reagent was added dropwise to a suspension of [W(CO)$_6$] (8.445 g, 24.00 mmol) in diethyl ether (60 mL) until IR spectroscopy indicated no [W(CO)$_6$] remained. The red solution was cooled to –78 °C before trifluoroacetic anhydride (3.40 mL, 24.3 mmol) was added dropwise over a period of 10 min, resulting in a yellow precipitate. After stirring for 30 min at –78 °C, 4-picoline (6.0 mL, 62 mmol) was added. The suspension was allowed to slowly warm to room temperature and stirred overnight. The yellow precipitate was isolated via cannula filtration and extracted with dichloromethane (60 mL) and the combined extracts were filtered through diatomaceous earth, followed by washing with further dichloromethane until the extracts were colourless (total volume 200 mL). The solvent volume was reduced to *ca* 40 mL under reduced pressure before slow dilution with hexane (300 mL) to precipitate a yellow-orange solid that was freed of supernatant via cannula filtration. Hexane (80 mL) was added, and the suspension was ultrasonically triturated for 10 min to remove residual [W(CO)$_6$]. The yellow-orange solid was collected on a sinter, washed with further hexane (20 mL) and dried under high vacuum (13.094 g, 21.446 mmol, 89% isolated yield).

IR (CH$_2$Cl$_2$, cm^{-1}): 1986 vs. ν_{CO}, 1897 vs. ν_{CO}. **IR** (ATR, cm^{-1}): 1970 vs. ν_{CO}, 1881 vs. ν_{CO}. **^{1}H NMR** (600 MHz, CD$_2$Cl$_2$, 298 K) δ_H = 8.91 [d, 4 H, $^3J_{HH}$ = 7, H^{2,6}(pic)], 7.25 [d, 2 H, $^3J_{HH}$ = 8, H^{2,6}(C$_6$H$_4$)], 7.13 [d, 4 H, $^3J_{HH}$ = 7, H^{3,5}(pic)], 7.09 [d, 2 H, $^3J_{HH}$ = 9, H^{3,5}(C$_6$H$_4$)], 2.38 [s, 6 H, pic-CH$_3$], 2.29 [s, 3 H, tolyl-CH$_3$]. **^{13}C{^{1}H} NMR** (151 MHz, CD$_2$Cl$_2$, 298 K)

δ_C = 263.9 [$^1J_{WC}$ = 203 Hz, W≡C], 221.4 [$^1J_{WC}$ = 169 Hz, CO], 153.3 [$C^{2,6}$(pic)], 151.0 [C^4(pic)], 146.9 [$C^1(C_6H_4)$], 138.5 [$C^4(C_6H_4)$], 129.4 [$C^{2,6}(C_6H_4)$], 129.1 [$C^{3,5}(C_6H_4)$], 126.3 [$C^{3,5}$(pic)], 21.8 [tolyl-CH_3], 21.3 [pic-CH_3]. **MS** (ESI, +ve ion, m/z): Found: 609.0375. Calcd for $C_{22}H_{22}N_2O_2{}^{79}Br^{184}W$ [M + H]$^+$: 609.0374. Crystals suitable for structural determination were grown from liquid diffusion of diethyl ether into a saturated dichloromethane solution of the sample at -20 °C. **Crystal Data** for $C_{22}H_{21}BrN_2O_2W\cdot(OEt_2)_{0.5}$ (M_w = 646.23 gmol^{-1}): monoclinic, space group $C2/c$ (no. 15), a = 23.3477(7) Å, b = 12.5106(2) Å, c = 17.9409(5) Å, β = 110.628(3) °, V = 4904.4(2) Å^3, Z = 8, T = 150.0(1) K, μ(Mo Kα) = 6.364 mm^{-1}, D_{calc} = 1.750 Mgm^{-3}, 37431 reflections measured (7.422° $\leq 2\Theta \leq$ 64.280°), 8075 unique which were used in all calculations. The final R_1 was 0.0311 ($I > 2\sigma(I)$) and wR_2 was 0.0671 (all data) with 291 refined parameters with one restraint, CCDC 2305468. The molecular geometry in the solid state is depicted in Figure 9.

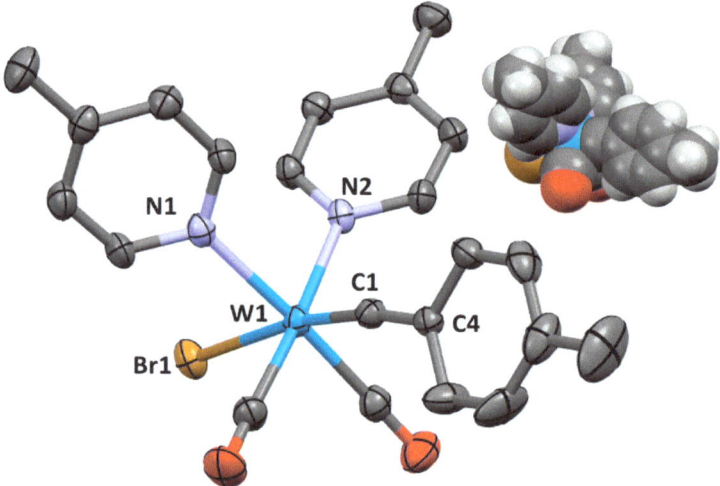

Figure 9. Molecular structure of cis,cis,trans-[W(≡CC_6H_4Me-4)(CO)$_2$(pic)$_2$(Br)] in a crystal (50% displacement ellipsoids, hydrogen atoms omitted for clarity).

*3.3. Synthesis of [Mo(≡CC_6H_4Me-4)(CO)$_2$(pic)$_2$(Br)] (**2b**)*

A solution of 4-bromotoluene (6.842 g, 40.00 mmol) in diethylether (50 mL) was cooled to 0 °C before lithium (1.3 g, 190 mmol, hammered and cut wire) was added. This was stirred vigorously at 0 °C for 30 min before being allowed to slowly warm to room temperature and being stirred for a further 3.5 h. The lithium reagent was added dropwise to a suspension of [Mo(CO)$_6$] (6.338 g, 24.01 mmol) in diethyl ether (60 mL) until negligible [Mo(CO)$_6$] remained, as indicated by in situ IR spectroscopy. The red solution was cooled to –78 °C before trifluoroacetic anhydride (3.40 mL, 24.3 mmol) was added dropwise over a period of 10 min. After being stirred for 45 min at –78 °C, 4-picoline (6.0 mL, 62 mmol) was added. The suspension was allowed to slowly warm to room temperature and stirred overnight. The yellow precipitate was isolated via cannula filtration and extracted with dichloromethane (50 mL) and the extracts filtered through diatomaceous earth, followed by washing with further dichloromethane (6 × 5 mL). The volume was reduced to 50 mL under reduced pressure before slow dilution with hexane (120 mL) to precipitate a yellow solid that was freed of supernatant via cannula filtration and dried under high vacuum (8.473 g, 16.25 mmol, 68% isolated yield).

IR (CH_2Cl_2, cm^{-1}): 2000 vs. ν_{CO}, 1918 vs. ν_{CO}. **IR** (ATR, cm^{-1}): 1986 vs. ν_{CO}, 1913 vs. ν_{CO}. **^{1}H NMR** (600 MHz, CD_2Cl_2, 298 K) δ_H = 8.87 [d, 4 H, $^3J_{HH}$ = 6, H^{2,6}(pic)], 7.36 [d, 2 H, $^3J_{HH}$ = 8, H$^{2,6}(C_6H_4)$], 7.08-7.14 [m, 6 H, H^{3,5}(pic) and H$^{3,5}(C_6H_4)$ over-lapped], 2.37 [s, 6 H, pic-CH_3], 2.32 [s, 3 H, tolyl-CH_3]. **^{13}C{^{1}H} NMR** (151 MHz, CH_2Cl_2, 298 K)

δ_C = 276.8 [Mo≡C], 224.4 [CO], 152.9 [$C^{2,6}$(pic)], 150.6 [C^4(pic)], 143.5 [$C^1(C_6H_4)$], 139.5 [$C^4(C_6H_4)$], 129.3 [$C^{3,5}(C_6H_4)$], 129.2 [$C^{2,6}(C_6H_4)$], 125.9 [$C^{3,5}$(pic)], 21.8 [tolyl-CH_3], 21.26 [pic-CH_3)]. **MS** (ESI, +ve ion, *m/z*): Found: 443.0659. Calcd for $C_{22}H_{22}N_2O_2{}^{98}Mo$ [M − Br]$^+$: 443.0662.

3.4. Synthesis of [Tris(1-methylimidazolium)borate] Bis(hexafluorophosphate) ([1](PF$_6$)$_2$)

A 1 L three-necked flask was fitted with a stirrer bar, water-cooled reflux condenser, pressure-equalizing dropping funnel and a gas outlet leading to a NaOH scrubber. The entire apparatus was flushed with nitrogen for 30 min before trimethylamine-borane complex (7.32 g, 100 mmol) was added, followed by 150 mL degassed chlorobenzene. To the dropping was added 50 mL chlorobenzene and bromine (7.8 mL, 85 mmol Br$_2$). The bromine solution was added to the flask at a rate of about one drop/second whilst the reaction was flushed with a gentle stream of nitrogen. This reaction is initially very exothermic and the rate of bromine addition should be adjusted accordingly; caution should also be exercised, since hydrogen gas is also liberated at this stage. After approximately half of the bromine had been added, the exothermicity was less pronounced and rate of addition of the remainder could be increased safely. The mixture was then stirred for 3 h at ambient temperature, during which time the orange colour of bromine slowly faded to a pale yellow. Hydrogen bromide was liberated during this time as nitrogen was continuously swept over the reaction. *N*-methylimidazole (28 mL, 330 mmol) was added to the mixture, and the apparatus was carefully transferred to a heating mantle, where it was brought to reflux for 4-6 h. Upon heating, a white crystalline solid precipitated from the reaction mixture; extended heating is to be discouraged, as this leads to the formation of tarry yellow materials and poor yields of product. The chlorobenzene layer was decanted from the solids while warm, and the flask was then rinsed with 3 × 100 mL portions of toluene; the washings were subsequently discarded. The white solid was dissolved into 100 mL methanol and added slowly to a vigorously stirred solution of NaPF$_6$ (35 g, 210 mmol; NaBF$_4$ may also be used) in 100 mL methanol, from which the product precipitated as a fluffy white solid. The white solids were collected via filtration, washed with 3 × 50 mL portions each of methanol and Et$_2$O and dried under suction. Purity was sufficient for synthetic purposes, though an analytically pure sample was obtained via re-crystallisation from acetone/Et$_2$O (vapour diffusion). Isolated yield 11.50 g (21 mmol, 21%) as the PF$_6$ salt or 12.90 g (30 mmol, 30%) as the BF$_4$ salt.

IR (THF, cm^{-1}): 2454 w ν_{BH}. **IR** (ATR, cm^{-1}): 2455 vs. ν_{BH}, 827 vs. ν_{PF}. **^{1}H NMR** (400 MHz, CD$_3$CN, 25 °C): δ_H = 8.17 (s, 3 H, N$_2$CH), 7.38 (t, $^3J_{HH}$ = 1.7 Hz, 3 H, NCHCH), 7.17 (t, $^3J_{HH}$ = 1.7 Hz, 3 H, NCHCH), 3.87, (s.br, 1 H, BH) 2.18 (s, 9 H, NCH$_3$). **^{13}C{^{1}H} NMR** (101 MHz, CD$_3$CN, 25 °C): δ_C= 139.9 (N$_2$C), 125.6 (NCC), 124.2 (NCC), 36.4 (NCH$_3$). **^{11}B{^{1}H}** NMR (128 MHz, CD$_3$CN, 25 °C): δ_B = −3.50 (BH). **^{11}B NMR** (128 MHz, CD$_3$CN, 25 °C): δ_B = −3.42 (d, $^1J_{BH}$ = 121.7 Hz, BH). ^{19}F NMR (376 MHz, CD$_3$CN, 25 °C): δ_F = −72.9 (d, $^1J_{PF}$ = 708 Hz, PF$_6$). **^{31}P{^{1}H} NMR** (162 MHz, CD$_3$CN, 25 °C): δ_P = −144.6 (sept, $^1J_{PF}$ = 700 Hz, PF$_6$). **MS** (ESI, +ve ion, *m/z*): Found: 257.1685. Calcd for $C_{12}H_{18}N_6{}^{11}B$. [M−H]$^+$: 257.1686. **Crystal Data** for $C_{12}H_{19}BF_{12}N_6P_2$ (M_w = 548.08 gmol^{-1}): monoclinic, space group $P2_1/n$ (no. 14), a = 20.7403(2) Å, b = 10.10590(10) Å, c = 20.7879(2) Å, β = 97.4530(10)°, V = 4320.32(7) Å^3, Z = 8, T = 150.2(1) K, μ(CuKα) = 2.945 mm^{-1}, D_{calc} = 1.685 Mgm^{-3}, 54277 reflections measured (5.664° ≤ 2Θ ≤ 156.216°), 9112 unique (R_{int} = 0.0625, R_{sigma} = 0.0400), which were used in all calculations. The final R_1 was 0.0603 ($I > 2\sigma(I)$) and wR_2 was 0.1725 (all data) for 711 refined parameters with 296 restraints. CCDC 2305467.

*3.5. Synthesis of [W(≡CC$_6$H$_4$Me-4)(CO)$_2${HB(ImMe)$_3$}] (**4**)*

Tris(1-methylimidazolium)borate bis(hexafluorophosphate) ([1](PF$_6$)$_2$: 0.400 g, 0.730 mmol) was dissolved in tetrahydrofuran (30 mL) and cooled (dry ice/propanone). A solution of *n*-butyllithium (1.40 mL, 1.6 M, 2.20 mmol, hexanes) was added dropwise at −78 °C. While being stirred for 90 min at this temperature, the solution became pale yellow, at which point solid

[W(≡CC$_6$H$_4$Me-4)(CO)$_2$(γ-pic)$_2$(Br)] (**2a**: 0.45 g, 0.70 mmol) was added. After it was stirred at this temperature for 15 min, the mixture was allowed to warm to room temperature and stirred for a further 3 h and then freed of volatiles under reduced pressure. The residual black tar was dissolved in a minimum of dichloromethane (~5mL) and subjected to flash column chromatography (silica gel, N$_2$, CH$_2$Cl$_2$). The orange band that eluted first was collected, and the solvent was removed under reduced pressure to give (**4**) a bright orange powder. Yield: 0.11 g (0.18 mmol, 20%).

IR (CH$_2$Cl$_2$, cm^{-1}): 2453 w ν$_{BH}$, 1958 vs. ν$_{CO}$, 1873 vs. ν$_{CO}$. **IR** (ATR, cm^{-1}): 2442 w ν$_{BH}$, 1949 vs. ν$_{CO}$, 1867 vs. ν$_{CO}$. **^{1}H NMR** (800 MHz, CD$_2$Cl$_2$, 25 °C): δ$_H$ = 7.36 [d, $^3J_{HH}$ = 7.4 Hz, 2 H, H^{2,6}(C$_6$H$_4$)], 7.11 [d, $^3J_{HH}$ = 1.4 Hz, 2 H, NCHCH], 7.08 (d, $^3J_{HH}$ = 1.5 Hz, 1 H, NCCH], 7.07 [d, $^3J_{HH}$ = 7.0 Hz, 2 H, H^{3,5}(C$_6$H$_4$)], 6.84 [d, $^3J_{HH}$ = 1.4 Hz, 2 H, NCCH], 6.76 [d, $^3J_{HH}$ = 1.3 Hz, 1 H, NCCH], 3.80 [s, 3 H, NCH$_3$], 3.79 [s, 6 H, NCH$_3$], 2.26 [s, 3 H, CCH$_3$]. **^{13}C{^{1}H} NMR** (201 MHz, CD$_2$Cl$_2$, 25 °C): δ$_C$ = 280.7 [$^1J_{CW}$ = 171.3 Hz, W≡C], 223.3 [$^1J_{CW}$ = 132.1 Hz, CO], 192.5 [$^1J_{CW}$ = 95.0 Hz, NCN)], 181.7 [$^1J_{CW}$ = 44.7 Hz, NCN], 151.5 [$^2J_{CW}$ = 39.1 Hz, C^4(C$_6$H$_4$)], 136.9 [C^{2,6}(C$_6$H$_4$)], 129.0 [C^{3,5}(C$_6$H$_4$)], 128.8 [C^4(C$_6$H$_4$)], 124.3 [C^5(C$_3$N$_2$H$_2$)], 123.7 [C^5(C$_3$N$_2$H$_2$)], 120.7 [C^4(C$_3$N$_2$H$_2$)], 120.3 [C^4(C$_3$N$_2$H$_2$)], 38.8 [NCH$_3$], 38.1 [NCH$_3$], 21.8 [CCH$_3$]. **^{11}B{^{1}H} NMR** (128 MHz, CDCl$_3$, 25 °C): δ$_B$ = −1.41 (BH). **^{11}B NMR** (128 MHz, CDCl$_3$, 25 °C): δ$_B$ = −1.35 (d, $^1J_{BH}$ = 97.9 Hz, BH). **MS** (ESI, +ve ion, m/z): Found: 599.1572. Calcd for C$_{22}$H$_{24}$^{11}BN$_6$O$_2$^{184}W. [M + H]$^+$: 599.1563. **CV** (CH$_2$Cl$_2$): $E_{\frac{1}{2}}$ = 0.00 V vs. [Fe(C$_5$H$_5$)$_2$]/ [Fe(C$_5$H$_5$)$_2$]$^+$. See Figure 8 for computationally optimized molecular structure.

3.6. Synthesis of [W(≡CC$_6$H$_4$Me-4)(CO)$_2$(Tp)] (5)*

The complex has been previously described via the reaction of the thermolabile intermediate [W(≡CC$_6$H$_4$Me-4)Br(CO)$_4$] (from [W{=C(OMe)C$_6$H$_4$Me-4}(CO)$_5$] and BBr$_3$) and K[Tp*] (80%) [17]. The present synthesis follows a similar approach to the synthesis of [W(≡CC$_6$H$_3$Me$_2$-2,6)(CO)$_2$(Tp)] [17]. Sodium hydrotris(3,5-dimethyl-1-pyrazolyl) borate (0.183 g, 0.544 mmol) was dissolved in dichloromethane (15 mL) and added to a solution of [W(≡CC$_6$H$_4$Me-4)(CO)$_2$(γ-pic)$_2$(Br)] (**2a**: 0.307 g, 0.506 mmol) in dichloromethane (20 mL) with stirring overnight. The solution slowly darkened from pale orange to dark red, and this transition was visible after 3 h. Solvent and picoline were removed under reduced pressure, and the residue was redissolved in a minimum of dichloromethane (~2mL) and purified via flash column chromatography using a 1:2 DCM to petroleum spirits 60–80 eluent (silica gel, N$_2$). The orange fraction which eluted first was collected, and solvent was removed under reduced pressure to give (**5**) a bright orange powder. Yield: 217 mg (0.339 mmol, 67%). **IR** (CH$_2$Cl$_2$, cm^{-1}): 2554w ν$_{BH}$, 1971 vs. ν$_{CO}$, 1879 vs. ν$_{CO}$. **IR** (ATR, cm^{-1}): 2550 w ν$_{BH}$, 1954 vs. ν$_{CO}$, 1861 vs. ν$_{CO}$. **^{1}H NMR** (800 MHz, CDCl$_3$, 25 °C): δ$_H$ = 7.36 [d, $^3J_{HH}$ = 7.4 Hz, 2 H, H^{2,6}(C$_6$H$_4$)], 7.10 [d, $^3J_{HH}$ = 7.9 Hz, 2 H, H^{3,5}(C$_6$H$_4$)], 5.89 [s, 2 H, H^4(pz)], 5.79 [s, 1 H, H^4(pz)], 2.52 [s, 6 H, pzCH$_3$], 2.47 [s, 3 H, pzCH$_3$], 2.38 [s, 6 H, pzCH$_3$], 2.35 [s, 3 H, pzCH$_3$], 2.31 [s, 3 H, C$_6$H$_4$CH$_3$]. **^{13}C{^{1}H} NMR** (201 MHz, CDCl$_3$, 25 °C): δ$_C$ = 279.2 [$^1J_{CW}$ = 186.3 Hz, W≡C], 224.0 [$^1J_{CW}$ = 166.5 Hz, CO], 152.4 [C^5(pz)], 152.1 [C^5(pz)], 148.0 [$^2J_{CW}$ = 42.5 Hz, C^1(C$_6$H$_4$)], 145.7 [C^3(pz)], 144.5 [C^3(pz)], 137.8 [C^{2,6}(C$_6$H$_4$)], 129.3 [C^{3,5}(C$_6$H$_4$)], 128.8 [C^4(C$_6$H$_4$)], 106.7 [C^4(pz)], 106.5 [C^4(pz)], 21.8 [C$_6$H$_4$CH$_3$], 16.7 [pzCH$_3$], 15.4 [pzCH$_3$], 12.8 [pzCH$_3$], 12.8 [pzCH$_3$]. **^{11}B{^{1}H} NMR** (128 MHz, CDCl$_3$, 25 °C): δ$_B$ = −9.17 (BH). **^{11}B NMR** (128 MHz, CDCl$_3$, 25 °C): δ$_B$ = −9.15 (br, BH). **MS** (ESI, +ve ion, m/z): Found: 641.2039. Calcd for C$_{25}$H$_{30}$^{11}BN$_6$O$_2$^{184}W 641.2033. [M + H]$^+$. **CV** (CH$_2$Cl$_2$): $E_{\frac{1}{2}}$ = 0.18 V vs. [Fe(C$_5$H$_5$)$_2$]/ [Fe(C$_5$H$_5$)$_2$]$^+$.

3.7. Synthesis of [WAu(μ$_2$-CC$_6$H$_4$Me-4)Cl(CO)$_2${HB(ImMe)$_3$}] (6)

To a solution of [W(≡CC$_6$H$_4$Me-4)(CO)$_2${HB(ImMe)$_3$}] (**4**: 20 mg, 0.033 mmol) in dichloromethane (5 mL) was added [AuCl(SMe$_2$)] (10 mg, 0.034 mmol) with stirring, whereupon the solution turned from bright orange to dark red. After 30 min, a further equivalent of [AuCl(SMe$_2$)] (10 mg, 0.034 mmol) was added to the reaction, which was stirred for a further 15 min (longer reaction times resulted in gold mirror formation). After this time, the resulting solution was subjected to flash column chromatography (diatomaceous earth, CH_2Cl_2, N_2) to collect the bright orange fraction, from which solvent was removed under reduced pressure. The resulting dark orange powder was suspended in n-hexane (10 mL) and then collected via vacuum filtration, washed with n-hexane (3 × 5 mL) and dried in vacuo for 45 min, to give a brown-gold powder of (**6**). Yield: 14 mg (8.7 μmol, 54%). IR (CH_2Cl_2, cm^{-1}): 2455 w ν_{BH}, 1986 vs. ν_{CO}, 1911 vs. ν_{CO}. IR (ATR, cm^{-1}): 2454 w ν_{BH}, 1983 vs. ν_{CO}, 1879 vs. ν_{CO}. **^{1}H NMR** (600 MHz, CD_2Cl_2, 25 °C): δ_H = 7.87 [d, $^3J_{HH}$ = 8.1 Hz, 2 H, H^{2,6}(C_6H_4)], 7.24 [d, $^3J_{HH}$ = 7.8 Hz, 2 H, H^{3,5}(C_6H_4)], 7.20 [d, $^3J_{HH}$ = 1.6 Hz, 2 H, NCCH], 7.14 [d, $^3J_{HH}$ = 1.6 Hz, 1 H, NCCH], 6.91 [d, $^3J_{HH}$ = 1.6 Hz, 2 H, NCCH], 6.90 [d, $^3J_{HH}$ = 1.6 Hz, 1 H, NCCH), 3.95 [s, 3 H, NCH$_3$], 3.72 [s, 6 H, NCH$_3$), 2.36[s, 3 H, CCH$_3$]. **^{13}C{^{1}H} NMR** (151 MHz, CD_2Cl_2, 25 °C): δ_C = 277.7 [$^1J_{CW}$ = 85 Hz, W≡C], 216.0 [$^1J_{CW}$ = 119 Hz, CO], 187.7 [$^1J_{CW}$ = 90 Hz, NCN], 173.7 [$^1J_{CW}$ = 71 Hz, NCN], 149.6 [C^{2,6}(C_6H_4)], 140.4 [$^2J_{CW}$ = 97 Hz, C^1(C_6H_4)], 130.3 [C^{3,5}(C_6H_4)], 129.5 [C^4(C_6H_4)], 124.8 [C^3($C_3N_2H_2$)], 124.5 [C^3($C_3N_2H_2$)], 122.0 [C^4($C_3N_2H_2$)], 121.5 [C^4($C_3N_2H_2$)], 39.4 [NCH$_3$], 38.0 [NCH$_3$], 21.81 [$C_6H_4CH_3$]. **^{11}B{^{1}H} NMR** (128 MHz, CD_2Cl_2, 25 °C): δ_B = −1.60 (BH). **^{11}B NMR** (128 MHz, CD_2Cl_2, 25 °C): δ_B = −1.47 [d, $^1J_{BH}$ = 101.2 Hz, BH]. **MS** (ESI, +ve ion, m/z): Found: 853.0721. Calcd for $C_{22}H_{23}Au^{11}B^{35}ClN_6O_2^{184}W$ 853.0731. [M + Na]$^+$. See Figure 8 for computationally optimized molecular geometry.

4. Conclusions

The first examples of mononuclear and binuclear carbyne complexes ligated by poly(imidazolylidenyl)borates have been isolated. Spectroscopic data for these add to the growing evidence that poly(imidazolylidenyl)borates are particularly strong net donor ligands. These data are contextualised by comparison with those having a wide range of more familiar κ^3, η^5 and η^6 facially capping ligands, with recourse to two parameters k_{CO} and $\%Vol_{bur}$. Reminiscent of the steric/electronic map presented by Tolman to describe the coordinative features of phosphines, a similar map based on k_{CO} and $\%Vol_{bur}$ suggests that HB(Im^R)$_3$ ligands occupy a sparsely populated region of ligand space, associated with potent net basicity and significant (but variable) steric encumbrance.

The first of these parameters, k_{CO} (a Cotton–Kraihanzel force constant), is given by

$$k_{CO} \text{ [Ncm}^{-1}\text{]} = 1.7426 \times 10^{-6} \text{ Ncm} \times (\nu_s^2 + \nu_{as}^2)$$

where ν_s and ν_{as} are the uncorrected frequencies (in cm^{-1}) calculated at the ωB97X-D/6-31G*/LANL2Dζ level of theory for the complexes [W(≡CC$_6$H$_4$Me-4)(CO)$_2$(L)] in the gas phase.

The second of these, $\%Vol_{bur}$, is obtained using the *SambVca* protocol [35] applied to either the computationally optimised geometries or, where available, the experimentally determined atomic coordinates with hydrogen atoms included based on a sphere of 3.5Å radius centred on tungsten. Because this approach may be applied to hypothetical as well as real molecules, the method may enjoy predictive value with limited computational expense. For comparison of calculated and experimentally determined infrared data in the region of ν_{CO}-associated vibrations (1850–2100 cm^{-1}), an anharmonic scaling factor of 0.9297 is recommended for the combination of the ωB97X-D functional and 6-31G* basis set.

Supplementary Materials: The following supporting information can be downloaded at: https://www.mdpi.com/article/10.3390/molecules28237761/s1, IR, ^{1}H, ^{13}C{^{1}H} and ^{11}B NMR spectra for new compounds.

Author Contributions: Conceptualization, A.F.H.; methodology, A.F.H.; formal analysis, C.M.I., R.A.M., R.M.K., M.D.S., L.J.W. and A.F.H.; investigation, C.M.I., M.S., R.A.M., M.D.S., R.M.K. and L.J.W.; resources, A.F.H.; data curation, writing, supervision, project administration, and funding acquisition, A.F.H. All authors have read and agreed to the published version of the manuscript.

Funding: This research was funded by the Australian Research Council, grant number DP230199215.

Data Availability Statement: The data presented in this study are available in the accompanying electronic supporting information.

Conflicts of Interest: The authors declare no conflict of interest.

References

1. Trofimenko, S. Boron-pyrazole Chemistry. IV. Carbon- and Boron-Substituted Poly[(1-pyrazolyl) borates]. *J. Am. Chem. Soc.* **1967**, *89*, 6288–6294. [CrossRef]
2. Trofimenko, S. *Scorpionates: The Coordination Chemistry of Polypyrazolylborate Ligands*; Imperial College Press: London, UK, 1999.
3. Pettinari, C. *Scorpionates II: Chelating Borate Ligands*; Imperial College Press: London, UK, 2008.
4. Hopkinson, M.; Richter, C.; Schedler, M.; Glorius, F. An Overview of N-Heterocyclic Carbenes. *Nature* **2014**, *510*, 485–496. [CrossRef] [PubMed]
5. Bellotti, P.; Koy, M.; Glorius, F. Recent Advances in the Chemistry and Applications of N-heterocyclic Carbenes. *Nat. Rev. Chem.* **2021**, *5*, 711–725. [CrossRef] [PubMed]
6. Smith, C.A.; Narouz, M.R.; Lummis, P.A.; Singh, I.; Nazemi, A.; Li, C.-H.; Crudden, C.M. N-Heterocyclic Carbenes in Materials Chemistry. *Chem. Rev.* **2019**, *119*, 4986–5056. [CrossRef] [PubMed]
7. Kernbach, U.; Ramm, M.; Luger, P.; Fehlhammer, W.P. A Chelating Triscarbene and its Hexacarbene Iron Complex. *Angew. Chem. Int. Ed.* **1996**, *35*, 310–312. [CrossRef]
8. Frankel, R.; Kniczek, J.; Ponikwar, W.; Noth, H.; Polborn, K.; Fehlhammer, W.P. Bis(imidazolin-2-ylidene-1-yl)borate Complexes of Palladium(II), Platinum(II) and Gold(I). *Inorg. Chim. Acta* **2001**, *312*, 23–39. [CrossRef]
9. Frankel, R.; Kernbach, U.; Bakola-Christianopoulou, M.; Plaia, U.; Suter, M.; Ponikwar, W.; Noth, H.; Moinet, C.; Fehlhammer, W.P. Hexacarbene Complexes. *J. Organomet. Chem.* **2001**, *617–618*, 530–545. [CrossRef]
10. Biffis, A.; Tubaro, C.; Scattolin, E.; Basato, M.; Papini, G.; Santini, C.; Alvarez, E.; Conejero, S. Trinuclear Copper(I) Complexes with Triscarbene Ligands: Catalysis of C–N and C–C Coupling Reactions. *Dalton Trans.* **2009**, *35*, 7223–7229. [CrossRef]
11. Nieto, I.; Cervantes-Lee, F.; Smith, J.M. A New Synthetic Route to Bulky "Second Generation" Tris(imidazol-2-ylidene)borate Ligands: Synthesis of a Four Coordinate Iron(ii) Complex. *Chem. Commun.* **2005**, *30*, 3811–3813. [CrossRef]
12. Lu, Z.; Williams, T.J. Di(carbene)-Supported Nickel Systems for CO_2 Reduction Under Ambient Conditions. *ACS Catal.* **2016**, *6*, 6670–6673. [CrossRef]
13. Meihaus, K.R.; Minasian, S.G.; Lukens, W.W., Jr.; Kozimor, S.A.; Shuh, D.K.; Tyliszczak, T.; Long, J.R. Influence of Pyrazolate vs N-Heterocyclic Carbene Ligands on the Slow Magnetic Relaxation of Homoleptic Trischelate Lanthanide(III) and Uranium(III) Complexes. *J. Am. Chem. Soc.* **2014**, *136*, 6056–6068. [CrossRef] [PubMed]
14. Hickey, A.K.; Muñoz, S.B.; Lutz, S.A.; Pink, M.; Chen, C.-H.; Smith, J.M. Arrested α-hydride migration activates a phosphido ligand for C–H insertion. *Chem. Commun.* **2017**, *53*, 412–415. [CrossRef] [PubMed]
15. Elpitiya, G.R.; Malbrecht, B.J.; Jenkins, D.M. A Chromium(II) Tetracarbene Complex Allows Unprecedented Oxidative Group Transfer. *Inorg. Chem.* **2017**, *56*, 14101–14110. [CrossRef] [PubMed]
16. Bass, H.M.; Cramer, S.A.; McCullough, A.S.; Bernstein, K.J.; Murdock, C.R.; Jenkins, D.M. Employing Dianionic Macrocyclic Tetracarbenes to Synthesize Neutral Divalent Metal Complexes. *Organometallics* **2013**, *32*, 2160–2167. [CrossRef]
17. Isbill, S.B.; Chandrachud, P.P.; Kern, J.L.; Jenkins, D.M.; Roy, S. Elucidation of the Reaction Mechanism of $C_2 + N_1$ Aziridination from Tetracarbene Iron Catalysts. *ACS Catal.* **2019**, *9*, 6223–6233. [CrossRef]
18. Chandrachud, P.P.; Bass, H.M.; Jenkins, D.M. Synthesis of Fully Aliphatic Aziridines with a Macrocyclic Tetracarbene Iron Catalyst. *Organometallics* **2016**, *35*, 1652–1657. [CrossRef]
19. Anneser, M.R.; Elpitiya, G.R.; Townsend, J.; Johnson, E.J.; Powers, X.B.; DeJesus, J.F.; Vogiatzis, K.D.; Jenkins, D.M. Unprecedented Five-Coordinate Iron(IV) Imides Generate Divergent Spin States Based on the Imide R-Groups. *Angew. Chem. Int. Ed.* **2019**, *58*, 8115–8118. [CrossRef]
20. Anneser, M.R.; Elpitiya, G.R.; Powers, X.B.; Jenkins, D.M. Toward a Porphyrin-Style NHC: A 16-Atom Ringed Dianionic Tetra-NHC Macrocycle and Its Fe(II) and Fe(III) Complexes. *Organometallics* **2019**, *38*, 981–987. [CrossRef]
21. DeJesus, J.F.; Jenkins, D.M. A Chiral Macrocyclic Tetra- N -Heterocyclic Carbene Yields an "All Carbene" Iron Alkylidene Complex. *Chem. Eur. J.* **2020**, *26*, 1429–1435. [CrossRef]
22. Nieto, I.; Bontchev, R.P.; Smith, J.M. Synthesis of a Bulky Bis(carbene)borate Ligand—Contrasting Structures of Homoleptic Nickel(II) Bis(pyrazolyl)borate and Bis(carbene)borate Complexes. *Eur. J. Inorg. Chem.* **2008**, *2008*, 2476–2480. [CrossRef]
23. Arrowsmith, M.; Hill, M.S.; Kociok-Köhn, G. Bis(imidazolin-2-ylidene-1-yl)borate Complexes of the Heavier Alkaline Earths: Synthesis and Studies of Catalytic Hydroamination. *Organometallics* **2009**, *28*, 1730–1738. [CrossRef]
24. Kaufhold, S.; Rosemann, N.W.; Chábera, P.; Lindh, L.; Losada, I.B.; Uhlig, J.; Pascher, T.; Strand, D.; Wärnmark, K.; Yartsev, A.; et al. Microsecond Photoluminescence and Photoreactivity of a Metal-Centered Excited State in a Hexacarbene–Co(III) Complex. *J. Am. Chem. Soc.* **2021**, *143*, 1307–1312. [CrossRef]
25. Kjær, K.S.; Kaul, N.; Prakash, O.; Chábera, P.; Rosemann, N.W.; Honarfar, A.; Gordivska, O.; Fredin, L.A.; Bergquist, K.E.; Häggström, L.; et al. Luminescence and reactivity of a charge-transfer excited iron complex with nanosecond lifetime. *Science* **2019**, *363*, 249–253. [CrossRef] [PubMed]

26. Forshaw, A.P.; Smith, J.M.; Ozarowski, A.; Krzystek, J.; Smirnov, D.; Zvyagin, S.A.; Harris, T.D.; Karunadasa, H.I.; Zadrozny, J.M.; Schnegg, A.; et al. Low-Spin Hexacoordinate Mn(III): Synthesis and Spectroscopic Investigation of Homoleptic Tris(pyrazolyl)borate and Tris(carbene)borate Complexes. *Inorg. Chem.* **2013**, *52*, 144–159. [CrossRef]
27. Prakash, O.; Chábera, P.; Rosemann, N.W.; Huang, P.; Häggström, L.; Ericsson, T.; Strand, D.; Persson, P.; Bendix, J.; Lomoth, R.; et al. A Stable Homoleptic Organometallic Iron(IV) Complex. *Chem. Eur. J.* **2020**, *26*, 12728–12732. [CrossRef] [PubMed]
28. Forshaw, A.P.; Bontchev, R.P.; Smith, J.M. Oxidation of the Tris(carbene)borate Complex PhB(MeIm)$_3$Mn^I(CO)$_3$ to Mn^{IV}[PhB(MeIm)$_3$]$_2$(OTf)$_2$. *Inorg. Chem.* **2007**, *46*, 3792–3794. [CrossRef]
29. Chen, F.; Wang, G.-F.; Li, Y.-Z.; Chen, X.-T.; Xue, Z.-L. Syntheses, structures and electrochemical properties of homoleptic ruthenium(III) and osmium(III) complexes bearing two tris(carbene)borate ligands. *Inorg. Chem. Commun.* **2012**, *21*, 88–91. [CrossRef]
30. Chen, F.; Wang, G.-F.; Li, Y.-Z.; Chen, X.-T.; Xue, Z.-L. Synthesis and characterization of rhodium(I) and iridium(I) carbonyl phosphine complexes with bis(N-heterocyclic carbene)borate ligands. *J. Organomet. Chem.* **2010**, *710*, 36–43. [CrossRef]
31. Chen, F.; Sun, J.-F.; Li, T.-Y.; Chen, X.-T.; Xue, Z.-L. Iridium(I) and Rhodium(I) Carbonyl Complexes with the Bis(3-*tert*-butylimidazol-2-ylidene)borate Ligand and Unusual B−H Fluorination. *Organometallics* **2011**, *30*, 2006–2011. [CrossRef]
32. Nishiura, T.; Takabatake, A.; Okutsu, M.; Nakazawa, J.; Hikichi, S. Heteroleptic cobalt(iii) acetylacetonato complexes with N-heterocyclic carbine-donating scorpionate ligands: Synthesis, structural characterization and catalysis. *Dalton Trans.* **2019**, *48*, 2564–2568. [CrossRef]
33. Fostvedt, J.I.; Lohrey, T.D.; Bergman, R.G.; Arnold, J. Structural diversity in multinuclear tantalum polyhydrides formed via reductive hydrogenolysis of metal–carbon bonds. *Chem. Commun.* **2019**, *55*, 13263–13266. [CrossRef]
34. Fostvedt, J.I.; Boreen, M.A.; Bergman, R.G.; Arnold, J. A Diverse Array of C–C Bonds Formed at a Tantalum Metal Center. *Inorg. Chem.* **2021**, *60*, 9912–9931. [CrossRef] [PubMed]
35. Nieto, I.; Bontchev, R.P.; Ozarowski, A.; Smirnov, D.; Krzystek, J.; Telser, J.; Smith, J.M. Synthesis and spectroscopic investigations of four-coordinate nickel complexes supported by a strongly donating scorpionate ligand. *Inorganica Chim. Acta* **2009**, *362*, 4449–4460. [CrossRef]
36. Papini, G.; Bandoli, G.; Dolmella, A.; Lobbia, G.G.; Pellei, M.; Santini, C. New homoleptic carbene transfer ligands and related coinage metal complexes. *Inorg. Chem. Commun.* **2008**, *11*, 1103–1110. [CrossRef]
37. Muñoz, S.B.; Foster, W.K.; Lin, H.-J.; Margarit, C.G.; Dickie, D.A.; Smith, J.M. Tris(carbene)borate Ligands Featuring Imidazole-2-ylidene, Benzimidazol-2-ylidene, and 1,3,4-Triazol-2-ylidene Donors. Evaluation of Donor Properties in Four-Coordinate {NiNO}^{10} Complexes. *Inorg. Chem.* **2012**, *51*, 12660–13668. [CrossRef] [PubMed]
38. Xu, W.-F.; Li, X.-W.; Li, Y.-Z.; Chen, X.-T.; Xue, Z.-L. Synthesis and structures of π-allylpalladium(II) complexes containing bis(1,2,4-triazol-5-ylidene-1-yl)borate ligands. An unusual tetrahedral palladium complex. *J. Organomet. Chem.* **2011**, *696*, 3800–3806. [CrossRef]
39. Huang, J.; Stevens, E.D.; Nolan, S.P.; Petersen, J.L. Olefin Metathesis-Active Ruthenium Complexes Bearing a Nucleophilic Carbene Ligand. *J. Am. Chem. Soc.* **1999**, *121*, 2674–2678. [CrossRef]
40. Scholl, M.; Trnka, T.M.; Morgan, J.P.; Grubbs, R.H. Increased Ring Closing Metathesis Activity of Ruthenium-Based Olefin Metathesis Catalysts Coordinated with Imidazolin-2-ylidene Ligands. *Tetrahedron Lett.* **1999**, *40*, 2247–2250. [CrossRef]
41. Ackermann, L.; Fürstner, A.; Weskamp, T.; Kohl, F.J.; Herrmann, W.A. Ruthenium carbene complexes with imidazolin-2-ylidene ligands allow the formation of tetrasubstituted cycloalkenes by RCM. *Tetrahedron Lett.* **1999**, *40*, 4787–4790. [CrossRef]
42. Scholl, M.; Ding, S.; Lee, C.W.; Grubbs, R.H. Synthesis and Activity of a New Generation of Ruthenium-Based Olefin Metathesis Catalysts Coordinated with 1,3-Dimesityl-4,5-dihydroimidazol-2-ylidene Ligands. *Org. Lett.* **1999**, *1*, 953–956. [CrossRef]
43. Cardin, D.J.; Doyle, M.J.; Lappert, M.F. Rhodium(I)-catalysed dismutation of electron-rich olefins: Rhodium(I) carbene complexes as intermediates. *J. Chem. Soc. Chem. Commun.* **1972**, *16*, 927–928. [CrossRef]
44. Shao, M.; Zheng, L.; Qiao, W.; Wang, J.; Wang, J. A Unique Ruthenium Carbyne Complex: A Highly Thermo-endurable Catalyst for Olefin Metathesis. *Adv. Synth. Catal.* **2012**, *354*, 2743–2750. [CrossRef]
45. Ledoux, N.; Drozdzak, R.; Allaert, B.; Linden, A.; Van Der Voort, P.; Verpoort, F. Exploring new synthetic strategies in the development of a chemically activated Ru-based olefin metathesis catalyst. *Dalton Trans.* **2007**, *44*, 5201–5210. [CrossRef] [PubMed]
46. Buil, M.L.; Cardo, J.J.F.; Esteruelas, M.A.; Oñate, E. Square-Planar Alkylidyne–Osmium and Five-Coordinate Alkylidene–Osmium Complexes: Controlling the Transformation from Hydride-Alkylidyne to Alkylidene. *J. Am. Chem. Soc.* **2016**, *138*, 9720–9728. [CrossRef] [PubMed]
47. Castarlenas, R.; Esteruelas, M.A.; Lalrempuia, R.; Oliván, M.; Oñate, E. Osmium−Allenylidene Complexes Containing an N-Heterocyclic Carbene Ligand. *Organometallics* **2008**, *27*, 795–798. [CrossRef]
48. Buil, M.L.; Cardo, J.J.F.; Esteruelas, M.A.; Oñate, E. Dehydrogenative Addition of Aldehydes to a Mixed NHC-Osmium-Phosphine Hydroxide Complex: Formation of Carboxylate Derivatives. *Organometallics* **2016**, *35*, 2171–2173. [CrossRef]
49. Castarlenas, R.; Esteruelas, M.A.; Oñate, E. Preparation and Structure of Alkylidene−Osmium and Hydride−Alkylidyne−Osmium Complexes Containing an N-Heterocyclic Carbene Ligand. *Organometallics* **2007**, *26*, 2129–2132. [CrossRef]
50. Buil, M.L.; Cardo, J.J.F.; Esteruelas, M.A.; Fernández, I.; Oñate, E. Hydroboration and Hydrogenation of an Osmium–Carbon Triple Bond: Osmium Chemistry of a Bis-σ-Borane. *Organometallics* **2015**, *34*, 547–550. [CrossRef]

51. Fuchs, J.; Irran, E.; Hrobárik, P.; Klare, H.F.T.; Oestreich, M. Si–H Bond Activation with Bullock's Cationic Tungsten(II) Catalyst: CO as Cooperating Ligand. *J. Am. Chem. Soc.* **2019**, *141*, 18845–18850. [CrossRef]
52. Koy, M.; Elser, I.; Meisner, J.; Frey, W.; Wurst, K.; Kästner, J.; Buchmeiser, M.R. High Oxidation State Molybdenum *N*-Heterocyclic Carbene Alkylidyne Complexes: Synthesis, Mechanistic Studies, and Reactivity. *Chem. Eur. J.* **2017**, *23*, 15484–15490. [CrossRef]
53. Hauser, P.M.; van der Ende, M.; Groos, J.; Frey, W.; Wang, D.; Buchmeiser, M.R. Cationic Tungsten Alkylidyne *N*-Heterocyclic Carbene Complexes: Synthesis and Reactivity in Alkyne Metathesis. *Eur. J. Inorg. Chem.* **2020**, *2020*, 3070–3082. [CrossRef]
54. Elser, I.; Groos, J.; Hauser, P.M.; Koy, M.; van der Ende, M.; Wang, D.; Frey, W.; Wurst, K.; Meisner, J.; Ziegler, F.; et al. Molybdenum and Tungsten Alkylidyne Complexes Containing Mono-, Bi-, and Tridentate *N*-Heterocyclic Carbenes. *Organometallics* **2019**, *38*, 4133–4146. [CrossRef]
55. Groos, J.; Koy, M.; Musso, J.; Neuwirt, M.; Pham, T.; Hauser, P.M.; Frey, W.; Buchmeiser, M.R. Ligand Variations in Neutral and Cationic Molybdenum Alkylidyne NHC Catalysts. *Organometallics* **2022**, *41*, 1167–1183. [CrossRef]
56. Groos, J.; Hauser, P.M.; Koy, M.; Frey, W.; Buchmeiser, M.R. Highly Reactive Cationic Molybdenum Alkylidyne *N*-Heterocyclic Carbene Catalysts for Alkyne Metathesis. *Organometallics* **2021**, *40*, 1178–1184. [CrossRef]
57. Hauser, P.M.; Musso, J.V.; Frey, W.; Buchmeiser, M.R. Cationic Tungsten Oxo Alkylidene *N*-Heterocyclic Carbene Complexes via Hydrolysis of Cationic Alkylidyne Progenitors. *Organometallics* **2021**, *40*, 927–937. [CrossRef]
58. Caldwell, L.M. Alkylidyne Complexes Ligated by Poly(pyrazolyl)borates. *Adv. Organomet. Chem.* **2008**, *56*, 1–94. [CrossRef]
59. Mathur, M.A.; Moore, D.A.; Popham, R.E.; Sisler, H.H.; Dolan, S.; Shore, S.G. Trimethylamine-Tribromoborane. *Inorg. Synth.* **2007**, *29*, 51–54. [CrossRef]
60. Dossett, S.J.; Hill, A.F.; Jeffery, J.C.; Marken, F.; Sherwood, P.; Stone, F.G.A. Synthesis and Reactions of the Alkylidynemetal Complexes [M(CR)(CO)$_2$(η-C$_5$H$_5$)](R = C$_6$H$_3$Me$_2$-2,6, M = Cr, Mo, or W; R = C$_6$H$_4$Me-2, C$_6$H$_4$OMe-2, or C$_6$H$_4$NMe$_2$-4, M = Mo); Crystal Structure of the Compound [MoFe(μ-CC$_6$H$_3$Me$_2$-2,6)(CO)$_5$(η-C$_5$H$_5$)]. *J. Chem. Soc. Dalton Trans.* **1988**, *9*, 2453–2465. [CrossRef]
61. Anderson, S.; Hill, A.F.; Nasir, B.A. Steric and "Indenyl" Effects in the Chemistry of Alkylidyne Complexes of Tungsten and Molybdenum. *Organometallics* **1995**, *14*, 2987–2992. [CrossRef]
62. Hill, A.F.; Malget, J.M.; White, A.J.P.; Williams, D.J. Dihydrobis(pyrazolyl)borate Alkylidyne Complexes of Tungsten. *Eur. J. Inorg. Chem.* **2004**, *2004*, 818–828. [CrossRef]
63. Ghadwal, R.S.; Rottschäfer, D.; Andrada, D.M.; Frenking, G.; Schürmann, C.J.; Stammler, H.-G. Normal-to-abnormal rearrangement of an N-heterocyclic carbene with a silylene transition metal complex. *Dalton Trans.* **2017**, *46*, 7791–7799. [CrossRef] [PubMed]
64. Jeffery, J.C.; Gordon, F.; Stone, A.; Williams, G.K. Synthesis of alkylidyne tungsten complexes with hydrotris(pyrazol-1-yl)borate ligands. *Polyhedron* **1991**, *10*, 215–219. [CrossRef]
65. Anderson, S.; Cook, D.J.; Hill, A.F.; Malget, J.M.; White, A.J.P.; Williams, D.J. Reactions of Tungsten Alkylidynes with Thionyl Chloride. *Organometallics* **2004**, *23*, 2552–2557. [CrossRef]
66. Wadepohl, H.; Arnold, U.; Pritzkow, H.; Calhorda, M.J.; Veiros, L.F. Interplay of ketenyl and nitrile ligands on d6-transition metal centres. Acetonitrile as an end-on (two-electron) and a side-on (four-electron) ligand. *J. Organomet. Chem.* **1999**, *587*, 233–243. [CrossRef]
67. Foreman, M.R.S.-J.; Hill, A.F.; White, A.J.P.; Williams, D.J. Hydrotris(methimazolyl)borato Alkylidyne Complexes of Tungsten. *Organometallics* **2003**, *22*, 3831–3840. [CrossRef]
68. Kläui, W.; Hamers, H. Molybdän- und wolfram-dicarbonyl-carbin-komplexe stabilisiert durch dreizähnige sauerstoff-liganden. *J. Organomet. Chem.* **1988**, *345*, 287–298. [CrossRef]
69. Mayr, A.; Ahn, S. Oxidatively induced insertion of an alkylidyne unit into the tungsten tris(pyrazolyl)borate cage. *Inorganica Chim. Acta* **2000**, *300–302*, 406–413. [CrossRef]
70. Delaney, A.R.; Frogley, B.J.; Hill, A.F. Metal coordination to a dimetallaoctatetrayne. *Dalton Trans.* **2019**, *48*, 13674–13684. [CrossRef]
71. Manzano, R.A.; Hill, A.F. Fluorocarbyne complexes via electrophilic fluorination of carbido ligands. *Chem. Sci.* **2023**, *14*, 3776–3781. [CrossRef]
72. Tolman, C.A. Steric effects of phosphorus ligands in organometallic chemistry and homogeneous catalysis. *Chem. Rev.* **1977**, *77*, 313–348. [CrossRef]
73. Gusev, D.G. Electronic and Steric Parameters of 76 N-Heterocyclic Carbenes in Ni(CO)$_3$(NHC). *Organometallics* **2009**, *28*, 6458–6461. [CrossRef]
74. Dorta, R.; Stevens, E.D.; Hoff, C.D.; Nolan, S.P. Stable, three-coordinate Ni(CO)$_2$(NHC)(NHC= N-heterocyclic carbene) complexes enabling the determination of Ni−NHC bond energies. *J. Am. Chem. Soc.* **2003**, *125*, 10490–10491. [CrossRef] [PubMed]
75. Dorta, R.; Stevens, E.D.; Scott, N.M.; Costabile, C.; Cavallo, L.; Hoff, C.D.; Nolan, S.P. Steric and Electronic Properties of N-Heterocyclic Carbenes (NHC): A Detailed Study on Their Interaction with Ni(CO)$_4$. *J. Am. Chem. Soc.* **2005**, *127*, 2485–2495. [CrossRef] [PubMed]
76. Huynh, H.V. Electronic Properties of N-Heterocyclic Carbenes and Their Experimental Determination. *Chem. Rev.* **2018**, *118*, 9457–9492. [CrossRef] [PubMed]

77. Abernethy, R.J.; Foreman MR, S.t.-J.; Hill, A.F.; Tshabang, N.; Willis, A.C.; Young, R.D. Similar arguments have been applied to complexes of the form [M(NO)(CO)$_2$(L)] and [M(κ^3-C$_3$H$_5$)(CO)$_2$(L)] (M = Mo, W): (b) Poly(methimazolyl)borato Nitrosyl Complexes of Molybdenum and Tungsten. *Organometallics* **2008**, *27*, 4455–4463. [CrossRef]
78. Abernethy, R.J.; Foreman, M.R.S.-J.; Hill, A.F.; Smith, M.K.; Willis, A.C. Relative hemilabilities of H2B(az)2 (az = pyrazolyl, dimethylpyrazolyl, methimazolyl) chelates in the complexes [M(η-C3H5)(CO)2{H2B(az)2}] (M = Mo, W). *Dalton Trans.* **2020**, *49*, 781–796. [CrossRef] [PubMed]
79. Green, M.; Howard, J.A.K.; James, A.L.; Nunn, C.M.; Stone, F.G.A. Alkylidyne(carbaborane)tungsten-gold and -Rhodium Complexes; Crystal Structures of [AuW(μ-CR)(CO)$_2$(PPh$_3$)(η^5-C$_2$B$_9$H$_9$Me$_2$)], [RhW(μ-CR)(CO)$_2$(PPh$_3$)$_2$(η^5-C$_2$B$_9$H$_9$Me$_2$)], and [RhW(μ-CR)(CO)$_2$(PPh$_3$)$_2${η^5-C$_2$B$_9$(C$_7$H$_9$)H$_8$Me$_2$}](R = C$_6$H$_4$Me-4). *J. Chem. Soc. Dalton Trans.* **1987**, *1*, 61–72. [CrossRef]
80. Lee, F.-W.; Chan, M.C.-W.; Cheung, K.-K.; Che, C.-M. Carbyne complexes of the group 6 metals containing 1,4,7-triazacyclononane and its 1,4,7-trimethyl derivative. *J. Organomet. Chem.* **1998**, *552*, 255–263. [CrossRef]
81. Lee, F.-W.; Chan, M.C.-W.; Cheung, K.-K.; Che, C.-M. Synthesis, crystal structures and spectroscopic properties of cationic carbyne complexes of molybdenum and tungsten supported by tripodal nitrogen, phosphorus and sulphur donor ligands. *J. Organomet. Chem.* **1998**, *563*, 191–200. [CrossRef]
82. Fischer, E.O.; Lindner, T.L.; Kreissl, F.R. Übergangsmetall-carbin-komplexe: XVI. π-Cyclopentadienyl(dicarbonyl)carbin-komplexe des wolframs. *J. Organomet. Chem.* **1976**, *112*, C27–C30. [CrossRef]
83. Fischer, E.O.; Lindner, T.L.; Huttner, G.; Friedrich, P.; Kreißl, F.R.; Besenhard, J.O. Übergangsmetall-Carbin-Komplexe, XXVII. Dicarbonyl(π-cyclopentadienyl)carbin-Komplexe des Wolframs. *Eur. J. Inorg. Chem.* **1977**, *110*, 3397–3404. [CrossRef]
84. Green, M.; Howard, J.A.K.; James, A.P.; Nunn, C.M.; Stone, F.G.A. Synthesis of Alkylidynetungsten Complexes with (Pyrazol-1-yl)borato Ligands; Crystal Structures of [W(CC$_6$H$_4$Me-4)(CO)$_2${B(pz)$_4$}], [Rh$_2$W(μ_3-CC$_6$H$_4$Me-4)(μ-CO)(CO)$_2$(η-C$_9$H$_7$)$_2${HB(pz)$_3$}]·CH$_2$Cl$_2$ and [FeRhW(μ_3-CC$_6$H$_4$Me-4)(μ-MeC$_2$Me)(CO)$_4$(η-C$_9$H$_7$){HB(pz)$_3$}]. *J. Chem. Soc. Dalton Trans.* **1986**, *1*, 187–197. [CrossRef]
85. Green, M.; Howard, J.A.K.; James, A.P.; Jelfs, A.N.d.M.; Nunn, C.M.; Stone, F.G.A. Alkylidyne[pyrazol-1-yl)borato]tungsten complexes:metal–carbon triple bonds as four-electron donors. *J. Chem. Soc. Chem. Commun.* **1984**, *24*, 1623–1625. [CrossRef]
86. Delgado, E.; Hein, J.; Jeffery, J.C.; Ratermann, A.L.; Stone, F.A. Addition of Methylene Groups to the Unsaturated Complex [FeW(μ-CC$_6$H$_4$Me-4)(CO)$_5$(η-C$_5$Me$_5$)]: μ-Alkenyl Ligand Rearrangements at a Dimetal Centre. *J. Organomet. Chem.* **1986**, *307*, C23–C26. [CrossRef]
87. Byers, P.K.; Stone, F.G.A. Alkylidyne Tungsten and Molybdenum Complexes with Pyrazolylmethane Ligands. *J. Chem. Soc. Dalton Trans.* **1990**, *11*, 3499–3505. [CrossRef]
88. Devore, D.D.; Henderson, S.J.; Howard, J.A.; Gordon, F.; Stone, A. Docosahedral carbaboranetungsten-alkylidyne complexes: Synthesis of compounds with heteronuclear metal-metal bonds. *J. Organomet. Chem.* **1988**, *358*, C6–C10. [CrossRef]
89. Crennell, S.J.; DeVore, D.D.; Henderson, S.J.B.; Howard, J.A.K.; Stone, F.G.A. Docosahedral Carbaborane(alkylidyne)tungsten Complexes as Reagents for the Synthesis of Compounds with Heteronuclear Metal–metal Bonds: Crystal Structures of [NEt$_4$][W(CC$_6$H$_6$Me$_2$-2,6)(CO)$_2$(η^6-C$_2$B$_{10}$H$_{10}$Me$_2$)] and [NEt$_4$][WFe(μ-CC$_6$H$_3$Me$_2$-2,6)(CO)$_4$(η^6-C$_2$B$_{10}$H$_{10}$Me$_2$)]. *J. Chem. Soc. Dalton Trans.* **1989**, *7*, 1363–1374. [CrossRef]
90. Cotton, F.A.; Kraihanzel, C.S. Vibrational Spectra and Bonding in Metal Carbonyls. I. Infrared Spectra of Phosphine-substituted Group VI Carbonyls in the CO Stretching Region. *J. Am. Chem. Soc.* **1962**, *84*, 4432–4438. [CrossRef]
91. Parker, D.J. Solvent effects on the infrared active CO stretching frequencies of some metal carbonyl complexes—I. Dimanganese decarbonyl and dirhenium decarbonyl. *Spectrochim. Acta Part A Mol. Spectrosc.* **1983**, *39*, 463–476. [CrossRef]
92. Poater, A.; Cosenza, B.; Correa, A.; Giudice, S.; Ragone, F.; Scarano, V.; Cavallo, L. SambVca: A Web Application for the Calculation of the Buried Volume of N-Heterocyclic Carbene Ligands. *Eur. J. Inorg. Chem.* **2009**, *2009*, 1759–1766. [CrossRef]
93. Falivene, L.; Cao, Z.; Petta, A.; Serra, L.; Poater, A.; Oliva, R.; Scarano, V.; Cavallo, L. Towards the online computer-aided design of catalytic pockets. *Nat. Chem.* **2019**, *11*, 872–879. [CrossRef]
94. Clavier, H.; Nolan, S.P. Percent buried volume for phosphine and N-heterocyclic carbene ligands: Steric properties in organometallic chemistry. *Chem. Commun.* **2010**, *46*, 841–861. [CrossRef] [PubMed]
95. Chai, J.-D.; Head-Gordon, M. Systematic optimization of long-range corrected hybrid density functionals. *J. Chem. Phys.* **2008**, *128*, 084106. [CrossRef] [PubMed]
96. Chai, J.D.; Head-Gordon, M. Long-range Corrected Hybrid Density Functionals with Damped Atom–atom Dispersion Corrections. *Phys. Chem. Chem. Phys.* **2008**, *10*, 6615–6620. [CrossRef]
97. Hehre, W.J.; Ditchfield, R.; Pople, J.A. Self—Consistent Molecular Orbital Methods. XII. Further Extensions of Gaussian—Type Basis Sets for Use in Molecular Orbital Studies of Organic Molecules. *J. Chem. Phys.* **1972**, *56*, 2257–2261. [CrossRef]
98. Hay, P.J.; Wadt, W.R. Ab initio effective core potentials for molecular calculations. Potentials for the transition metal atoms Sc to Hg. *J. Chem. Phys.* **1985**, *82*, 270–283. [CrossRef]
99. Hay, P.J.; Wadt, W.R. Ab initio effective core potentials for molecular calculations. Potentials for K to Au including the outermost core orbitals. *J. Chem. Phys.* **1985**, *82*, 299–310. [CrossRef]
100. Wadt, W.R.; Hay, P.J. Ab initio effective core potentials for molecular calculations. Potentials for main group elements Na to Bi. *J. Chem. Phys.* **1985**, *82*, 284–298. [CrossRef]
101. *SPARTAN20*®, Wavefunction Inc.: Irvine, CA, USA, 2022.

102. The National Institute of Standards and Technology Computational Chemistry Comparison and Benchmark Data Base. Recommends a scaling factor on 0.9485 for the B97X-D/6-31G* combination. 2022. Available online: https://cccbdb.nist.gov/vibscalejust.asp. (accessed on 20 November 2023).
103. Merrick, J.P.; Moran, D.; Radom, L. An Evaluation of Harmonic Vibrational Frequency Scale Factors. *J. Phys. Chem. A* **2007**, *111*, 11683–11700. [CrossRef]
104. Lin, C.Y.; George, M.W.; Gill, P.M.W. EDF2: A Density Functional for Predicting Molecular Vibrational Frequencies. *Aust. J. Chem.* **2004**, *57*, 365–370. [CrossRef]
105. Halls, M.D.; Velkovski, J.; Schlegel, H.B. Harmonic frequency scaling factors for Hartree-Fock, S-VWN, B-LYP, B3-LYP, B3-PW91 and MP2 with the Sadlej pVTZ electric property basis set. *Theor. Chem. Accounts* **2001**, *105*, 413–421. [CrossRef]
106. Jacobsen, R.L.; Johnson, R.D.; Irikura, K.K.; Kacker, R.N. Anharmonic Vibrational Frequency Calculations Are Not Worthwhile for Small Basis Sets. *J. Chem. Theory Comput.* **2013**, *9*, 951–954. [CrossRef] [PubMed]
107. Barone, V.; Cossi, M. Quantum Calculation of Molecular Energies and Energy Gradients in Solution by a Conductor Solvent Model. *J. Phys. Chem. A* **1998**, *102*, 1995–2001. [CrossRef]
108. Marenich, A.V.; Cramer, C.J.; Truhlar, D.G. Universal Solvation Model Based on Solute Electron Density and on a Continuum Model of the Solvent Defined by the Bulk Dielectric Constant and Atomic Surface Tensions. *J. Phys. Chem. B* **2009**, *113*, 6378–6396. [CrossRef]
109. Klamt, A.; Schüürmann, G. COSMO: A new approach to dielectric screening in solvents with explicit expressions for the screening energy and its gradient. *J. Chem. Soc. Perkin Trans.* **1993**, *2*, 799–805. [CrossRef]
110. Klamt, A. Conductor-like Screening Model for Real Solvents: A New Approach to the Quantitative Calculation of Solvation Phenomena. *J. Phys. Chem.* **1995**, *99*, 2224–2235. [CrossRef]
111. Sokolenko, T.M.; Petko, K.I.; Yagupolskii, L.M. N-Trifluoromethylazoles. *Chem. Heterocyclic Comp.* **2009**, *45*, 430–435. [CrossRef]
112. Maffett, L.S.; Gunter, K.L.; Kreisel, K.A.; Yap, G.P.; Rabinovich, D. Nickel nitrosyl complexes in a sulfur-rich environment: The first poly(mercaptoimidazolyl)borate derivatives. *Polyhedron* **2007**, *26*, 4758–4764. [CrossRef]
113. Green, J.C.; Underwood, C. Pentamethylcyclopentadienylnickelnitrosyl: Synthesis and photoelectron spectrum. *J. Organomet. Chem.* **1997**, *528*, 91–94. [CrossRef]
114. Landry, V.K.; Pang, K.; Quan, S.M.; Parkin, G. Tetrahedral nickel nitrosyl complexes with tripodal [N3] and [Se3] donor ancillary ligands: Structural and computational evidence that a linear nitrosyl is a trivalent ligand. *Dalton Trans.* **2007**, *8*, 820–824. [CrossRef]
115. MacBeth, C.E.; Thomas, J.C.; Betley, T.A.; Peters, J.C. The Coordination Chemistry of "[BP₃]NiX" Platforms: Targeting Low-Valent Nickel Sources as Promising Candidates to L₃NiE and L₃Ni⋮E Linkages. *Inorg. Chem.* **2004**, *43*, 4645–4662. [CrossRef] [PubMed]
116. Tomson, N.C.; Crimmin, M.R.; Petrenko, T.; Roseburgh, L.E.; Sproules, S.; Boyd, W.C.; Bergman, R.G.; DeBeer, S.; Toste, F.D.; Wieghardt, K. A Step Beyond the Feltham-Enemark Notation: Spectroscopic and Correlated Ab Initio Computational Support for an Antiferromagnetically Coupled M(II)(NO) Description of Tp*M(NO)(M = Co, Ni). *J. Am. Chem. Soc.* **2011**, *133*, 18785–18801. [CrossRef] [PubMed]
117. Hoffmann, R.; Chen, M.M.L.; Elian, M.; Rossi, A.R.; Mingos, D.M.P. Pentacoordinate nitrosyls. *Inorg. Chem.* **1974**, *13*, 2666–2675. [CrossRef]
118. Rerek, M.E.; Ji, L.-N.; Basolo, F. The indenyl ligand effect on the rate of substitution reactions of Rh(η-C$_9$H$_7$)(CO)$_2$ and Mn(η-C$_9$H$_7$)(CO)$_3$. *J. Chem. Soc. Chem. Commun.* **1983**, *21*, 1208–1209. [CrossRef]
119. Hart-Davis, A.J.; Mawby, R.J. Reactions of π-indenyl complexes of transition metals. Part I. Kinetics and mechanisms of reactions of tricarbonyl-π-indenylmethylmolybdenum with phosphorus(III) ligands. *J. Chem. Soc. A Inorg. Phys. Theor.* **1969**, 2403–2407. [CrossRef]
120. Bartik, T.; Weng, W.; Ramsden, J.A.; Szafert, S.; Falloon, S.B.; Arif, A.M.; Gladysz, J.A. New Forms of Coordinated Carbon: Wirelike Cumulenic C$_3$ and C$_5$ sp Carbon Chains that Span Two Different Transition Metals and Mediate Charge Transfer. *J. Am. Chem. Soc.* **1998**, *120*, 11071–11081. [CrossRef]
121. Manzano, R.A.; Hill, A.F. For a review of tricarbido complexes see Propargylidyne and tricarbido complexes. *Adv. Organomet. Chem.* **2019**, *72*, 103–171. [CrossRef]
122. Cairns, G.A.; Carr, N.; Green, M.; Mahon, M.F. Reaction of [W(η^2-PhC$_2$Ph)$_3$(NCMe)] with o-diphenylphosphino-styrene and -allylbenzene; evidence for novel carbon–carbon double and triple bond cleavage and alkyne insertion reactions. *Chem. Commun.* **1996**, *21*, 2431–2432. [CrossRef]
123. Yeh, W.-Y. C$_{60}$-induced alkyne–alkyne coupling and alkyne scission reactions of a tungsten tris(diphenylacetylene) complex. *Chem. Commun.* **2011**, *47*, 1506–1508. [CrossRef]
124. Buss, J.A.; Agapie, T. Four-electron deoxygenative reductive coupling of carbon monoxide at a single metal site. *Nature* **2016**, *529*, 72–75. [CrossRef]
125. Buss, J.A.; Agapie, T. Mechanism of Molybdenum-Mediated Carbon Monoxide Deoxygenation and Coupling: Mono- and Dicarbyne Complexes Precede C−O Bond Cleavage and C−C Bond Formation. *J. Am. Chem. Soc.* **2016**, *138*, 16466–16477. [CrossRef]
126. Buss, J.A.; Bailey, G.A.; Oppenheim, J.; VanderVelde, D.G.; Goddard, W.A.; Agapie, T. CO Coupling Chemistry of a Terminal Mo Carbide: Sequential Addition of Proton, Hydride, and CO Releases Ethenone. *J. Am. Chem. Soc.* **2019**, *141*, 15664–15674. [CrossRef] [PubMed]

127. Bailey, G.A.; Buss, J.A.; Oyala, P.H.; Agapie, T. Terminal, Open-Shell Mo Carbide and Carbyne Complexes: Spin Delocalization and Ligand Noninnocence. *J. Am. Chem. Soc.* **2021**, *143*, 13091–13102. [CrossRef] [PubMed]
128. Ortin, Y.; Lugan, N.; Mathieu, R. Subtle reactivity patterns of non-heteroatom-substituted manganese alkynyl carbene complexes in the presence of phosphorus probes. *Dalton Trans.* **2005**, *9*, 1620–1636. [CrossRef] [PubMed]
129. Casey, C.P.; Kraft, S.; Kavana, M. Intramolecular CH Insertion Reactions of (Pentamethylcyclopentadienyl)Rhenium Alkynylcarbene Complexes. *Organometallics* **2001**, *20*, 3795–3799. [CrossRef]
130. Fischer, E.O.; Chen, J. Nucleophilic Cleavage of the Carbon-Fluorine Bond in a Fluorocarbene Complex of Manganese. *Z. Naturforsch.* **1983**, *38*, 580–581. [CrossRef]
131. Fischer, E.O.; Chen, J.; Scherzer, K. π-Cyclopentadienyl(dicarbonyl)[aryl(halogen)-carben]-komplexe des Mangans und Rheniums. *J. Organomet. Chem.* **1983**, *253*, 231–241. [CrossRef]
132. Orama, O.; Schubert, U.; Kreissl, F.R.; Fischer, E.O. Reaction of a Cationic Carbyne Complex with a Carbonylmetalate. The Phenylketenyl Group as a Bridging Ligand. *Z. Naturforsch.* **1980**, *35*, 82–85. [CrossRef]
133. Young, R.D.; Lawes, D.J.; Hill, A.F.; Ball, G.E. Observation of a Tungsten Alkane σ-Complex Showing Selective Binding of Methyl Groups Using FTIR and NMR Spectroscopies. *J. Am. Chem. Soc.* **2012**, *134*, 8294–8297. [CrossRef]
134. Carter, T.J.; Kampf, J.W.; Szymczak, N.K. Reduction of Borazines Mediated by Low-Valent Chromium Species. *Angew. Chem. Int. Ed.* **2012**, *51*, 13168–13349. [CrossRef]
135. Prinz, R.; Werner, H. Tricarbonylhexamethylborazolechromium. *Angew. Chem. Int. Ed. Engl.* **1967**, *6*, 91–92. [CrossRef]
136. Kreickmann, T.; Arndt, S.; Schrock, R.R.; Müller, P. Imido Alkylidene Bispyrrolyl Complexes of Tungsten. *Organometallics* **2007**, *26*, 5702–5711. [CrossRef]
137. Mathey, F. The Organic Chemistry of Phospholes. *Chem. Rev.* **1988**, *88*, 429–453. [CrossRef]
138. Carmichael, D.; Mathey, F. New Trends in Phosphametallocene Chemistry. *Top. Curr. Chem.* **2002**, *220*, 27–51.
139. Mills, D.P.; Evans, P. f-Block Phospholyl and Arsolyl Chemistry. *Chem. Eur. J.* **2021**, *27*, 6645–6665. [CrossRef] [PubMed]
140. Abel, E.W.; Clark, N.; Towers, C. η-Tetraphenylphospholyl and η-tetraphenylarsolyl derivatives of manganese, rhenium, and iron. *J. Chem. Soc. Dalton Trans.* **1979**, 1552–1556. [CrossRef]
141. Kirk, R.M.; Hill, A.F. Arsolyl-supported intermetallic dative bonding. *Chem. Sci.* **2022**, *13*, 6830–6835. [CrossRef] [PubMed]
142. Kirk, R.M.; Hill, A.F. Free and coordinated biarsolyls. *Dalton Trans.* **2023**, *52*, 13235–13243. [CrossRef]
143. Kirk, R.M.; Hill, A.F. Bridging arsolido complexes. *Dalton Trans.* **2023**, *52*, 10190–10196. [CrossRef]
144. Byers, P.K.; Carr, N.; Stone, F.G.A. Chemistry of polynuclear metal complexes with bridging carbene or carbyne ligands. Part 106. Synthesis and reactions of the alkylidyne complexes [M(CR)(CO)₂{(C₆F₅)AuC(pz)₃}](M = W or Mo, R = alkylaryl, pz = pyrazol-1-yl); crystal structure of [WPtAu(C₆F₅)(μ₃-CMe)(CO)₂(PMe₂Ph)₂{(C₆F₅)AuC(pz)₃}] and 105 preceding parts in the series. *J. Chem. Soc. Dalton Trans.* **1990**, *12*, 3701–3708. [CrossRef]
145. Clark, G.; Cochrane, C.; Roper, W.; Wright, L. The interaction of an osmium-carbon triple bond with copper(I), silver(I) and gold(I) to give mixed dimetallocyclopropene species and the structures of Os(AgCl)(CR)Cl(CO)(PPh₃)₂. *J. Organomet. Chem.* **1980**, *199*, C35–C38. [CrossRef]
146. Carr, N.; Gimeno, M.C.; Goldberg, J.E.; PIlotti, M.U.; Stone, F.G.A.; Topiloglu, I. Synthesis of Mixed-metal Compounds via the Salts [NEt₄][Rh(CO)L(η⁵-C₂B₉H₉R₂)](L = PPh₃, R = H; L = CO, R = Me); Crystal Structures of the Complexes [WRhAu(μ-CC₆H₄Me-4)(CO)₃(PPh₃)(η-C₅H₅)(η⁵-C₂B₉H₁₁)] and [WRh₂Au₂(μ₃-CC₆H₄Me-4)(CO)₆(η-C₅H₅)(η⁵-C₂B₉H₉Me₂)₂]·0.5CH₂Cl₂. *J. Chem. Soc. Dalton Trans.* **1990**, *7*, 2253–2261. [CrossRef]
147. Carriedo, G.A.; Riera, V.; Sánchez, G.; Solans, X. Synthesis of new heteronuclear complexes with bridging carbyne ligands between tungsten and gold. X-Ray crystal structure of [AuW(μ-CC₆H₄Me-4)(CO)₂(bipy)(C₆F₅)Br]. *J. Chem. Soc. Dalton Trans.* **1988**, *7*, 1957–1962. [CrossRef]
148. Strasser, C.E.; Cronje, S.; Raubenheimer, H.G. Fischer-type tungsten acyl (carbeniate), carbene and carbyne complexes bearing C5-attached thiazolyl substituents: Interaction with gold(i) fragments. *New J. Chem.* **2010**, *34*, 458–469. [CrossRef]
149. Zhou, X.; Li, Y.; Shao, Y.; Hua, Y.; Zhang, H.; Lin, Y.-M.; Xia, H. Reactions of Cyclic Osmacarbyne with Coinage Metal Complexes. *Organometallics* **2018**, *37*, 1788–1794. [CrossRef]
150. Borren, E.S.; Hill, A.F.; Shang, R.; Sharma, M.; Willis, A.C. A Golden Ring: Molecular Gold Carbido Complexes. *J. Am. Chem. Soc.* **2013**, *135*, 4942–4945. [CrossRef] [PubMed]
151. Frogley, B.J.; Hill, A.F.; Onn, C.S.; Watson, L.J. Bi- and Polynuclear Transition-Metal Carbon Tellurides. *Angew. Chem. Int. Ed.* **2019**, *58*, 15349–15353. [CrossRef] [PubMed]
152. Frogley, B.J.; Hill, A.F. Synthesis of pyridyl carbyne complexes and their conversion to N-heterocyclic vinylidenes. *Chem. Commun.* **2019**, *55*, 15077–15080. [CrossRef] [PubMed]
153. Onn, C.S.; Hill, A.F.; Olding, A. Metal coordination of phosphoniocarbynes. *Dalton Trans.* **2020**, *49*, 12731–12741. [CrossRef]
154. Hill, A.F.; Manzano, R.A. A [C1+C2] Route to Propargylidyne Complexes. *Dalton Trans.* **2019**, *48*, 6596–6610. [CrossRef] [PubMed]
155. Burt, L.K.; Hill, A.F. Heterobimetallic μ₂-Halocarbyne Complexes. *Dalton Trans.* **2022**, *51*, 12080–12099. [CrossRef]
156. Green, M.L.H.; Parkin, G. Application of the Covalent Bond Classification Method for the Teaching of Inorganic Chemistry. *J. Chem. Educ.* **2014**, *91*, 807–816. [CrossRef]
157. Wilford, J.B.; Powell, H.M. The crystal and molecular structure of tricarbonyl-π-cyclopentadienyltungstiotriphenylphosphinegold. *J. Chem. Soc. A Inorganic Phys. Theor.* **1969**, 8–15. [CrossRef]

158. Fischer, P.J.; Krohn, K.M.; Mwenda, E.T.; Young, V.G. (2-(Dimethylammonium)ethyl)cyclopentadienyltricarbonylmetalates: Group VI Metal Zwitterions. Attenuation of the Brønsted Basicity and Nucleophilicity of Formally Anionic Metal Centers. *Organometallics* **2005**, *24*, 5116–5126. [CrossRef]
159. CrysAlisPro. *Oxford Diffraction*; Revision 5.2; Agilent Technologies UK Ltd.: Yarnton, UK, 2013.
160. Dolomanov, O.V.; Bourhis, L.J.; Gildea, R.J.; Howard, J.A.K.; Puschmann, H. OLEX2: A complete structure solution, refinement and analysis program. *J. Appl. Cryst.* **2009**, *42*, 339–341. [CrossRef]
161. Macrae, C.F.; Bruno, I.J.; Chisholm, J.A.; Edgington, P.R.; McCabe, P.; Pidcock, E.; Rodriguez-Monge, L.; Taylor, R.; van de Streek, J.; Wood, P.A. Mercury CSD 2.0—New features for the visualization and investigation of crystal structures. *J. Appl. Crystallogr.* **2008**, *41*, 466–470. [CrossRef]
162. Kuveke, R.E.H.; Barwise, L.; van Ingen, Y.; Vashisht, K.; Roberts, N.; Chitnis, S.S.; Dutton, J.L.; Martin, C.D.; Melen, R. An International Study Evaluating Elemental Analysis. *ACS Cent. Sci.* **2022**, *8*, 855–863. [CrossRef]

Disclaimer/Publisher's Note: The statements, opinions and data contained in all publications are solely those of the individual author(s) and contributor(s) and not of MDPI and/or the editor(s). MDPI and/or the editor(s) disclaim responsibility for any injury to people or property resulting from any ideas, methods, instructions or products referred to in the content.

Article

The Covalent Linking of Organophosphorus Heterocycles to Date Palm Wood-Derived Lignin: Hunting for New Materials with Flame-Retardant Potential

Daniel J. Davidson [1,2], Aidan P. McKay [1], David B. Cordes [1], J. Derek Woollins [1,3] and Nicholas J. Westwood [1,2,*]

[1] School of Chemistry, University of St Andrews and EaStCHEM, North Haugh, St Andrews KY16 9ST, UK; dd68@st-andrews.ac.uk (D.J.D.); apm31@st-andrews.ac.uk (A.P.M.); jdw3@st-andrews.ac.uk (J.D.W.)
[2] Biomedical Sciences Research Complex, University of St Andrews, North Haugh, St Andrews KY16 9ST, UK
[3] Department of Chemistry, Khalifa University, Abu Dhabi 127788, United Arab Emirates
* Correspondence: njw3@st-andrews.ac.uk

Abstract: Environmentally acceptable and renewably sourced flame retardants are in demand. Recent studies have shown that the incorporation of the biopolymer lignin into a polymer can improve its ability to form a char layer upon heating to a high temperature. Char layer formation is a central component of flame-retardant activity. The covalent modification of lignin is an established technique that is being applied to the development of potential flame retardants. In this study, four novel modified lignins were prepared, and their char-forming abilities were assessed using thermogravimetric analysis. The lignin was obtained from date palm wood using a butanosolv pretreatment. The removal of the majority of the ester groups from this heavily acylated lignin was achieved via alkaline hydrolysis. The subsequent modification of the lignin involved the incorporation of an azide functional group and copper-catalysed azide–alkyne cycloaddition reactions. These reactions enabled novel organophosphorus heterocycles to be linked to the lignin. Our preliminary results suggest that the modified lignins had improved char-forming activity compared to the controls. ^{31}P and HSQC NMR and small-molecule X-ray crystallography were used to analyse the prepared compounds and lignins.

Keywords: organophosphorus; heterocycles; lignin; biomass pretreatment; deacylation; click reaction; flame retardants; X-ray crystallography; NMR analysis

Citation: Davidson, D.J.; McKay, A.P.; Cordes, D.B.; Woollins, J.D.; Westwood, N.J. The Covalent Linking of Organophosphorus Heterocycles to Date Palm Wood-Derived Lignin: Hunting for New Materials with Flame-Retardant Potential. *Molecules* **2023**, *28*, 7885. https://doi.org/10.3390/molecules28237885

Academic Editor: Yves Canac

Received: 13 October 2023
Revised: 19 November 2023
Accepted: 24 November 2023
Published: 1 December 2023

Copyright: © 2023 by the authors. Licensee MDPI, Basel, Switzerland. This article is an open access article distributed under the terms and conditions of the Creative Commons Attribution (CC BY) license (https://creativecommons.org/licenses/by/4.0/).

1. Introduction

Historically, flame-retardant compounds have been toxic and persistent in the environment, with polyhalogenated/polybrominated flame retardants being a well-documented issue [1]. These compounds are now largely banned or heavily restricted; therefore, replacements are required. Organophosphorus flame retardants (OPFRs) [2–5] are being developed as less toxic and less harmful alternatives, although this class of compounds is not concern-free [6,7]. The main effect of OPFRs likely occurs in the condensed phase via the degradation of the phosphorus motif and polymerisation of resulting free phosphoric acid-containing units to form a char layer. This layer insulates the flammable substrate from the required oxygen, disrupting the fire triangle. The incorporation of a nitrogen-containing functional group to form a P-N bond can provide additional gas-phase flame-retardant character. The nitrogen-containing gases, formed upon decomposition at elevated temperatures, dilute the oxygen content in the vicinity of the fire and therefore inhibit flame growth [8,9]. Many OPFRs are physically blended as small molecules into polymers to produce flame-retardant materials (for example, the extensive use of DOPO [10]). More recently, rather than just blending, the chemical attachment of a OPFR to the polymer has been demonstrated, either by covalent [11] or reversible dynamic bonds [12].

Lignin is a renewable biopolymer isolated from biomass alongside cellulose and hemicellulose. A wide range of different pretreatments are used to obtain lignin, including organosolv methods [13,14]. We, and others, have focused on the use of butanol as a sustainably sourced organic solvent that delivers high-quality lignin via a butanosolv pretreatment [15–21]. Recently, we have extended the butanosolv methodology to enable the use of more unusual biomasses, including cocoa pod husks, a by-product from chocolate manufacturing (Figure 1A) [22].

Importantly, in the context of this work, the simple addition of unmodified lignin to a polymer is known to enhance the flame-retardant properties of the polymer. This is proposed to result from the degradation of the lignin, leading to improved char layer formation [23]. Studies have shown that modification by covalently linking OPFRs to the lignin can lead to materials with flame-retardant properties (Figure 1A and others) [22,24,25]. Butanosolv lignin is highly suited to selective covalent modification as it is soluble in most organic solvents enabling the use of standard reaction sequences. For example, butanosolv (and other) lignins have been used as substrates for grafting on small molecules using click chemistry [22,26,27]. Increasingly, researchers are interested in enhancing the inherent flame-retardant properties of lignin through its covalent modification with OPFRs.

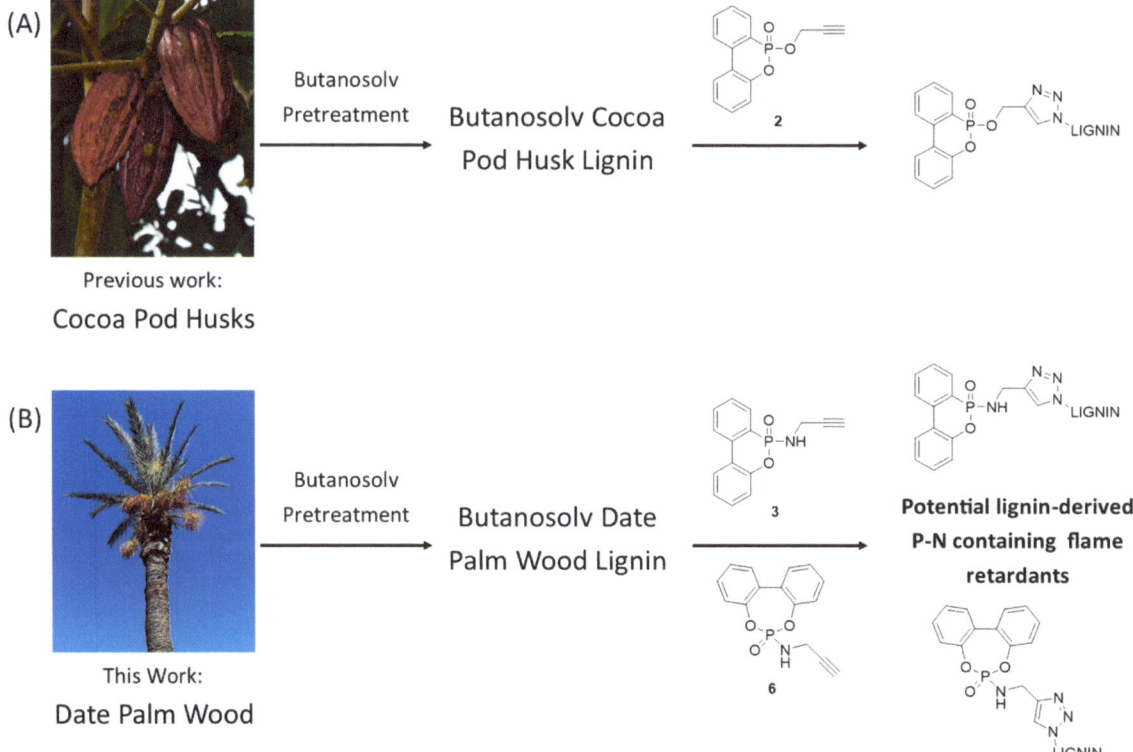

Figure 1. Combining potential organophosphorus flame-retardants (OPFRs) and lignin. (A) Previous work in which the known OPFR O-DOPO 1 (Scheme 1) was attached to a cocoa pod husk lignin using the alkyne analogue 2 [22]. (B) This work in which a novel application of the known DOPO analogue 3 is provided and a novel potential OPFR 6 with a lignin prepared from date palm waste is described.

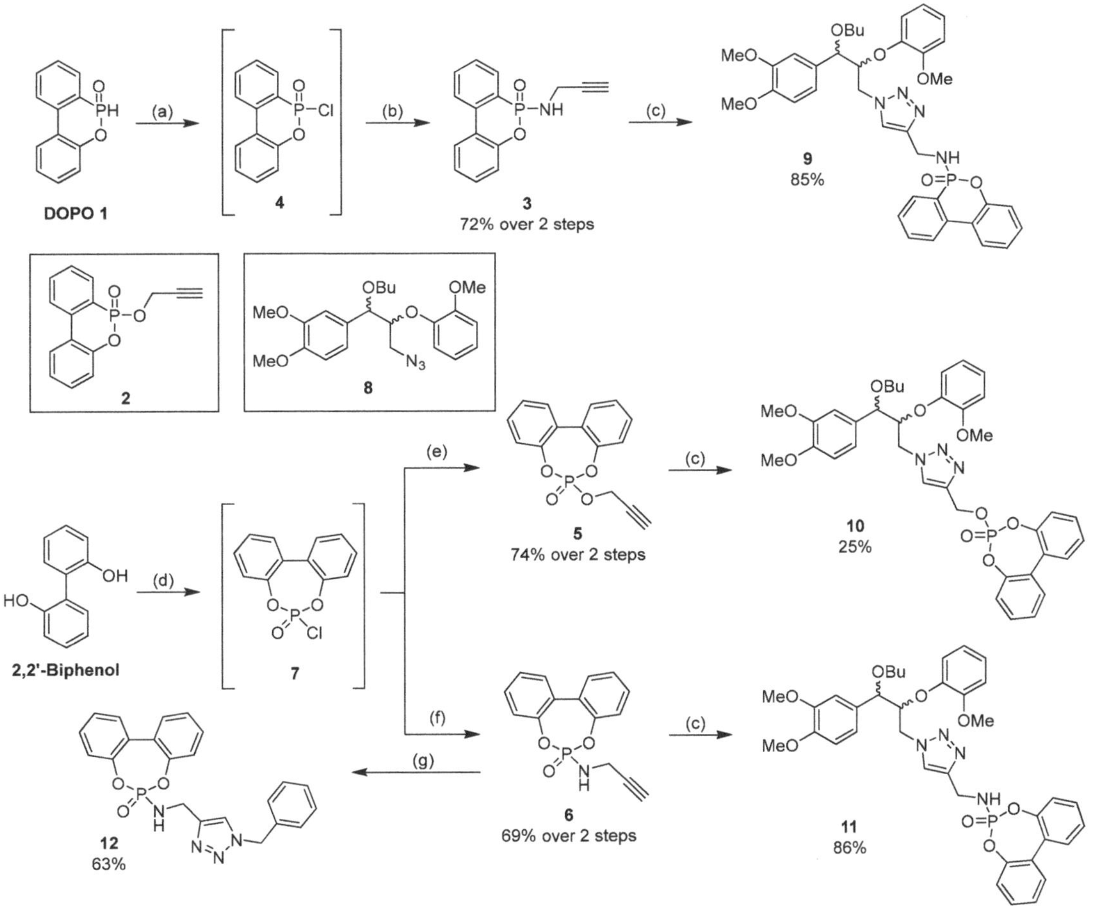

Scheme 1. OPFR and model compound synthesis. Conditions: (**a**) N-chlorosuccinimide, DCM, 0 °C to rt, 16 h; (**b**) propargylamine, NEt₃, DCM, 0 °C to rt, 16 h; (**c**) **8**, sodium ascorbate, CuSO₄·5H₂O, MeOH, rt, 16 h; (**d**) POCl₃, NEt₃, THF, 0 °C to rt, 2 h; (**e**) propargyl alcohol, NEt₃, THF, 0 °C to rt, 16 h; (**f**) propargylamine, NEt₃, THF, 0 °C to rt, 16 h; (**g**) benzyl azide, sodium ascorbate, CuSO₄·5H₂O, MeOH, rt, 16 h.

The work presented here is dedicated to our excellent colleague at the University of St Andrews, Professor Derek Woollins. Derek's interests continue to be wide-ranging and include the synthesis of phosphorus-, selenium-, or tellerium-containing heterocycles [28–30]. Here, we present the synthesis of novel P-heterocycles and the structural analysis of three of these through the use of small-molecule X-ray crystallography. In addition, as a direct result of a collaboration with Derek, we gained access to a relatively understudied biomass source, date palm wood (Figure 1B). We show that an interesting lignin can be obtained by subjecting date palm wood to butanosolv pretreatment, complementing previous work on this lignin type [31]. Through the use of ^{31}P NMR spectroscopy methods, a technique frequently used by Professor Woollins [32,33], this lignin was characterised before and after modification with the novel P-heterocycles. Our preliminary assessment of the flame-retardant potential of the novel lignin–OPFR conjugates will guide future work in this developing research area. We would like to thank Professor Woollins for his scientific inspiration and leadership skills.

2. Results and Discussion

2.1. Phosphorus-Containing Heterocycle Synthesis

The flame-retardant properties of the DOPO motif **1** (Scheme 1) are well known [34–36], and we have previously reported that after the attachment of O-propargyl DOPO **2** to lignin, the resulting product demonstrates potential flame-retardant properties (Figure 1A) [22]. Based on previous reports [8], the use of N-propargyl DOPO analogue **3** may enable additional gas-phase cooperative flame-retardant activity in this system.

The synthesis and/or use of **3** has been reported in the context of electrode additives [37] and bioactive compound synthesis [38]; however, a slightly modified approach to **3** was used here to convert DOPO **1** to **3** via **4** (Scheme 1). A small-molecule X-ray crystallographic analysis of **3** was carried out (Figure 2).

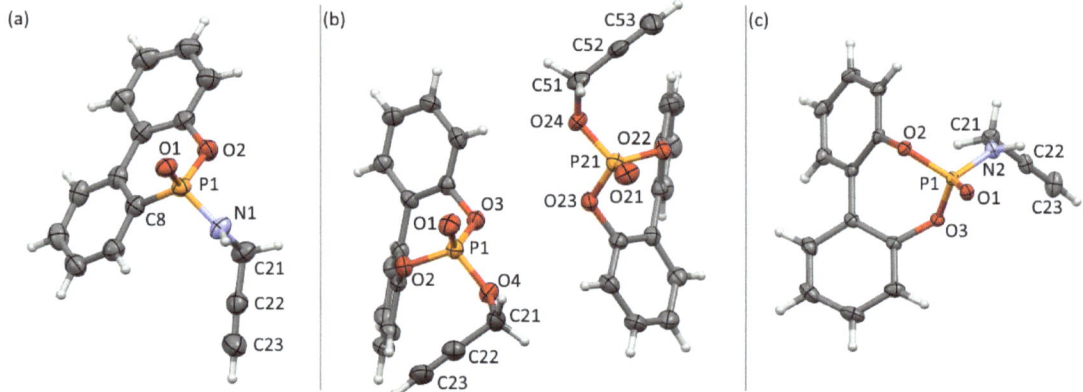

Figure 2. Thermal ellipsoid plots (50% probability ellipsoids) of the structures of: (**a**) **3**, selected bond lengths (Å), angles (°), and torsions (°): P1-O1 1.478(3), P1-O2 1.600(3), P1-N1 1.616(3), P1-C8 1.780(4), C21-C22 1.456(6), C22-C23 1.182(6), O2-P1-C8 101.90(17), O1-P1-N1 111.46(17), C21-C22-C23 176.4(5), P1-N1-C21-C22 -122.6(4). (**b**) **5**, selected bond lengths (Å), angles (°), and torsions (°): P1-O1 1.457(4), P1-O2 1.585(4), P1-O3 1.575(4), P1-O4 1.569(4), C21-C22 1.501(15), C22-C23 1.156(15), P2-O21 1.453(4), P2-O22 1.590(4), P2-O23 1.580(4), P2-O24 1.567(3), C51-C52 1.429(15), C52-C53 1.200(17), O2-P1-O3 104.75(19). O1-P1-O4 116.7(2), O22-P2-O23 104.46(19), O21-P2-O24 116.5(2), C21-C22-C23 176.5(9), C51-C52-C53 177.9(8), P1-O4-C21-C22 73.7(5), P2-O24-C51-C52 75.8(5). (**c**) **6**, selected bond lengths (Å), angles (°), and torsions (°): P1-O1 1.4666(8), P1-O2 1.5975(8), P1-O3 1.5898(8), P1-N2 1.6109(10), C21-C22 1.4637(18), C22-C23 1.1834(19), O2-P1-O3 102.49(4). O1-P1-N2 113.25(5), C21-C22-C23 178.95(15), P1-N2-C21-C22 -110.21(10).

It has been proposed that dibenzo[d,f][1,3,2]-dioxaphosphepine 6-oxide (BPPO)-derived phosphorus heterocycles should also demonstrate flame-retardant properties [39,40]. Novel compounds O- and N-propargyl BPPO **5** and **6**, respectively, were therefore prepared via **7** (Scheme 1). The preliminary testing of the use of O-propargyl BPPO **5** in copper-catalysed alkyne–azide click reactions (CuAAC) identified several issues on both models and lignin (see SI for more detail, Figure S1); therefore, the main focus of this study became the modification of lignin by N-propargyl DOPO **3** and N-propargyl BPPO **6**.

2.2. X-ray Crystallography

Crystals of **3**, **5**, and **6** suitable for X-ray analysis were grown from ethanol, dichloromethane, or isopropanol solutions of the respective compounds. The compounds crystallised in the monoclinic $P2_1/c$, orthorhombic $Pca2_1$, and triclinic $P\bar{1}$ space groups, respectively, and contain either one (**3** and **6**) or two (**5**) molecules in the asymmetric units (Figure 2). The two aryl groups in **3** are nearly co-planar with a slight twist of 9.77(14)°, with the six membered oxophosphinine ring forming a slightly distorted hexagon (O2-P1-C8 101.90(17)°). In contrast, the two aryl groups in both **5** and **6** show moderate twists of 46.2(2)°, 44.1(2)°, and 42.61(4)°, respectively, and have slightly puckered dioxophosphepine rings (endo-cyclic O-P-O 104.75(19)°, 104.46(19)°, and 102.49(4)°).

Compounds **3** and **6** form hydrogen bonded chains down [1 0 0] and [0 1 0], respectively, through $C(7)\left[R_2^2(8)R_4^2(14)\right]$ motifs composed of both moderate strength NH···O (H···O 1.88(2) and 2.017(14) Å, N···O 2.846(4) and 2.9101(13) Å) and non-classical C^{sp}H···O (H···O 2.278(3) and 2.2846(8) Å, C···O 3.208(6) and 3.2049(16) Å) hydrogen bonds (Figure 3 for **3**). When viewed down [1 0 0], the hydrogen bonded chains of **3** form a herringbone arrangement. A combination of weaker CH···O (H···O 2.557(3) and 2.707(3) Å, C···O 3.409(5) and 3.617(6) Å) and π-stacking (C···centroid 3.708(4) Å) interactions leads to the formation of sheets in the (1 0 0) plane. The chains of **6** do not adopt a herringbone arrangement and form sheets in the (0 1 0) plane through weak CH···O interactions (H···O 2.5870(8) Å and 2.8830(8) Å, C···O 3.5218(14) and 3.6209(14) Å).

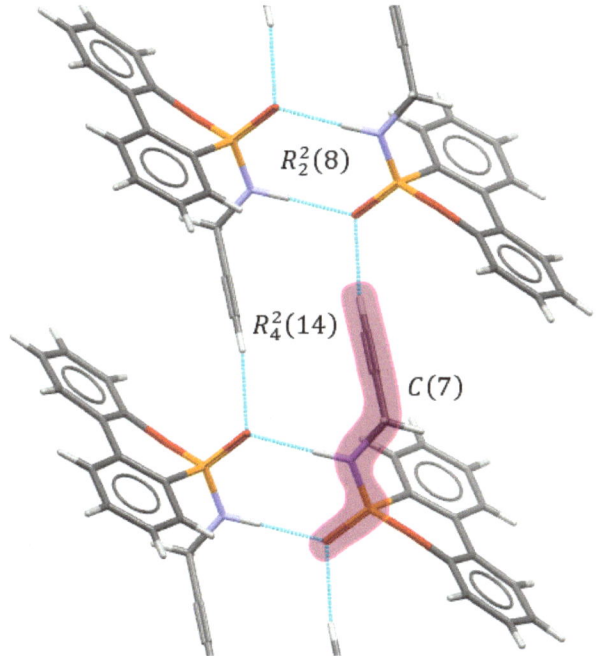

Figure 3. View of part of the X-ray crystal structure of **3** showing the packing of molecules into chains along [1 0 0] and the components of the $C(7)\left[R_2^2(8)R_4^2(14)\right]$ motif. Hydrogen bonds are shown via the light blue dashed lines.

In the structure of **5**, each molecule forms $C(7)\left[R_3^3(14)\right]$ chains down [1 0 0] through non-classical C^{sp}H···O (H···O 2.577(4) and 2.544(4) Å, C···O 3.446(13) and 3.437(12) Å) and weaker CH···O (H···O 2.376(4) and 2.378(4) Å, C···O 3.362(10) and 3.353(10) Å) hydrogen bonds, supported by CH···π interactions (H···centroid 3.001(3) Å, C···centroid 3.821(9) Å) (Figure 4). These chains form sheets in the (1 0 0) plane through weak CH···O (H···O 2.588(4)–2.678(4) Å, C···O 3.216(8)–3.489(7) Å), CH···π (H···centroid 2.888(3) Å, C···centroid 3.712(8) Å), and π···π (centroid···centroid 3.723(2) Å) interactions.

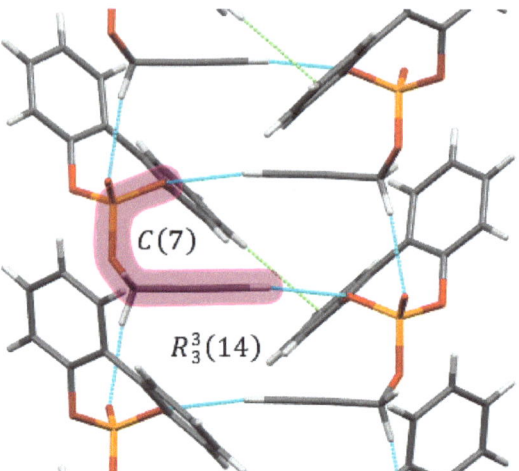

Figure 4. View of part of the X-ray crystal structure of **5** showing the packing of molecules into chains along [1 0 0] and the components of the $C(7)\left[R_3^3(14)\right]$ motif. Hydrogen bonds are shown via the light blue dashed lines, and CH···π interactions are shown via the light green dashed lines.

2.3. Model Compound Synthesis

Due to the complex heterogeneous structure of lignin [41], the assignment of structural features via NMR analysis is aided by the preparation and NMR analysis of simplified model compounds [42]. We have previously prepared models of the butoxylated β-O-4 linkage modified with various functional groups at the γ-position, including **8**, which contains an azide functionality that can be utilised in copper-catalysed azide–alkyne cycloaddition (CuAAC) click reactions (Scheme 1) [43]. Novel model compounds **9**, **10**, and **11** were prepared from **8** and characterised for comparison with the modified lignins (Scheme 1).

2.4. Lignin Substrate Preparation

Using a procedure previously described in the literature (optimised for unusual biomasses [22]), a sample of date palm wood (DPW) was processed using a butanol pretreatment to prepare a date palm wood lignin (**DPW Lignin**) in good yield (Figure S2). This butanosolv lignin was initially characterised by 2D HSQC NMR and quantitative ^{31}P NMR after phosphitylation using a procedure previously described in the literature [44,45] (Figures 5A–C and S3A and Tables S1 and S2). Whilst the aliphatic OH content of this lignin was reasonably high (6.8 mmol/g, Figure 5A,B), it was observed that many of the potentially modifiable β-O-4 sites were acylated (Figure 5C).

These acyl groups were expected based on previous reports that have shown that date palm lignins contain a range of ester pendant groups at the γ-position of the β-O-4 units, including abundant benzoate and *p*-hydroxybenzoate esters, as well as minor components such as vanillic and syringic esters [31]. Inspired by well-established methods of hydrolysing ester units in lignin [31,46,47], aqueous sodium hydroxide solution was used with a sample of **DPW lignin** to produce a deacylated lignin (**DeAcyl Lignin**). Following this reaction, there was a nearly 40% increase in aliphatic OH content (from 6.8 to 9.4 mmol/g, Figure 5B), with the acylated β-O-4 linkage content (labelled *p*-BH in Figure 5) having decreased. Detailed HSQC and HMBC NMR analyses of the aqueous component after hydrolysis allowed for identification of the free benzoic, *p*-hydroxybenzoic, and syringic acids that were cleaved from the lignin (Figure S4). The identification of the free acids facilitated the assignment of the corresponding ester moieties in the aromatic region of the HSQC NMR spectra of the lignin (Figure S3). Whilst each of these ester moieties have been identified in palm lignins before, the expected relative abundance differed compared to previous reports [22]. No *p*-coumarate or ferulate esters were detected in the HSQC NMR analysis, possibly suggesting that these esters were more facile to hydrolyse and may have been removed earlier in the processing of the biomass, as observed with acetyl groups in a previous work [22].

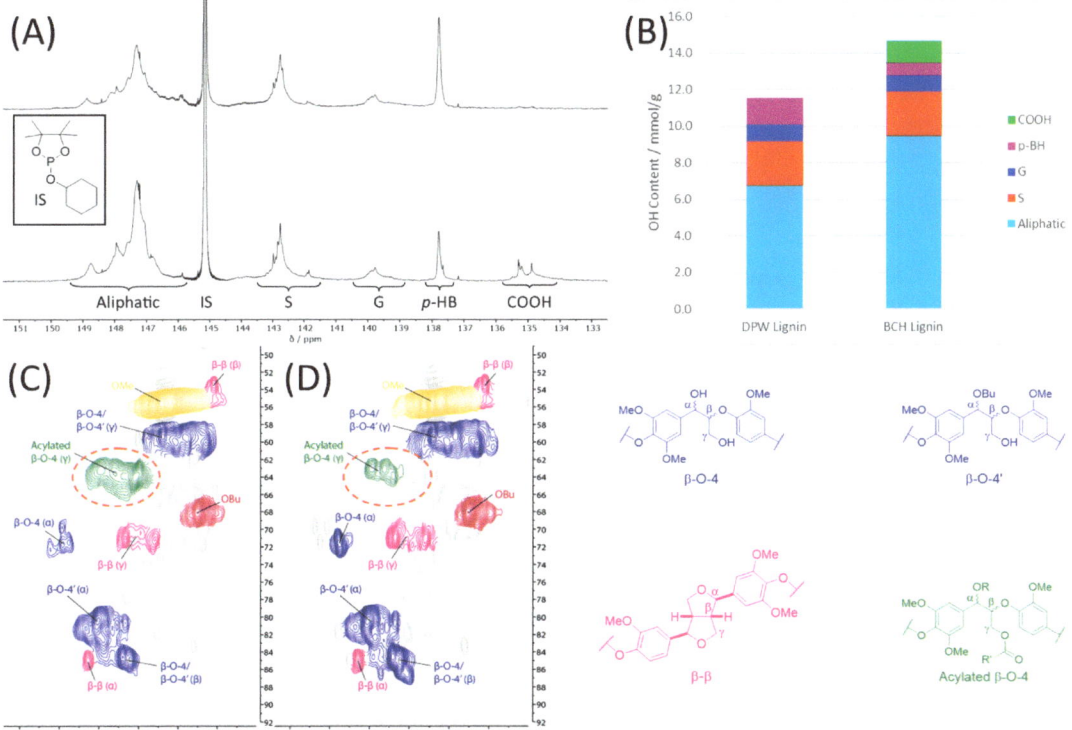

Figure 5. (**A**) Quantitative ^{31}P NMR analysis after the phosphitylation [44,45] of the starting date palm wood lignin (**DPW Lignin**, top) and the deacylated lignin (**DeAcyl Lignin**, bottom) obtained after aqueous hydroxide treatment with integrals corresponding to relevant structural features; (**B**) calculated hydroxyl content of **DPW Lignin** and **DeAcyl Lignin**; HSQC NMR (700 MHz, DMSO-d_6) analysis of the linkage region of (**C**) **DPW lignin** and (**D**) **DeAcyl lignin**, with the region corresponding to the C$\underline{H}_2$OAcyl in the γ-position of acylated β-O-4 units highlighted in the red circle (see structure of Acylated β-O-4 unit in Figure 5 right hand side). Colour-coded structures that correspond to regions in the HSQC NMR spectra are shown. The corresponding aromatic regions are shown in Figure S3.

2.5. Lignin Modification

Having obtained the required date palm lignins in sufficient quantities, subsequent modification to incorporate azide functional groups was carried out (Scheme 2) based on previously established methods [43]. This culminated in the synthesis of **DPW Lignin N$_3$** and **DeAcyl Lignin N$_3$**, which were characterised by HSQC NMR and IR at each stage to confirm successful modification (see SI for further details; Figures S5 and S6).

The modified lignins containing an azide functional handle at the γ-position were then reacted with OPFRs **3** or **6** under CuAAC click conditions to prepare grafted lignins. These modified lignin samples were precipitated and then further purified via column chromatography on silica gel to give final lignins **DPW-3** (118 wt% yield), **DPW-6** (137 wt%), **DeAcyl-3** (130 wt%), and **DeAcyl-6** (137 wt%). Some challenges were encountered at this stage due to the polar nature of both the OPFRs and the final modified lignin (see below and SI for a more detailed discussion).

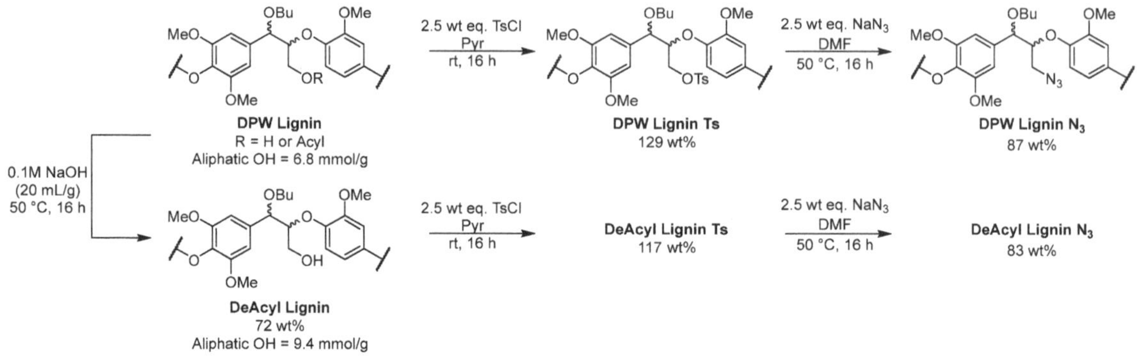

Scheme 2. Lignin modification to give azide-containing lignins.

2.6. OPFR-Grafted Lignin Characterisation

The ^{31}P NMR and HSQC NMR analyses of model compound **9** (structure in Scheme 1) were compared with the NMR spectra of the OPFR grafted lignins obtained from reaction with **3** to determine if the click reaction had been successful. The broad signal in the ^{31}P NMR spectrum corresponding to the final DPW lignin (**DPW-3**) obtained upon the CuAAC reaction of **DPW lignin N$_3$** with **3** showed good alignment with the signals for model compound **9** (diastereomeric mixture, Figure 6A). However, a sharp signal corresponding to free **3** that was contaminating **DPW-3** was also observed. In addition, overlay of the HSQC NMR analysis of **9** with the final deacylated lignin (**DeAcyl-3**) obtained upon the CuAAC reaction of **DeAcyl lignin N$_3$** with **3** also supported a successful reaction (Figure 6B). For example, a signal at ^{1}H 4.30-3.85/^{13}C 37.2-34.1 corresponded to the methylene hydrogens adjacent to the newly formed triazole ring (Figure 6B). This shows perfect overlay with the analogous signal in **9**. However, the presence of unreacted OPFR **3** was also observed in the HSQC NMR spectra (Figure S7). Analogous results were obtained for the other possible combinations of the lignin azides with the OPFRs (for **DeAcyl lignin N$_3$** and **6**, see Figure 6C, and for all other combinations, see Figure S7). Whilst it was gratifying that the CuAAC reaction was successful for all combinations tested, it was disappointing that, despite purifying the final lignins via column chromatography, it was not possible to remove all of the starting small-molecule OPFRs. This observation was in contrast to a previous report on how one can successfully purify OPFR-grafted lignins when using OPFR **2** (Figure 1A and Scheme 1) and cocoa pod husk lignin [22]. Presumably, the incorporation of the NHR motif into the OPFR structures (in **3** and **6** compared to **2**) meant that the OPFRs co-eluted with the lignin during purification. Attempts to solve this problem will be the subject of future work.

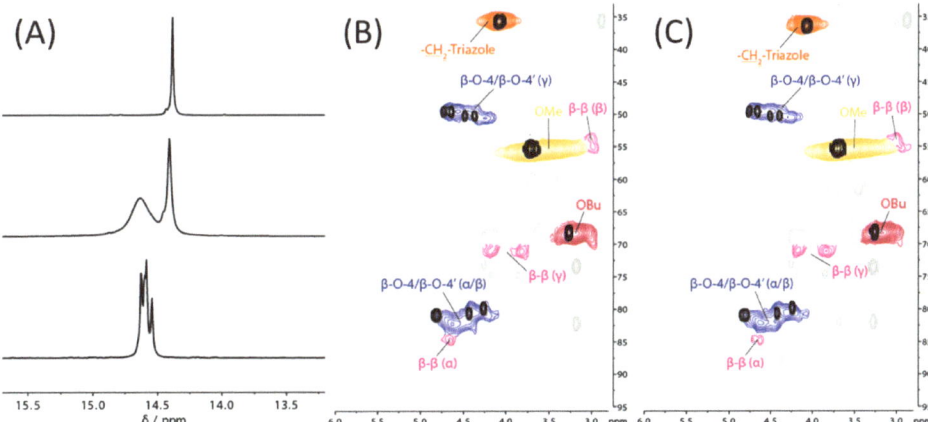

Figure 6. (**A**) ^{31}P NMR spectra of OPFR **3** (top), grafted **DWP Lignin** (**DPW-3**, middle), and model compound **9** (bottom) emphasising the successful modification of the **DWP lignin** N_3 on reaction with **3** under CuAAC conditions. The final lignin sample was contaminated with unreacted **3** (Figure S7); (**B**) HSQC NMR (700 MHz, DMSO-d$_6$) analysis of the linkage region of **DeAcyl Lignin** grafted with **3** (overlaid with the analysis of **9**, black); (**C**) HSQC NMR (700 MHz, DMSO-d$_6$) analysis of the linkage region of **DeAcyl Lignin** grafted with **6** (overlaid with the analysis of **11**, black). See Scheme 1 for the structures of **9** and **11**.

2.7. Thermogravimetric Analysis of OPFR-Grafted Lignins

Despite the presence of small molecular impurities in the final samples of the OPFR-grafted lignins, it was decided to complete this study by carrying out a thermogravimetric analysis (TGA) of the four final lignins (**DPW-3**, **DPW-6**, **DeAcyl-3**, and **DeAcyl-6**). It was proposed that a control TGA experiment would also be carried out, in which a physical mixture (blend) of non-modified **DPW Lignin** and the model compound **12** (Scheme 1) would be used. Small molecule **12** represents a compound in which the triazole ring formed in a CuAAc reaction is present, hence enabling the impact of the triazole ring to also be controlled for. This mixture is referred to as the **Control Mixture** below. It was decided that a 5:1 w/w ratio of **DPW Lignin**/**12** should be used as it was felt that this corresponded to a higher level of small molecule contaminant **12** compared to the amounts of small-molecule OPFRs likely present in the final lignin samples. Any difference in the TGA results (Figure 7) of the final lignins from this **Control Mixture** must be due to the presence of the grafted OPFRs.

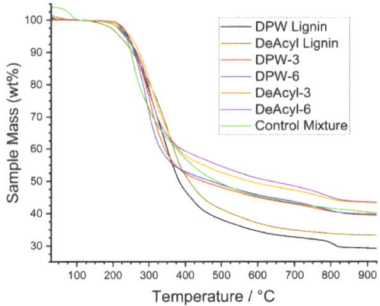

Figure 7. TGA curves of the original (**DPW** and **DeAcyl lignin**) and modified lignins (**DPW-3**, **DPW-6**, **DeAcyl-3**, **DeAcyl-6**) and of the **Control Mixture**. Curves were normalised to set 100 wt% at 110 °C after the drying isotherm.

A key factor in assessing the potential of a material for use in flame-retardant applications is char formation. This is determined by comparing the mass of sample remaining (as char) after heating the sample to temperatures nearing 1000 °C against suitable controls. Here, three control samples were used: (i) the starting date palm wood lignin (**DPW lignin**), (ii) the starting deacylated DPW lignin (**DeAcyl lignin**), and (iii) the **Control Mixture** discussed above. These controls were compared to the four test lignins: **DPW-3**, **DPW-6**, **DeAcyl-3**, and **DeAcyl-6**. In brief, the two starting lignins (**DPW and DeAcyl lignins**) did lead to some char formation, as expected, but this was lower than the amount of char formed by the test lignins. Interestingly, the two best performing lignins were **DeAcyl-3** and **DeAcyl-6**, which are believed to contain a greater amount of OPFRs covalently bonded to the lignin compared to **DPW-3** and **DPW-6**. In addition, both **DeAcyl-3** and **DeAcyl-6** formed an increased amount of char compared to the **Control Mixture**, suggesting that the covalent attachment of the OPFRs to the lignin may provide an advantage over just simply physically mixing OPFRs with lignin. While further work is clearly required to assess the full potential of these materials, one possible explanation for the observed differences is that by holding the OPFR motif closer to the lignin through the use of a covalent bond, the initial degradation reaction is more likely to lead to the intertwining of the lignin- and OPFR-derived chars. This may provide additional structure to the forming char, ultimately improving its formation and therefore the overall char-forming ability of the bulk material.

3. Materials and Methods

For a detailed discussion of the lignin experimental procedures, lignin model compounds synthesis, and general experimental considerations, see the Supplementary Materials.

3.1. 6-(prop-2-yn-1-ylamino)dibenzo[c,e][1,2]oxaphosphinine 6-oxide 3

DOPO **1** (2.02 g, 9.33 mmol, 1.00 eq.) was dissolved in DCM (25 mL) and cooled to 0 °C under a N_2 atmosphere. N-chlorosuccinimide (1.37 g, 10.3 mmol, 1.10 eq.) was added slowly portionwise over 10 min, and the resulting mixture was warmed to rt and stirred under N_2 for 16 h. The resulting suspension was filtered, and the solvent was removed under reduced pressure to afford intermediate **4**, which was used immediately in the next step. Crude **4** was dissolved in fresh DCM (25 mL) and cooled to 0 °C under N_2, and propargylamine (1.56 mL, 11.2 mmol, 1.20 eq.) and NEt_3 (0.72 mL, 11.2 mmol, 1.20 eq.) were added slowly dropwise over 10 min then warmed to rt and stirred for 16 h under N_2. The resulting suspension was filtered, and the filtrate diluted with aq. sat. $NaHCO_3$ (20 mL) and extracted with DCM (3 × 15 mL). The combined organic extracts were washed with brine (20 mL) and dried over anhydrous $MgSO_4$, and the solvent was removed under reduced pressure. The crude product was purified via column chromatography on silica gel eluting with EtOAc/hexane (0–95%) to afford 6-(prop-2-yn-1-ylamino)dibenzo[c,e][1,2]oxaphosphinine 6-oxide 3 (1.82 g, 72%) as a yellow solid. ^{1}H NMR (500 MHz, DMSO-d6) δ 3.18 (1H, t, J = 2.5 Hz, H16), 3.67–3.74 (2H, m, H14), 6.24 (1H, dt, J = 13.3, 6.8 Hz, H13), 7.27 (1H, dd, J = 8.1, 1.3 Hz, H4), 7.29–7.34 (1H, m, H2), 7.42–7.48 (1H, m, H3), 7.56–7.62 (1H, m, H10), 7.74–7.80 (1H, m, H11), 7.82–7.88 (1H, m, H9), 8.16–8.23 (2H, m, H1/12). ^{13}C NMR (126 MHz, DMSO-d6) δ 29.59 (C14), 73.55 (C16), 82.30 (d, J = 5.4 Hz, C15), 120.23 (d, J = 5.9 Hz, C4), 121.97 (d, J = 11.5 Hz, C6), 124.12 (d, J = 10.7 Hz, C12), 124.42 (C2), 125.42 (d, J = 162.6 Hz, C8), 125.48 (C8), 128.41 (d, J = 14.1 Hz, C10), 129.66 (d, J = 10.0 Hz, C9), 130.48 (C3), 132.91 (d, J = 2.3 Hz, C11), 135.96 (d, J = 6.8 Hz,

C7), 149.36 (d, J = 7.0 Hz, C5). ^{31}P NMR (202 MHz, DMSO-d6) δ 14.39. IR (ATR) 3229, 3167, 2893, 1597, 1477, 1444, 1213, 1168, 922, 752. mp 147–148 °C. HRMS (ESI) calculated for $C_{15}H_{12}O_2NPNa$ [M + Na]+ 292.0503; found 292.0495.

3.2. 6-(prop-2-yn-1-yloxy)dibenzo[d,f][1,3,2]dioxaphosphepine 6-oxide 5

2,2′-biphenol (2.00 g, 10.7 mmol, 1.00 eq.) was dissolved in dry THF (40 mL) and cooled to 0 °C under N_2. $POCl_3$ (1.00 mL, 10.7 mmol, 1.00 eq.) was added, followed by the dropwise addition of NEt_3 (3.00 mL, 21.5 mmol, 2.00 eq.), and then warmed to rt and stirred under N_2 for 3 h. The resulting suspension was filtered, and the solvent was removed under reduced pressure to afford intermediate **7**, which was used immediately in the next step. Crude **7** was dissolved in fresh dry THF (25 mL) and cooled to 0 °C under N_2. Propargyl alcohol (0.69 mL, 11.9 mmol, 1.10 eq.) and NEt_3 (1.64 mL, 11.77 mmol, 1.1 eq.) were dissolved in dry THF (5 mL), and the amine solution was added slowly dropwise over 10 min, then warmed to rt, and subsequently stirred for 16 h. The resulting suspension was filtered, and the solvent was removed from the filtrate under reduced pressure, and the crude product was purified via column chromatography on silica gel eluting with EtOAc/hexane (0–75%) to afford 6-(prop-2-yn-1-yloxy)dibenzo[d,f][1,3,2]dioxaphosphepine 6-oxide **5** (2.26 g, 74%) as an off-white solid. ^{1}H NMR (500 MHz, DMSO-d_6) δ 3.90 (1H, t, J = 2.4 Hz, H9), 5.02 (2H, dd, J = 11.4, 2.5 Hz, H7), 7.40–7.43 (2H, m, H6), 7.47–7.51 (2H, m, H4), 7.58 (2H, dddd, J = 8.2, 7.4, 1.7, 0.9 Hz, H5), 7.71 (2H, dd, J = 7.7, 1.7 Hz, H3). ^{13}C NMR (126 MHz, DMSO-d_6) δ 56.75 (d, J = 4.3 Hz, C7), 77.65 (d, J = 6.6 Hz, C8), 80.12 (C9), 121.31 (d, J = 4.4 Hz, C6), 127.10 (C4), 127.45 (C2), 130.31 (C11), 130.66 (C5), 146.91 (d, J = 9.1 Hz, C1). ^{31}P NMR (202 MHz, DMSO-d_6) δ 1.82. IR (ATR) 3289, 1476, 1437, 1381, 1292, 1182, 1028, 945, 758. mp 103–105 °C. HRMS (ESI) calculated for $C_{15}H_{11}O_4PNa$ [M + Na]$^+$ 309.0293; found 309.0279.

3.3. 6-(prop-2-yn-1-ylamino)dibenzo[d,f][1,3,2]dioxaphosphepine 6-oxide 6

2,2′-biphenol (2.00 g, 10.7 mmol, 1.00 eq.) was dissolved in dry THF (40 mL) and cooled to 0 °C under N_2. $POCl_3$ (1.00 mL, 10.7 mmol, 1.00 eq.) was added, followed by the dropwise addition of NEt_3 (3.00 mL, 21.5 mmol, 2.00 eq.), and then warmed to rt and stirred under N_2 for 3 h. The resulting suspension was filtered, and the solvent was removed from the filtrate under reduced pressure to afford intermediate **7**, which was used immediately in the next step. Crude **7** was dissolved in fresh dry THF (25 mL) and cooled to 0 °C under N_2. Propargylamine (0.76 mL, 11.9 mmol, 1.10 eq.) and NEt_3 (1.67 mL, 12.0 mmol, 1.10 eq.) were dissolved in dry THF (5 mL), and the amine solution was added slowly dropwise over 10 min, then warmed to rt, and subsequently stirred for 16 h. The resulting suspension was filtered, and the solvent was removed from the filtrate under reduced pressure, and the crude product was purified via column chromatography on silica gel eluting with EtOAc/hexane (0–80%) to afford 6-(prop-2-yn-1-ylamino)dibenzo[d,f][1,3,2]dioxaphosphepine 6-oxide 6 (2.16 g, 69%) as an orange solid. ^{1}H NMR (500 MHz, DMSO-d6) δ 3.28 (1H, t, J = 2.5 Hz, H10), 3.69 (2H, ddd, J = 14.5, 6.9, 2.5 Hz, H8), 6.47 (1H, dt, J = 13.9, 6.9 Hz, H7), 7.32–7.36 (2H, m, H6), 7.41–7.46 (2H, m, H4), 7.54 (2H, dddd, J = 8.1, 7.4, 1.7, 0.8 Hz, H5), 7.68 (2H, dd, J = 7.7, 1.7 Hz, H3). ^{13}C NMR (126 MHz, DMSO-d6) δ 30.21 (C8), 73.77 (C10), 82.11 (d, J = 4.4 Hz, C19), 121.77 (d, J = 3.7 Hz, C6), 126.38 (C4), 128.04 (C2), 129.91 (C3), 130.22 (C5), 147.45 (d, J = 9.3 Hz, C1). ^{31}P NMR (202 MHz, DMSO-d6) δ 13.89. IR (ATR) 3248, 3225, 2918, 1477, 1437, 1246, 1184, 999, 918, 758. mp 182–183 °C (decomp.). HRMS (ESI) calculated for $C_{15}H_{12}O_3NPNa$ [M + Na]+ 308.0452; found 308.0439.

3.4. 6-(((1-benzyl-1H-1,2,3-triazol-4-yl)methyl)amino)dibenzo[d,f][1,3,2]dioxaphosphepine 6-oxide 12

6 (53.8 mg, 0.19 mmol, 1.05 eq.), benzyl azide (23.8 mg, 0.18 mmol, 1.00 eq.), sodium ascorbate (7.70 mg, 0.04 mmol, 0.20 eq.), and copper sulfate pentahydrate (9.40 mg, 0.04 mmol, 0.20 eq.) were dissolved in MeOH (3 mL) and stirred at rt for 12 h. The reaction was diluted with water (10 mL); extraction was carried out with DCM (3 × 5 mL), and the combined organic extracts were washed with aq. sat. NaHCO$_3$ (10 mL) and brine (10 mL) before being dried over anhydrous MgSO$_4$, and the solvent was removed under reduced pressure. The crude product was purified via column chromatography on silica gel eluting with EtOAc/hexane (0–90%) to afford6-(((1-benzyl-1H-1,2,3-triazol-4-yl)methyl)amino)dibenzo[d,f][1,3,2]dioxaphosphepine 6-oxide 12 (46.8 mg, 63%) as a white solid. ^{1}H NMR (500 MHz, DMSO-d6) δ 4.11 (2H, dd, J = 13.5, 6.9 Hz, H8), 5.61 (2H, s, H11), 6.46 (1H, dt, J = 13.9, 6.9 Hz, H7), 7.13–7.18 (2H, m, H6), 7.31–7.47 (9H, m, H4/5/13/14/15), 7.64 (2H, dd, J = 7.5, 1.9 Hz, H3), 8.01 (1H, s, H10). ^{13}C NMR (126 MHz, DMSO-d6) δ 36.47 (C8), 52.76 (C11), 121.64 (d, J = 3.6 Hz, C6), 122.94 (C10), 126.26 (C4), 128.02 (C13), 128.04 (C2), 128.16 (C15), 128.78 (C14), 129.87 (C3), 130.11 (C5), 136.21 (C12), 146.46 (d, J = 4.9 Hz, C9), 147.55 (d, J = 9.3 Hz, C1). ^{31}P NMR (202 MHz, DMSO-d6) δ 13.99. IR (ATR) 2931, 2870, 1593, 1500, 1437, 1251, 1093, 1024, 935, 785, 754. mp 177–178 °C HRMS (ESI) calculated for $C_{22}H_{19}O_3N_4PNa$ [M + Na]+ 441.1092; found 441.1078.

3.5. X-ray Crystallography

X-ray diffraction data for **5** were collected at 173 K using a Rigaku SCXmini CCD diffractometer with a SHINE monochromator [Mo Kα radiation (λ = 0.71073 Å)]. Intensity data were collected using ω steps accumulating area detector images spanning at least a hemisphere of reciprocal space. X-ray diffraction data for **6** were collected at 125 K using a Rigaku FR-X Ultrahigh Brilliance Microfocus RA generator/confocal optics with a XtaLAB P200 diffractometer [Mo Kα radiation (λ = 0.71073 Å)], and data for **3** were collected at 173 K using a Rigaku MM-007HF High Brilliance RA generator/confocal optics with XtaLAB P100 diffractometer [Cu Kα radiation (λ = 1.54187 Å)]. Data for **5** were collected using CrystalClear [48], and for **6** and **3**, data were collected using CrysAlisPro [49]; all data were processed (including correction for Lorentz, polarisation, and absorption) using CrysAlisPro. Structures were solved using dual-space methods (SHELXT) [50] and refined using full-matrix least squares against F^2 (SHELXL-2019/3) [51]. Non-hydrogen atoms were refined anisotropically, and hydrogen atoms were refined using a riding model, except for the hydrogen atoms on N2 (in both **3** and **6**), which were located from the difference

Fourier map and refined isotropically subject to a distance restraint. All calculations were performed using the Olex2 [52] interface. The structure of **5** is in the polar space group $Pca2_1$ and has an ambiguous flack x parameter (0.13(7)). With the lack of chiral directing groups, the crystal is considered to likely be a racemate. Selected crystallographic data are presented in Tables S3 and S4. CCDC 2300932–2300934 contains the supplementary crystallographic data for this paper. These data can be obtained free of charge from The Cambridge Crystallographic Data Centre via the following link: www.ccdc.cam.ac.uk/structures.

4. Conclusions

The development of novel flame-retardant materials is important. Here, the potential impact that novel organophosphorus-containing heterocycles bonded to lignin could have in the context of the development of novel flame-retardant materials was assessed. The study began with the synthesis of the phosphorus-containing heterocycles that were analysed using small-molecule X-ray crystallography. The preparation of two different lignins from date palm was then achieved, and both ^{31}P and ^{1}H-^{13}C HSQC NMR methods were used to determine the lignin's structure. The results of our thermogravimetric analysis revealed that by covalently linking the novel heterocycles to lignin, an increased amount of char was formed compared to lignin alone or a physically mixed control.

Supplementary Materials: The following supporting information can be downloaded at: https://www.mdpi.com/article/10.3390/molecules28237885/s1, Figure S1: Analysis of **DPW-5**; Figure S2: Pretreatment mass balance; Figure S3: Aromatic regions of HSQC NMR of **DPW Lignin** and **DeAcyl Lignin**; Table S1: Quantitative ^{31}P NMR integrals; Table S2: Hydroxyl content of lignin samples; Figure S4: HSQC and HMBC NMR analysis of hydrolysis; Figure S5: HSQC NMR analysis of modified lignins; Figure S6: IR analysis of modified lignins; Figure S7: HSQC NMR analysis of final test lignins; Table S3: Selected crystallographic data; Table S4: Hydrogen bond geometry lengths (Å) and angles (°); Figures S8–S54: NMR spectra of compounds and lignins.

Author Contributions: Conceptualisation, N.J.W., D.J.D. and J.D.W.; methodology, N.J.W., D.J.D., A.P.M. and D.B.C.; software, D.J.D., A.P.M. and D.B.C.; validation, N.J.W., D.J.D., A.P.M., D.B.C. and J.D.W.; formal analysis, N.J.W., D.J.D., A.P.M., D.B.C. and J.D.W.; investigation, D.J.D., A.P.M. and D.B.C.; resources, N.J.W.; data curation, N.J.W., D.J.D., A.P.M. and D.B.C.; writing—original draft preparation, N.J.W., D.J.D., A.P.M. and D.B.C.; writing—review and editing, N.J.W., D.J.D., A.P.M., D.B.C. and J.D.W.; visualisation, N.J.W., D.J.D., A.P.M. and D.B.C.; supervision, N.J.W.; project administration, N.J.W.; funding acquisition, N.J.W. All authors have read and agreed to the published version of the manuscript.

Funding: This research article was funded by EaSI-CAT at the University of St Andrews (studentship to D.J.D.).

Institutional Review Board Statement: Not applicable.

Informed Consent Statement: Not applicable.

Data Availability Statement: The research data underpinning this publication can be accessed at https://doi.org/10.17630/e0b856df-ea8f-443a-94fc-18bb38436b01.

Acknowledgments: We would like to thank David Francis at Leonardo UK Ltd. (Leonardo, UK) for supplying a sample of DPW collected from date farms near Al Ain in the UAED, Gavin Peters for supporting the running of TGA measurements, and Caroline Horsburgh for helping with the mass spectrometry analysis.

Conflicts of Interest: The authors declare no conflict of interest.

References

1. Chang, C.J.; Terrell, M.L.; Marcus, M.; Marder, M.E.; Panuwet, P.; Ryan, P.B.; Pearson, M.; Barton, H.; Barr, D.B. Serum Concentrations of Polybrominated Biphenyls (PBBs), Polychlorinated Biphenyls (PCBs) and Polybrominated Diphenyl Ethers (PBDEs) in the Michigan PBB Registry 40 Years after the PBB Contamination Incident. *Environ. Int.* **2020**, *137*, 105526. [CrossRef] [PubMed]
2. Woodward, G.; Harris, C.; Manku, J. Design of New Organophosphorus Flame Retardants. *Phosphorus Sulfur Silicon Relat. Elem.* **1999**, *144*, 25–28. [CrossRef]
3. Özer, M.S.; Gaan, S. Recent Developments in Phosphorus Based Flame Retardant Coatings for Textiles: Synthesis, Applications and Performance. *Prog. Org. Coat.* **2022**, *171*, 107027. [CrossRef]
4. Wendels, S.; Chavez, T.; Bonnet, M.; Salmeia, K.A.; Gaan, S. Recent Developments in Organophosphorus Flame Retardants Containing P-C Bond and Their Applications. *Materials* **2017**, *10*, 784. [CrossRef]
5. Nazir, R.; Gaan, S. Recent Developments in P(O/S)–N Containing Flame Retardants. *J. Appl. Polym. Sci.* **2020**, *137*, 47910. [CrossRef]
6. Yang, J.; Zhao, Y.; Li, M.; Du, M.; Li, X.; Li, Y. A Review of a Class of Emerging Contaminants: The Classification, Distribution, Intensity of Consumption, Synthesis Routes, Environmental Effects and Expectation of Pollution Abatement to Organophosphate Flame Retardants (OPFRs). *Int. J. Mol. Sci.* **2019**, *20*, 2874. [CrossRef] [PubMed]
7. Blum, A.; Behl, M.; Birnbaum, L.S.; Diamond, M.L.; Phillips, A.; Singla, V.; Sipes, N.S.; Stapleton, H.M.; Venier, M. Organophosphate Ester Flame Retardants: Are They a Regrettable Substitution for Polybrominated Diphenyl Ethers? *Environ. Sci. Technol. Lett.* **2019**, *6*, 638–649. [CrossRef] [PubMed]
8. Shen, J.; Liang, J.; Lin, X.; Lin, H.; Yu, J.; Wang, S. The Flame-Retardant Mechanisms and Preparation of Polymer Composites and Their Potential Application in Construction Engineering. *Polymers* **2022**, *14*, 82. [CrossRef]
9. Camino, G.; Costa, L. Performance and Mechanisms of Fire Retardants in Polymers—A Review. *Polym. Degrad. Stab.* **1988**, *20*, 271–294. [CrossRef]
10. Bifulco, A.; Varganici, C.D.; Rosu, L.; Mustata, F.; Rosu, D.; Gaan, S. Recent Advances in Flame Retardant Epoxy Systems Containing Non-Reactive DOPO Based Phosphorus Additives. *Polym. Degrad. Stab.* **2022**, *200*, 109962. [CrossRef]
11. Jian, R.K.; Ai, Y.F.; Xia, L.; Zhao, L.J.; Zhao, H.B. Single Component Phosphamide-Based Intumescent Flame Retardant with Potential Reactivity towards Low Flammability and Smoke Epoxy Resins. *J. Hazard Mater.* **2019**, *371*, 529–539. [CrossRef]
12. Abdur Rashid, M.; Liu, W.; Wei, Y.; Jiang, Q. Review of Reversible Dynamic Bonds Containing Intrinsically Flame Retardant Biomass Thermosets. *Eur. Polym. J.* **2022**, *173*, 111263. [CrossRef]
13. vom Stein, T.; Grande, P.M.; Kayser, H.; Sibilla, F.; Leitner, W.; Domínguez de María, P. From Biomass to Feedstock: One-Step Fractionation of Lignocellulose Components by the Selective Organic Acid-Catalyzed Depolymerization of Hemicellulose in a Biphasic System. *Green Chem.* **2011**, *13*, 1772–1777. [CrossRef]
14. Grande, P.M.; Viell, J.; Theyssen, N.; Marquardt, W.; Domínguez De María, P.; Leitner, W. Fractionation of Lignocellulosic Biomass Using the OrganoCat Process. *Green Chem.* **2015**, *17*, 3533–3539. [CrossRef]
15. Del Rio, L.F.; Chandra, R.P.; Saddler, J.N. The Effect of Varying Organosolv Pretreatment Chemicals on the Physicochemical Properties and Cellulolytic Hydrolysis of Mountain Pine Beetle-Killed Lodgepole Pine. *Appl. Biochem. Biotechnol.* **2010**, *161*, 1–21. [CrossRef] [PubMed]
16. Mesa, L.; González, E.; Cara, C.; González, M.; Castro, E.; Mussatto, S.I. The Effect of Organosolv Pretreatment Variables on Enzymatic Hydrolysis of Sugarcane Bagasse. *Chem. Eng. J.* **2011**, *168*, 1157–1162. [CrossRef]
17. Teramura, H.; Sasaki, K.; Oshima, T.; Matsuda, F.; Okamoto, M.; Shirai, T.; Kawaguchi, H.; Ogino, C.; Hirano, K.; Sazuka, T.; et al. Organosolv Pretreatment of Sorghum Bagasse Using a Low Concentration of Hydrophobic Solvents Such as 1-Butanol or 1-Pentanol. *Biotechnol. Biofuels* **2016**, *9*, 27. [CrossRef] [PubMed]
18. Lancefield, C.S.; Panovic, I.; Deuss, P.J.; Barta, K.; Westwood, N.J. Pre-Treatment of Lignocellulosic Feedstocks Using Biorenewable Alcohols: Towards Complete Biomass Valorisation. *Green Chem.* **2017**, *19*, 202–214. [CrossRef]
19. Schmetz, Q.; Teramura, H.; Morita, K.; Oshima, T.; Richel, A.; Ogino, C.; Kondo, A. Versatility of a Dilute Acid/Butanol Pretreatment Investigated on Various Lignocellulosic Biomasses to Produce Lignin, Monosaccharides and Cellulose in Distinct Phases. *ACS Sustain. Chem. Eng.* **2019**, *7*, 11069–11079. [CrossRef]
20. Zijlstra, D.S.; Analbers, C.A.; de Korte, J.; Wilbers, E.; Deuss, P.J. Efficient Mild Organosolv Lignin Extraction in a Flow-Through Setup Yielding Lignin with High β-O-4 Content. *Polymers* **2019**, *11*, 1913. [CrossRef]
21. Viola, E.; Zimbardi, F.; Morgana, M.; Cerone, N.; Valerio, V.; Romanelli, A. Optimized Organosolv Pretreatment of Biomass Residues Using 2-Methyltetrahydrofuran and n-Butanol. *Processes* **2021**, *9*, 2051. [CrossRef]
22. Davidson, D.J.; Lu, F.; Faas, L.; Dawson, D.M.; Warren, G.P.; Panovic, I.; Montgomery, J.R.D.; Ma, X.; Bosilkov, B.G.; Slawin, A.M.Z.; et al. Organosolv Pretreatment of Cocoa Pod Husks: Isolation, Analysis, and Use of Lignin from an Abundant Waste Product. *ACS Sustain. Chem. Eng.* **2023**, *11*, 14323–14333. [CrossRef] [PubMed]
23. Zhang, D.; Zeng, J.; Liu, W.; Qiu, X.; Qian, Y.; Zhang, H.; Yang, Y.; Liu, M.; Yang, D. Pristine Lignin as a Flame Retardant in Flexible PU Foam. *Green Chem.* **2021**, *23*, 5972–5980. [CrossRef]
24. Zhang, Y.M.; Zhao, Q.; Li, L.; Yan, R.; Zhang, J.; Duan, J.C.; Liu, B.J.; Sun, Z.Y.; Zhang, M.Y.; Hu, W.; et al. Synthesis of a Lignin-Based Phosphorus-Containing Flame Retardant and Its Application in Polyurethane. *RSC Adv.* **2018**, *8*, 32252–32261. [CrossRef] [PubMed]
25. Yang, H.; Yu, B.; Xu, X.; Bourbigot, S.; Wang, H.; Song, P. Lignin-Derived Bio-Based Flame Retardants toward High-Performance Sustainable Polymeric Materials. *Green Chem.* **2020**, *22*, 2129–2161. [CrossRef]

26. Jawerth, M.; Johansson, M.; Lundmark, S.; Gioia, C.; Lawoko, M. Renewable Thiol-Ene Thermosets Based on Refined and Selectively Allylated Industrial Lignin. *ACS Sustain. Chem. Eng.* **2017**, *5*, 10918–10925. [CrossRef]
27. Jedrzejczyk, M.A.; Madelat, N.; Wouters, B.; Smeets, H.; Wolters, M.; Stepanova, S.A.; Vangeel, T.; Van Aelst, K.; Van den Bosch, S.; Van Aelst, J.; et al. Preparation of Renewable Thiol-Yne "Click" Networks Based on Fractionated Lignin for Anticorrosive Protective Film Applications. *Macromol. Chem. Phys.* **2022**, *223*, 2100461. [CrossRef]
28. Bhattacharyya, P.; Woollins, J.D. Selenocarbonyl Synthesis Using Woollins Reagent. *Tetrahedron Lett.* **2001**, *42*, 5949–5951. [CrossRef]
29. Gray, I.P.; Bhattacharyya, P.; Slawin, A.M.Z.; Woollins, J.D. A New Synthesis of (PhPSe2)2 (Woollins Reagent) and Its Use in the Synthesis of Novel P–Se Heterocycles. *Chem. Eur. J.* **2005**, *11*, 6221–6227. [CrossRef]
30. Hua, G.; Fuller, A.L.; Slawin, A.M.Z.; Woollins, J.D. Formation of New Organoselenium Heterocycles and Ring Reduction of 10-Membered Heterocycles into Seven-Membered Heterocycles. *Polyhedron* **2011**, *30*, 805–808. [CrossRef]
31. Karlen, S.D.; Smith, R.A.; Kim, H.; Padmakshan, D.; Bartuce, A.; Mobley, J.K.; Free, H.C.A.; Smith, B.G.; Harris, P.J.; Ralph, J. Highly Decorated Lignins in Leaf Tissues of the Canary Island Date Palm *Phoenix Canariensis*. *Plant Physiol.* **2017**, *175*, 1058–1067. [CrossRef] [PubMed]
32. Hua, G.; Du, J.; Surgenor, B.A.; Slawin, A.M.Z.; Woollins, J.D. Novel Fluorinated Phosphorus-Sulfur Heteroatom Compounds: Synthesis and Characterization of Ferrocenyl-and Aryl-Phosphonofluorodithioic Salts, Adducts, and Esters. *Molecules* **2015**, *20*, 12175–12197. [CrossRef] [PubMed]
33. Sanhoury, M.A.; Mbarek, T.; Slawin, A.M.Z.; Ben Dhia, M.T.; Khaddar, M.R.; Woollins, J.D. Synthesis, Characterization and Structures of Cadmium(II) and Mercury(II) Complexes with Bis(Dipiperidinylphosphino)Methylamine Dichalcogenides. *Polyhedron* **2016**, *119*, 106–111. [CrossRef]
34. Salmeia, K.A.; Gaan, S. An Overview of Some Recent Advances in DOPO-Derivatives: Chemistry and Flame Retardant Applications. *Polym. Degrad. Stab.* **2015**, *113*, 119–134. [CrossRef]
35. Chi, Z.; Guo, Z.; Xu, Z.; Zhang, M.; Li, M.; Shang, L.; Ao, Y. A DOPO-Based Phosphorus-Nitrogen Flame Retardant Bio-Based Epoxy Resin from Diphenolic Acid: Synthesis, Flame-Retardant Behavior and Mechanism. *Polym. Degrad. Stab.* **2020**, *176*, 109151. [CrossRef]
36. Lu, X.; Yu, M.; Wang, D.; Xiu, P.; Xu, C.; Lee, A.F.; Gu, X. Flame-Retardant Effect of a Functional DOPO-Based Compound on Lignin-Based Epoxy Resins. *Mater. Today Chem.* **2021**, *22*, 100562. [CrossRef]
37. Lee, K.; Hwang, J.; Jeong, P.H.; Park, J.; Lee, K.; Ko, J.M. New Quinone-Based Electrode Additives Electrochemically Polymerized on Activated Carbon Electrodes for Improved Pseudocapacitance. *Macromol. Res.* **2023**, *31*, 171–179. [CrossRef]
38. Oh, J.; Park, J.Y.; Park, K.C.; Hwang, J.H.; Park, J.H. Phosphonamidate Compounds for Butyrylcholinesterase Selective Inhibitors. *Bull. Korean Chem. Soc.* **2020**, *41*, 1153–1160. [CrossRef]
39. Januszewski, R.; Dutkiewicz, M.; Maciejewski, H.; Marciniec, B. Synthesis and Characterization of Phosphorus-Containing, Silicone Rubber Based Flame Retardant Coatings. *React. Funct. Polym.* **2018**, *123*, 1–9. [CrossRef]
40. Denis, M.; Coste, G.; Sonnier, R.; Caillol, S.; Negrell, C. Influence of Phosphorus Structures and Their Oxidation States on Flame-Retardant Properties of Polyhydroxyurethanes. *Molecules* **2023**, *28*, 611. [CrossRef]
41. Ralph, J.; Lapierre, C.; Boerjan, W. Lignin Structure and Its Engineering. *Curr. Opin. Biotechnol.* **2019**, *56*, 240–249. [CrossRef] [PubMed]
42. Lahive, C.W.; Kamer, P.C.J.; Lancefield, C.S.; Deuss, P.J. An Introduction to Model Compounds of Lignin Linking Motifs; Synthesis and Selection Considerations for Reactivity Studies. *ChemSusChem* **2020**, *13*, 4238–4265. [CrossRef] [PubMed]
43. Panovic, I.; Montgomery, J.R.D.; Lancefield, C.S.; Puri, D.; Lebl, T.; Westwood, N.J. Grafting of Technical Lignins through Regioselective Triazole Formation on β-O-4 Linkages. *ACS Sustain. Chem. Eng.* **2017**, *5*, 10640–10648. [CrossRef]
44. Granata, A.; Argyropoulos, D.S. 2-Chloro-4,4,5,5-Tetramethyl-1,3,2-Dioxaphospholane, a Reagent for the Accurate Determination of the Uncondensed and Condensed Phenolic Moieties in Lignins. *J. Agric. Food Chem.* **1995**, *43*, 1538–1544. [CrossRef]
45. Jiang, Z.-H.; Argyropoulos, D.S.; Granata, A. Correlation Analysis of ^{31}P NMR Chemical Shifts with Substituent Effects of Phenols. *Magn. Reason. Chem.* **1995**, *33*, 375–382. [CrossRef]
46. Smith, D.C.C. Ester Groups in Lignin. *Nature* **1955**, *176*, 267–268. [CrossRef]
47. Zijlstra, D.S.; de Korte, J.; de Vries, E.P.C.; Hameleers, L.; Wilbers, E.; Jurak, E.; Deuss, P.J. Highly Efficient Semi-Continuous Extraction and In-Line Purification of High β-O-4 Butanosolv Lignin. *Front. Chem.* **2021**, *9*, 655983. [CrossRef] [PubMed]
48. *CrystalClear-SM Expert*, v2.1.; Rigaku Americas: The Woodlands, TX, USA; Rigaku Corporation: Tokyo, Japan, 2015.
49. *CrysAlisPro, v1.171.42.94a Rigaku Oxford Diffraction*; Rigaku Corporation: Tokyo, Japan, 2023.
50. Sheldrick, G.M. SHELXT—Integrated space-group and crystal structure determination. *Acta Crystallogr. Sect. A Found. Adv.* **2015**, *71*, 3–8. [CrossRef]
51. Sheldrick, G.M. Crystal structure refinement with SHELXL. *Acta Crystallogr. Sect. C Struct. Chem.* **2015**, *71*, 3–8. [CrossRef]
52. Dolomanov, O.V.; Bourhis, L.J.; Gildea, R.J.; Howard, J.A.K.; Puschmann, H. OLEX2: A complete structure solution, refinement and analysis program. *J. Appl. Crystallogr.* **2009**, *42*, 339–341. [CrossRef]

Disclaimer/Publisher's Note: The statements, opinions and data contained in all publications are solely those of the individual author(s) and contributor(s) and not of MDPI and/or the editor(s). MDPI and/or the editor(s) disclaim responsibility for any injury to people or property resulting from any ideas, methods, instructions or products referred to in the content.

Article

Thermal Rearrangement of Thiocarbonyl-Stabilised Triphenylphosphonium Ylides Leading to (Z)-1-Diphenylphosphino-2-(phenylsulfenyl)alkenes and Their Coordination Chemistry

R. Alan Aitken *, Graham Dawson, Neil S. Keddie, Helmut Kraus, Heather L. Milton, Alexandra M. Z. Slawin, Joanne Wheatley and J. Derek Woollins

EaStCHEM School of Chemistry, University of St Andrews, North Haugh, St. Andrews, Fife KY16 9ST, UK
* Correspondence: raa@st-and.ac.uk; Tel.: +44-1334-463865

Abstract: While thiocarbonyl-stabilised phosphonium ylides generally react upon flash vacuum pyrolysis by the extrusion of Ph_3PS to give alkynes in an analogous way to their carbonyl-stabilised analogues, two examples with a hydrogen atom on the ylidic carbon are found to undergo a quite different process. The net transfer of a phenyl group from P to S gives (Z)-configured 1-diphenylphosphino-2-(phenylsulfenyl)alkenes in a novel isomerisation process via intermediate λ^5-1,2-thiaphosphetes. These prove to be versatile hemilabile ligands with a total of seven complexes prepared involving five different transition metals. Four of these are characterised by X-ray diffraction with two involving the bidentate ligand forming a five-membered ring metallacycle and two with the ligand coordinating to the metal only through phosphorus.

Keywords: flash vacuum pyrolysis; phosphonium ylide; phosphine; hemilabile ligand; transition metal complex; X-ray structure

Citation: Aitken, R.A.; Dawson, G.; Keddie, N.S.; Kraus, H.; Milton, H.L.; Slawin, A.M.Z.; Wheatley, J.; Woollins, J.D. Thermal Rearrangement of Thiocarbonyl-Stabilised Triphenylphosphonium Ylides Leading to (Z)-1-Diphenylphosphino-2-(phenylsulfenyl)alkenes and Their Coordination Chemistry. *Molecules* 2024, 29, 221. https://doi.org/10.3390/molecules29010221

Academic Editor: Fabio Marchetti

Received: 13 December 2023
Revised: 28 December 2023
Accepted: 30 December 2023
Published: 31 December 2023

Copyright: © 2023 by the authors. Licensee MDPI, Basel, Switzerland. This article is an open access article distributed under the terms and conditions of the Creative Commons Attribution (CC BY) license (https:// creativecommons.org/licenses/by/ 4.0/).

1. Introduction

The thermal extrusion of Ph_3PO from carbonyl-stabilised triphenylphosphonium ylides **1** is a well-established synthetic route to functionalised alkynes **2** (Scheme 1) [1–3]. The process proceeds particularly well using flash vacuum pyrolysis (FVP), and we have found that the phosphorus to carbonyl coupling constant $^2J_{P-CO}$ provides a diagnostic parameter for the likely success of the reaction, with ylides for which $^2J_{P-CO} < 11$ Hz usually providing the alkynes in high yield [4]. By way of contrast, the thermal behaviour of the corresponding thiocarbonyl-stabilised ylides **3** has only been examined in a few cases and Bestmann and Schaper found that heating the ylides **3** above their melting point resulted in a bimolecular process with the loss of Ph_3P and Ph_3PS to give tetrasubstituted thiophenes **4** [5]. Some time ago, we described a preliminary study in which FVP of the ylides **5** was found to proceed as expected by analogy with the carbonyl analogues **1** with the loss of Ph_3PS to give alkynes **2** for $R^1 \neq H$, but when R^1 was hydrogen, a quite different process was observed: rearrangement with the transfer of a phenyl group from P to S giving the potentially useful bidentate proligands **6** [6]. In this paper, we describe in more detail the synthesis and structure of these novel phosphinovinyl sulfides as well as their coordination chemistry.

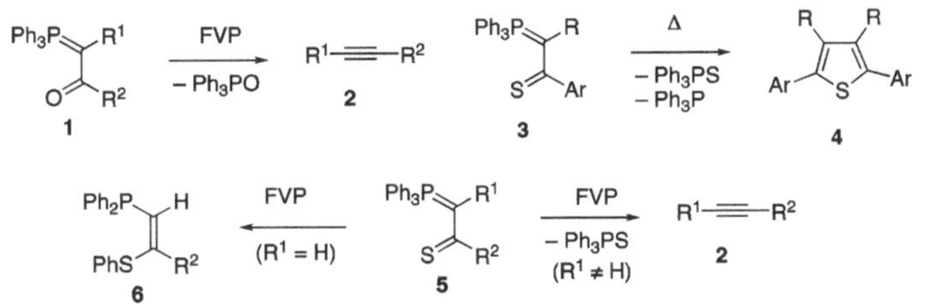

Scheme 1. Previously reported thermal reactivity of carbonyl- and thiocarbonyl-stabilised phosphonium ylides.

2. Results and Discussion

Synthetic access to thiocarbonyl-stabilised ylides **5** is available using various methods including the treatment of non-stabilised ylides with dithioesters or dithiocarbonates [7], or activation of the corresponding carbonyl-stabilised ylides with triflic anhydride [8] or POCl$_3$ [9] followed by treatment with sodium sulfide. For the current study, we used the direct reaction of carbonyl-stabilised ylides **1** with Lawesson's reagent introduced by Capuano and coworkers [10], and in this way, we prepared the five examples **7–11** (Figure 1) from their carbonyl analogues.

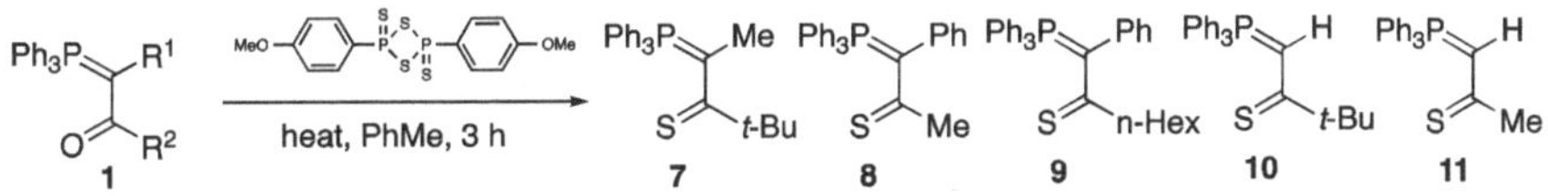

Figure 1. Thiocarbonyl ylides used in this study.

Compounds **7–9** are already known [8] and gave analytical and spectroscopic data in agreement with the reported values. The new compounds **10** and **11** were fully characterised and showed distinctive NMR signals confirming the presence of P=CH–C=S [**10** δ_P +5.0; δ_H 5.22 (d, ^2J 34 Hz, P=CH); δ_C 81.3 (d, ^1J 118 Hz, P=CH), 214.4 (d, ^2J 4 Hz, C=S). **11** δ_P +8.1; δ_H 5.18 (d, ^2J 32 Hz, P=CH); δ_C 84.1 (d, ^1J 113 Hz, P=CH), 200.5 (d, ^2J 4 Hz, C=S)]. In fact, the phosphorus coupling extended throughout the structures with all carbon signals except the CH$_3$ of *t*-butyl observed as doublets in the ^{13}C NMR spectra. This is discussed further below in the context of a comparison of the coupling pattern before and after the thermal rearrangement. The structure of **10** was also confirmed by X-ray diffraction (CSD RefCode: AJOMUI) and this was described in our earlier communication [6]. We might also note at this point that compound **10** has a particularly pungent and unpleasant smell. Although it is not very volatile, it is extremely persistent and inadvertent contact with equipment or work surfaces contaminated with **10**, even after several months, results in the release of its characteristic smell. We speculate that this may be due to slow hydrolysis of the P=C bond to give Ph$_3$PO and release thiopinacolone, *t*-BuC(=S)Me.

When the thiocarbonyl ylides **7**, **8** and **9** were subjected to FVP, there was complete reaction at a furnace temperature of 650 °C to give Ph$_3$PS (δ_P +43) at the furnace exit and the expected alkynes **2**, 2,2-dimethylpent-3-yne, 1-phenylpropyne and 1-phenyloctyne, in the cold trap. Thus, these compounds behave in a similar way to their carbonyl analogues but react more readily than the latter, which require a temperature of 750 °C for complete reaction [11]. The higher reactivity of thiocarbonyl- as compared to carbonyl-stabilised ylides is a feature that we have already noted and quantified in a series of kinetic studies on the pyrolysis of carbamoyl and thiocarbamoyl ylides [12–14].

When the two ylides **10** and **11** were subjected to FVP, the reaction was also complete at 650 °C, but the process involved turned out to be completely different. In each case, only a single main product was obtained, which was isomeric with the starting material. In the case of **10**, the product **12** was obtained in good yield and in a pure form as a crystalline solid after preparative TLC. This showed a ^{31}P NMR signal at −19.8 ppm, in the region expected for an alkyldiphenylphosphine, and the single P–CH= hydrogen gave a ^{1}H NMR singlet at 6.94 ppm (Scheme 2).

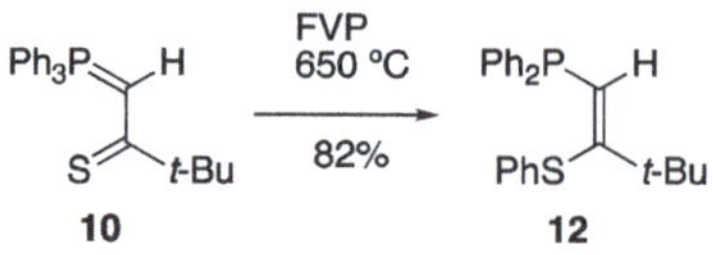

Scheme 2. FVP of **10** to give **12**.

The ^{13}C NMR spectrum, and particularly the pattern of P–C coupling, was particularly informative and showed major changes from the values in **10** (Figure 2).

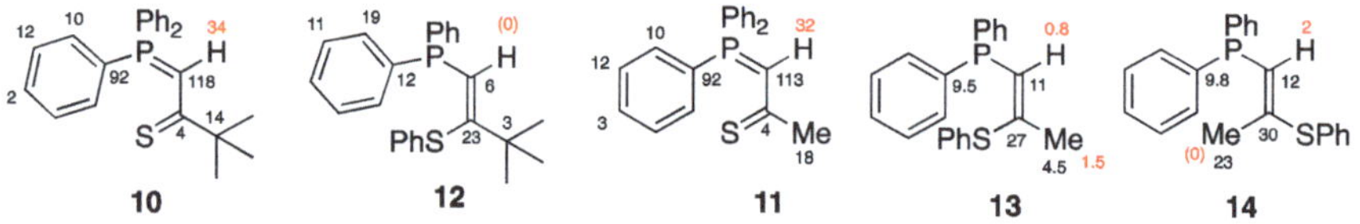

Figure 2. Magnitude of P–H (red) and P–C (black) coupling constants (Hz).

The values of J_{P-C} for the thiocarbonyl ylides **10** and **11** are consistent with those well established for the corresponding carbonyl ylides [11] and carbamoyl/thiocarbamoyl ylides [12–14]. However, in the rearranged product **12**, the values around P-Ph are more similar to those in Ph$_3$P, and the much higher coupling to P–C=C as compared to P–C=C as well as the absence of coupling to P–CH are surprising features. As already mentioned in our preliminary communication [6], the structure of **12** was confirmed by X-ray diffraction (CSD RefCode: AJOMOC).

When we examined the corresponding pyrolysis of the thioacetyl ylide **11**, the corresponding reaction occurred at the same temperature, but the product was now a liquid formed in lower yield and containing some impurities including Ph$_3$P, Ph$_3$PO and Ph$_3$PS. More interestingly, while it was predominantly (6.5:1) the (Z)-isomer **13** (δ_P −22.7), signals attributed to the (E)-isomer **14** (δ_P −25.4) were also apparent (Scheme 3).

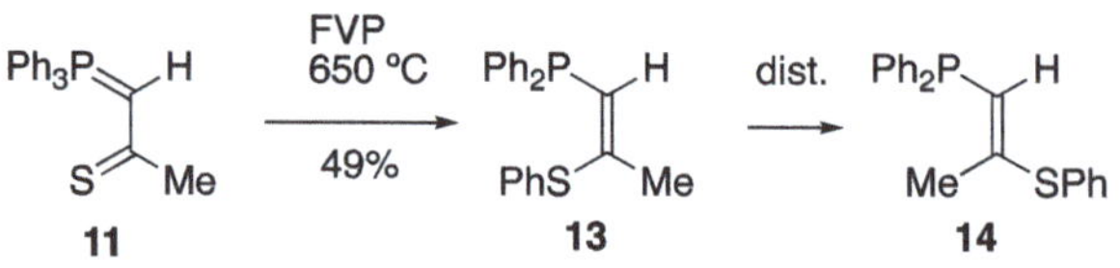

Scheme 3. FVP of **11** to give **13** and **14**.

Attempts to purify this by repeated Kugelrohr distillation under reduced pressure led instead to isomerisation to give a 1:1 mixture of **13** and **14**. At first sight, it might seem surprising that the product is obtained mainly as the less thermodynamically stable isomer from pyrolysis at 650 °C but then isomerises to the more stable isomer upon simple distillation at 90 °C, but this only serves to emphasise the mild nature of the flash vacuum pyrolysis technique. In addition, it has been shown that under FVP conditions, (Z)-alkenes do not normally isomerise to the (E)-isomer to any great extent at temperatures as low as 650 °C with the degree of conversion of (Z)- to (E)-stilbene, for example, being determined as 12%, in good agreement with our results [15]. The presence of aromatic impurities even after distillation prevented full assignment of the ^{13}C NMR spectra for **13** and **14**, but the key signals and the values of the phosphorus coupling constants (Figure 2) were in good agreement with those for **12** while also showing significant differences between the two isomers. The detailed form of the signals for P–CH and P–C=C–Me in the two isomers was at first sight surprising. However, this could be explained by coincidental equivalence of some H–H and P–H coupling constants, and the observed patterns (see Supplementary Materials, Figure S14) could be satisfactorily simulated using the values shown in Figure 3.

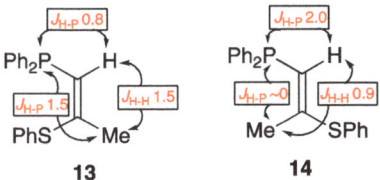

Figure 3. Coupling constants in ^{1}H NMR spectra of **13** and **14**.

Although the 1,2-arrangement of phosphine and sulfide functions on an alkene double bond, particularly in the (*Z*)-configuration, gives potentially valuable "hemilabile" proligands, few such compounds seem to be known (Figure 4). Chlorinated compounds such as **15** [16], **16** [17] and **17** [18] have been prepared as mixtures of (*E*)- and (*Z*)-isomers by the addition of phosphorus compounds to alkynes. The simpler disubstituted alkenes **18** [19], **19** [20] and **20** [21] have also been prepared but these are the (*E*)-isomers as shown. More recently, the (*Z*)-configured vinylphosphonates **21** containing sulfide, selenide and telluride functions [22] as well as the tellurovinylphosphine oxides **22** [23] have also been reported.

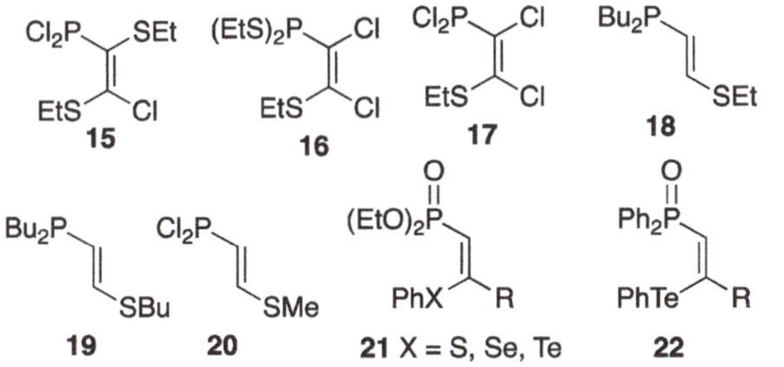

Figure 4. Some previously reported P–C=C–S proligands.

In terms of the mechanism of this new thermal rearrangement, we envisage attack of the nucleophilic sulfur at phosphorus to give a transient λ^5-1,2-thiaphosphete, which is of course the same intermediate involved in the extrusion of Ph$_3$PS to give alkynes as observed for **7–9**. This behaviour is also consistent with that of the isolable thiaphosphete **23**, which fragments with the loss of an alkyne to give the benzoxaphosphole *P*-sulfide [24]. However, perhaps due to the relief of steric congestion, the thiaphosphetes derived from **10** and **11** instead undergo what is effectively a reductive elimination at phosphorus to give the (*Z*)-phosphinovinyl sulfides **12** and **13** (Scheme 4).

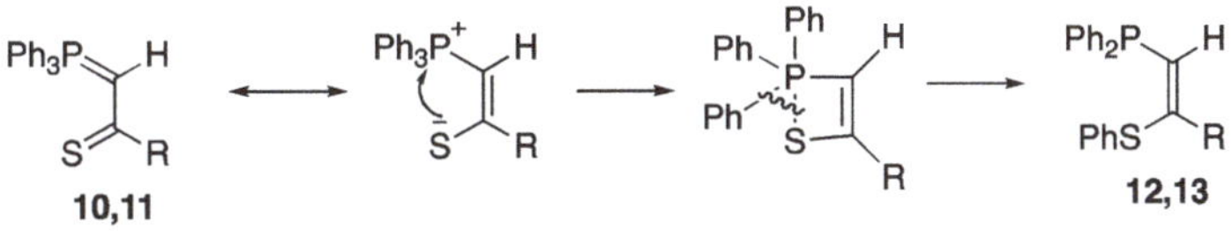

Scheme 4. Mechanism proposed for the thermal rearrangement of **10** and **11**.

Somewhat similar processes are the transfer of Ph from Se to O in acylselenonium ylides such as **24** to give **25** observed by Rakitin [25] (Scheme 5), the transformation of **26** into **27** postulated by Zbiral in the interaction of ylides Ph$_3$P=CHR with benzyne [26] and the rearrangement of the ylide-containing N-heterocyclic carbene **28** via **29** to give 3-phosphinoindole **30** [27].

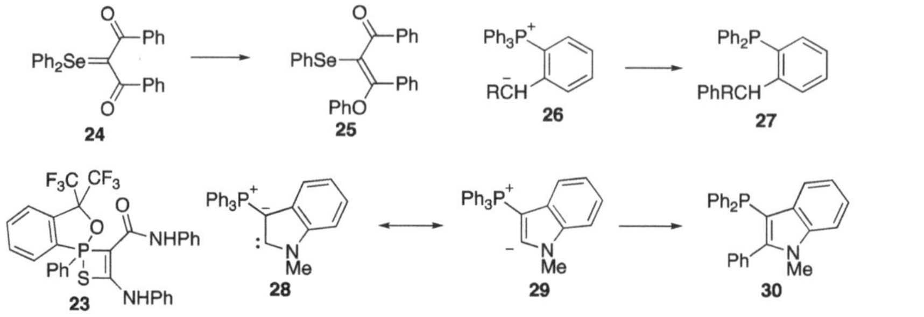

Scheme 5. Some mechanistic precedents.

With the two new hemilabile proligands in hand, we now set about exploring their coordination chemistry. In our preliminary communication [6], the formation of the square planar platinum complex **31** from **12** was described along with its X-ray structure determination (CSD RefCode: AJONAP). We were also successful in obtaining complexes of **12** with a wide range of other standard transition metal reagents (Scheme 6).

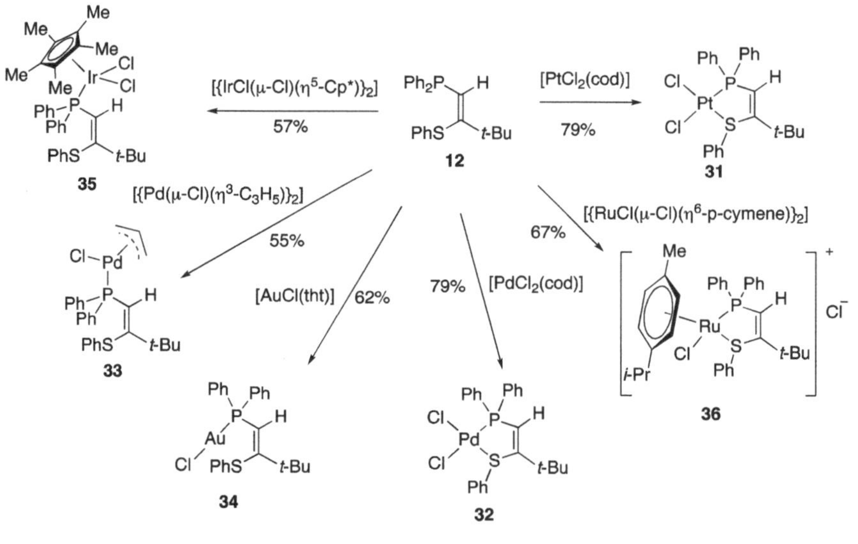

Scheme 6. Formation of transition metal complexes from **12**.

Reaction of the starting materials in CH$_2$Cl$_2$ followed by partial evaporation and precipitation with diethyl ether gave the new complexes **32–36** in moderate to good yield as crystalline solids, giving the expected microanalytical data and ^{1}H and ^{31}P NMR spectra. From the analytical data, it was clear that the complexes had formed with the expected stoichiometry according to the metal source employed, with **12** acting as a bidentate ligand in the square planar platinum(II) and palladium(II) complexes **31** and **32** and the cationic ruthenium(II) complex **36**, but as a monodentate ligand through the more strongly donating phosphorus atom in palladium(II) complex **33**, gold(I) complex **34** and iridium(III) complex **35**. In addition to the compound **31** already confirmed by X-ray diffraction [6], we were able to obtain X-ray structures for gold complex **34** and iridium complex **35** (see below).

Although the methyl proligand **13** was available in lower quantity and purity, we were able to prepare its palladium dichloride complex **37** analogous to **32** and also determined its X-ray structure (Scheme 7).

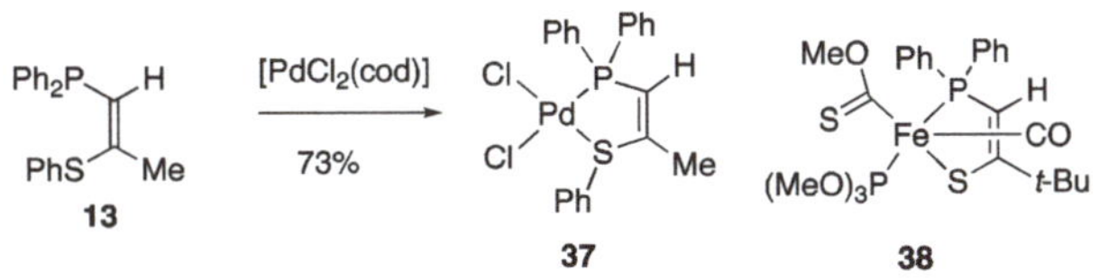

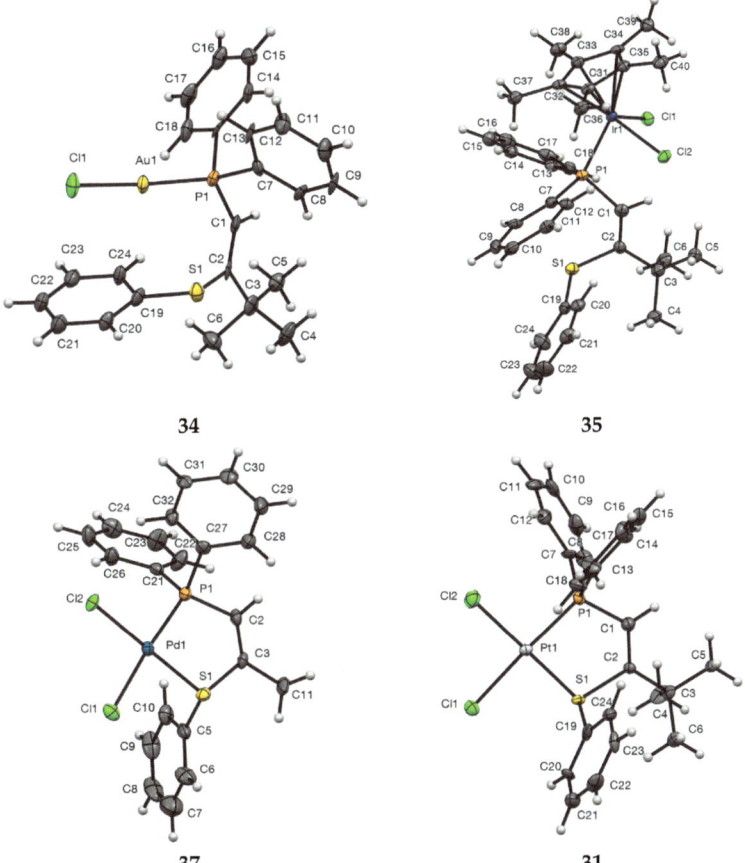

Scheme 7. Formation of palladium complex **37** and structure of a related iron complex.

The structures of the complexes **34**, **35** and **37** together with the numbering systems used are shown below (Figure 5) with that of **31** [6] also included for comparison.

Figure 5. X-ray structures of complexes **34**, **35**, **37** and **31** showing probability ellipsoids at 50% level and numbering systems used.

For comparison, we also show (Figure 6) the previously reported structures of **10** [6] and **12** [6], and the iron(II) complex **38** [28], which, although made in a quite different way, contains the phosphinovinylthiolate corresponding to **12** as an anionic bidentate ligand.

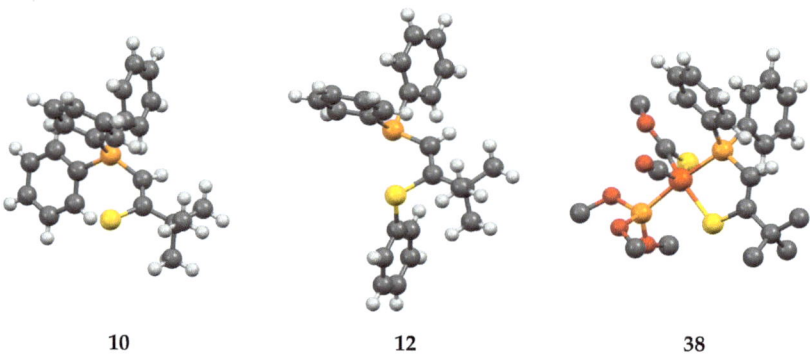

Figure 6. Published structures of **10**, **12** and **38**.

The key structural parameters for complexes **34**, **35** and **37** and, for comparison, those for **10**, **12**, **31** and **38** are presented in Table 1. In all seven structures, the P–C–C–S fragment is quite accurately planar with a torsion angle of <10° in every case.

Table 1. Geometric parameters for **10** [6], **12** [6], **31** [6], **34**, **35**, **37** and, for comparison, **38** [28].

Compound	Bond lengths (Å)				
	P–CH	CH=C	=C–S	P–M	M–S
10	1.739(2)	1.373(3)	1.708(2)	—	—
12	1.818(4)	1.316(6)	1.788(4)	—	—
31	1.773(14)	1.34(2)	1.825(14)	2.216(6)	2.259(5)
34	1.795(12)	1.332(17)	1.782(13)	2.229(3)	—
35	1.824(5)	1.318(7)	1.787(5)	2.3076(12)	—
37	1.798(8)	1.322(11)	1.796(8)	2.227(2)	2.252(2)
38	1.770(6)	1.346(8)	1.764(5)	2.256(1)	2.307(2)

Compound	Angles (°)				
	P–C=C	C=C–S	=C–P–M	P–M–S	M–S–C=
10	124.05(16)	122.46(16)	—	—	—
12	125.5(3)	115.1(3)	—	—	—
31	121.4(11)	118.0(10)	106.6(6)	88.59(19)	105.3(5)
34	128.7(10)	118.9(9)	121.2(4)	—	—
35	131.7(4)	119.8(4)	109.27(15)	—	—
37	119.6(6)	119.1(6)	107.2(3)	87.07(8)	106.8(3)
38	118.3(2)	120.7(2)	108.4(1)	85.1(0)	106.3(1)

If we first compare **10** and **12**, the major change in geometry associated with the transformation of P=C–C=S into P–C=C–S is clear. However, comparing the structural parameters of **12** with those of its complexes **31**, **34** and **35** as well as the related thiolate complex **38** shows a remarkable degree of consistency. As expected, the complexes **34** and **35** where the metal binds only to phosphorus have bond lengths around the =C–SPh that are relatively unaffected, while the bidentate binding to platinum in **31** results in the significant lengthening of C–S and shortening of C–P. While the smaller size of AuCl means the ligand **12** can retain the orientation of the PPh$_2$ group, coordination to the much larger Cp*IrCl$_2$ requires the PPh$_2$ group to rotate, placing the phenyl groups facing towards SPh. The similarity between the geometry of the neutral ligand **12** in complexes such as **31** and the anionic enethiolate in **38** is also notable, with only the =C–S length being significantly shorter in the latter. The angles within the five-membered ring in complexes **31**, **37** and **38** are also remarkably consistent.

3. Experimental Section

3.1. General Experimental Details

NMR spectra were recorded on solutions in CDCl$_3$ unless otherwise stated using Bruker instruments, and chemical shifts are given in ppm to high frequency from Me$_4$Si for ^{1}H and ^{13}C and H$_3$PO$_4$ for ^{31}P with coupling constants J in Hz. The ^{13}C NMR spectra are referenced to the solvent signal at 77.0 (CDCl$_3$). IR spectra were recorded on a Perkin Elmer 1420 instrument. Elemental analysis was conducted using a Carlo Erba CHNS analyser. Mass spectra were obtained using a Micromass instrument and the ionisation method used is noted in each case. Preparative TLC was carried out using 1.0 mm layers of Merck alumina 60 G containing 0.5% Woelm fluorescent green indicator on glass plates. Melting points were recorded on a Gallenkamp 50W melting point apparatus or a Reichert hot-stage microscope.

Flash vacuum pyrolysis (FVP) was carried out in a conventional flow system by subliming the starting material through a horizontal quartz tube (30 × 2.5 cm) externally heated by a tube furnace to 650 °C and maintained at a pressure of 2–5 × 10^{-2} Torr by a rotary vacuum pump. Products were collected in a liquid N$_2$ cooled U-shaped trap and purified as noted.

General organic and inorganic reagents and solvents were obtained from standard suppliers and used as received. Dry THF was prepared by storage over sodium wire. Starting transition metal complexes [AuCl(tetrahydrothiophene)] [29], [PdCl$_2$(cyclooctadiene)] [30], [PtCl$_2$(cyclooctadiene)] [31], [{RuCl(μ-Cl)(η^6-p-MeC$_6$H$_4$^{i}Pr)}$_2$] [32], [{IrCl(μ-Cl)(η^5-C$_5$Me$_5$)}$_2$] [33] and [{Pd(m-Cl)(η^3-C$_3$H$_5$)}$_2$] [34] were prepared by the reported methods.

3.2. Preparation of Thiocarbonyl Ylides

3.2.1. Preparation of Thiopivaloylmethylenetriphenylphosphorane 10

A solution of pivaloylmethylenetriphenylphosphorane (5.0 g, 13.9 mmol) and Lawesson's reagent (2.81 g, 6.9 mmol) in toluene (300 mL) was heated under reflux under nitrogen for 3 h. The mixture was allowed to cool to RT and the solution was poured off leaving an insoluble oily residue and evaporated. Recrystallisation of the resulting solid from ethyl acetate gave the product (2.86 g, 55%) as pale-yellow crystals, mp 200–202 °C; (Found: C, 76.5; H, 6.4. C$_{24}$H$_{25}$PS requires C, 76.6; H, 6.7%); v_{max}/cm^{-1} 1572, 1260, 1205, 1160, 1105, 978, 880, 792, 751, 722, 713, 691 and 620; ^{1}H NMR (300 MHz) δ$_H$ 1.40 (9H, s), 5.22 (1H, d, J 34, CH=P), 7.40–7.48 (6H, m), 7.48–7.52 (3H, m) and 7.67–7.80 (6H, m); ^{13}C NMR (75 MHz) δ$_C$ 31.3 (3C), 43.2 (d, J 14, CMe$_3$), 81.3 (d, J 118, P=C), 125.3 (d, J 92, C-1 of Ph), 128.5 (d, J 12, C-3 of Ph), 131.6 (d, J 2, C-4 of Ph), 132.8 (d, J 10, C-2 of Ph) and 214.4 (d, J 4, C=S); ^{31}P NMR (121 MHz) δ$_P$ +5.0; MS (EI) m/z 376 (M$^+$, 16%), 343 (9), 319 (100), 294 (7), 262 (12) and 183 (23).

3.2.2. Preparation of Thioacetylmethylenetriphenylphosphorane 11

A solution of acetylmethylenetriphenylphosphorane (8.0 g, 25 mmol) and Lawesson's reagent (5.1 g, 12.6 mmol) in toluene (300 mL) was heated under reflux under nitrogen for 3 h. The mixture was allowed to cool to RT and the solution was poured off leaving an insoluble oily residue and evaporated. Recrystallisation of the resulting solid from ethyl acetate gave the product (4.62 g, 55%) as pale-yellow crystals, mp 172–174 °C; (Found: C, 75.2; H, 5.3. C$_{21}$H$_{19}$PS requires C, 75.4; H, 5.7%); v_{max}/cm^{-1} 1585, 1270, 1175, 1106, 993, 872, 763, 747, 723, 686 and 660; ^{1}H NMR (300 MHz) δ$_H$ 2.63 (3H, s), 5.18 (1H, d, J 32, CH=P), 7.45–7.55 (6H, m), 7.55–7.65 (3H, m) and 7.70–7.80 (6H, m); ^{13}C NMR (75 MHz) δ$_C$ 36.8 (d, J 18, Me), 84.1 (d, J 113, P=C), 124.6 (d, J 92, C-1 of Ph), 128.9 (d, J 12, C-3 of Ph), 132.3 (d, J 3, C-4 of Ph), 133.3 (d, J 10, C-2 of Ph) and 200.5 (d, J 4, C=S); ^{31}P NMR (121 MHz) δ$_P$ +8.1; MS (EI) m/z 334 (M$^+$, 85%), 319 (16), 301 (40), 262 (14), 225 (30), 183 (38) and 43 (100).

3.3. Thermal Rearrangement of Thiocarbonyl Ylides

3.3.1. Preparation of (Z)-1-Diphenylphosphino-3,3-dimethyl-2-phenylthiobut-1-ene 12

FVP of the ylide **10** (0.50 g, 1.33 mmol) was performed at 650 °C and 3.8×10^{-2} Torr and was complete within 1 h. Preparative TLC (silica, diethyl ether) of the crude material gave the product **12** (0.41 g, 82%) as pale-yellow plates, mp 121–123 °C; (Found: C, 76.3; H, 6.7; S, 8.2. $C_{24}H_{25}PS$ requires C, 76.6; H, 6.7; S, 8.5%); ν_{max}/cm^{-1} 1582, 1551, 1303, 1181, 1118, 1027, 960, 740, 720 and 694; ^1H NMR (300 MHz) δ_H 1.21 (9H, s), 6.94 (1H, s), 7.05–7.25 (5H, m) and 7.25–7.40 (10H, m); ^{13}C NMR (75 MHz) δ_C 29.8 (3CH$_3$), 41.5 (d, J 3, CMe$_3$), 125.2 (C-4 of SPh), 127.5 (C-3 of SPh), 128.24 (C-4 of PPh), 128.25 (d, J 11, C-3 of PPh), 128.6 (C-2 of SPh), 132.6 (d, J 19, C-2 of PPh), 137.2 (C-1 of SPh), 137.8 (d, J 6, P–CH=), 139.4 (d, J 12, C-1 of PPh) and 158.8 (d, J 23, S–C=); ^{31}P NMR (121 MHz) δ_P −19.8; MS (CI) m/z 377 (M+H$^+$, 100%), 319 (9) and 279 (10).

3.3.2. Preparation of (Z)-1-Diphenylphosphino-2-phenylthiopropene 13

FVP of the ylide **11** (0.50 g, 78.8 µmol) was performed at 650 °C and 3.8×10^{-2} Torr and was complete within 1 h. NMR analysis of the crude product (0.245 g, 49%) showed a 6.5:1 ratio of (Z)- and (E)-**13**. Repeated Kugelrohr distillation of this (bp 90 °C/0.1 Torr) in an attempt to remove trace impurities of Ph$_3$P, Ph$_3$PO and Ph$_3$PS resulted in isomerisation to afford a 1:1 ratio of (Z) and (E)-**13**. By comparing the NMR data before and after distillation, the following assignments could be made (owing to peak overlap, definite assignment of the remaining aromatic ^{13}C NMR signals was not possible):

(Z)-**13**: ^1H NMR (300 MHz) δ_H 2.02 (3H, t, J 1.5), 6.33 (1H, qd, J 1.5, 0.8) and 7.25–7.75 (15H, m); ^{13}C NMR (75 MHz) δ_C 148.8 (d, J 27, =C–S), 139.0 (d, J 9.5, PPh C-1), 132.2 (d, J 11, P–CH) and 25.7 (d, J 4.5, CH$_3$); ^{31}P NMR (121 MHz) δ_P −22.7.

(E)-**13**: ^1H NMR (300 MHz) δ_H 2.23 (3H, d, J 0.9), 5.94 (1H, dq, J 2.0, 0.9) and 7.25–7.75 (15H, m); ^{13}C NMR (75 MHz) δ_C 149.8 (d, J 30, =C–S), 138.9 (d, J 9.8, PPh C-1), 122.0 (d, J 12, P–CH) and 20.9 (d, J 23, CH$_3$); ^{31}P NMR (121 MHz) δ_P −25.4.

For the isomer mixture (Found: C, 75.2; H, 5.7. $C_{21}H_{19}PS$ requires C, 75.4; H, 5.7%); ν_{max}/cm^{-1} 1584, 1478, 1435, 1184, 1109, 1026, 999, 743, 719 and 694; HRMS (EI) m/z calcd for $C_{21}H_{19}PS$ (M$^+$) 334.0945. Found 334.0960.

3.4. Formation of Transition Metal Complexes

3.4.1. (Z)-1-Diphenylphosphino-3,3-dimethyl-2-phenylthiobut-1-ene Platinum Dichloride Complex 31

A solution of [PtCl$_2$(cod)] (66 mg, 0.18 mmol) in CH$_2$Cl$_2$ (5 mL) was stirred while a solution of (Z)-1-diphenylphosphino-3,3-dimethyl-2-phenylthiobut-1-ene **12** (66 mg, 0.18 mmol) in CH$_2$Cl$_2$ (5 mL) was added dropwise over 30 min. After 1 h, the mixture was reduced to 1 mL by evaporation, and the addition of diethyl ether (15 mL) led to precipitation of the product as an off-white solid (89 mg, 79%), which was isolated by filtration. (Found: C, 44.8; H, 3.6; S, 4.8. $C_{24}H_{25}Cl_2PPtS$ requires C, 44.9; H, 3.9; S, 5.0%); ν_{max}/cm^{-1} 1665, 1437, 295; ^1H NMR (300 MHz, CD$_2$Cl$_2$) δ_H 8.0–7.5 (15H, m), 6.70 (1H, dd, $^3J_{Pt-H}$ 67, $^2J_{P-H}$ 10) and 1.20 (9H, s); ^{31}P NMR (121 MHz, CD$_2$Cl$_2$) δ_P +29.4 (d, $^1J_{P-Pt}$ 3524); MS (ESI$^-$) m/z 641 (M–H).

3.4.2. (Z)-1-Diphenylphosphino-3,3-dimethyl-2-phenylthiobut-1-ene Palladium Dichloride Complex 32

A solution of [PdCl$_2$(cod)] (38 mg, 0.13 mmol) in CH$_2$Cl$_2$ (5 mL) was stirred while a solution of (Z)-1-diphenylphosphino-3,3-dimethyl-2-phenylthiobut-1-ene **12** (50 mg, 0.13 mmol) in CH$_2$Cl$_2$ (5 mL) was added dropwise over 30 min. After 2 h, the mixture was reduced to 1 mL by evaporation, and the addition of diethyl ether (15 mL) led to precipitation of the product as a yellow solid (63 mg, 79%), which was isolated by filtration. (Found: C, 49.7; H, 4.4. $C_{24}H_{25}Cl_2PPdS\bullet 0.5$ CH$_2$Cl$_2$ requires C, 49.4; H, 4.4%); ν_{max}/cm^{-1} 1575, 1436, 289; ^1H NMR (300 MHz, CD$_2$Cl$_2$) δ_H 8.0–7.5 (15H, m), 6.70 (1H, d, $^2J_{P-H}$ 8) and 1.20 (9H, s); ^{31}P NMR (121 MHz, CD$_2$Cl$_2$) δ_P +50.4; MS (ESI$^+$) m/z 553 (M).

3.4.3. (Z)-1-Diphenylphosphino-3,3-dimethyl-2-phenylthiobut-1-ene η³-allyl Palladium Chloride Complex 33

A solution of [Pd(η³-allyl)Cl] (32 mg, 0.22 mmol) in CH$_2$Cl$_2$ (5 mL) was stirred while a solution of (Z)-1-diphenylphosphino-3,3-dimethyl-2-phenylthiobut-1-ene **12** (66 mg, 0.18 mmol) in CH$_2$Cl$_2$ (5 mL) was added dropwise over 30 min. After 2 h, the mixture was reduced to 0.5 mL by evaporation, and the addition of diethyl ether (10 mL) led to precipitation of the product as a yellow solid (54 mg, 55%), which was isolated by filtration. (Found: C, 56.7; H, 4.6. C$_{27}$H$_{30}$ClPPdS•0.4 CH$_2$Cl$_2$ requires C, 56.4; H, 5.3%); ν_{max}/cm^{-1} 1599, 1435, 296; ^{1}H NMR (300 MHz, CD$_2$Cl$_2$) δ_H 8.0–7.5 (15H, m), 6.70 (1H, d, $^2J_{P-H}$ 8) and 1.20 (9H, s); ^{31}P NMR (121 MHz, CD$_2$Cl$_2$) δ_P +40.9; MS (ESI$^+$) m/z 523 (M–Cl).

3.4.4. (Z)-1-diphenylphosphino-3,3-dimethyl-2-phenylthiobut-1-ene Gold Chloride Complex 34

A solution of [Au(tht)Cl] (18 mg, 0.06 mmol) and (Z)-1-diphenylphosphino-3,3-dimethyl-2-phenylthiobut-1-ene **12** (21 mg, 0.06 mmol) in CH$_2$Cl$_2$ (2 mL) was stirred for 18 h. The mixture was reduced to 0.5 mL by evaporation, and the addition of diethyl ether (10 mL) led to precipitation of the product as a white solid (21 mg, 62%), which was isolated by filtration. (Found: C, 47.2; H, 4.1. C$_{24}$H$_{25}$AuClPS requires C, 47.3; H, 4.1%); ν_{max}/cm^{-1} 1577, 1436, 253; ^{1}H NMR (300 MHz, CD$_2$Cl$_2$) δ_H 7.5–7.0 (15H, m), 6.70 (1H, d, $^2J_{P-H}$ 12) and 1.20 (9H, s); ^{31}P NMR (121 MHz, CDCl$_3$) δ_P +18.2; MS (ESI$^+$) m/z 631 (M+Na).

3.4.5. (Z)-1-Diphenylphosphino-3,3-dimethyl-2-phenylthiobut-1-ene Pentamethylcyclopentadienyl Iridium Dichloride Complex 35

A solution of [{IrCl(μ-Cl)(η^5-C$_5$Me$_5$)}$_2$] (75 mg, 0.1 mmol) in CH$_2$Cl$_2$ (5 mL) was stirred while a solution of (Z)-1-diphenylphosphino-3,3-dimethyl-2-phenylthiobut-1-ene **12** (71 mg, 0.19 mmol) in CH$_2$Cl$_2$ (5 mL) was added dropwise over 30 min. After 2 h, the mixture was reduced to 0.5 mL by evaporation, and the addition of diethyl ether (20 mL) led to precipitation of the product as a yellow solid (84 mg, 57%), which was isolated by filtration. (Found: C, 48.1; H, 4.65. C$_{34}$H$_{40}$Cl$_2$IrPS•1.25 CH$_2$Cl$_2$ requires C, 48.1; H, 4.9%); ν_{max}/cm^{-1} 1648, 1437, 290; ^{1}H NMR (300 MHz, CD$_2$Cl$_2$) δ_H 8.0–7.5 (15H, m), 6.70 (1H, d, $^2J_{P-H}$ 8), 1.20 (9H, s) and 1.00 (15H, s); ^{31}P NMR (121 MHz, CD$_2$Cl$_2$) δ_P −8.7; MS (ESI$^+$) m/z 739 (M).

3.4.6. (Z)-1-Diphenylphosphino-3,3-dimethyl-2-phenylthiobut-1-ene p-Cymene Ruthenium Dichloride Complex 36

A solution of [{RuCl(μ-Cl)(η^6-p-MeC$_6$H$_4^i$Pr)}$_2$] (26 mg, 0.04 mmol) in CH$_2$Cl$_2$ (5 mL) was stirred while a solution of (Z)-1-diphenylphosphino-3,3-dimethyl-2-phenylthiobut-1-ene **12** (32 mg, 0.08 mmol) in CH$_2$Cl$_2$ (5 mL) was added dropwise over 30 min. After 18 h, the mixture was reduced to 0.5 mL by evaporation, and the addition of diethyl ether (10 mL) led to precipitation of the product as an orange solid (39 mg, 67%), which was isolated by filtration. (Found: C, 56.9; H, 4.0. C$_{34}$H$_{39}$Cl$_2$PRuS•0.5 CH$_2$Cl$_2$ requires C, 57.1; H, 5.6%); ν_{max}/cm^{-1} 1637, 1436, 291; ^{1}H NMR (300 MHz) δ_H 8.0–7.5 (19H, m), 6.70 (1H, d, $^2J_{P-H}$ 8), 2.50 (1H, m), 1.80 (3H, s), 1.20 (9H, s) and 0.70 (6H, m); ^{31}P NMR (121 MHz) δ_P +14.6; MS (ESI$^+$) m/z 647 (M–Cl).

3.4.7. (Z)-1-Diphenylphosphino-2-phenylthiopropene Palladium Dichloride Complex 37

A solution of [PdCl$_2$(cod)] (33 mg, 0.1 mmol) in CH$_2$Cl$_2$ (5 mL) was stirred while a solution of (Z)-1-diphenylphosphino-2-phenylthiopropene **13** (64 mg, 0.2 mmol) in CH$_2$Cl$_2$ (5 mL) was added dropwise over 30 min. After 2 h, the mixture was reduced to 0.5 mL by evaporation, and the addition of diethyl ether (10 mL) led to precipitation of the product as a yellow solid (43 mg, 73%), which was isolated by filtration. (Found: C, 49.5; H, 3.2. C$_{21}$H$_{19}$ClPPdS requires C, 49.3; H, 3.7%); ν_{max}/cm^{-1} 1576, 1435, 296; ^{1}H NMR (300 MHz, CD$_2$Cl$_2$) δ_H 8.0–7.5 (15H, m), 6.30 (1H, d, $^2J_{P-H}$ 8) and 2.00 (3H, s); ^{31}P NMR (121 MHz, CD$_2$Cl$_2$) δ_P +52.4; MS (ESI$^+$) m/z 532 (M+Na).

3.5. X-ray Structure Determination of Complexes

Data were collected on a Bruker SMART diffractometer using graphite monochromated Mo Kα radiation λ = 0.71075 Å. The data were deposited at the Cambridge Crystallographic Data Centre and can be obtained free of charge via http://www.ccdc.cam.ac.uk/getstructures (accessed on 13 December 2023). The structure was solved by direct methods and refined by full-matrix least-squares against F^2 (SHELXL, Version 2018/3 [35]).

3.5.1. (Z)-1-Diphenylphosphino-3,3-dimethyl-2-phenylthiobut-1-ene Gold Chloride Complex **34**

Crystal data for $C_{24}H_{25}AuClPS$, M = 608.91, colourless prism, crystal dimensions 0.13 × 0.03 × 0.03 mm, monoclinic, space group $P2_1/c$ (No. 14), a = 18.290(6), b = 7.007(2), c = 17.660(6) Å, β = 96.966(6)°, V = 2246.7(13) Å^3, Z = 4, D_c = 1.800 g cm^{-3}, T = 125 K, $R1$ = 0.0622, $Rw2$ = 0.1417 for 2869 reflections with $I > 2\sigma(I)$ and 253 variables. CCDC 2298238.

3.5.2. (Z)-1-Diphenylphosphino-3,3-dimethyl-2-phenylthiobut-1-ene Pentamethylcyclopentadienyl Iridium Dichloride Complex **35**

Crystal data for $C_{34}H_{40}Cl_2IrPS$, M = 774.85, yellow prism, crystal dimensions 0.30 × 0.20 × 0.20 mm, triclinic, space group P-1 (No. 2), a = 10.2041(15), b = 10.2862(15), c = 16.724(3) Å, α = 80.903(2), β = 82.655(2), γ = 65.666(2)°, V = 1575.5(4) Å^3, Z = 2, D_c = 1.633 g cm^{-3}, T = 125 K, $R1$ = 0.0263, $Rw2$ = 0.0717 for 4283 reflections with $I > 2\sigma(I)$ and 354 variables. CCDC 2298239.

3.5.3. (Z)-1-Diphenylphosphino-2-phenylthiopropene Palladium Dichloride Complex **37**

Crystal data for $C_{21}H_{19}Cl_2PPdS$, M = 511.69, orange prism, crystal dimensions 0.30 × 0.15 × 0.10 mm, triclinic, space group P-1 (No. 2), a = 8.677(3), b = 11.063(4), c = 11.665(4) Å, α = 76.460(6), β = 87.468(6), γ = 71.174(5)°, V = 1029.8(6) Å^3, Z = 2, D_c = 1.650 g cm^{-3}, T = 125 K, $R1$ = 0.0519, $Rw2$ = 0.1335 for 2660 reflections with $I > 2\sigma(I)$ and 235 variables. CCDC 2298237.

4. Conclusions

While thiocarbonyl ylides with other groups on the ylidic carbon undergo thermal extrusion of Ph$_3$PS upon FVP at 650 °C, two examples, **10** and **11**, with hydrogen on the ylidic carbon instead undergo a novel isomerisation under the same conditions to afford useful (Z)-configured 1-diphenylphosphino-2-phenylsulfenylalkenes. The *t*-butyl compound **12** is obtained in good yield as the pure (Z)-isomer and behaves well as a ligand, forming a range of transition metal complexes with both bidentate binding via P and S and monodentate binding via only P. The methyl compound is obtained in lower yield mainly as the (Z)-isomer **13** but with a significant proportion of (E)-**14**, which increases upon distillation. A more limited study of its coordination chemistry resulted in the isolation of a bidentate bonded palladium complex. It is clear that while seven new complexes involving the two ligands have been isolated and characterised, including in four cases by X-ray diffraction, much more work needs to be carried out to fully exploit the potential of these simple yet versatile proligands, which are now readily available thanks to this unusual thermal rearrangement.

Supplementary Materials: The following are available online at https://www.mdpi.com/article/10.3390/molecules29010221/s1, Figures S1–S16: ^{1}H, ^{13}C and ^{31}P NMR spectra of compounds **10**, **11**, **12** and **13**.

Author Contributions: G.D., N.S.K. and H.K. carried out the organic synthesis work; J.W. prepared the metal complexes; H.L.M. and J.W. collected the X-ray data and solved the structures; A.M.Z.S. supervised the X-ray structure determination and optimised the structures; J.D.W. supervised the preparation of metal complexes; R.A.A. designed the study, supervised the organic synthesis work, analysed the data and wrote the paper. All authors have read and agreed to the published version of the manuscript.

Funding: This work received no external funding.

Institutional Review Board Statement: Not applicable.

Informed Consent Statement: Not applicable.

Data Availability Statement: The data presented in this study are available in supplementary material here.

Conflicts of Interest: The authors declare no conflicts of interest.

References

1. Aitken, R.A.; Atherton, J.I. A new general synthesis of aliphatic and terminal alkynes: Flash vacuum pyrolysis of β-oxoalkylidenetriphenylphosphoranes. *J. Chem. Soc. Chem. Commun.* **1985**, 1140–1141. [CrossRef]
2. Aitken, R.A.; Thomas, A.W. Pyrolysis involving compounds with C=C, C=O and C=N double bonds. In *Chemistry of the Functional Groups*; Patai, S., Ed.; Wiley: Chichester, UK, 1997; Suppl. A3, pp. 473–536. [CrossRef]
3. Eymery, F.; Iorga, B.; Savignac, P. The usefulness of phosphorus compounds in alkyne synthesis. *Synthesis* **2000**, 185–213. [CrossRef]
4. Aitken, R.A.; Boubalouta, Y.; Chang, D.; Cleghorn, L.P.; Gray, I.P.; Karodia, N.; Reid, E.J.; Slawin, A.M.Z. The Value of $^2J_{P-CO}$ as a Diagnostic Parameter for the Structure and Thermal Reactivity of Carbonyl-Stabilised Phosphonium Ylides. *Tetrahedron* **2017**, *73*, 6275–6285. [CrossRef]
5. Bestmann, H.J.; Schaper, W. Reaktionen von Thioacylalkylidentriphenylphosphoranen-eine neue Thiophensynthese. *Tetrahedron Lett.* **1979**, *20*, 243–244. [CrossRef]
6. Aitken, R.A.; Dawson, G.; Keddie, N.S.; Kraus, H.; Slawin, A.M.Z.; Wheatley, J.; Woollins, J.D. Thermal rearrangement of thiocarbonyl-stabilised triphenylphosphonium ylides leading to (Z)-1-diphenylphosphino-2-phenylsulfenylalkenes. *Chem. Commun.* **2009**, 7381–7383. [CrossRef] [PubMed]
7. Yoshida, H.; Matsuura, H.; Ogata, T.; Inokawa, S. α-Thiocarbonyl-stabilized phosphonium ylides: Preparation, structure, and alkylation reactions. *Bull. Chem. Soc. Jpn.* **1975**, *48*, 2907–2910. [CrossRef]
8. Bestmann, H.J.; Pohlschmidt, A.; Kumar, K. Eine Methode zur Überführung von Acylalkylidentriphenylphosphoranen in Thioacylalkylidentriphenylphosphorane. *Tetrahedron Lett.* **1992**, *33*, 5955–5958. [CrossRef]
9. Pasenok, S.; Appel, W. Process for the Preparation of Novel Stabilised Phosphorus Ylides. European Patent 741138 A2, 6 November 1996.
10. Capuano, L.; Drescher, S.; Huch, V. Neue Synthesen mit 1,3-ambident-nucleophilen Phosphor-Yliden, VII. Heterocyclische Triphenylphosphonium-chloride, Triphenylphosphonio-olate, acyclische Triphenylphosphonio-thiolate und ihre Wittig-Derivate. *Liebigs Ann. Chem.* **1993**, *1993*, 125–129. [CrossRef]
11. Aitken, R.A.; Atherton, J.I. Flash vacuum pyrolysis of stabilised phosphorus ylides. Part 1. Preparation of aliphatic and terminal alkynes. *J. Chem. Soc. Perkin Trans. 1* **1994**, 1281–1284. [CrossRef]
12. Aitken, R.A.; Al-Awadi, N.A.; Dawson, G.; El-Dusouqi, O.M.E.; Farrell, D.M.M.; Kaul, K.; Kumar, A. Synthesis, thermal reactivity and kinetics of substituted [(benzoyl)(phenylcarbamoyl)methylene]triphenylphosphoranes and their thiocarbamoyl analogue. *Tetrahedron* **2005**, *61*, 129–135. [CrossRef]
13. Aitken, R.A.; Al-Awadi, N.A.; El-Dusouqi, O.M.E.; Farrell, D.M.M.; Kumar, A. Synthesis, thermal reactivity and kinetics of stabilized phosphorus ylides, part 2: [(arylcarbamoyl)(cyano)methylene]triphenylphosphoranes and their thiocarbamoyl analogues. *Int. J. Chem. Kinet.* **2006**, *38*, 496–502. [CrossRef]
14. Aitken, R.A.; Al-Awadi, N.A.; Dawson, G.; El-Dusouqi, O.M.E.; Kaul, K.; Kumar, A. Kinetic and mechanistic study on the thermal reactivity of stabilized phosphorus ylides, part 3: [(acetyl)(arylcarbamoyl)methylene]triphenylphosphoranes and [(alkoxycarbonyl)(arylcarbamoyl)methylene]triphenylphosphoranes and their thiocarbamoyl analogues. *Int. J. Chem. Kinet.* **2007**, *39*, 6–16. [CrossRef]
15. Hickson, C.L.; McNab, H. E–Z Isomerization of alkenes by flash vacuum pyrolysis. *J. Chem. Res. (S)* **1989**, 176–177.
16. Seredkina, S.G.; Kolbina, V.E.; Rozinov, V.G.; Mirskova, A.N.; Donskikh, V.I.; Voronkov, M.G. Phosphorylation of bis(organothio)acetylenes with phosphorus pentachloride. *J. Gen. Chem. USSR* **1982**, *52*, 2375–2379, *Zh. Obshch. Khim.* **1982**, *52*, 2694–2698.
17. Sinyashin, O.G.; Zubanov, V.A.; Musin, R.Z.; Batyeva, E.S.; Pudovik, A.N. Reaction of thioesters of trivalent phosphorus acids with dichloroacetylene. *J. Gen. Chem. USSR* **1989**, *59*, 454–458, *Zh. Obshch. Khim.* **1989**, *59*, 512–516.
18. Seredkina, S.G.; Mirskova, A.N.; Bannikova, O.B.; Dolgushin, G.V. Phosphorylation of organylthiochloroacetylenes by phosphorus pentachloride. *J. Gen. Chem. USSR* **1991**, *61*, 983–988, *Zh. Obshch. Khim.* **1991**, *61*, 1084–1090.
19. Voskuil, W.; Arens, J.F. Chemistry of acetylenic ethers LXII. Tertiary phosphines with an acetylene-phosphorus bond. *Recl. Trav. Chim. Pays-Bas* **1962**, *81*, 993–1008. [CrossRef]

20. Kolomiets, A.F.; Fokin, A.V.; Rudnitskaya, L.S.; Krolevets, A.A. Alkenyldichlorophosphines. *Bull. Acad. Sci. USSR Div. Chem. Sci.* **1976**, *25*, 171–173, *Izv. Akad. Nauk. SSSR Ser. Khim.* **1976**, 181–183. [CrossRef]
21. Kolomiets, A.F.; Fokin, A.V.; Krolevets, A.A.; Bronnyi, O.V. Reactions of alkenes with phosphorus pentachloride and trichlorosilane. *Bull. Acad. Sci. USSR Div. Chem. Sci.* **1976**, *25*, 200–201, *Izv. Akad. Nauk. SSSR, Ser. Khim.* **1976**, 207–209. [CrossRef]
22. Braga, A.L.; Alves, E.F.; Silveira, C.C.; de Andrade, L.H. Stereoselective addition of sodium organyl chalcogenolates to alkynylphosphonates: Synthesis of diethyl 2-(organyl)-2-(organochalcogenyl)vinylphosphonates. *Tetrahedron Lett.* **2000**, *41*, 161–163. [CrossRef]
23. Braga, A.L.; Vargas, F.; Zeni, G.; Silveira, C.C.; de Andrade, L.H. Synthesis of β-organotelluro vinylphosphine oxides by hydrotelluration of 1-alkynylphosphine oxides and their palladium-catalyzed cross-coupling with alkynes. *Tetrahedron Lett.* **2002**, *43*, 4399–4402. [CrossRef]
24. Kawashima, T.; Iijima, T.; Kikuchi, H.; Okazaki, R. Synthesis of the first stable pentacoordinate 1,2-thiaphosphetene. *Phosphorus Sulfur Silicon Relat. Elem.* **1999**, *144–146*, 149–152. [CrossRef]
25. Magdesieva, N.N.; Kyandzhetsian, R.A.; Rakitin, O.A. Rearrangements of selenonium ylides with two electron withdrawing groups. *J. Org. Chem. USSR* **1975**, *11*, 2636–2641, *Zh. Org. Khim.* **1975**, *11*, 2562–2567.
26. Zbiral, E. Phosphororganische Verbindungen III. Zum Mechanismus der durch Arine an Alkylenphosphoranen ausgelösten Umlagerung. *Tetrahedron Lett.* **1964**, *5*, 3963–3967. [CrossRef]
27. Nakafuji, S.; Kobayashi, J.; Kawashima, T. Generation and coordinating properties of a carbene bearing a phosphorus ylide: An intensely electron-donating ligand. *Angew. Chem. Int. Ed.* **2008**, *47*, 1141–1144. [CrossRef] [PubMed]
28. Robert, P.; Le Bozec, H.; Dixneuf, P.H.; Hartstock, F.; Taylor, N.J.; Carty, A.J. Chemistry of η^2-CS_2 complexes. Mononuclear iron compounds containing alkoxythiocarbonyl and chelating $Ph_2PCH=C(R)S$ ligands vis coupling of coordinated CS_2 and phosphinoacetylenes: X-ray structure of $Fe(CO)[P(OMe)_3][Ph_2PCH=C(t-Bu)S][CS(OMe)]$. *Organometallics* **1982**, *1*, 1148–1154. [CrossRef]
29. Uson, R.; Laguna, A.; Laguna, M.; Briggs, D.A.; Murray, H.H.; Fackler, J.P., Jr. (Tetrahydrothiophene)gold(I) or gold(II) complexes. *Inorg. Synth.* **1989**, *26*, 85–91. [CrossRef]
30. Drew, D.; Doyle, J.R.; Shaver, A.G. Cyclic diolefin complexes of platinum and palladium. *Inorg. Synth.* **1991**, *28*, 346–349. [CrossRef]
31. McDermott, J.X.; White, J.F.; Whitesides, G.M. Thermal decomposition of bis(phosphone)platinum(II) metallocycles. *J. Am. Chem. Soc.* **1976**, *60*, 6521–6528. [CrossRef]
32. Bennett, M.A.; Huang, T.-N.; Matheson, T.W.; Smith, A.K.; Ittel, S.; Nickerson, W. (η^6-Hexamethylbenzene)ruthenium complexes. *Inorg. Synth.* **1982**, *21*, 74–78. [CrossRef]
33. White, C.; Yates, A.; Maitlis, P.M.; Heinekey, D.M. (η^5-Pentamethylcyclopentadienyl)rhodium and -iridium compounds. *Inorg. Synth.* **1992**, *29*, 228–234. [CrossRef]
34. Tatsuno, Y.; Yoshida, T.; Seiotsuka; Al-Salem, N.; Shaw, B.L. (η^3-Allyl)Palladium(II) complexes. *Inorg. Synth.* **1979**, *19*, 220–223. [CrossRef]
35. Sheldrick, G.M. A short history of SHELXL. *Acta Crystallogr. Sect. A* **2008**, *64*, 112–122. [CrossRef] [PubMed]

Disclaimer/Publisher's Note: The statements, opinions and data contained in all publications are solely those of the individual author(s) and contributor(s) and not of MDPI and/or the editor(s). MDPI and/or the editor(s) disclaim responsibility for any injury to people or property resulting from any ideas, methods, instructions or products referred to in the content.

molecules

Article

Ligand Hydrogenation during Hydroformylation Catalysis Detected by In Situ High-Pressure Infra-Red Spectroscopic Analysis of a Rhodium/Phospholene-Phosphite Catalyst [†]

José A. Fuentes [1], Mesfin E. Janka [2], Aidan P. McKay [1], David B. Cordes [1], Alexandra M. Z. Slawin [1], Tomas Lebl [1] and Matthew L. Clarke [1,*]

[1] EaStCHEM School of Chemistry, University of St Andrews, Purdie Building, North Haugh, St Andrews KY16 9ST, UK; jaf14@st-andrews.ac.uk (J.A.F.); apm31@st-andrews.ac.uk (A.P.M.); dbc21@st-andrews.ac.uk (D.B.C.); amzs@st-andrews.ac.uk (A.M.Z.S.); tl12@st-andrews.ac.uk (T.L.)
[2] Eastman Chemical Company, 200 South Wilcox Drive, Kingsport, TN 37660, USA
* Correspondence: mc28@st-andrews.ac.uk
[†] This paper is dedicated to Professor J. Derek Woollins on the elongated event of his retirement, a great colleague who was still amusing to be around and spoke his mind in a straightforward way- even when working in senior management.

Abstract: Phospholane-phosphites are known to show highly unusual selectivity towards branched aldehydes in the hydroformylation of terminal alkenes. This paper describes the synthesis of hitherto unknown unsaturated phospholene borane precursors and their conversion to the corresponding phospholene-phosphites. The relative stereochemistry of one of these ligands and its Pd complex was assigned with the aid of X-ray crystal structure determinations. These ligands were able to approach the level of selectivity observed for phospholane-phosphites in the rhodium-catalysed hydroformylation of propene. High-pressure infra-red (HPIR) spectroscopic monitoring of the catalyst formation revealed that whilst the catalysts showed good thermal stability with respect to fragmentation, the C=C bond in the phospholene moiety was slowly hydrogenated in the presence of rhodium and syngas. The ability of this spectroscopic tool to detect even subtle changes in structure, remotely from the carbonyl ligands, underlines the usefulness of HPIR spectroscopy in hydroformylation catalyst development.

Keywords: hydroformylation; phosphacycles; homogeneous catalysis; rhodium; in situ spectroscopy; regioselectivity

1. Introduction

Phosphacycles are widely applied ligands in several areas of homogenous catalysis [1–11], and hence new examples of phosphacycles are important in both main group chemistry and catalysis disciplines. Rhodium complexes of enantiomerically pure 2,5-disubstituted phospholanes, such as Me-DUPHOS (Figure 1), are produced at the kilogram scale and are applied industrially, whilst other related bis-phospholanes have many applications as catalysts in asymmetric synthesis [2,8,9]. The 2,5-diarylphospholano motif is present in the widely used ligand Ph-BPE (Ph-BPE = phenyl, bis-phospholano-ethane, Figure 1) [10,11] and additionally in phospholane-phosphite ligands, such as Bobphos (Figure 1) [11]. The latter confer very unusual branched regioselectivity up to 6:1 in the hydroformylation of a range of unbiased terminal 'alkyl-alkenes', a type of substrate that normally forms linear aldehydes [12–15]. Phospholane-phosphites of this type have also recently been found to be preferred ligands for certain enantioselective arylation reactions using arylboron reagents [16,17].

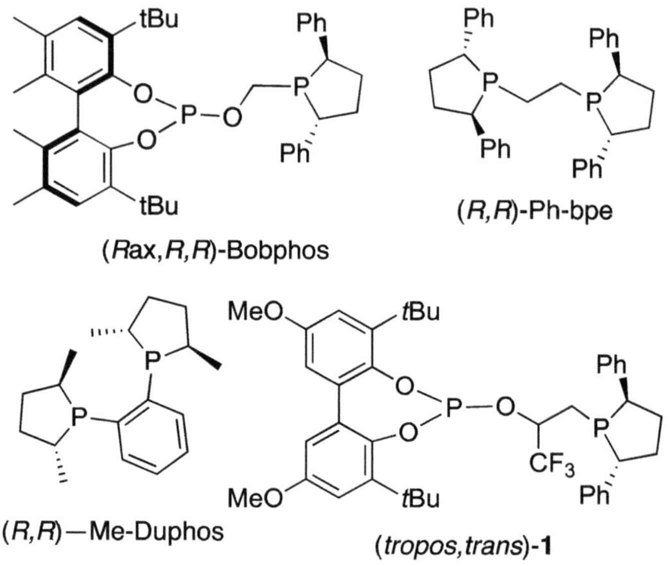

Figure 1. Structures of some phospholanes used in homogeneous catalysis.

These catalysts were originally developed for producing high-value, branched, enantiomerically pure aldehydes for pharmaceutical and fine chemical synthesis, but more recently we have been working on a programme exploring the possibility of developing a large-scale selective route to *iso*-butanal from propene. We reported that several milestones were reached, namely, (i) the use of unusual process conditions to improve the *iso*-butanal regioselectivity observed in the initial screening from around 60% up to around 80%, maintaining a selectivity of at least 65% even at high temperatures, at which industrially acceptable rates were observed [14]; (ii) the redesign of the ligand to obtain structures such as **1** (Figure 1) to confer stability at high temperatures for elongated periods of time [18]. These solutions to significant hurdles delivered stable and selective catalysts that operated with no decomposition in experiments lasting over several days, producing kilogram amounts of products [18]. One issue, however, is that for such a high-volume speciality chemical like isobutanal, with a market of many thousands of tonnes, the ligand structure is quite complex, needing eight synthetic steps; simpler ligand structures that can produce catalysts with similar performance are of significant interest.

Considering the synthetic route to Bobphos (Scheme 1) and the proposed origin of selectivity [19], it seemed plausible that an unsaturated *phospholene*-phosphite could be obtained with two fewer synthetic steps. It was unclear what the effect of the more rigid unsaturated ring and the different relative stereochemistry of the two phenyl groups would be, but we hoped that the molecule might be sufficiently similar to **1** to still favour branched aldehyde formation. Here, we describe the synthesis of this type of ligand and its precursors, before reporting on a study of the stability of their Rh catalysts. In situ high-pressure infra-red (HPIR) spectroscopy was discovered to be a sufficiently sensitive tool to detect the hydrogenation of the C=C bond in the phospholene under conditions relevant to hydroformylation catalysis.

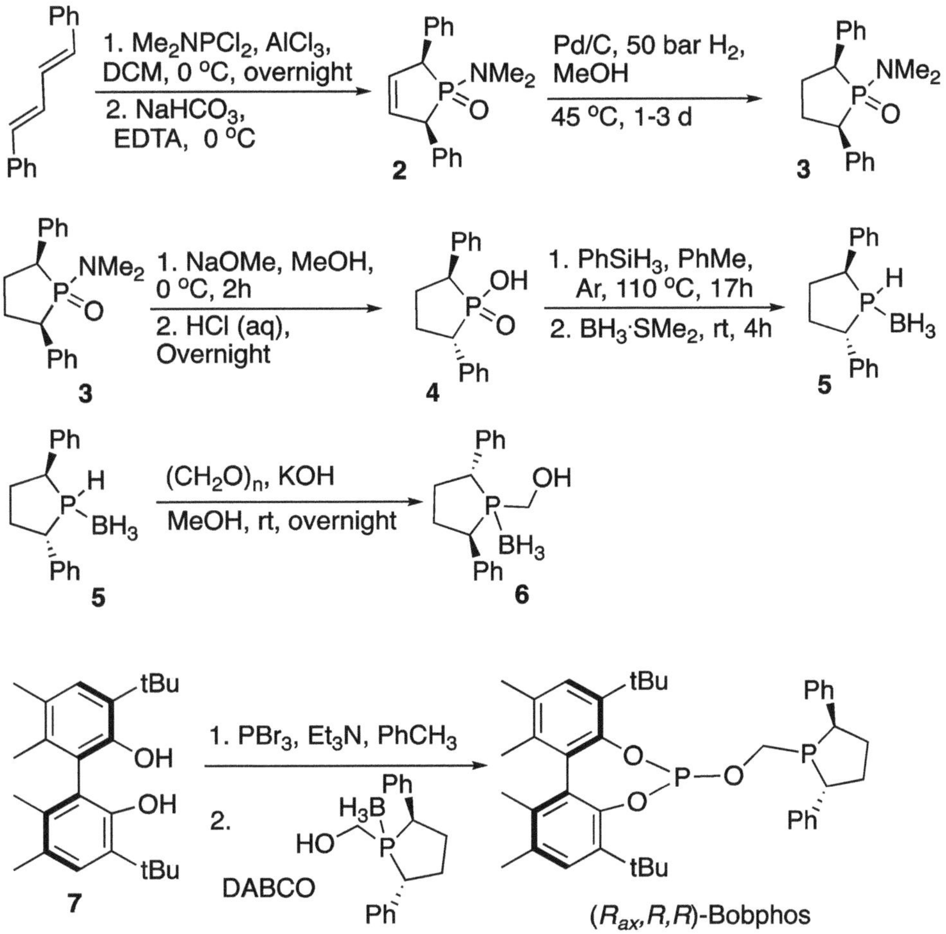

Scheme 1. Synthesis of (R_{ax},R,R)-Bobphos.

2. Results and Discussion

To the best of our knowledge, phospholene ligands are very scarcely applied in transition metal catalysis or as ligands in general [20,21]. Phosphorous chemists have prepared a variety of these molecules over the years [20–27], primarily using a McCormack cycloaddition between dienes and phosphenium cations. Their main utility has been as precursors to other phosphacycles [22–26]. A secondary phospholene borane precursor to produce 2,5-diarylsubstituted phospholenes was consequently unknown, but was readily prepared in this study by simply omitting the C=C hydrogenation and cis-trans isomerisation steps that were previously used to prepare secondary phospholane borane, **5** [10,22]. The synthesis is shown in Scheme 2 and started from amino-phosphine oxide **2** obtained in one step from diphenylbutadiene, as shown in Scheme 1. The cis–trans isomerization and hydrogenation steps for the synthesis of **1** were omitted, making phospholenic acid **8** available in just two steps in place of four steps needed to make phospholanic acid **4**.

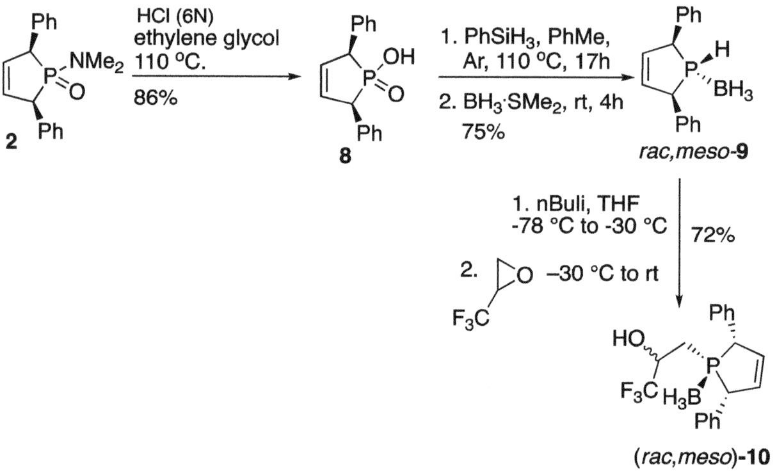

Scheme 2. Synthesis of secondary phospholene borane **9** and the precursor to phospholene-phosphite ligands. (See Scheme 1 for synthesis of **2**).

Phospholenic acid **8** was already described in the literature [24]. Reduction of acid **8** followed by protection with borane-dimethylsulfide afforded the corresponding secondary phospholene **9**. The preparation of the phospholene-containing fragment was achieved by deprotonation of secondary phospholene **9** with *n*-BuLi and attack of the resulting anion on the less substituted carbon of the commercially available 2-(trifluoromethyl)oxirane to obtain adduct **10** (Scheme 2). The NMR data suggested the presence of a single, racemic diastereomer and a single regioisomer in solution for compounds **9** and **10** [27]. The relative stereochemistry in phospholene borane **9** was investigated using 1D gs-NOESY. An NOE between hydrogens, three bonds apart in a five-membered ring, cannot be reliably used to assign the relative stereochemistry without a known standard to compare to. This is not only due to the likelihood of zero-quantum coherence effect (ZQC), but also due to a high chance of seeing an NOE signal even in the case of *anti* geometry. Whilst such an NOE would be expected to be weaker for an *anti* relationship between P-H and C-H, in the case of only one isomer, it would not be possible to assign the relative stereochemistry without knowing the NOE for the other isomer. Fortunately, in this case, the previously synthesised phospholane borane **5** has C-H bonds with both *syn* and *anti* relationships with the P-H bond; a 1D gs-NOESY spectrum of phospholane borane **5** (Supplementary Materials) showed the presence of both NOE effects and H *syn* to P-H and H *anti* to P-H relationships. There was a strong NOE signal for one of the CH α to P and a weak antiphase (ZQC) signal for the other CH, as expected (see Supplementary Materials). We found that 1D gs-NOESY of phospholene borane **9** only showed an antiphase (ZQC) artefact signal when irradiating *H*-P, due to scalar coupling between those spins ($J = 8.2$ Hz). Furthermore, a significant NOE of the *H*-P with one of the aromatic hydrogens (see Supplementary Materials) was observed, suggesting *anti* stereochemistry, as depicted in Scheme 2.

The final step in the synthesis was the coupling between **10** and a chloro-phosphite. In addition to the main target **13a**, a bulkier phosphite fragment, **13b**, and a less electron-donating phosphite fragment, **13c**, were targeted for synthesis. The requisite diol, **11b**, for the bulkier ligand was not commercially available and was prepared using a classical oxidative coupling in the presence of MnO_2 (Scheme 3).

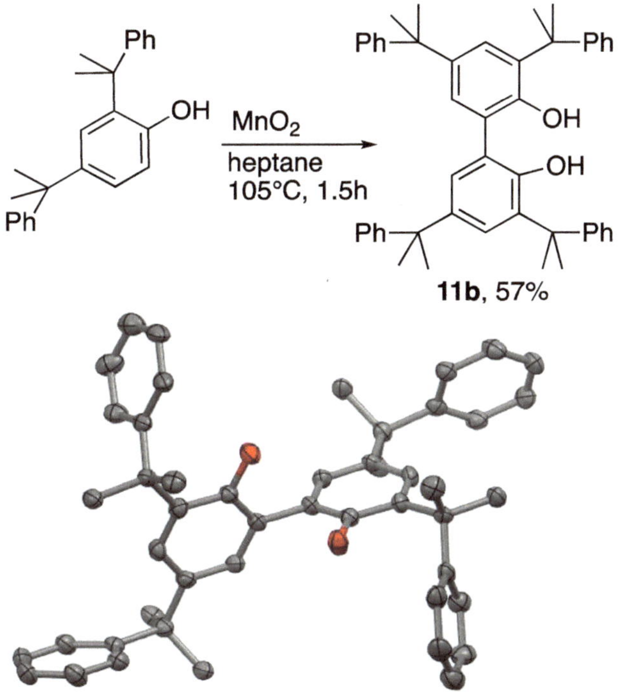

Scheme 3. Synthesis and X-ray crystal structure of **11b** (thermal ellipsoid plot, 50% probability ellipsoids). Carbon atoms shown grey, oxygen red). Hydrogen atoms omitted for clarity.

Compound **11b** was reported previously in the literature [28]. Our sample of diol **11b**, prepared as in Scheme 3, resembled a pure compound with the expected NMR spectrum, but the NMR data did not fully match the literature [28], as our NMR spectrum contained significantly fewer peaks. Esguerra et al. attributed the unexpectedly high number of observed signals in their data to the presence of two conformational isomers. For whatever reason, the sample of **11b** we prepared did *not* show extra signals, possibly assignable to the unexplained lack of an extra conformational isomer or to a major impurity in the compound reported in reference [28] To completely confirm that our NMR data indicated the proposed structure of compound **11b**, we determined the structure of diol **11b** using X-ray crystallography. The crystal structure is shown in Scheme 3, confirming the structure of the diol prepared in Scheme 3. One feature that merits discussion concerns the twists between the rings. We found that biphenol **11b** possessed a smaller twist of 79.1(4)° than the one observed for diol **11a**, of 89.84(8)° [29]. In comparison, unsubstituted [1,1′-biphenyl]-2,2′-diol displayed a much smaller torsion angle of 48.71(5)°, probably due to the lack of substituents and the presence of intramolecular hydrogen bonding [30].

Phosphite coupling to prepare the ligands **13a–13c** was achieved by synthesising the corresponding chlorophosphites from *tropos* diols **11a–11c**, and these were reacted directly with precursor **10** in the presence of 1,4-diazabicyclo-[2,2,2]-octane (DABCO, Scheme 4).

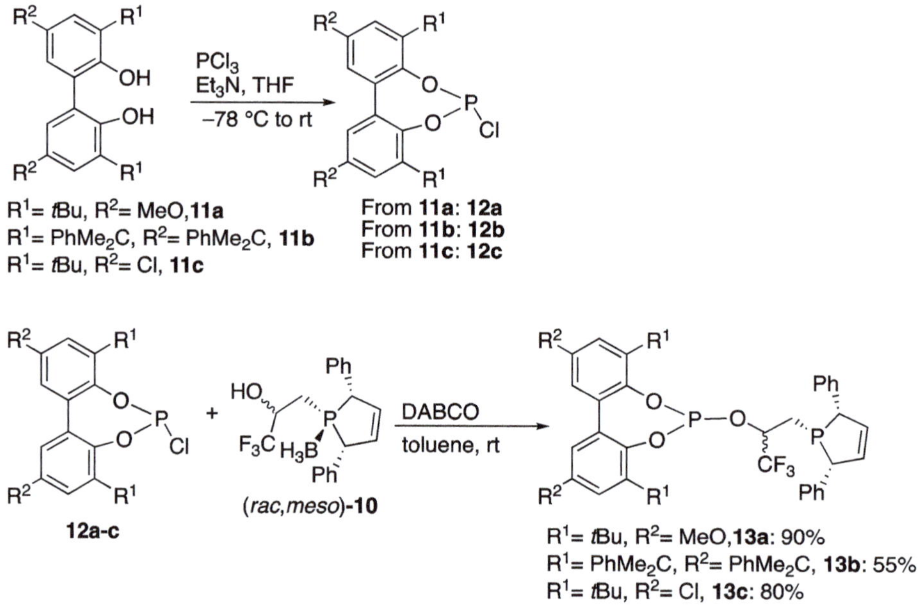

Scheme 4. Phosphite coupling and deprotection gives the final phospholene-phosphite ligands.

In order to confirm the relative configuration of ligand **13a**, an X-ray crystal structure was obtained (Figure 2). In the structure, it can be observed that the lone pair on phosphorous is *anti* relative to the *meso-cis* phenyl groups. This relative stereochemistry for alkyl or P-H relative to phenyl was similar to the stereochemical assignment made for **9**. It is quite possible that, rather than the deprotonation–alkylation step occurring with retention of configuration at phosphorus, the deprotonated form of **9** might be configurationally unstable, with the most thermodynamically stable and/or most reactive stereoisomer of the anion leading to the observed stereochemistry in precursor **10**. Retaining the same stereochemistry in **10** and **13a** is as expected. Ligand **13a** (in racemic form) was, as expected, a mixture of the (S_c, *meso-cis*) and (R_c, *meso-cis*) isomers.

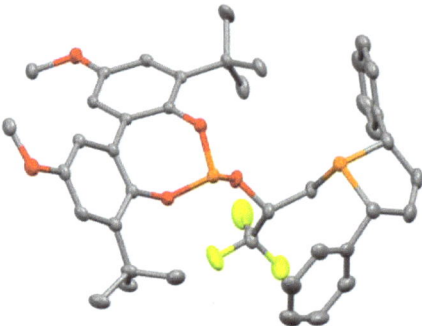

Figure 2. X-ray crystal structure of **13a** (thermal ellipsoid plot, 50% probability ellipsoids). Carbon atoms shown grey, oxygen red, fluorine green, phosphorus orange). Hydrogen atoms omitted for clarity.

It is worth mentioning that ligand **13a** contains a *tropos* diol, which, in solution, displays a rapid interconversion through the planar conformation. However, in the solid state, the diol settles in a preferred *atropos* conformation that depends on the chirality of the stereocentre with the CF$_3$ substituent. The two isomers (enantiomers) observed were, therefore, (S_c, pseudo-R_{ax}, meso-cis$_{ring}$) and (R_c, pseudo-S_{ax}, meso-cis$_{ring}$), both with the P lone pair *anti* to the phenyl rings.

Crystals of a rhodium complex derived from ligands **13a–c** were not available; so, one example of a Pd complex was prepared, which had the additional desirable feature of being comparable to the analogous Pd(L)Cl$_2$ complex of Bobphos that was structurally characterised [19]. Complex **14** was prepared by reacting ligand **13a** with [PdCl$_2$(PhCN)$_2$]. and single crystals suitable for analysis were grown from chloroform. The X-ray crystal structure of [PdCl$_2$(**13a**)]·3CHCl$_3$ (**14**) (Scheme 5, Table 1) revealed a slightly distorted square planar geometry about palladium. The bidentate ligand **13a** occupied two coordination sites, with a P-Pd-P crystallographic bite angle of 96.22(2)°, enlarged by about 6° over the preferred 90° for this type of complex. This crystallographic bite angle was over 10° larger than the one in [PdCl$_2${(S_{ax},S,S)-Bobphos}]·2CHCl$_3$ [19], which was largely due to the introduction of an extra carbon in the linker of **13a**. The Pd-Cl bonds were slightly shorter than in [PdCl$_2$(S_{ax},S,S)-Bobphos)], and contrary to the Pd/Bobphos complex, the Pd-Cl bond *trans* to the phosphite, 2.3385(5) Å, was slightly longer than the Pd-Cl bond *trans* to the phosphine, 2.3271(5) Å.

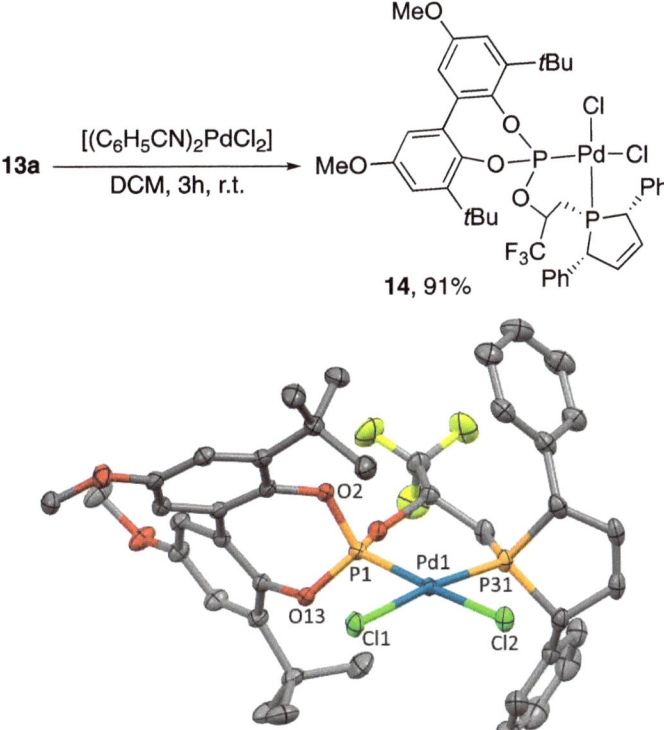

Scheme 5. Synthesis and X-ray crystal structure of **14** (thermal ellipsoid plot, 50% probability ellipsoids). Carbon atoms shown grey, oxygen red, fluorine light green, phosphorus orange, chlroine green, palladium turquoise). Hydrogen atoms and chloroform solvates omitted for clarity.

Table 1. Comparison of selected bond lengths (Å), angles (°) and torsions (°) for [PdCl$_2$(**13a**)]·3CHCl$_3$ (**14**) and [PdCl$_2$(S_{ax},S,S)-Bobphos·2CHCl$_3$ (**NEKXEJ**).

	13a	14	NEKXEJ
Pd-P(1)		2.2023 (5)	2.183 (2)
Pd-P(31)		2.2426 (2)	2.225 (2)
Pd-Cl(2)trans phosphite		2.3385 (5)	2.342 (3)
Pd-Cl(1)trans phosphine		2.3271 (5)	2.348 (2)
P(1)-Pd-P(31)		96.22 (2)	86.13 (9)
Cl(1)-Pd-Cl(2)		92.50 (1)	93.8 (1)
O(2)-P1-O(13)		106.47 (7)	106.5 (4)
C(2)-C(1)-C(12)-C(13)		51.0 (2)	65 (1)
C(6)-C(1)-C(12)-C(17)		45.0 (2)	62 (1)
Phenol-phenol twist	53.05 (4)	50.07 (6)	66.8 (4)

The twist observed between the phenol groups of 53.05(4)° in the free ligand **13a** was most similar to that in the unsubstituted biphenol, with the phospholene ring restricting the rotation between the phenol moieties, and remained quite far from perpendicular in the Pd complex at 50.07(6)° This was a narrower twist than that observed in the analogous complex derived from an atropos biphenol, [PdCl$_2$(S_{ax},S,S)-Bobphos·2CHCl$_3$], which was 66.8(4)°.

Hydroformylation catalysis continues to attract academic research interest and promote new discoveries in the industry [31–34]. Whilst the reaction is of interest across several sectors, the largest scale hydroformylation reaction practised is the Rh-catalysed hydroformylation of propene. Whilst linear selective reactions produce *n*-butanal, needed on a large scale, the formation of *iso*-butanal is of more recent interest, both as a conceptual challenge to reverse the innate preference of these reactions and since *iso*-butanal now has a very significant and increasing market [33–35]. This reaction has been the focus of a long-standing project in our laboratories, and branched selective hydroformylations (and hydroformylation with unusually low *n/iso* ratios) have been studied quite widely [36]. Until recently, class-leading results were generally to tilt the *n/iso* ratio; so, the *iso*-product was slightly in excess. Some reactions more similar to *iso*-selective reactions have been reported recently [14,18,35–37]. The performances of these new, more readily synthesised *phospholene-phosphite* ligands, **13a–13c**, were studied in the hydroformylation of propene and were compared to class-leading *phospholane-phosphite* **1** (Table 2). We previously reported the use of *n*-dodecane, a cheap non-volatile solvent, indicating it as a good solvent that can be used as an alternative to more expensive but preferred fluorinated solvents; dodecane was the solvent used in this study [18]. The catalysts originated from these ligands were preactivated by mixing [Rh(acac)(CO)$_2$] and the corresponding ligand in dodecane and syngas at 90 °C before the vessel was brought to reaction temperature and then filled with the propene/syngas mixture, as we described in previous papers. Complete conversion was 1450 TON for reactions filled at 90 °C; so, the turnover measurements were performed at <50% conversion after 1 h, and the values obtained can be considered to be the average TOF for the first stage of the reaction. The time taken for this activation step was separately studied by in situ HPIR spectroscopy (Supplementary Materials) to ensure that the bands associated with the catalyst had grown to full intensity within the chosen activation times used in the catalysis experiments (30–45 min). The activation times for the phospholene-phosphite ligands **11a–11c** were found to be similar to those for ligand **1**. We were pleased to find that Rh/**13a** generated a branched selective catalyst, affording *iso*-butanal with a selectivity of 67.5% at 75 °C (Table 2, entry 2). The *iso*-selectivity of the new complex was lower than that of the complex obtained with Rh/**1** at the same temperature, (74.6%, Table 2, entry 3), but not by a huge margin. The stereochemical change to a *meso-cis* arrangement of the two phenyl groups seemed only to have a marginal impact on regioselectivity; this was far from obvious, given the various subtleties we observed during our work with phospholanes [14].

Table 2. Hydroformylation of propene catalysed by Rhodium complexes of new phospholane-phosphite ligands.

$$[Rh(acac)(CO)_2] \ (0.25 \text{ mM}), \text{Ligand (L/Rh 2:1)}, H_2/CO, \text{ solvent}, 1h \longrightarrow i\text{-CHO} + n\text{-CHO}$$

Entry	Ligand [a]	T [°C]	TON (in 1 h)	Iso (%)
1 [b]	13a	50	204	71.0
2	13a	75	117	67.5
3	1	75	121	74.6
4	13a	90	333	65.7
5 [c]	1	90	397	70.9
6	13a	105	751	64.5
7	1	105	782	67.2
8	13b	75	78	67.0
9	13b	90	184	64.8
10	13b	105	502	62.3
11	13c	75	181	65.3
12	13c	90	462	63.3
13	13c	105	993	61.0

[a] Catalyst preformed from [Rh(acac)(CO)$_2$] (5.12 × 10^{-3} mmol) and ligand (10.24 × 10^{-3} mmol) by stirring at 20 bar of CO/H$_2$, at 90 °C for 40 min (**13a**), 45min (**13b**), 30min (**13c**), and 30 min (**1**) in the desired solvent (20 mL) and then increasing or decreasing the reaction T prior to running the reaction using a gas feed of propene/CO/H$_2$ at a 1:4.5:4.5 ratio. Rh concentration = 2.52 × 10^{-4} mol dm^{-3}. Product determined by GC using 1-methylnaphthalene as an internal standard. The TON values can be treated as average TOF values for the first part of the reaction; see the text for a discussion. [b] Reaction time, 16 h. Average TOF over 16 h = 13. [c] Ligand **1** (major isomer) from reference [17].

Improved turnover frequencies (TOFs) were obtained at 90 and 105 °C, and the loss of selectivity with temperature compared with that for Rh/**1** was also lower, allowing Rh/**13a** to have a similar *iso*-butanal selectivity to Rh/**1** at 105 °C (64.5% vs. 67.2%, Table 2, entries 6 and 7). The bulkier phosphite-phospholene **13b** did not improve the branched selectivity in the reaction, and there was a drop in activity (Table 2, entries 8–10). The introduction of a Cl atom in the backbone of the diol moiety in the phosphite (ligand **13c**), in order to make the phosphorus atom of the phosphite less basic, afforded the expected increase in the reaction rate (Table 2, entries 11–13). However, this did come at the cost of a drop in selectivity in the reaction.

Initially, an HPIR study was carried out to characterise the catalyst resting state and, as previously discussed, to investigate the time taken for the activation step (For discussion on the HPIR set-up, see ref. [38]). The presence of two main bands in the spectrum at 2019 and 1977 cm^{-1} and the asymmetric nature of the bands were consistent with an equatorial–axial (*ea*) coordination mode (Figure 3) [18,39]. We were very pleased to find that the stability, with respect to the decomposition of a range of unselective, unidentifiable species, was good: after 5 days at 90 °C, the IR bands for [RhH(CO)$_2$(**13a**)] were characteristic of an eq–ax isomer of a complex of type [RhH(CO)$_2$(L)] (Figure 3). However, a small shift was observed for both bands, from 2019 to 2017 and from 1977 to 1975 cm^{-1}, after 24 h (Figure 3).

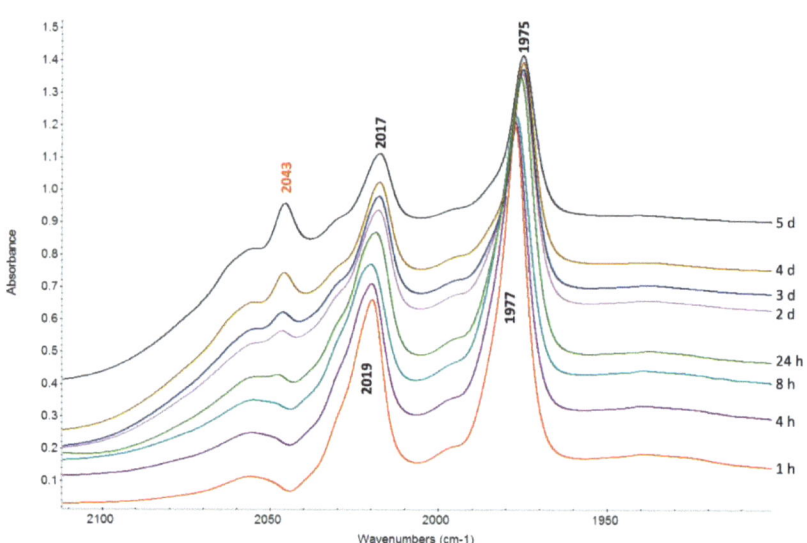

Figure 3. HPIR spectra of [RhH(CO)$_2$**13a**]. Conditions: Rh/L = 1:1.25 (C$_{Rh}$ 1 mM in dodecane), T = 90 °C, P = 20 bar, CO:H2 1:1. Colours represent spectra obtained at times given on right of figure.

Given the lack of data on phospholene ligands in homogeneous catalysis, one aspect of interest to us at the outset of this study was if the alkene function in the phospholene was stable in hydroformylation conditions and whether in situ HPIR spectroscopy was a sensitive-enough tool to observe relatively remote changes in the catalyst relative to the carbonyl ligands on rhodium. It is possible the alkene could undergo hydroformylation, polymerisation, isomerisation or hydrogenation. Polymerisation was relatively unlikely from a reactivity perspective and would be revealed by changes in physical properties. Hydroformylation should lead to new carbonyl bands in the IR and formyl protons when investigated by NMR. Detection of the catalyst resting state by NMR spectroscopy was therefore performed.

We previously found all of the [RhH(CO)$_2$(**L**)] complexes from bidentate ligands related to Bobphos (and to diphosphites) to be stable for several hours, enabling NMR characterisation under 1 atm of syngas; so, the desired complex, [RhH(CO)$_2$(**13a**)] was generated in a pressure vessel and then sampled for NMR interrogation. For a broader discussion on the expected fluxionality for a Rh complex of an asymmetrically substituted bidentate ligand, see References [18,40]. However, the key parameter is the magnitude of $^2J_{P-H}$, since $^2J_{P-H}$ for H-*trans*-P is known to be much larger (100–250 Hz) than for a *cis* coupling (10–20 Hz) (Figure 4). In addition, phosphites have significantly larger coupling constants than phosphines (by a factor of nearly 2; hence, a complex with phosphite *trans* to H would have $^2J_{P-H}$ of around 200 Hz). Intermediate values (e.g., $^2J_{P-H cis}$ of around 50 Hz), can sometimes be observed when there is a rapidly interconverting mixture of complexes with phosphite *trans* to H and complexes with phosphine *trans* to H, but this was not significant here. Thus, it was possible to measure two $^2J_{P-H}$ coupling constants from the hydride region of the ^{1}H NMR spectrum, which were ~115 Hz and either 10 or 12 Hz ($^1J_{Rh-H}$ was also 10 or 12 Hz and was not distinguishable from the very similar $^2J_{P-H cis}$). Comparing the ^{31}P NMR and ^{31}P{^{1}H} NMR spectra clearly showed that the phosphite only possessed a small $^2J_{P-H}$ coupling constant (measured as 11 Hz, i.e., 10 or 12 Hz), whereas the phospholene region had a large coupling constant. These data are clearly supportive of the structure shown in Figure 4, with the phospholene in the axial position, displaying the H-*trans*-P coupling. The small size of the $^2J_{P-H cis}$ coupling constant in the phosphite region indicated that a rapidly interconverting mixture of isomers was either not observed or contained a very high ratio of isomers favouring the structure in Figure 4. After 2 h in the pressure vessel, the alkene protons in the phospholene ring were visible in the complex (see Supplementary Materials, HSQC).

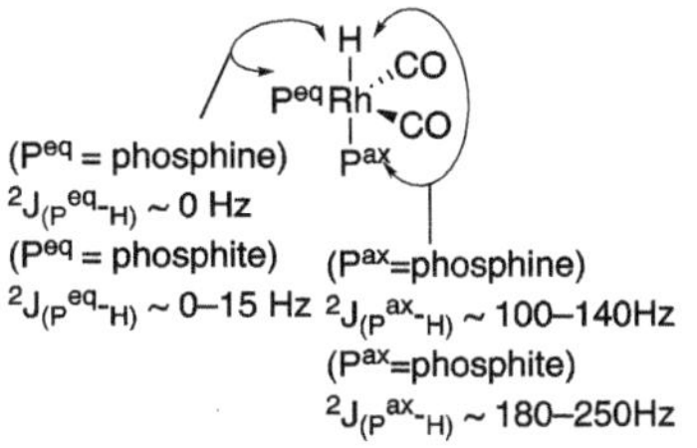

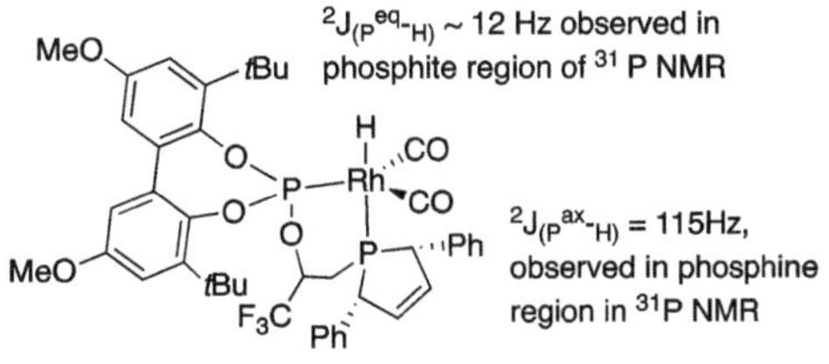

Figure 4. (**Left**) Diagnostic coupling ranges for phosphorus ligands bound in axial and equatorial sites in trigonal bipyramidal rhodium hydride complexes. (**Right**) Proposed structure and coupling constants observed for [RhH(CO)$_2$(**13a**)].

Carrying out the reaction of ligand, Rh source and syngas, while stirring for elongated reaction times (7 days) to ensure that whatever change that was implied by the shifting bands in the HPIR spectrum occurred, led to a different compound in comparison with that obtained in the 2 h reaction. There was no extensive fragmentation of the catalyst, but rather a different rhodium–hydride–dicarbonyl complex formed. This complex showed quite different peak shape and shift in the hydride region of the ^{1}H-NMR spectrum relative to both [RhH(CO)$_2$(**13a**)] and [RhH(CO)$_2$(**1**)] (see comparative spectra in Supplementary Materials). The HSQC spectra showed that in the region where the alkene protons of [RhH(CO)$_2$(**13a**)] were visible, there was now no resonance. There was no sign of any other alkene protons or multiple overlapping peaks for each signal, nor any change in physical properties, and there was no signal in the aldehyde region of the ^{1}H-NMR spectrum. This seemed to rule out polymerisation, hydroformylation, or isomerisation, whilst the fact that the species was not [RhH(CO)$_2$(**1**)] ruled out hydrogenation and epimerisation at one of the C-HPh centres. The most likely structure by far was then a simple hydrogenation of the C=C bond, with no other significant changes. One aspect of the NMR spectrum that was not fully explained was that there were twice as many peaks in the spectrum of the catalyst derived from the hydrogenated ligand. More specifically, these were consistent with two very similar isomers formed in a 50/50 ratio, both of which had similar spectral features as those of the single isomer of [RhH(CO)$_2$(**13a**)]. Both sets of peaks contained a large $^2J_{P-H}$ coupling constant for the phospholene ligand, indicating an e-a isomer with phospholene in the apical site. There are two likely explanations for this. One is that [RhH(CO)$_2$(**13a**)] and [RhH(CO)$_2$(**1**)] had a phosphite unit derived from a *tropos* biphenol that froze to one atropisomer in the complex. This was observed in various forms using *tropos* compounds that either were coordinated to metals or had further chiral centres within their structure. It is possible that the 7 days of heating and ligand hydrogenation led to the hydrogenated ligand product interconverting to both atropisomers, which were both detectable by NMR. Alternatively, it is also possible that the relative stereochemistry of the P-alkyl bond, which was always *syn* to the Ph group for **10**, **13a** and **14**, interconverted by pyramidalization at phosphorous [27,41–44]. This is a known reaction for a phosphacycle but is normally accomplished at higher temperatures; here, however, the reaction time was very long. The former explanation seems, by far, most likely; in any case, the data are most consistent with a ligand hydrogenation reaction to produce a *meso-cis* phospholane-based ligand. The small shift of the IR bands to lower wavenumbers is consistent with a slight increase in the donor strength of the phosphorous ligand, which is consistent with the basicity of saturated heterocycles versus that of unsaturated heterocycles.

3. Conclusions

The observations in this study should be meaningful to various groups of chemists, ranging from those interested in the synthesis and reactivity of phosphocycles to those who study and develop selective hydroformylation catalysts. A new nucleophilic precursor, secondary phospholene borane, was synthesised. This could prove a useful synthon for further studies on phospholene ligands. There is a preference, most likely thermodynamic, in borane precursors (secondary phospholene and its anion) towards one stereoisomer for the final tertiary phospholenes, as confirmed by both NMR and X-ray crystallographic studies.

Whilst outside of the scope of this project, the phospholene C=C bond might provide useful for the further remote functionalisation of a phosphacycle ligand for various purposes and would certainly be of interest in the future. The stereochemical preferences and synthetic routes show that the start of such a project should be relatively straightforward. Phospholene-phosphite ligands have many similarities to the corresponding phospholanes and can readily act as bidentate ligands for Pd and Rh. Activation of H_2 along with coordination of CO occurred readily to form complexes of the [RhH(CO)$_2$(**13**)] type, where the phospholene portion is in the apical site and *trans* to hydride. These complexes were tested in a reaction of industrial interest: the formation of *iso*-butanal with a low *n/iso* ratio (i.e., with some branched selectivity). They appeared very nearly as good as [RhH(CO)$_2$(**1**)] in terms of selectivity, reaction rate, or stability with respect to fragmentation/total decomposition. The synthesis of phospholene here described is two-step shorter than that of the analogous phospholanes, including the elimination of one step using a relatively expensive Pd/C catalyst. This is a technical improvement with likely reduced ligand cost, although we note that a catalyst that would clearly justify its incorporation into an established and efficient industrial-scale reaction would desirably be even simpler to access. It was pleasing that a relatively subtle and remote change within a ligand could be detected by in situ HPIR spectroscopic monitoring. The phospholene ligand was cleanly hydrogenated within a few days of operation at 90 °C and 20 bar syngas. This was longer than the operation time of some laboratory-scale batch catalysis reactions, but is relevant to processes that run for a long time. It is probably desirable that phospholene be hydrogenated rather than hydroformylated, since aldehydes are very reactive functional groups, allowing various other reactions to be triggered. In the low-pressure Rh-catalysed hydroformylation of alkyl-alkenes such as cyclopentenes, C=C hydrogenation is rare. Alkene hydrogenation under hydroformylation conditions is only regularly observed with reactive compounds such as unsaturated esters or alkynes. In this case, it seems likely that the C=C bond in the phospholane was able to insert into a Rh-hydride, but due to the coordination of another phosphorous moiety (probably the phosphite), this Rh alkyl was not mobile enough to undergo another migratory insertion reaction with Rh-CO. Instead, hydrogenolysis occurred to produce the saturated ring. In any case, the observation of this possible side reaction could be useful if further studies of phospholenes as ligands in homogeneous catalysis are carried out in the future. Whilst there has now been significant progress towards *iso*-selective hydroformylation reactions, in the case of the simplest but most important example, converting propene to *iso*-butanal, more streamlined ligands would be desirable, as would also be higher *iso*-selectivity.

4. Materials and Methods

4.1. Safety Note

Hydroformylations make use of hydrogen and carbon monoxide gases. Both are flammable, and CO is toxic. Reactions should only be carried out by trained personnel in pressure vessels designed for high-pressure reactions. The dispensing of CO should be carried out using a controllable cylinder head with a secondary method for stopping the flow of CO. Carbon monoxide detectors should be warned, and adequate signage and control of the laboratory to prevent access from non-trained personnel should be ensured.

4.2. General Information

All reactions were performed under an inert atmosphere of nitrogen or argon using standard Schlenk techniques, unless otherwise stated. All glassware used was flame-dried. Dry and degassed solvents were obtained from a solvent still or a solvent purification system (SPS). Commercially purchased anhydrous solvents were degassed before use by the freeze–pump–thaw method or by purging with inert gas. Triethylamine and $CDCl_3$ were dried and degassed before use. All chemicals, unless specified, were purchased commercially and used as received. CO/H_2 and propylene/CO/H_2 (10/45/45%) were obtained pre-mixed from BOC. NMR spectra were recorded on a Bruker Avance 300, 400 or 500 MHz instrument. Proton chemical shifts are referenced to internal residual solvent protons. Carbon chemical shifts are referenced to the carbon signal of the deuterated solvent. Signal multiplicities are provided as s (singlet), d (doublet), t (triplet), q (quartet), m (multiplet) or a combination of the above. Where appropriate, coupling constants (*J*) are quoted in Hz and are reported to the nearest 0.1 Hz. All spectra were recorded at r.t. (unless otherwise stated), and the solvent for a particular spectrum is indicated in parentheses. NMR spectra of compounds containing phosphorus were recorded under an inert atmosphere in dry and degassed solvent. Gas chromatography was performed on an Agilent Technologies 7820A machine. Mass spectrometry was performed on a Micromass GCT spectrometer, a Micromass LCT spectrometer and on Waters ZQ4000, Thermofisher LTQ Orbitrap XL or Finnigan MAT 900 XLT instruments. Flash column chromatography was performed using Merck Geduran Si 60 (40–63 µm) silica gel. Thin-layer chromatographic (TLC) analyses were carried out using POLYGRAM SIL G/UV254 or POLYGRAM ALOX N/UV254 plastic plates. TLC plates were visualised using a UV visualizer or stained using potassium permanganate dip followed by gentle heating. The synthesis and characterisation of compounds not described below, experimental protocols and spectra can be found in the Supplementary Materials. Crude unprocessed NMR data are available in a data archive [45]. High pressure infrared spectroscopy was performed in a Parr high pressure IR CSTR vessel constructed from Hastelloy C, fitted with CaF_2 windows and rated to 275 bar. The adjustable pathlength was set to 4 mm. The high-pressure IR spectra were recorded using an Avatar 360 FT-IR. Further discussion of the HPIR set up is available in the Supplementary Materials, and in reference [38]. The presence of 'unmodified catalysts is ruled out as discussed, in agreement with previous data [38,46].

4.3. X-ray Crystallography

X-ray diffraction data for compound **11b** were collected at 125 K using Rigaku MM-007HF high-brilliance RA generator/confocal optics with a XtaLAB P200 diffractometer [Cu Kα radiation (λ = 1.54187 Å)] using Crystal Clear [47]. Intensity data were collected using ω steps, accumulating area detector images spanning at least a hemisphere of reciprocal space. X-ray diffraction data for compounds **13a** and **14** were collected at 125 K using Rigaku FR-X ultrahigh-brilliance Microfocus RA generator/confocal optics with a XtaLAB P200 diffractometer [Mo Kα radiation (λ = 0.71073 Å)] using CrysAlisPro [48]. Data for all compounds analysed were processed (including correction for Lorentz, polarization and absorption) using CrysAlisPro. The structures were solved by dual-space methods (SHELXT [49]) and refined by full-matrix least squares against F^2 (SHELXL-2019/3) [50]. Non-hydrogen atoms were refined anisotropically, and hydrogen atoms were refined using a riding model, except for the hydrogen atom on O2 in the structure of **11b**, which was located from the difference Fourier map and refined isotropically subject to a distance restraint. The calculations were performed using either the CrystalStructure [51] or the Olex2 [52] interfaces. Selected crystallographic data are presented in Table 3. CCDC 2310520-2310522 contains the supplementary crystallographic data for this paper. These data can be obtained free of charge from the Cambridge Crystallographic Data Centre via www.ccdc.cam.ac.uk/structures.

Table 3. Selected crystallographic data.

	11b	13a	14
formula	$C_{48}H_{50}O_2$	$C_{41}H_{45}O_5F_3P_2$	$C_{44}H_{48}O_5F_3P_2Cl_{11}Pd$
fw	658.92	736.71	1272.11
crystal description	colourless plate	colourless block	yellow prism
crystal size [mm^3]	0.1 × 0.1 × 0.02	0.12 × 0.11 × 0.04	0.1 × 0.09 × 0.05
space group	$P3_221$	$P2_1/c$	$P2_1/n$
a [Å]	9.9188 (3)	18.5498 (3)	14.94065 (19)
b [Å]		10.34298 (16)	19.3961 (2)
c [Å]	32.7080 (13)	20.2110 (3)	18.4869 (2)
β [°]		103.8332 (16)	95.1095 (11)
vol [Å]^3	2786.78 (16)	3765.21 (11)	5336.02 (11)
Z	3	4	4
ρ (calc) [g/cm^3]	1.178	1.300	1.583
μ [mm^{-1}]	0.534	0.174	1.011
F(000)	1062.0	1552.0	2568.0
reflections collected	12302	74481	115446
independent reflections (R_{int})	3712 (0.1250)	9058 (0.0333)	13030 (0.0352)
parameters, restraints	235/1	468/27	753/166
GoF on F^2	1.225	1.028	1.031
R_1 [$I > 2\sigma(I)$]s	0.0586	0.0386	0.0291
wR_2 (all data)	0.2517	0.0997	0.0682
largest diff. peak/hole [e/Å^3]	0.44/−0.30	0.30/−0.33	0.55/−0.40

4.4. General Procedure for the Rhodium-Catalysed Hydroformylation of Propene

The hydroformylation reactions of propene were performed in a Parr 4590 Micro Reactor fitted with a gas entrainment stirrer; comprising holes which allowed for better gas dispersion throughout the reaction mixture. The vessel had a volume capacity of 0.1 L, an overhead stirrer with gas entrainment head (set to 1200 r.p.m.), temperature controls, a pressure gauge and the ability to be connected to a gas cylinder. The ligand (10.24 μmol (Rh/L 1:2)) was added to a Schlenk tube, which was then purged with nitrogen (or argon). The internal standard 1-methylnaphthalene (0.1 mL) was then added. The mixture was dissolved in a stock solution of [Rh(acac)(CO)$_2$] in toluene (2 mg/mL, 0.65 mL, 5.12 μmol of [Rh(acac)(CO)$_2$]), followed by the addition of the designated solvent (19.35 mL). The solution was transferred via a syringe to the pressure vessel (which had been purged with CO/H$_2$) through the injection port. CO/H$_2$ (1:1) (20 bar) was added, and the heating jacket was set to the desired temperature while stirring. Once the desired temperature was reached, the reaction was stirred for the required time to fully activate the catalyst. Then, pressure was slowly released, and repressurisation was achieved with propene/CO/H$_2$. The reaction was then run for the time specified in the tables. After this time, stirring was stopped, and the reaction was cooled by placing the vessel in a basin of cold water. The pressure was released, and the crude sample was analysed immediately by GC (in toluene). The GC method was run on a HP-5 Agilent column with a length of 30 m, a diameter of 0.250 mm and a film of 0.25 μm. The oven was initially held at 25 °C for 6 min, and then the temperature was increased to 60 °C at a rate of 10 °C per minute. The ramp was then increased to 20 °C per minute until the temperature reached 300 °C. The following products with the indicated retention times could be identified: *iso*-butyraldehyde (1.02 min); *n*-butyraldehyde (1.15 min); and 1-methylnaphthalene (13.50 min). The GC was calibrated for propene hydroformylation using (1-methylnaphthalene) as an internal standard. Both the linear (*n*-butyraldehyde) and the branched (*iso*-butyraldehyde) products were calibrated against the internal standard and against each other. Caution: The hydroformylation protocol should be carried out in an adequate vessel for the pressures encountered, and the use of a CO detector is recommended when handling syngas (a poisonous and highly flammable gas).

4.4.1. Synthesis of (*Meso*)-2,5-cis-diphenylphospholene borane Adduct 9: Borane-Protected (*meso*)-2,5-diphenyl-2,5-dihydro-1H-phosphole, 9

The compound (*meso*)-1-hydroxy-2,5-diphenyl-2,5-dihydrophosphole 1-oxide (**8**) (2.0 g, 7.4 mmol) was suspended in dry and degassed toluene (16 mL) under an inert atmosphere. Phenyl silane (1.83 mL, 14.8 mmol, 2 eq.) was added slowly to the reaction mixture using a syringe. The mixture was then heated to 110 °C and stirred for 17 h. After that time, the reaction mixture was cooled to around 5 °C using an ice bath, and the borane–dimethylsulfide complex (0.783 mL, 8.14 mmol, 1.1 eq.) was added over 1 min. The reaction mixture was then allowed to warm to room temperature and stirred for 4 h. The resulting solution was then filtered through a plug of silica and eluted with toluene (40 mL), and the solvent was removed under reduced pressure to leave a 'sticky' colourless solid. The solid was stirred in toluene/heptane 1:4 (10 mL) for 30 min, filtered, washed with toluene/heptane 1:4 (1 × 2 mL) and dried under vacuum to afford the desired product **9** as a white solid (0.654 g, 2.59 mmol, 35%). The organic fractions from the trituration and the washes were concentrated in vacuo. Purification by flash chromatography on silica gel (9:1 hexane/EtOAc) yielded more of the desired product **9** (0.757 g, 3.00 mmol, 40.6%) as a white solid. Combined isolated yield: 1.411 g, 5.59 mmol, 75.6%. ^{1}H NMR (CDCl$_3$, 500 MHz) δ 7.37–7.26 (10H, m, ArH), 6.25 (2H, d, $^3J_{P-H}$ = 17.9 Hz, CH=CH), 4.33–4.31 (2H, m, P-CH), 4.29 (1H, dm, $^1J_{P-H}$ = 367.9 Hz, P-H), 0.84 (3H, q, J = 87.5 Hz, BH$_3$). ^{31}P{^{1}H} NMR (CDCl$_3$, 202 MHz) δ 57.1 (br d, J = 42.8 Hz). ^{31}P NMR (CDCl$_3$, 202 MHz) δ 57.1 (br dm, $^1J_{P-H}$ = 367.9 Hz). ^{13}C NMR (CDCl$_3$, 126 MHz) δ 138.54 (d, J = 7.2 Hz 2 × ArC), 133.63 (d, J = 2.5 Hz, CH=CH), 129.38 (d, J = 2.0 Hz, 4 × ArCH), 127.74 (d, J = 2.4 Hz, 2 × ArCH), 127.35 (d, J = 4.2 Hz, 4 × ArCH), 48.66 (d, $^1J_{C-P}$ = 27.7 Hz, 2 × P-CH). HRMS (ES$^+$) C$_{16}$H$_{18}$BPNa [MNa]$^+$ m/z: 275.1128 found, 275.1131 required.

4.4.2. Borane-protected 3-((*meso*)-2,5-diphenyl-2,5-dihydro-1H-phosphol-1-yl)-1,1,1-trifluoropropan-2-ol, 10

To a stirred solution of (*meso*)-2,5-*cis*-diphenylphospholene borane adduct **9** (0.770 g, 3.06 mmol) in THF (13 mL) at −78 °C, under an atmosphere of nitrogen, a 1.58 M solution of *n*-BuLi in hexanes (1.94 mL, 3.06 mmol) was added dropwise via a syringe. The reaction was then allowed to slowly warm to −30 °C, and after stirring for 3 h, a solution of 2-(trifluoromethyl)oxirane (0.29 mL, 3.36 mmol) in THF (4 mL) was added dropwise via a syringe. Once the addition was complete, the reaction was allowed to warm to room temperature and stirred for 2 h. The reaction was quenched by the slow addition of saturated NaHCO$_3$ (aq) (5 mL) and water (5 mL) and diluted with diethyl ether (10 mL), and the organic layer was separated. The aqueous layer was extracted with diethyl ether (3 × 10 mL). The organic fractions were combined, dried (MgSO$_4$), filtered and concentrated in vacuo to yield a white solid. Purification by flash chromatography on silica gel (3:1 hexane/Et$_2$O) yielded the desired product **10** (0.802 g, 2.20 mmol, 72%) as a white solid. ^{1}H NMR (CDCl$_3$, 500 MHz) δ 7.44–7.24 (10H, m, ArH), 6.32–6.24 (2H, m, CH=CH), 4.63–4.60 (2H, m, P-CH), 3.03–2.96 (1H, m, CH-O), 2.74 (1H, br s, OH), 1.59–0.70 (5H, m, P-CH$_2$, BH$_3$). ^{31}P{^{1}H} NMR (CDCl$_3$, 202 MHz) δ 46.9 (br d, J = 54.7 Hz). ^{19}F NMR (CDCl$_3$, 470 MHz) δ −80.85 (s). ^{13}C NMR (CDCl$_3$, 126 MHz) δ 134.46 (d, J = 7.6 Hz ArC), 134.33 (d, J = 7.4 Hz ArC), 132.87 (d, J = 3.2 Hz, CH=CH), 132.34 (d, J = 3.3 Hz, CH=CH), 129.51 (d, J = 2.0 Hz, 2 × ArCH), 129.37 (d, J = 2.0 Hz, 2 × ArCH), 128.27 (d, J = 2.5 Hz, ArCH), 128.16 (d, J = 2.5 Hz, ArCH), 127.82 (d, J = 3.6 Hz, 2 × ArCH), 127.73 (d, J = 3.5 Hz, 2 × ArCH), 123.47 (qd, $^1J_{C-F}$ = 281.2 Hz, J = 16.3 Hz, CF$_3$), 65.74 (q, $^2J_{C-F}$ = 32.7 Hz, OCH), 50.10 (d, $^1J_{C-P}$ = 28.1 Hz, P-CH), 50.06 (d, $^1J_{C-P}$ = 27.1 Hz, P-CH), 20.42 (d, $^1J_{C-P}$ = 28.8 Hz, P-CH$_2$). HRMS (ES$^+$) C$_{19}$H$_{21}$OBF$_3$NaP [MNa]$^+$ m/z: 387.1256 found, 387.1267 required.

4.4.3. 4,8-di-tert-butyl-6-((3-((meso)-2,5-diphenyl-2,5-dihydro-1H-phosphol-1-yl)-1,1,1-trifluoropropan-2-yl)oxy)-2,10-dimethoxydibenzo[d,f][1,3,2]dioxaphosphepine, 13a

3,3'-di-*tert*-butyl-5,5'-dimethoxy-[1,1'-biphenyl]-2,2'-diol [53] (0.271 g, 0.755 mmol) was placed in a Schlenk tube and dissolved in 3 mL of THF. The resulting solution was cooled to −78 °C, and PCl$_3$ (0.086 mL, 0.982 mmol) was added slowly. NEt$_3$ (0.315 mL, 2.265 mmol) was also added to the reaction mixture, which was then stirred and allowed to reach room temperature over 1 h and then stirred for another hour. The suspension was filtered using a frit under an inert atmosphere, and the filtrate was evaporated using a Schlenk line and dried under vacuum to remove any residual PCl$_3$. The crude ^{31}P{^{1}H} NMR (202.4 MHz, C$_6$D$_6$) spectrum showed a single peak at δ 172.0 ppm, corresponding to the chlorophosphite. The product was used in the next step without further purification. To a Schlenk flask containing a solution of the chlorophosphite from the previous step in toluene (6 mL) a solution of (*rac,meso*)-phospholene 10 (0.250 g, 0.687 mmol) in toluene (6 mL) was added, followed by a solution of 1,4-diazabicyclo-[2,2,2]-octane (DABCO) (0.462 g, 4.12 mmol, 6 eq.) in toluene (5.5 mL). The reaction mixture was then stirred at room temperature overnight (20 h). The resulting suspension was filtered through silica gel (previously dried overnight in an oven) under an inert atmosphere, using dry toluene to compact and wash SiO$_2$ after filtration. Purification of (*tropos,meso*)-13a was achieved by recrystallisation. Heptane (2 mL) was added to a flask containing the reaction mixture; then, the flask was gently warmed with a heat gun, causing the solid to dissolve. The resulting solution was left standing at room temperature, which led to the formation of crystals (0.453 g, 0.615 mmol, 89.5%). ^{1}H NMR (C$_6$D$_6$, 500 MHz) δ 7.17–7.02 (12H, m, ArH), 6.71 (1H, d, J = 3.0 Hz, ArH), 6.61 (1H, d, J = 3.0 Hz, ArH), 5.85–5.82 (2H, m, CH=CH), 4.39–4.28 (2H, m, P-CH), 3.34 (3H, s, OCH$_3$), 3.29 (3H, s, OCH$_3$), 3.19–3.08 (1H, m, CH-O), 1.82–1.79 (1H, m, P-CH$_2$), 1.47 (9H, s, 3 × CH3), 1.44 (9H, s, 3 × CH3), 1.18–1.12 (1H, m, P-CH$_2$). ^{31}P{^{1}H} NMR (C$_6$D$_6$, 202 MHz) δ 143.1 (ap dd, J_{P-P} = 53.6 Hz, J_{P-F} = 3.2 Hz), 5.0 (ap dd, J_{P-P} = 53.6 Hz, J_{P-F} = 3.2 Hz). ^{19}F NMR (C$_6$D$_6$, 470 MHz) δ −77.73 (ap t, J_{P-F} = 4.2 Hz). ^{13}C NMR (C$_6$D$_6$, 126 MHz) δ 156.82 (ArC), 156.00 (ArC), 143.25 (ArC), 143.11 (d, J = 11.3 Hz ArC), 142.77 (ArC), 141.21 (ArC), 138.14 (d, J = 3.8 Hz ArC), 137.98 (d, J = 3.5 Hz ArC), 135.31 (d, J = 4.8 Hz ArC), 133.99 (ArC), 135.55 (d, J = 3.2 Hz, CH=CH), 134.04 (d, J = 3.0 Hz, CH=CH), 129.42 (2 × ArCH), 128.91 (2 × ArCH), 128.69 (d, J = 1.7 Hz, 2 × ArCH), 128.16 (d, J = 1.8 Hz, 2 × ArCH), 127.09 (ArCH), 126.97 (ArCH), 124.46 (qm, $^1J_{C-F}$ = 281.6 Hz, CF$_3$), 115.01 (ArCH), 114.82 (ArCH), 113.65 (ArCH), 112.87 (ArCH), 71.0–69.86 (m, OCH), 55.14 (OCH$_3$), 54.08 (OCH$_3$), 51.95 (dd, $^1J_{C-P}$ = 22.2, J = 3 Hz, P-CH), 51.68 (d, $^1J_{C-P}$ = 22.3 Hz, P-CH), 35.67 (C(CH$_3$)$_3$), 35.61 (C(CH$_3$)$_3$), 31.49 (C(CH$_3$)$_3$), 31.24 (d, J_{C-P} = 3.7 Hz, C(CH$_3$)$_3$), 21.64 (d, $^1J_{C-P}$ = 30.6 Hz, P-CH$_2$). HRMS (ES$^+$) C$_{41}$H$_{46}$O$_5$F$_3$P$_2$ [MH]$^+$ m/z: 737.2767 found, 737.2753 required. Recrystallisation from heptane afforded X-ray-quality crystals to determine the relative configuration of the ligand (in racemic form)

The synthesis and characterisation of other ligands and intermediates discussed in this paper can be found in the Supplementary Materials, along with all relevant NMR spectra.

Supplementary Materials: The following supporting information can be downloaded at: https://www.mdpi.com/article/10.3390/molecules29040845/s1.

Author Contributions: J.A.F. carried out all the experimental work and contributed creatively, including providing a significant contribution to the production of the manuscript. M.E.J. contributed to the management and experimental design. A.P.M. obtained the X-ray crystallographic data, made the main contribution to solving the structures of 13a and 14 and provided helpful contributions to the preparation of the manuscript. D.B.C. provided supervision and contributions to solving the structures of 13a and 14, as well as helpful contributions to the preparation of the manuscript. A.M.Z.S. obtained the data and solved the structure of 11b and provided helpful contributions to the preparation of the manuscript. T.L. assisted with the NMR spectroscopic determination of the stereochemistry of the phospholenes and provided helpful contributions to the preparation of the manuscript. M.L.C. devised and managed the project, made contributions to the experimental design and characterisation and led the writing of the manuscript. All authors have read and agreed to the published version of the manuscript.

Funding: This research was funded by Eastman Chemical Company.

Institutional Review Board Statement: Not applicable.

Informed Consent Statement: Not applicable.

Data Availability Statement: Crude unprocessed NMR data are available in a data archive [45].

Acknowledgments: We thank all the technical staff in the School of Chemistry for ongoing support. We thank Kevin J. Fontenot and Jody Rodgers and other colleagues at Eastman Chemical Company for fruitful discussions throughout this project.

Conflicts of Interest: Mesfin E. Janka was employed by Eastman Chemical Company. The other authors declare no conflicts of interest.

References

1. Clark, T.P.; Landis, C.R.; Freed, S.L.; Klosin, J.; Abboud, K.A. Highly active, regioselective, and enantioselective hydroformylation catalysts ligated by Bis-3,3-diazaphospholanes. *J. Am. Chem. Soc.* **2005**, *127*, 5040–5042. [CrossRef]
2. Zhang, W.; Chi, Y.; Zhang, X. Developing chiral ligands for asymmetric hydrogenation. *Acc. Chem. Res.* **2007**, *40*, 1278. [CrossRef]
3. Xu, G.; Senayake, C.H.; Tang, W. P-chiral phosphorous ligands based on a 2.3-dihydrobenzo[d][1,3]oxaphosphole motif for asymmetric catalysis. *Acc. Chem. Res.* **2019**, *52*, 1101. [CrossRef]
4. Ramazanova, K.; Chakrabortty, S.; Kallmeier, F.; Kretzschmar, N.; Tin, S.; Lönnecke, P.; De Vries, J.G.; Hey-Hawkins, E. The continued interest in chiral phosphacycles is evidenced by one of the papers in this special issue also being in this field. *Molecules* **2023**, *28*, 6210. [CrossRef]
5. Axtell, A.T.; Klosin, J.; Abboud, K.A. Evaluation of asymmetric hydrogenation ligands in asymmetric hydroformylation reactions. Highly enantioselective ligands based on bis-phosphacycles. *Organometallics* **2006**, *25*, 5003–5009. [CrossRef]
6. Carreira, M.; Charernsuk, M.; Eberhard, M.; Fey, N.; van Ginkel, R.; Hamilton, A.; Mul, W.P.; Orpen, A.G.; Phetmung, H.; Pringle, P.G. Anatomy of phobanes, Diasteroselective synthesis of three isomers of *n*-butylphobane and a comparison of their donor properties. *J. Am. Chem. Soc.* **2009**, *131*, 3078–3092. [CrossRef]
7. Coles, N.T.; Abels, A.S.; Leitl, J.; Wolf, R.; Grützmacher, H.; Müller, C. Phosphinine-based ligands: Recent development in coordination chemistry and applications. *Coord. Chem. Rev.* **2021**, *433*, 213729. [CrossRef]
8. Cobley, C.J.; Johnson, N.B.; Lennon, I.C.; McCague, R.; Ramsden, J.A.; Zanotti-Gerosa, A. Chap. 2 in *Asymmetric Catalysis on Industrial Scale: Challenges, Approaches and Solutions*; Blaser, H.U., Schmidt, E., Eds.; Wiley-VCH: Weinheim, Germany, 2004.
9. Ager, D.J.; de Vries, A.H.M.; de Vries, J.G. Asymmetric homogeneous hydrogenations at scale. *Chem. Soc. Rev.* **2012**, *41*, 3340–3380. [CrossRef]
10. Pilkington, C.J.; Zanotti-Gerosa, A. Expanding the family of phospholane-based ligands: 1,2-Bis(2,5-diphenylphospholano)ethane. *Org. Lett.* **2003**, *5*, 1273–1275. [CrossRef]
11. Axtell, A.T.; Cobley, C.J.; Klosin, J.; Whiteker, G.T.; Zanotti-Gerosa, A.; Abboud, K.A. Highly Regio- and Enantioselective Asymmetric Hydroformylation of Olefins Mediated by 2,5-Disubstituted Phospholane Ligand. *Angew. Chem. Int. Ed.* **2004**, *44*, 5834. [CrossRef]
12. Noonan, G.M.; Fuentes, J.A.; Cobley, C.J.; Clarke, M.L. An Asymmetric Hydroformylation Catalyst that Delivers Branched Aldehydes from Alkyl Alkenes. *Angew. Chem. Int. Ed.* **2012**, *51*, 2477–2480. [CrossRef]
13. Pittaway, R.; Fuentes, J.A.; Clarke, M.L. Diastereoselective and branched-aldehyde-selective tandem hydroformylation–hemiaminal formation: Synthesis of functionalized piperidines and amino alcohols. *Org. Lett.* **2017**, *19*, 2845–2848. [CrossRef]
14. Iu, L.; Fuentes, J.A.; Janka, M.E.; Fontenot, K.J.; Clarke, M.L. High *iso* aldehyde selectivity in the hydroformylation of short-chain alkenes. *Angew. Chem. Int. Ed.* **2019**, *58*, 2120–2124. [CrossRef]
15. Herle, B.; Späth, G.; Schreyer, L.; Fürstner, A. Total Synthesis of Mycinolide IV and Path-Scouting for Aldgamycin N. *Angew. Chem. Int. Ed.* **2021**, *60*, 7893–7899. [CrossRef]
16. Gilbert, S.H.; Fuentes, J.A.; Cordes, D.B.; Slawin, A.M.Z.; Clarke, M.L. Phospholane-phosphites for Rh catalyzed conjugate addition: Unsually reactive catalysts for challenging couplings. *Eur. J. Org. Chem.* **2020**, *20*, 3071–3076. [CrossRef]
17. Ortiz, K.G.; Dotson, J.J.; Robinson, D.J.; Singman, M.S.; Karimov, R.R. Catalyst-controlled Enantioselective and Regiodivergent Addition of Aryl Boron Nucleophiles to N-Alkyl Nicotinate Salts. *J. Am. Chem. Soc.* **2023**, *145*, 11781–11788. [CrossRef] [PubMed]
18. Fuentes, J.A.; Janka, M.E.; Rogers, J.; Fontenot, K.J.; Bühl, M.; Slawin, A.M.Z.; Clarke, M.L. Effect of ligand backbone on the selectivity and stability of rhodium hydroformylation catalysts derived from phospholane-phosphites. *Organometallics* **2021**, *40*, 3966–3978. [CrossRef]
19. Dingwall, P.; Fuentes, J.A.; Crawford, L.; Slawin, A.M.Z.; Bühl, M.; Clarke, M.L. Understanding a Hydroformylation catalyst that produces branched aldehydes from alkyl alkenes. *J. Am. Chem. Soc.* **2017**, *139*, 15921–15932. [CrossRef]

20. Bagi, P.; Kovács, T.; Szilvási, T.; Pongrácz, P.; Kollár, L.; Drahos, L.; Fogassy, E.; Keglevich, G. Platinum,(II) complexes incorporating racemic and optically active 1-alkyl-3-phospholene P-ligands: Synthesis, stereostructure, NMR properties and catalytic activity. *J. Organomet. Chem.* **2014**, *751*, 306–313. [CrossRef]
21. Leca, F.; Réau, R. 2-Pyridyl-2-phospholene: New P,N ligands for the palladium-catalyzed isoprene telomerisation. *J. Catal.* **2006**, *238*, 425–429. [CrossRef]
22. Lin, J.; Coles, N.T.; Dettling, L.; Steiner, L.; Felix-Witte, J.; Paulus, N.M.; Müller, C. Phospholenes from Phosphabenzenes by Selective Ring Contraction. *Chem. Eur. J.* **2022**, *28*, e202203406. [CrossRef]
23. Lim, K.M.-H.; Hayashi, T. Dynamic Kinetic Resolution in Rhodium-Catalyzed Asymmetric Arylation of Phospholene oxides. *J. Am. Chem. Soc.* **2017**, *139*, 8122–8125. [CrossRef]
24. Hintermann, L.; Schmitz, M. Enantioselective Synthesis of Phospholene via Asymmetric Organocatalytic Alkene Isomerization. *Adv. Synth Catal.* **2008**, *350*, 1469–1473. [CrossRef]
25. Hintermann, L.; Schmitz, M.; Maltsev, O.V.; Naumov, P. Organocatalytic stereoisomerization versus alkene isomerization: Catalytic asymmetric synthesis of 1-hydroxy-trans-2, 5-diphenylphospholane 1-oxide. *Synthesis* **2013**, *45*, 308–325. [CrossRef]
26. Guillen, F.; Rivard, M.; Toffano, M.; Legros, J.-Y.; Daran, J.-C.; Fiaud, J.-C. Synthesis and first applications of a new family of chiral monophosphine ligand: 2,5-diphenylphospholanes. Tetrahedron. *Tetrahedron* **2002**, *58*, 5895–5904. [CrossRef]
27. Quin, L.D.; Barket, T.P. Stereoisomerism in some derivatives of the 2-substituted 3-phospholene system. *J. Am. Chem. Soc.* **1970**, *92*, 4303–4308. [CrossRef]
28. Esguerra, K.V.N.; Fall, Y.; Petitjean, L.; Lumb, J.-P. Controlling the catalytic aerobic oxidation of phenols. *J. Am. Chem. Soc.* **2014**, *136*, 7662–7668. [CrossRef]
29. Byrne, J.J.; Chavant, P.Y.; Averbuch-Pouchot, M.-T.; Vallee, Y. 2,2′Biphenol. *Acta Crystallogr. Sect. C Cryst. Struct. Commun.* **1998**, *54*, 1154. [CrossRef]
30. Elsler, B.; Schollmeyer, D.; Waldvogel, S.R. Synthesis of iodobiaryls and dibenzofurans by direct coupling at BDD anodes. *Faraday Discuss.* **2014**, *172*, 413–420. [CrossRef]
31. Franke, R.; Selent, D.; Börner, A. Applied hydroformylation. *Chem. Rev.* **2012**, *112*, 5675–5732. [CrossRef]
32. Börner, A.; Franke, R. *Hydroformylation. Fundamentals, Processes, and Applications in Organic Synthesis*; Börner, A., Franke, R., Eds.; Wiley-VCH: Weinheim, Germany, 2016.
33. Puckette, T. Hydroformylation catalysis at Eastman chemicals. *Top. Catal.* **2012**, *55*, 421–425. [CrossRef]
34. Guo, L.; Sun, L.; Huo, Y.X. Towards bioproduction of oxo chemicals from C1 feedstocks using isobutyraldehyde as an example. *Biotechnol. Biofuels* **2022**, *15*, 80. [CrossRef]
35. Ibrahim, M.Y.S.; Bennett, J.A.; Mason, D.; Rodgers, J.; Abolhasani, M. Flexible homogeneous hydroformylation: On-demand tuning of aldehyde branching with a cyclic fluorophosphite ligand. *J. Catal.* **2022**, *409*, 105–117. [CrossRef]
36. Wang, X.; Nurttila, S.; Czik, W.I.; Becker, R.; Rodgers, J.; Reek, J.N.H. Tuning the porphyrin building block in self assembled cages for branched selective hydroformylation of propene. *Chem. Eur. J.* **2017**, *23*, 14769–14777. [CrossRef]
37. Sigrist, M.; Zhang, Y.; Antheame, C.; Dydio, P. Isoselective Hydroformylation by Iodide-Assisted Palladium Catalysis. *Ange. Chem. Int. Ed.* **2022**, *61*, e202116406. [CrossRef]
38. How, R.C.; Dingwall, P.; Hembre, R.T.; Ponasik, J.A.; Tolleson, G.S.; Clarke, M.L. Composition of catalysts resting states of hydroformylation catalysts derived from bulky mono-phosphorous ligands, rhodium dicarbonyl acetylacetonate and syngas. *Mol. Catal.* **2017**, *434*, 116–122. [CrossRef]
39. Chikkali, S.H.; van der Vlugt, J.I.; Reek, J.N.H. Hybrid diphosphorus ligands in rhodium catalysed asymmetric hydroformylation. *Coord. Chem. Rev.* **2014**, *262*, 1–15. [CrossRef]
40. Castillo-Molina, D.A.; Casey, C.P.; Müller, I.; Nozaki, K.; Jäkel, C. New low temperature NMR studies establish the presence of Second equatorial-apical isomer of [(R,S)-BINAPHOS](CO)$_2$RhH. C. *Organometallics* **2010**, *29*, 3362–3367. [CrossRef]
41. Hommer, H.; Gordillo, B. Synthesis and study of the thermal epimerization of r-2-Ethoxy-cis-4-cis-5-Dimethyl-1,3,2-3-Dioxaphospholane using 31 P NMR. *Phosphorous. Sulfur Silicon* **2002**, *177*, 465–470. [CrossRef]
42. Egan, W.; Tang, R.; Zon, G.; Mislow, K. Low barrier to pyramidal inversions in phospholes. Measure of aromaticity. *J. Am. Chem. Soc.* **1970**, *92*, 1442–1444. [CrossRef]
43. Cremer, S.E.; Chorvat, R.J.; Chang, C.H.; Davis, D.W. Pyramidal inversion in substituted phosphetanes. *Tetrahedron Lett.* **1968**, *55*, 5799–5802. [CrossRef]
44. Hoge, G. Stereoselective cyclization and pyramidal inversion strategies for P-chirogenic phospholane synthesis. *J. Am. Chem. Soc.* **2004**, *126*, 9920–9921. [CrossRef] [PubMed]
45. Fuentes, J.A.; Clarke, M.L. *Ligand Hydrogenation during Hydroformylation Catalysis Detected by In-Situ High Pressure Infra-Red Spectroscopic Analysis of a Rhodium/Phospholene-Phosphite Catalyst: Dataset*; University of St Andrews: St Andrews, UK, 2023.
46. Allian, A.D.; Garland, M. Spectral resolution of fluxional organometallics: The observation and FTIR characterization of all-terminal [Rh$_4$(CO)$_{12}$]. *Dalton. Trans.* **2005**, *2005*, 1957. [CrossRef]
47. *Crystal Clear-SM Expert*; Version 2.1; Rigaku Americas: The Woodlands, TX, USA; Rigaku Corporation: Tokyo, Japan, 2015.
48. *Crys Alis Pro*; Version 1.171.42.96a; Rigaku Oxford Diffraction; Rigaku Corporation: Tokyo, Japan, 2023.
49. Sheldrick, G.M. SHELXT–Integrated space-group and crystal structure determination. *Acta Crystallogr. Sect. A Found. Adv.* **2015**, *71*, 3–8. [CrossRef] [PubMed]

50. Sheldrick, G.M. Crystal structure refinement with SHELXL. *Acta Crystallogr. Sect. C Struct. Chem.* **2015**, *71*, 3–8. [CrossRef] [PubMed]
51. *Crystal Structure*; Version 4.3.0; Rigaku Americas: The Woodlands, TX, USA; Rigaku Corporation: Tokyo, Japan, 2018.
52. Dolomanov, O.V.; Bourhis, L.J.; Gildea, R.J.; Howard, J.A.K.; Puschmann, H. OLEX2: A complete structure solution, refinement and analysis program. *J. Appl. Crystallogr.* **2009**, *42*, 339–341. [CrossRef]
53. Li, C.; Xiong, K.; Yan, L.; Jiang, M.; Song, X.; Wang, T.; Chen, X.; Zhan, Z.; Ding, Y. Designing highly efficient Rh/CPOL-bp&PPh$_3$ heterogeneous catalysts for hydroformylation of internal and terminal olefins. *Catal. Sci. Technol.* **2016**, *6*, 2143–2149.

Disclaimer/Publisher's Note: The statements, opinions and data contained in all publications are solely those of the individual author(s) and contributor(s) and not of MDPI and/or the editor(s). MDPI and/or the editor(s) disclaim responsibility for any injury to people or property resulting from any ideas, methods, instructions or products referred to in the content.

Article

Synthesis and Structural Studies of *peri*-Substituted Acenaphthenes with Tertiary Phosphine and Stibine Groups †

Laurence J. Taylor [1], Emma E. Lawson [2], David B. Cordes [2], Kasun S. Athukorala Arachchige [3], Alexandra M. Z. Slawin [2], Brian A. Chalmers [2,*] and Petr Kilian [2,*]

[1] School of Chemistry, University of Nottingham, Nottingham NG7 2RD, UK
[2] EaStCHEM School of Chemistry, University of St Andrews, North Haugh, St Andrews, Fife KY16 9ST, UK
[3] Centre for Microscopy and Microanalysis, The University of Queensland, St Lucia, QLD 4072, Australia
* Correspondence: bac8@st-andrews.ac.uk (B.A.C.); pk7@st-andrews.ac.uk (P.K.)
† This paper is dedicated to Professor J. Derek Woollins on the occasion of his retirement, for his towering contribution and unwavering support of all things related to main group chemistry.

Abstract: Two mixed *peri*-substituted phosphine-chlorostibines, Acenap(P*i*Pr$_2$)(SbPhCl) and Acenap(P*i*Pr$_2$)(SbCl$_2$) (Acenap = acenaphthene-5,6-diyl) reacted cleanly with Grignard reagents or *n*BuLi to give the corresponding tertiary phosphine-stibines Acenap(P*i*Pr$_2$)(SbRR′) (R, R′ = Me, *i*Pr, *n*Bu, Ph). In addition, the Pt(II) complex of the tertiary phosphine-stibine Acenap(P*i*Pr$_2$)(SbPh$_2$) as well as the Mo(0) complex of Acenap(P*i*Pr$_2$)(SbMePh) were synthesised and characterised. Two of the phosphine-stibines and the two metal complexes were characterised by single-crystal X-ray diffraction. The *peri*-substituted species act as bidentate ligands through both P and Sb atoms, forming rather short Sb-metal bonds. The tertiary phosphine-stibines display through-space J(CP) couplings between the phosphorus atom and carbon atoms bonded directly to the Sb atom of up to 40 Hz. The sequestration of the P and Sb lone pairs results in much smaller corresponding J(CP) being observed in the metal complexes. QTAIM (Quantum Theory of Atoms in Molecules) and EDA-NOCV (Energy Decomposition Analysis employing Naturalised Orbitals for Chemical Valence) computational techniques were used to provide additional insight into a weak n(P)→σ*(Sb-C) intramolecular bonding interaction (pnictogen bond) in the phosphine-stibines.

Keywords: *peri*-substitution; phosphorus; antimony; NMR; single-crystal X-ray structures; synthesis; QTAIM; EDA-NOCV; pnictogen bond

Citation: Taylor, L.J.; Lawson, E.E.; Cordes, D.B.; Athukorala Arachchige, K.S.; Slawin, A.M.Z.; Chalmers, B.A.; Kilian, P. Synthesis and Structural Studies of *peri*-Substituted Acenaphthenes with Tertiary Phosphine and Stibine Groups. *Molecules* **2024**, *29*, 1841. https://doi.org/10.3390/molecules29081841

Academic Editor: Yves Canac

Received: 21 March 2024
Revised: 10 April 2024
Accepted: 15 April 2024
Published: 18 April 2024

Copyright: © 2024 by the authors. Licensee MDPI, Basel, Switzerland. This article is an open access article distributed under the terms and conditions of the Creative Commons Attribution (CC BY) license (https://creativecommons.org/licenses/by/4.0/).

1. Introduction

Tertiary amines and phosphines play a key role as tuneable ligands, with uses in transition metal catalysis and other applications. The heavier tertiary pnictines (ER$_3$, E = As, Sb, Bi, R = alkyl, aryl) also serve as L-type ligands in a number of complexes, although they generally display lower donor strength than corresponding N and P-based ligands [1,2].

Several unusual properties stemming from the close-proximity of two pnictine groups in *peri*-substituted scaffolds have been noted. The first of these was in the 1960's, when the remarkably high basicity of proton sponge 1,8-bis(dimethylamino)naphthalene (**A**, Figure 1) was reported by Alder [3]. Tertiary bis(phosphines) such as the phosphorus analogues of the proton sponge **B** (Figure 1) were reported shortly thereafter [4,5], as were several of their metal complexes [6–8].

Syntheses of *peri*-substituted tertiary bis(arsines), such as **C** (Figure 1), and their complexes were also reported as early as the 1960's [9]. However, no crystal structures of such ligands or complexes have appeared in the literature. Prototypical naphthalene bis(stibines) with dimethylstibino (**D1**) and diphenylstibino groups (**D2**) were synthesised by Reid, together with their Mo(0) and Pt(II) complexes [10]. The aryl species **D2** (both naphthalene and acenaphthene variants) and **D3** (naphthalene variant) were recently structurally characterised by Schulz, together with the two bis(bismuthines) **E** [11,12]. However, no structural data for any of the bis(stibine) or bis(bismuthine) metal complexes have been published to date.

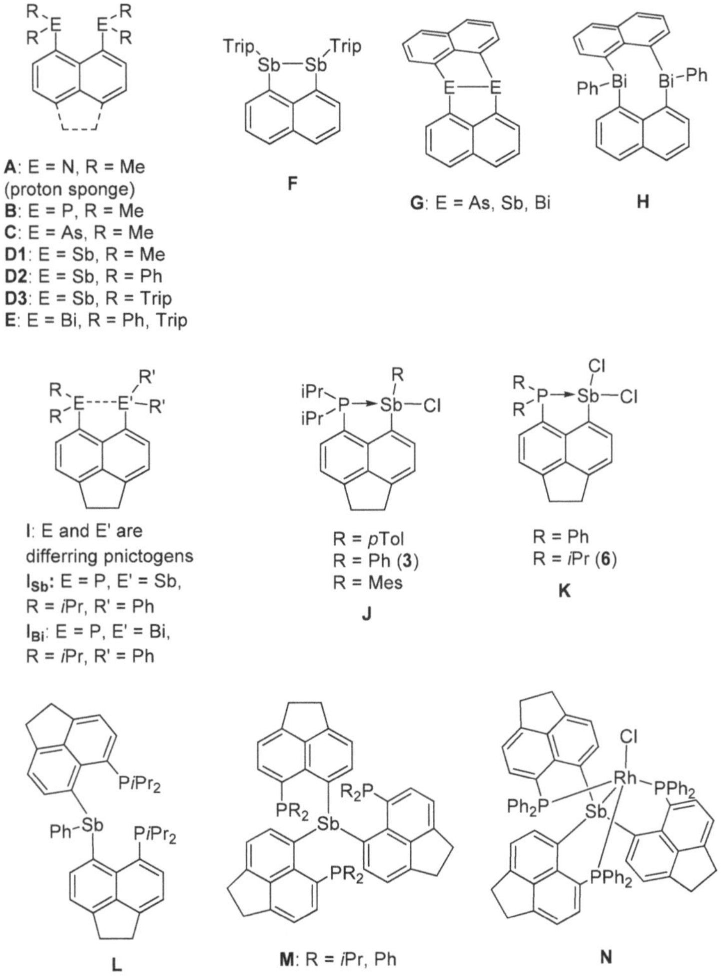

Figure 1. Literature Group 15 peri-substituted species mentioned in the introduction.

The related Sb–Sb bonded species **F** [12], as well as the doubly backboned species **G** [13–15] and **H** [11], have received significant attention recently, and a few transition metal complexes with these as ligands have also been reported [13].

Species **I** (Figure 1) with two differing Group 15 *peri*-atoms display intriguing dative bonding and NMR properties [16,17]. Surprisingly, only two bis(tertiary) phosphine-stibine and phosphine-bismuthine *peri*-substituted species have been reported to date: **I$_{Sb}$** [18] and **I$_{Bi}$** [19]. Both of these display repulsive interactions between the two pnictogen-centred groups, although their geometries indicate a weak pnictogen bond (nP→σ*(Pn-C)) is present, as indicated in Figure 1 by a dashed line. This is in contrast to the related E(III)−E(III) halophosphines, such as **J** [18,20] and **K** [18,21], which display strong dative pnictogen-pnictogen bonds.

Apart from the phosphine-stibine **I**$_{Sb}$ [18], the most closely related work to this paper are the geminally substituted bis- and tris(acenaphthene) species **L** and **M** [22,23]. Only one metal complex of these has been structurally characterised, the Rh(I) species **N** [23].

As a continuation of our synthetic, structural and bonding studies of *peri*-substituted species, we investigated the utility of the halostibines **J** and **K** (Figure 1), reported by us earlier [18], as synthons towards primary stibine functionalities. Prompted by the paucity of the literature data, we have also probed the coordination chemistry of the produced tertiary phosphine-stibines.

2. Results and Discussion

*2.1. Synthesis and Spectroscopic Properties of the Tertiary Stibines **4**, **5**, **7**, and **8***

The three major precursors for the syntheses reported in this paper were bis(aryl) stibine **2**, chloro(aryl) stibine **3** and dichlorostibine **6** (Scheme 1). Syntheses and structural information for these compounds, starting from **1**, have been reported by us [18].

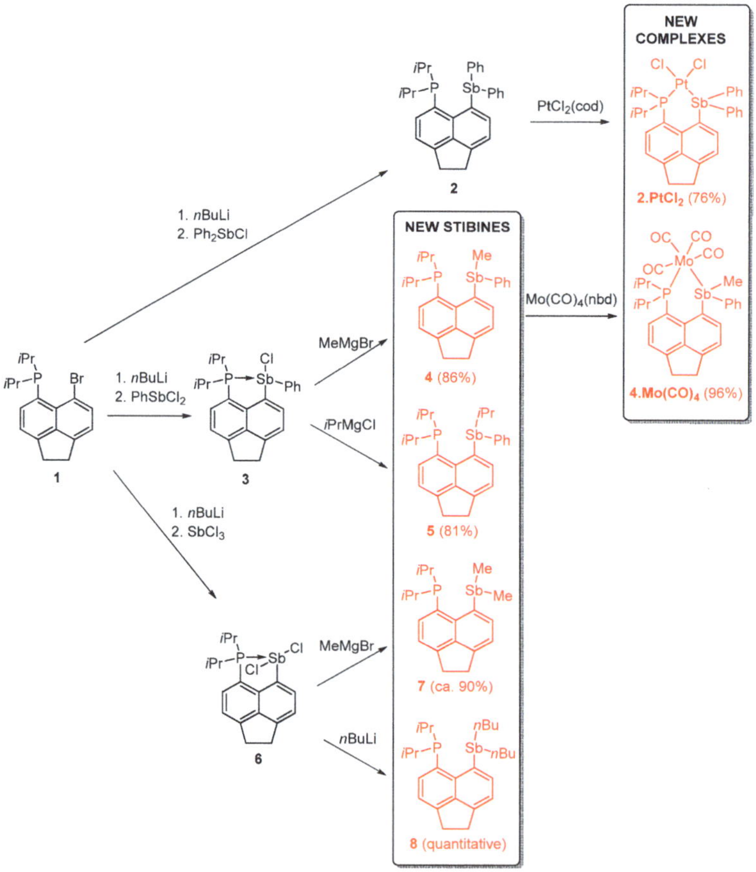

Scheme 1. Syntheses of the peri-substituted phosphine-stibines reported in this paper. Compounds highlighted in frames are newly synthesised here. Note: nbd = norbornadiene, cod = 1,5-cyclooctadiene.

The reactive Sb−Cl motifs in chlorostibine **3** and dichlorostibine **6** were used to form new Sb−C bonds via reactions with carbon nucleophiles. The chlorostibine **3** was reacted with one equivalent of alkyl-Grignard reagents, MeMgBr and *i*PrMgCl, to afford alkyl-aryl stibines **4** (86%) and **5** (81%), respectively.

The reaction of dichlorostibine **6** with two equivalents of MeMgBr afforded dimethylstibine **7** as an off-white solid (yield ca. 90%; exact yield determination was not possible as ^{1}H NMR indicated solvation by Et$_2$O). Reacting nBuLi with **6** also resulted in Sb−C bond formation; reaction with two equivalents of nBuLi afforded crude di-n-butylstibine **8** in quantitative yield (obtained as an oil). The crystallisation of **8** from common organic solvents was not successful. However, a small amount of crystalline **8** was obtained through the long standing of the oil at room temperature (see below).

All the Sb−C bond-forming reactions were remarkably clean, as judged by ^{31}P{^{1}H} and ^{1}H NMR spectroscopy. The newly prepared compounds were further characterised by ^{13}C DEPT-Q NMR, HRMS (peaks corresponding to (M + H)$^+$ with correct isotopic patterns were observed in all cases) and (for **7** and **8**) also by Raman spectroscopy. The purity of **5** was confirmed by CHN microanalysis. The novel tertiary stibines appear to be hydrolytically stable (in some cases, an aqueous wash was involved in the work-up); however they are oxidised in the presence of air. Both **7** and **8** decomposed in chloroform solutions within several days, indicating instability in halogenated solvents.

The reaction of **6** with one equivalent of nBuLi gave an oil after the workup. This oil was shown by ^{31}P{^{1}H} NMR to be a complex mixture, with a major peak at δ_P 18.5 ppm, corresponding to the doubly substituted species **8**, indicating that selective single substitution using an organolithium as a nucleophile may be difficult to achieve.

The ^{31}P{^{1}H} NMR spectra of the phosphine-stibines **4**, **5**, **7** and **8** display singlets within a narrow range of δ_P (−18.5 to −20.8 ppm). Notable through-space couplings (indicated by the TS superscript in the J notation, ^{TS}J) are observed in the ^{13}C{^{1}H} NMR spectra between the phosphorus atom and carbons attached to the antimony atom. In **4**, the *ipso*-C of the Sb-Ph moiety shows a $^{5TS}J_{CP}$ of 16.2 Hz. An even larger $^{5TS}J_{CP}$ of 34.1 Hz is observed for the Sb-CH$_3$ of **4**. Interestingly, the acenaphthene *ipso*-carbon atom shows no detectable coupling to the phosphorus atom, despite having a shorter bond path (formally $^3J_{CP}$).

A similar situation is observed in **5** ($^{5TS}J_{CP}$ = 17.3 Hz (*ipso*-Ph) and $^{5TS}J_{CP}$ = 36.8 Hz (Sb-CH), although in this case small magnitude splitting with the *ipso*-acenaphthene carbon (C1 in the numbering scheme shown in Figure 2) is observable ($^3J_{CP}$ = 2.4 Hz). Similar magnitudes of J_{CP} involving carbon atoms bonded directly to Sb atoms are also observed for **7** ($^{5TS}J_{CP}$ = 34.9 Hz (CH$_3$); $^3J_{CP}$ = 5.1 Hz (C1, Acenap)) and **8** ($^{5TS}J_{CP}$ = 30.6 Hz (CH$_2$); $^3J_{CP}$ = 5.0 Hz (C1, Acenap)). Observation of the through-space couplings in **4**, **5**, **7** and **8** is consistent with the significant overlap of P and Sb lone pairs as confirmed by single crystal X-ray diffraction (*vide infra*) and is in agreement with observations made in our previous study of P-Sb acenaphthenes [18].

2.2. Synthesis and Spectroscopic Properties of Tertiary Stibine Metal Complexes **2.PtCl$_2$** *and* **4.Mo(CO)$_4$**

Peri-substituted species **2** and **4**, bearing tertiary phosphine and tertiary stibine groups, were reacted with platinum(II) and molybdenum(0) motifs to explore their coordination chemistry. It was of interest to see if the phosphine-stibine species would act as bidentate ligands, with the metal coordinating through both phosphorus and antimony atoms.

[PtCl$_2$(cod)] was reacted with **2** in dichloromethane, giving **2.PtCl$_2$** as a yellow powder in a good yield (76%). Similarly, the reaction of [Mo(CO)$_4$(nbd)] with **4** in dichloromethane gave **4.Mo(CO)$_4$** as a brown powder in a near-quantitative yield (Scheme 1). Both complexes were stable to air in the solid and solution in the chlorinated solvents used to acquire their NMR spectra (CD$_2$Cl$_2$ and CDCl$_3$, respectively).

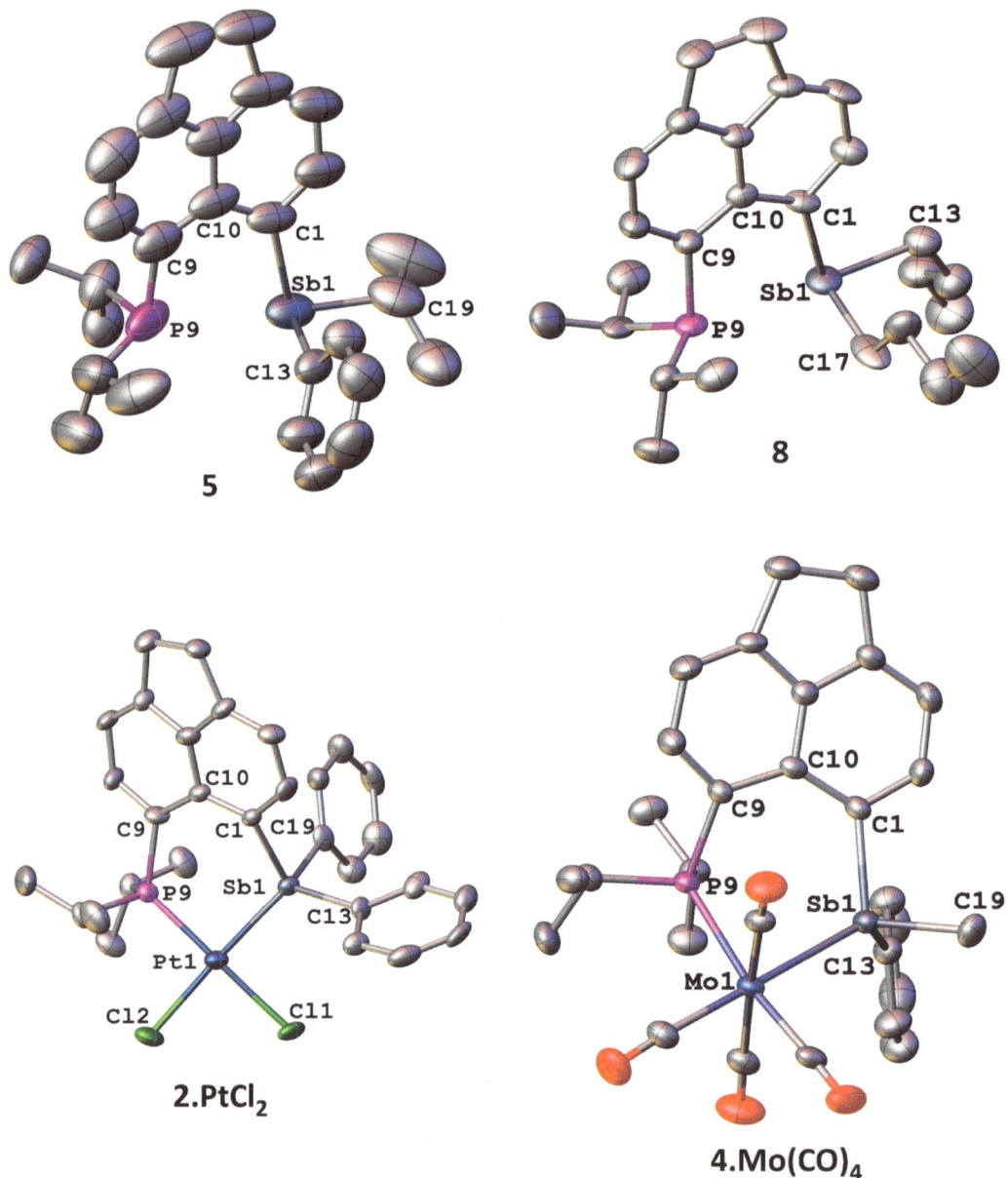

Figure 2. Molecular structures of **5**, **8**, **2.PtCl₂**, and **4.Mo(CO)₄**. Anisotropic displacement ellipsoids are plotted at the 50% probability level. Hydrogen atoms, solvent molecules and minor components of disorder are omitted for clarity.

The $^{31}P\{^{1}H\}$ NMR spectrum of **2.PtCl₂** consists of a singlet with a set of ^{195}Pt satellites (δ_P 7.8 ppm, $^{1}J_{PPt}$ = 3357 Hz), with the complementary doublet observed in the $^{195}Pt\{^{1}H\}$ NMR spectrum (δ_{Pt} −4541 ppm). The coordination of platinum centres resulted in a high-frequency shift (c.f. free ligand **2**, δ_P −21.9 ppm) [18] as well as loss of the through-space J_{CP} coupling (c.f. $^{5TS}J_{CP}$ 40.3 Hz for *ipso*-Ph carbon in **2**).

Coordination of the Mo(CO)$_4$ fragment to **4** resulted in an even more pronounced high-frequency shift for **4.Mo(CO)$_4$** (δ_P 43.4; c.f. δ_P −19.6 ppm in free ligand **4**). Similar to **2.PtCl$_2$**, the J_{CP} couplings between the phosphorus atom and the carbon atoms adjacent to the antimony atom are much smaller magnitudein **4.Mo(CO)$_4$** than **4**. This is notable as the through-bond coupling paths are shorter in the complex (formally $^3J_{CP}$, 2.3 and 2.9 Hz), compared to those in the free ligand **4** ($^{5TS}J_{CP}$, 16.2 and 34.1 Hz).

2.3. Structural Discussion

Two of the phosphine-stibines (**5** and **8**), as well as the two complexes **2.PtCl$_2$** and **4.Mo(CO)$_4$**, were subjected to the single crystal diffraction study. The structures are shown in Figures 2 and 3 and Table 1.

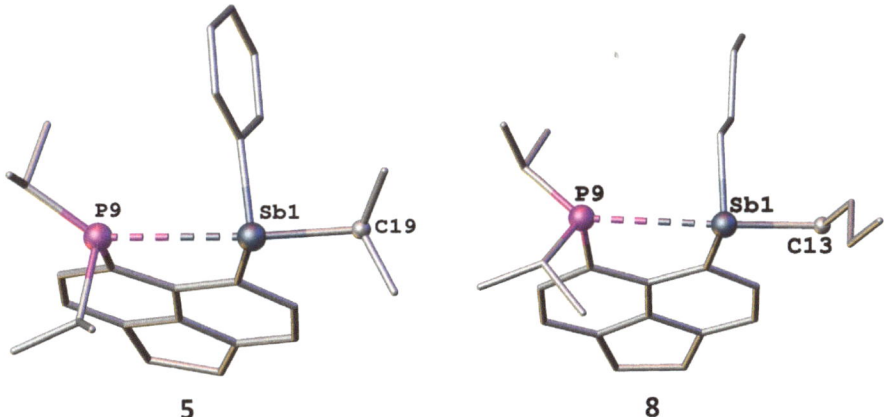

Figure 3. Molecular structures of **5** and **8** with molecules aligned approximately along the acenaphthene plane to show the quasi-linear P⋯Sb-C motifs. Hydrogen atoms and minor components of disorder are omitted for clarity.

Table 1. Selected bond distances, displacements, angles and torsion angles for the phosphine-stibines and their metal complexes.

Compound	5	8	2.PtCl$_2$·CH$_2$Cl$_2$	4.Mo(CO)$_4$
peri-region distances (Å)				
P9⋯Sb1	3.172(3)	3.218(2)	3.357(4)	3.3762(16)
P9–M1	-	-	2.248(4)	2.5432(16)
Sb1–M1	-	-	2.4570(10)	2.7007(6)
peri-region bond angles (°)				
P9⋯Sb1–C (quasi-linear)	168.48(19) [a]	167.14(16) [b]	-	-
P9–M1–Sb1	-	-	90.93(9)	80.10(4)
Splay [c]	15.1(12)	16.3(9)	16(2)	17.2(8)
Out-of-plane displacements (Å)				
P9	0.256(6)	0.202(5)	0.509(13)	0.571(6)
Sb1	0.064(6)	0.213(5)	0.788(13)	0.406(6)
M1	-	-	0.428(17)	1.007(7)
peri-region torsion angle (°)				
P9–C9⋯C1–Sb1	6.7(3)	11.2(3)	30.7(7)	24.0(3)

[a] for Sb-*i*Pr carbon C19; [b] for carbon C13; [c] Splay angle = sum of the bay region angles—360.

Crystals of **5** were grown from ethanol. The structure of **5** displays a moderately strained geometry, with a P9···Sb1 distance of 3.172(3) Å (129% of $\sum r_{covalent}$, 76% of $\sum r_{vdW}$) [24,25] and a splay angle of 15.1(12)°. These parameters indicate that, while the two functional groups in the *peri*-positions are forced into close proximity, the P···Sb interaction is primarily repulsive. However, a more detailed look at the *peri*-region geometry indicates the presence of a weak intramolecular pnictogen bond (n(P)→σ*(Sb–C$_{iPr}$)), which manifests through a quasi-linear arrangement of the P9···Sb1–C19 motif (168.5°, see Figure 3).

Crystals of **8** were obtained by prolonged standing of the crude oily product. The molecule of **8** in the structure displays a similar geometry to **5**, with a slightly larger P···Sb distance of 3.218(2) Å. In contrast to **5**, the "homoleptic" substitution pattern of the Sb atom in **8** allows direct comparison of the Sb-C bond lengths for the two *n*-butyl groups. This reveals that the Sb1–C13 bond length is significantly elongated compared to the Sb1–C17 bond length (2.197(6) vs. 2.100(7) Å). This indicates the donation of electron density (n(P9)) into the (antibonding) σ*(Sb1–C13) orbital, consistent with the formation of a (quasi-linear) pnictogen bond n(P9)→σ*(Sb1–C3), P9···Sb1–C13 angle 167.1°, see Figure 3.

Crystals of **2.PtCl$_2$** were grown from dichloromethane/hexane with a solvated molecule of dichloromethane. The platinum atom adopts a distorted square planar geometry, with the P and Sb atoms of ligand **2** bound in a *cis* fashion. Coordination of the PtCl$_2$ fragment results in elongation of the P···Sb distance to 3.357(12) Å (c.f. 3.191(1) Å in **2**) and significantly increased out-of-plane distortions within the acenaphthene ligand (see Table 1) [18]. While the P–Pt distance (2.248(4) Å) is as expected, the Sb–Pt distance in **2.PtCl$_2$** (2.4570(10) Å) is one of the shortest Sb–Pt bonds known, most likely due to the geometric constraints of the ligand. Of the 143 Sb–Pt bonds recorded in the Cambridge Structural Database to date, only 6 are shorter than the bond in compound **2.PtCl$_2$**. Those 6 examples are all Sb(V) species with highly electrophilic Sb centres [23,26–28], hence **2.PtCl$_2$** is the shortest Pt–Sb bond for a stibine (R$_3$Sb) ligand.

Crystals of **4.Mo(CO)$_4$** were grown from hexane. The molybdenum adopts a (distorted) octahedral geometry as expected, with ligand **4** attached in *cis* fashion. As above, the P-Mo distance is as expected; however, the Sb–Mo bond length of 2.7007(6) Å, is one of the shortest Sb–Mo bonds known. Of the 518 independent Sb–Mo bonds (in 97 compounds) recorded in the Cambridge Structural Database, only 8 are shorter than the bond in compound **4.Mo(CO)$_4$**. The three compounds showing the shortest distances (the shortest being 2.64386(19) Å) are all stibine (or halostibine) complexes, possessing a tridentate scaffold combining stibine and phosphine functionalities, with similar constraints as those seen in **4** [29].

2.4. Computational Analysis

DFT calculations were employed to further investigate the nP→σ*(Sb–C) interaction in **5** and **8**. Geometry optimisations were performed on these compounds (PBE0/SARC-ZORA-TZVP for Sb, PBE0/ZORA-def2-TZVP for all other atoms), and the resulting structures were in good agreement with the X-ray geometries. In particular, the Sb–C bond lengths in **8** were well reproduced, with the Sb–C bond opposite the P atom being elongated (Sb1–C13 2.197(6) Å experimental, 2.208 Å calculated; Sb1–C17 2.100(7) Å experimental, 2.175 Å calculated).

A Quantum Theory of Atoms in Molecules (QTAIM) [30,31] analysis was applied to **5** and **8**. Bond critical points (BCPs) were located between the Sb1 and P9 atoms for both **5** and **8**, indicative of a bonding interaction. Selected QTAIM parameters evaluated at BCPs for these molecules are summarized in Table 2. The Sb1···P9 BCPs all display a relatively low electron density (ρ_{BCP}) and a small and positive Laplacian ($\nabla^2 \rho_{BCP}$), which are typical of interactions between heavier elements [32,33].

Table 2. Selected properties of the electron density at bond critical points according to QTAIM analysis. $\rho(r)$ and $\nabla^2\rho(r)$ are given in standard atomic units.

Compound	Bond	$\rho(r)$	$\nabla^2\rho(r)$	E_i (kcal mol^{-1})	BD	$|V(r)|/G(r)$	ε
5	P9···Sb1	0.0272	0.0349	−4.6	−0.108	1.25	0.0446
	C1–Sb1	0.1053	0.0620	−34.9	−0.455	1.76	0.0901
	C9–P9	0.1600	−0.0631	−87.1	−0.916	2.12	0.1394
	Sb1–C13	0.1110	0.0652	−38.0	−0.472	1.76	0.0275
	Sb1–C19	0.1020	0.0278	−30.9	−0.450	1.87	0.0330
8	P9···Sb1	0.0246	0.0332	−4.0	−0.088	1.21	0.0536
	C1–Sb1	0.1038	0.0623	−34.2	−0.450	1.75	0.0933
	C9–P9	0.1603	−0.0637	−87.3	−0.917	2.12	0.1128
	Sb1–C13	0.1018	0.0401	−31.7	−0.447	1.82	0.0412
	Sb1–C17	0.1074	0.0347	−34.1	−0.466	1.85	0.0212

The bond degree parameter [32] (BD = H_{BCP}/ρ_{BCP}; H_{BCP} = energy density at BCP) [32,34] and the ratio of $|V_{BCP}|/G_{BCP}$ [32] (V_{BCP} = electronic potential energy at the BCP and G_{BCP} = electronic kinetic energy at the BCP) are two valuable metrics in QTAIM for analysing bonds between heavier elements. The BD indicates the amount of covalency in a bond, with larger negative values denoting a greater covalent interaction [32,34]. The P9···Sb1 interactions in **5** and **8** both show small, negative values, suggesting a weakly covalent interaction (Table 2). |BD| is smaller for **8** than **5**, suggesting less covalency in the P9···Sb1 interaction for **8**. This can be rationalised by the Sb(nBu)$_2$ moiety being more electron-rich than Sb(Ph)iPr, and thus a poorer electron acceptor. The interaction energy ($E_i = \frac{1}{2}V_{BCP}$) [35], which can be used as a rough estimate of bond strength, similarly indicates a weaker P9···Sb1 interaction in **8**.

The $|V_{BCP}|/G_{BCP}$ ratio differentiates between different bond types: purely closed-shell interactions such as van der Waals or ionic bonds exhibit $|V_{BCP}|/G_{BCP} < 1$, while fully covalent interactions show $|V_{BCP}|/G_{BCP} > 2$. Bonds with intermediate ratios ($1 < |V_{BCP}|/G_{BCP} < 2$) are termed transit closed-shell interactions, such bonds possess partial covalent character [32]. Both **5** and **8** display $1 < |V_{BCP}|/G_{BCP} < 2$, with a larger value for **5** than **8**. This once again suggests a more covalent P9···Sb1 in **5** than **8**. Crucially, the BD and $|V_{BCP}|/G_{BCP}$ suggest that the P9···Sb1 interaction in both **5** and **8** is not purely closed shell (i.e., Van der Waals), and that there is some degree of electron sharing between the P and Sb atoms, consistent with a nP→σ*(Sb–C) interaction. Also of note is the difference in QTAIM parameters for Sb1−C13 and Sb1−C17 in **8**. Sb−C13 shows a slightly reduced BD, $|V_{BCP}|/G_{BCP}$ and E_i compared with Sb−C17 (Table 2), consistent with a weakening of the Sb1−C13 bond due to donation into the Sb−C σ* orbital.

This P9···Sb1 interaction was further probed by an Energy Decomposition Analysis employing Naturalised Orbitals for Chemical Valence (EDA-NOCV) [36–39]. This allows the donor-acceptor interaction between P9 and Sb1 to be visualised and also allows for quantification of the interaction energy. For this analysis, the molecules were divided into two closed-shell fragments: an Acenap(PiPr$_2$)$^-$ anion and an SbR$_2$$^+$ cation. The total interaction energy between these fragments (ΔE_{int}) was computed and divided into terms for ΔE_{steric}, ΔE_{orb} and ΔE_{disp} (Table 3). ΔE_{orb} and ΔE_{disp} are the orbital and dispersion interaction energies, respectively. ΔE_{steric} is the combined electrostatic attraction and Pauli-repulsion energy terms [40]. In both **5** and **8**, ΔE_{steric} is negative, indicating a significant electrostatic attraction. This is a result of formally assigning the fragments as cationic and anionic. **8** is observed to have a slightly smaller ΔE_{int}, ΔE_{steric}, ΔE_{orb} and ΔE_{disp} than **5**, which can again be contributed to more electron rich groups on Sb weakening the donor-acceptor interaction.

Table 3. Energy decomposition analysis for compounds 5 and 8. All values are in kcal mol^{-1}. ΔE_{int} = $\Delta E_{steric} + \Delta E_{orb} + \Delta E_{disp}$.

Compound	ΔE_{int}	ΔE_{steric}	ΔE_{orb}	ΔE_{disp}
5	−226.67	−40.55	−173.78	−12.34
8	−221.92	−39.61	−171.04	−11.27

The ΔE_{orb} term can be broken down into pairs of natural orbitals for chemical valence (NOCVs), which represent the orbital interactions between the Acenap(PiPr$_2$)$^-$ and SbR$_2$$^+$ fragments. For each pair of NOCVs, a deformation density plot ($\Delta \rho_k$), which represents the flow of electrons between the molecular fragments, and its corresponding energy contribution to ΔE_{orb}, can be determined [36]. The first deformation density plots ($\Delta \rho_1$) for **5** and **8**, which have the largest energetic contribution to ΔE_{orb}, are dominated by electron flow from the (anionic) carbon of Acenap(PiPr$_2$)$^-$ to the Sb atom. However, $\Delta \rho_2$ and $\Delta \rho_3$ for **5** and **8** both appear to show electron donation from the P lone pair to a Sb−C σ* orbital (Figure 4). The energy contributions of these interactions are $\Delta \rho_2$ = −15.8 kcal mol^{-1}, $\Delta \rho_3$ = −12.1 kcal mol^{-1} for **5**; $\Delta \rho_2$ = −14.4 kcal mol^{-1}, $\Delta \rho_3$ = −11.0 kcal mol^{-1} for **8**. These values are likely a significant overestimate of the nP→σ*(Sb−C) interaction energy, as the deformation density plots also show significant contributions from the π-systems of **5** and **8**. However, these plots do strongly support the existence of donor-acceptor interactions between P9 and Sb1 in both compounds. Note that blue isosurface (an indicator of accepting electron density) is primarily observed on the Sb−C bond *opposite* the P-atom and not the other Sb−Ph (**5**) or Sb−nBu (**8**) bond (Figure 4).

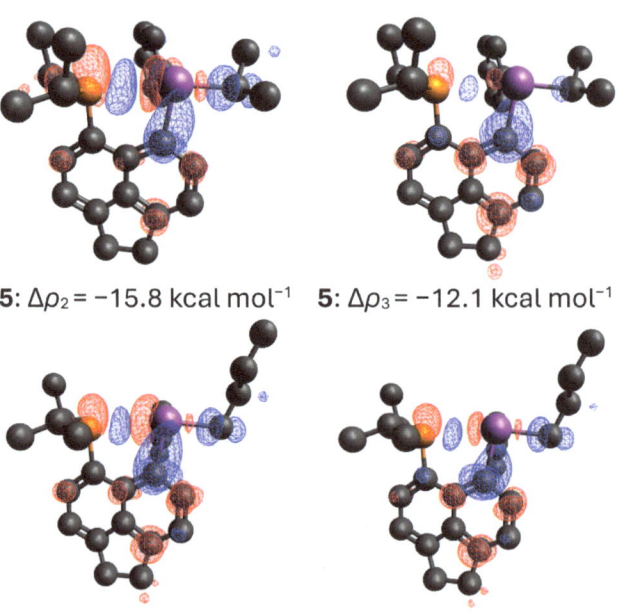

5: $\Delta \rho_2$ = −15.8 kcal mol^{-1} **5:** $\Delta \rho_3$ = −12.1 kcal mol^{-1}

8: $\Delta \rho_2$ = −14.4 kcal mol^{-1} **8:** $\Delta \rho_3$ = −11.0 kcal mol^{-1}

Figure 4. Deformation densities for **5** and **8** associated with nP→σ*(Sb-C) interaction. Charge flow is from the negative isosurface (red) to the positive isosurface (blue). All isosurfaces are plotted at 0.001 au.

3. Experimental Section

3.1. General Considerations

Unless otherwise stated, all experimental procedures were carried out under an atmosphere of dry nitrogen using standard Schlenk techniques or under an argon atmosphere in a Saffron glove box. Dry solvents were used unless otherwise stated and were either collected from an MBraun SPS-800 Solvent Purification System, or dried and stored according to literature procedures [41]. The peri-substituted acenapthene precursors **1** [42], **2**, **3** and **6** [18] were synthesised according to literature procedures. "*In vacuo*" refers to a pressure of ca. 2×10^{-2} mbar.

3.2. NMR Spectroscopy

All novel compounds were characterised where possible by ^{1}H, ^{13}C DEPTQ and ^{31}P{^{1}H} NMR spectroscopy, including measurements of ^{1}H{^{31}P}, H-H DQF COSY, H-C HSQC, H-C HMBC and H-P HMBC. ^{13}C{^{1}H} NMR spectra were recorded using the DEPTQ-135 pulse sequence with broadband proton decoupling. Measurements were performed at 20 °C using a Bruker Avance 300, Bruker Avance II 400 or Bruker Avance III 500 (MHz) spectrometer. For both ^{1}H and ^{13}C NMR, chemical shifts are relative to Me$_4$Si, which was used as an external standard. The residual solvent peaks were used for calibration (CHCl$_3$, δ_H 7.26, δ_C 77.16 ppm; CD$_2$Cl$_2$, δ_H 5.32, δ_C 53.84 ppm). For ^{31}P NMR, 85% H$_3$PO$_4$ in D$_2$O (δ_P 0 ppm) was used as an external standard. ^{195}Pt NMR was acquired for **2.PtCl$_2$**, and 1.2 M Na$_2$[PtCl$_6$] in D$_2$O (δ_{Pt} 0 ppm) was used as the external standard. The NMR numbering scheme is shown in Figure 5.

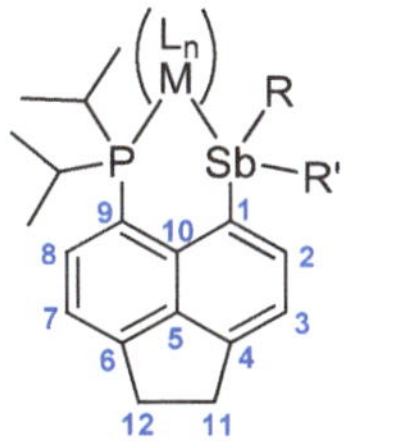

Figure 5. NMR numbering scheme.

3.3. Other Analyses

Elemental analyses (C, H and N) were performed at London Metropolitan University. High resolution mass spectrometry was performed by the EPSRC UK National Mass Spectrometry Facility (NMSF) at Swansea University using either a Waters Xevo G2-S (ASAP) or a Thermofisher LTQ Orbitrap XL (APCI) mass spectrometer. Electrospray ionisation (ES) spectra were acquired at the University of St Andrews Mass Spectrometry Facility using a Thermo Exactive Orbitrap Mass Spectrometer. Both IR and Raman spectra were collected on a Perkin Elmer 2000 NIR FT spectrometer. KBr tablets were used in IR measurements; powders in sealed glass capillaries were used for Raman spectra acquisitions. Melting (or decomposition) points were determined by heating solid samples in glass capillaries using a Stuart SMP30 melting point apparatus.

*3.4. [iPr$_2$P-Ace-SbPh$_2$]PtCl$_2$, **2.PtCl$_2$***

To a suspension of dichloro(1,5-cyclooctadiene)platinum(II) (72 mg, 170 µmol) in dichloromethane (4 mL), a solution of **2** (100 mg, 170 µmol) in dichloromethane (10 mL) was added dropwise. The solution was left to stir at room temperature overnight. The volatiles were removed *in vacuo* to give **2.PtCl$_2$** as a pale-yellow powder (104 mg, 76%). Crystals suitable for X-ray diffraction were grown from the vapour diffusion of hexane into a saturated solution of the compound in dichloromethane. M.p. 108 °C with decomposition.

^{1}H NMR: δ_H (500.1 MHz, CD$_2$Cl$_2$) 8.09 (1H, dd, $^3J_{HP}$ 11.0, $^3J_{HH}$ 7.6 Hz, H-8), 7.71 (1H, d, $^3J_{HH}$ 7.0 Hz, H-2), 7.67–7.63 (4H, m, *o*-Ph CH), 7.57 (1H, d, $^3J_{HH}$ 7.5 Hz, H-7), 7.54–7.51 (2H, m, *p*-Ph CH), 7.49–7.44 (5H, m, H-3, *m*-Ph CH), 3.55–3.45 (6H, m, H-11, 12, 2× *i*Pr CH), 1.36 (6H, dd, $^3J_{HP}$ 18.2, $^3J_{HH}$ 7.0 Hz, 2× *i*Pr CH$_3$), 1.17 (6H, dd, $^3J_{HP}$ 16.1, $^3J_{HH}$ 7.0 Hz, 2× *i*Pr CH$_3$).

^{13}C DEPTQ NMR: δ_C (125.8 MHz, CD$_2$Cl$_2$) 152.9 (s, qC-6), 152.7 (s, qC-4), 140.3 (d, $^3J_{CP}$ 7.2 Hz, qC-5), 139.8 (d, $^2J_{CP}$ 6.8 Hz, qC-10), 139.1 (s, C-2), 135.7 (d, $^2J_{CP}$ 4.3 Hz, C-8), 135.5 (s, o-Ph CH), 131.1 (s, m-Ph CH), 129.4 (s, p-Ph CH), 126.6 (s, i-Ph qC), 120.4 (s, C-3), 119.4 (d, $^3J_{CP}$ 9.4 Hz, C-7), 114.0 (d, $^1J_{CP}$ 48.4 Hz, qC-9), 111.2 (d, $^3J_{CP}$ 7.8 Hz, qC-1), 30.5 (s, C-11/12), 30.0 (s, C-11/12), 29.9 (d, $^1J_{CP}$ 34.7 Hz, 2× iPr CH), 19.4 (s, 2× iPr CH$_3$), 19.3 (s, 2 × iPr CH$_3$).
^{31}P{^{1}H} NMR: δ_P (202.5 MHz, CD$_2$Cl$_2$) 7.8 (s with ^{195}Pt satellites, $^1J_{PPt}$ 3357.0 Hz).
^{195}Pt{^{1}H} NMR: δ_{Pt} (107.0 MHz, CD$_2$Cl$_2$) −4541 (d, $^1J_{PtP}$ 3357.0 Hz).
IR (KBr disc, cm^{-1}) ν_{max} 3047m (ν_{Ar-H}), 2925m (ν_{C-H}), 1601s, 1479m, 1434vs, 1334m, 1254m, 1033m, 998w, 848m, 734vs, 693s, 451m, 270m, 241s.
Raman (glass capillary, cm^{-1}) ν_{max} 3052s (ν_{Ar-H}), 2926s (ν_{C-H}), 1604m, 1578m, 1444m, 1337m, 1000vs, 661s, 581m, 319s, 180vs.
MS (ES+): m/z (%) 775.06 (100) [M − Cl].

3.5. iPr$_2$P-Ace-Sb(Ph)Me, 4

To a stirred suspension of **3** (0.50 g, 0.99 mmol) in tetrahydrofuran (20 mL), cooled to −78 °C, methylmagnesium bromide (0.50 mL, 3.0 M solution in diethyl ether, 1.5 mmol) was added dropwise. The reaction mixture was allowed to warm to room temperature and stirred for 1 h, then cooled to 0 °C, and degassed water (0.5 mL) was added cautiously. Volatiles were removed *in vacuo*, and the resulting oil was redissolved in hexane (40 mL). The resulting suspension was filtered to remove insoluble impurities, and volatiles were removed *in vacuo* to afford **4** as a pale-yellow oil (0.41 g, 0.85 mmol, 86%).

^{1}H NMR δ_H (400 MHz, CDCl$_3$) 7.67–7.59 (4H, m, ArH-2, ArH-8, m-Ph CH), 7.32 (1H, d, $^3J_{HH}$ = 7.1 Hz, ArH-7), 7.29–7.25 (3H, m, o/p-Ph CH), 7.16 (1H, d, $^3J_{HH}$ = 7.1 Hz, ArH-3), 3.38 (4H, s, H-11, H-12, 2 × CH$_2$), 2.25 (1H, septd, $^3J_{HH}$ = 6.9 Hz, $^2J_{HP}$ = 5.1 Hz, iPr CH), 2.09 (1H, septd, $^3J_{HH}$ = 7.0 Hz, $^2J_{HP}$ = 3.2 Hz, iPr CH), 1.22 (3H, dd, $^3J_{HP}$ = 15.0 Hz, $^3J_{HH}$ = 6.9 Hz, iPr CH$_3$), 1.18 (3H, d, $^6J_{HP}$ = 1.8 Hz, Me(Sb)), 1.04 (3H, dd, $^3J_{HP}$ = 14.7 Hz, $^3J_{HH}$ = 6.9 Hz, iPr CH$_3$), 0.97 (3H, dd, $^3J_{HP}$ = 12.5 Hz, $^3J_{HH}$ = 7.0 Hz, iPr CH$_3$), 0.66 (3H, dd, $^3J_{HP}$ = 11.8 Hz, $^3J_{HH}$ = 7.0 Hz, iPr CH$_3$).
^{13}C{^{1}H} NMR δ_C (75 MHz, CDCl$_3$) 149.1 (s, qC-6), 147.2 (d, $^4J_{CP}$ = 1.6 Hz, qC-4), 146.0 (d, $^1J_{CP}$ = 40.5 Hz, qC-9), 142.1 (s, qC-1), 140.0 (d, $^3J_{CP}$ = 7.8 Hz, qC-5), 138.2 (s, C-2), 136.6 (s, m-Ph CH), 134.0 (s, qC-10), 133.9 (d, $^2J_{CP}$ = 2.4 Hz, C-8), 130.1 (d, $^5J_{CP}$ = 16.2 Hz, i-Ph qC), 128.4 (s, o-Ph CH), 127.6 (s, p-Ph CH), 120.0 (s, C-3), 119.0 (s, C-7), 30.3 (s, C-11/C-12), 30.0 (s, C-11/C-12), 26.1 (d, $^1J_{CP}$ = 12.5 Hz, iPr CH), 25.9 (d, $^1J_{CP}$ = 13.7 Hz, iPr CH), 20.6 (d, $^2J_{CP}$ = 17.7 Hz, iPr CH$_3$), 20.1 (d, $^2J_{CP}$ = 13.3 Hz, iPr CH$_3$), 20.0 (d, $^2J_{CP}$ = 7.3 Hz, iPr CH$_3$), 19.2 (d, $^2J_{CP}$ = 7.2 Hz, iPr CH$_3$), 4.74 (d, $^5J_{CP}$ = 34.1 Hz, Me(Sb)).
^{31}P NMR δ_P (109 MHz, CDCl$_3$) −19.7 (m).
^{31}P{^{1}H} NMR δ_P (109 MHz, CDCl$_3$) −19.6 (s).
MS (APCI+) m/z 390.05 (85%, M − Ph − Me), 405.07 (100, M − Ph), 467.09 (73, M − Me), 483.12 (11, M + H), 499.11 (4, M + O + H).
HRMS (APCI+) C$_{25}$H$_{31}$PSb (M + H)$^+$; calculated: 483.1196; found: 483.1195.

3.6. [iPr$_2$P-Ace-Sb(Ph)Me]Mo(CO)$_4$, 4.Mo(CO)$_4$

A solution of compound **4** (240 mg, 0.497 mmol) in dichloromethane (40 mL) was added to a stirred suspension of *cis*-tetracarbonyl(norbornadiene)molybdenum(0) (0.164 g, 0.546 mmol) in dichloromethane (10 mL), and the resulting suspension was stirred at room temperature for 3 days. Insoluble material was removed by filtration through celite, and volatiles were removed *in vacuo* to afford **4.Mo(CO)$_4$** as a pale brown solid (329 mg, 0.476 mmol, 96%). Crystals suitable for single crystal X-ray diffraction were grown from hexane at −25 °C.

^{1}H NMR δ_H (400 MHz, CDCl$_3$) 7.79 (1H, d, $^3J_{HH}$ = 6.9 Hz, ArH-2), 7.70 (1H, ≈ t, $^3J_{HP}$ = 7.8 Hz, $^3J_{HH}$ = 7.8 Hz, ArH-8), 7.56–7.50 (2H, m, m-Ph CH), 7.39–7.33 (4H, m, ArH-7, o/p-Ph CH), 7.30 (1H, d, $^3J_{HH}$ = 6.9 Hz, ArH-3), 3.40 (4H, s, H-11/H-12), 2.50–2.38 (1H, m, iPr CH), 2.35–2.25 (1H, m, iPr CH), 1.58 (s, 3H, Me(Sb)), 1.23 (3H, dd, $^3J_{PH}$ = 15.4 Hz,

$^3J_{HH}$ = 6.9 Hz, iPr CH$_3$), 1.10 (3H, dd, $^3J_{PH}$ = 15.8 Hz, $^3J_{HH}$ = 7.1 Hz, iPr CH$_3$), 1.04 (3H, dd, $^3J_{PH}$ = 15.6 Hz, $^3J_{HH}$ = 7.1 Hz, iPr CH$_3$), 0.99 (3H, dd, $^3J_{PH}$ = 15.0 Hz, $^3J_{HH}$ = 6.9 Hz, iPr CH$_3$).

^{13}C{^{1}H} NMR δ_C (101 MHz, CDCl$_3$) 218.4 (d, $^2J_{CP}$ = 8.1 Hz, CO), 216.4 (d, $^2J_{CP}$ = 20.3 Hz, CO), 211.1 (d, $^2J_{CP}$ = 9.3 Hz, CO), 210.8 (d, $^2J_{CP}$ = 9.1 Hz, CO), 150.6 (s, qC-6), 150.4 (d, $^4J_{CP}$ = 1.4 Hz, qC-4), 140.9 (d, $^3J_{CP}$ = 6.5 Hz, qC-5), 139.5 (d, $^2J_{CP}$ = 10.3 Hz, qC-10), 138.3 (s, C-2), 135.6 (d, $^3J_{CP}$ = 2.3 Hz, i-Ph qC), 133.9 (s, m-Ph CH), 133.3 (s, C-8), 129.4 (s, p-Ph CH), 129.0 (s, o-Ph CH), 124.6 (d, $^1J_{CP}$ = 20.4, qC-9), 123.4 (d, $^3J_{CP}$ = 2.4 Hz, qC-1), 119.8 (s, C-3), 118.9 (d, $^3J_{CP}$ = 5.7 Hz, C-7), 30.2 (s, C-11/C-12), 29.5 (s, C-11/C-12), 28.7 (d, $^1J_{CP}$ = 15.9 Hz, iPr CH), 28.1 (d, $^1J_{CP}$ = 14.9 Hz, iPr CH), 19.1 (d, $^2J_{CP}$ = 4.3 Hz, iPr CH$_3$), 19.0 (d, $^2J_{CP}$ = 4.1 Hz, iPr CH$_3$), 18.24 (d, $^2J_{CP}$ = 5.9 Hz, iPr CH$_3$), 18.15 (d, $^2J_{CP}$ = 4.4 Hz, iPr CH$_3$), 4.4 (d, $^3J_{CP}$ = 2.9 Hz, Me(Sb)).

^{31}P NMR δ_P (109 MHz, CDCl$_3$) 43.3 (m).

^{31}P{^{1}H} NMR δ_P (109 MHz, CDCl$_3$) 43.4 (s).

IR (KBr disk, cm^{-1}) ν_{max} 3007 (ν_{Ar-H}, w), 2963 (ν_{C-H}, m), 2014 ($\nu_{C\equiv O}$, vs), 1899 ($\nu_{C\equiv O}$, vs), 1603 (m), 1261 (m), 1084 (s), 1023 (s), 802 (m), 733 (m), 694 (m), 613 (m), 585 (m), 453 (m).

MS (APCI+) m/z 271.16 (100%, M − Mo(CO)$_4$ − Sb(Me)Ph + H), 287.16 (M − Mo(CO)$_4$ − Sb(Me)Ph + O + H), 637.02 (1, M − 2CO + H)

HRMS (APCI+) C$_{27}$H$_{31}$MoO$_2$PSb (M − 2CO + H)$^+$; calculated: 637.0149; found: 637.0148.

3.7. iPr$_2$P-Ace-Sb(Ph)iPr, **5**

To a stirred suspension of **3** (1.00 g, 1.98 mmol) in tetrahydrofuran (20 mL), cooled to −78 °C, a solution of isopropylmagnesium chloride (1.5 mL, 1.70 M solution in THF, 2.55 mmol) was added dropwise. The reaction mixture was allowed to warm to room temperature with stirring overnight, then cooled to 0 °C, and degassed water (0.5 mL) was added cautiously. Volatiles were removed *in vacuo* to give an oil. Hexane (50 mL) was added, and the resultant suspension was filtered to remove the insoluble salts. Volatiles were removed *in vacuo* to afford **5** as a yellow oil, which crystallised to a yellow solid on standing at room temperature for several days (0.824 g, 1.61 mmol, 81%). Crystals suitable for single crystal X-ray diffraction were grown from ethanol at −25 °C. M. p. 73–76 °C.

Elemental Analysis: C$_{27}$H$_{34}$PSb; calculated (%) C 63.43, H 6.70; found (%) C 63.31, H 6.73.

^{1}H NMR δ_H (500 MHz; CDCl$_3$) 7.85 (1H, d, $^3J_{HH}$ = 7.0 Hz, ArH-2), 7.61 (1H, dd, $^3J_{HH}$ = 7.1 Hz, $^3J_{HP}$ = 3.6 Hz, ArH-8), 7.58–7.54 (2H, m, m-Ph CH), 7.30 (1H, d, $^3J_{HH}$ = 7.1 Hz, ArH-3), 7.26 (1H, d, $^3J_{HH}$ = 7.1 Hz, ArH-7), 7.23–7.19 (3H, m, o/p-Ph CH), 3.39 (4H, s, H-11, H-12), 2.32 (1H, sept, $^3J_{HH}$ = 7.2 Hz, iPr(Sb) CH), 2.22 (1H, septd, $^3J_{HH}$ = 7.1 Hz, $^2J_{HP}$ = 3.4 Hz, iPr(P) CH), 2.04–1.93 (1H, m, iPr(P) CH), 1.36 (3H, d, $^3J_{HH}$ = 7.2 Hz, iPr(Sb) CH$_3$), 1.25–1.19 (6H, m, iPr(Sb) CH$_3$, iPr(P) CH$_3$), 0.99 (3H, dd, $^3J_{HP}$ = 11.9 Hz, $^3J_{HH}$ = 7.0 Hz, iPr(P) CH$_3$), 0.95 (3H, dd, $^3J_{HP}$ = 14.6 Hz, $^3J_{HH}$ = 6.9 Hz, iPr(P) CH$_3$), 0.42 (3H, dd, $^3J_{HP}$ = 12.4 Hz, $^3J_{HH}$ = 7.0 Hz, iPr(P) CH$_3$).

^{13}C{^{1}H} NMR δ_C (101 MHz; CDCl$_3$) 148.9 (s, qC-6), 147.3 (d, $^4J_{CP}$ = 1.7 Hz, qC-4), 144.2 (d, $^1J_{CP}$ = 28.2 Hz, qC-9), 142.2 (d, $^2J_{CP}$ = 27.2 Hz, qC-10), 140.1 (d, $^3J_{CP}$ = 7.8 Hz, qC-5), 138.1 (s, C-2), 136.7 (s, m-Ph CH), 134.0 (d, $^2J_{CP}$ = 2.4 Hz, C-8), 133.9 (d, $^3J_{CP}$ = 5.6 Hz, qC-1), 130.7 (d, $^5J_{CP}$ = 17.3 Hz, i-Ph qC), 128.2 (s, o-Ph CH), 127.4 (s, p-Ph CH), 120.1 (s, C-7), 119.0 (s, C-3), 30.2 (s, C-11/C-12), 30.0 (s, C-11/C-12), 26.4 (d, $^1J_{CP}$ = 14.8 Hz, iPr(P) CH), 26.2 (d, $^1J_{CP}$ = 14.0 Hz, iPr(P) CH), 25.1 (d, $^5J_{CP}$ = 36.8 Hz, iPr(Sb) CH), 22.6 (s, iPr(Sb) CH$_3$), 21.7 (s, iPr(Sb) CH$_3$), 20.7 (d, $^2J_{CP}$ = 16.8 Hz, iPr(P) CH$_3$), 20.1 (d, $^2J_{CP}$ = 8.1 Hz, iPr CH$_3$), 19.9 (d, $^2J_{CP}$ = 16.6 Hz, iPr(P) CH$_3$), 19.2 (d, $^2J_{CP}$ = 9.3 Hz, iPr(P) CH$_3$).

^{31}P NMR δ_P (109 MHz; CDCl$_3$) −20.8 (m).

^{31}P{^{1}H} NMR δ_P (109 MHz; CDCl$_3$) −20.8 (s).

Raman: (glass capillary, cm^{-1}) ν_{max} 3038s (ν_{Ar-H}), 2921vs (ν_{C-H}), 1601s, 1562s, 1442s, 1325vs, 1001vs, 657m, 582s, 490s.

MS (APCI+) m/z 390.05 (100%, M − iPr − Ph), 433.10 (60, M − Ph), 467.09 (50, M − iPr), 511.15 (16, M + H), 527.15 (4, M + O + H).

HRMS (APCI+): $C_{27}H_{35}PSb$ (M + H)$^+$; calculated: 511.1509; found: 511.1510.

3.8. iPr$_2$P-Ace-SbMe$_2$, 7

A solution of **6** (0.49 g, 1.06 mmol) in diethyl ether (40 mL) was cooled to −78 °C. A solution of methylmagnesium bromide in tetrahydrofuran (0.7 mL, 3.0 M solution, 2.10 mmol, diluted with diethyl ether, 4 mL) was added dropwise with stirring over one hour. The resulting suspension was stirred at -78 °C for a further 90 min before warming to ambient temperature overnight. The suspension was filtered, and volatiles were removed from the filtrate *in vacuo*, affording **7** as an off-white solid (0.42 g, 94%). The yield is approximate, as ^{1}H NMR spectra indicate the presence of solvated diethylether. M.p. 239 °C with decomposition.

^{1}H NMR: δ_H (400.1 MHz, CDCl$_3$) 7.77 (1H, d, $^3J_{HH}$ 7.0 Hz, H-2), 7.66 (1H, dd, $^3J_{HH}$ 7.1, $^3J_{HP}$ 3.6 Hz, H-8), 7.32 (1H, d, $^3J_{HH}$ 7.1 Hz, H-7), 7.26 (1H, d, $^3J_{HH}$ 7.0 Hz, H-3), 3.38 (4H, s, H-11, 12), 2.19 (2H, br sept, $^3J_{HH}$ 7.0 Hz, iPr CH), 1.17 (6H, dd, $^3J_{HP}$ 14.6, 3J_{HH} 6.9 Hz, iPr CH$_3$), 0.96 (6H, d, $^{6ts}J_{HP}$ 1.2 Hz, Sb-CH$_3$), 0.94 (6H, dd, $^3J_{HP}$ 12.3, $^3J_{HH}$ 7.0 Hz, iPr CH$_3$).

^{13}C DEPTQ NMR: δ_C (100.6 MHz, CDCl$_3$) 149.0 (s, qC-6), 147.0 (s, qC-4), 141.5 (d, $^2J_{CP}$ 4.3 Hz, qC-10), 139.8 (d, $^3J_{CP}$ 1.7 Hz, qC-5), 136.0 (s, C-2), 133.8 (d, $^2J_{CP}$ 2.5 Hz, C-8), 133.3 (d, $^3J_{CP}$ 5.1 Hz, qC-1), 130.3 (d, $^1J_{CP}$ 20.0 Hz, qC-9), 119.9 (s, C-3), 118.9 (s, C-7), 30.2 (s, C-11/12), 29.9 (s, C-11/12), 26.0 (d, $^1J_{CP}$ 13.5 Hz, iPr CH), 20.4 (d, $^2J_{CP}$ 16.7 Hz, iPr CH$_3$), 19.9 (d, $^2J_{CP}$ 9.0 Hz, iPr CH$_3$), 3.4 (d, $^{5ts}J_{CP}$ 34.9 Hz, Sb-CH$_3$).

^{31}P{^{1}H} NMR: δ_P (162.0 MHz, CDCl$_3$) −18.5 (s).

Raman: (glass capillary, cm^{-1}) ν_{max} 2933vs (ν_{C-H}), 2124w, 1601m, 1562m, 1447m, 1325vs, 577s, 509s, 482vs.

HRMS (ASAP+): m/z Calcd. for $C_{20}H_{29}PSb$ 421.1045, found 421.1042 [M + H]; Calcd. for $C_{19}H_{25}PSb$ 405.0732, found 405.0724 [M − Me].

3.9. iPr$_2$P-Ace-Sb(nBu)$_2$, 8

A solution of **6** (1.00 g, 2.16 mmol) in diethyl ether (40 mL) was cooled to −78 °C. To this, a solution of n-butyllithium in hexane (1.7 mL, 2.5 M solution, 4.25 mmol) was added dropwise with stirring over one hour. The resulting suspension was stirred at this temperature for a further hour before warming to ambient temperature overnight. The suspension was filtered, and the volatiles were removed from the filtrate *in vacuo*, affording **8** as a yellow oil (yield quantitative). A few crystals suitable for single-crystal X-ray diffraction formed spontaneously from the oil after long standing at room temperature. M. p. 53 °C.

^{1}H NMR: δ_H (400.1 MHz, CDCl$_3$) 7.73 (1H, d, $^3J_{HH}$ 7.0 Hz, H-2), 7.67 (1H, dd, $^3J_{HH}$ 7.1, $^3J_{HP}$ 3.6 Hz, C-8), 7.31 (1H, d, $^3J_{HH}$ 7.1 Hz, H-7), 7.25 (1H, d, $^3J_{HH}$ 7.0 Hz, H-3), 3.38 (4H, s, H-11, 12), 2.22 (2H, dsept, $^3J_{HH}$ 6.9, $^2J_{HP}$ 3.9 Hz, iPr CH), 1.71−1.48 (8H, m, SbCH$_2$ and SbCH$_2$CH$_2$), 1.44−1.36 (4H, m, Sb CH$_2$CH$_2$CH$_2$), 1.23 (6H, dd, $^3J_{HP}$ 14.4, $^3J_{HH}$ 6.9 Hz, 2× iPr CH$_3$), 0.95 (6H, dd, $^3J_{HP}$ 12.4, $^3J_{HH}$ 7.2 Hz, 2× iPr CH$_3$), 0.90 (6H, t, $^3J_{HH}$ 7.2 Hz, 2× n-Bu CH$_3$).

^{13}C DEPTQ NMR: δ_C (100.6 MHz, CDCl$_3$) 148.9 (s, qC-6), 146.8 (d, $^4J_{CP}$ 1.7 Hz, qC-4), 142.1 (d, $^2J_{CP}$ 26.9 Hz, qC-10), 139.9 (d, $^3J_{CP}$ 7.8 Hz, qC-5), 136.5 (s, C-2), 133.8 (d, $^2J_{CP}$ 2.4 Hz, C-8), 131.9 (d, $^3J_{CP}$ 5.0 Hz, qC-1), 130.9 (d, $^1J_{CP}$ 18.0 Hz, qC-9), 119.9 (s, C-3), 118.9 (s, C-7), 30.2 (s, SbCH$_2$CH$_2$), 29.9 (s, C-11, 12), 27.0 (s, SbCH$_2$CH$_2$CH$_2$), 26.4 (d, $^1J_{CP}$ 14.3 Hz, iPr CH), 20.7 (d, $^{5ts}J_{CP}$ 30.6 Hz, SbCH$_2$), 20.5 (d, $^2J_{CP}$ 16.9 Hz, iPr CH$_3$), 20.2 (d, $^2J_{CP}$ 9.5 Hz, iPr CH$_3$), 14.0 (s, n-Bu CH$_3$).

^{31}P{^{1}H} NMR: δ_P (162.0 MHz, CDCl$_3$) −18.5 (s)

Raman: (glass capillary, cm^{-1}) ν_{max} 2919vs (ν_{C-H}), 2868vs, 1556m, 1435m, 1325s, 1157w, 583m.

HRMS (ASAP+): m/z Calcd. for $C_{26}H_{41}PSb$ [M + H]: 505.1984, found 505.1986.

3.10. X-ray Diffraction

X-ray diffraction data for compound **2.PtCl2** were collected at 125 K using the St Andrews Automated Robotic Diffractometer (STANDARD) [43], consisting of a Rigaku sealed-tube X-ray generator equipped with a SHINE monochromator [Mo Kα radiation (λ = 0.71075 Å)], and a Saturn 724 CCD area detector, coupled with a Microglide goniometer head and an ACTOR SM robotic sample changer. Diffraction data for compounds **4.Mo(CO)$_4$**, **5** and **8** were collected at 173 K using a Rigaku FR-X Ultrahigh Brilliance Microfocus RA generator/confocal optics [Mo Kα radiation (λ = 0.71075 Å)] with an XtaLAB P200 diffractometer. Intensity data for all compounds were collected using ω steps, accumulating area detector images spanning at least a hemisphere of reciprocal space. Data for all compounds analysed were collected using CrystalClear [44] and processed (including correction for Lorentz, polarization and absorption) using either CrystalClear or CrysAlisPro [45]. Structures were solved by direct (SHELXS [46]), Pattterson (PATTY [47]) or charge-flipping (Superflip [48]) methods and refined by full-matrix least-squares against F^2 (SHELXL-2019/3 [49]). Non-hydrogen atoms were refined anisotropically, and hydrogen atoms were refined using a riding model. In **5**, both isopropyl groups bound to phosphorus were disordered over two positions. Atoms were split and refined with partial occupancies, and restraints to bond distances were required. Crystals of **8** were affected by pseudo-merohedral twinning, showing a twin law of [−0.9999 0.0198 0.0009 −0.0032 0.9992 −0.0203 −0.0007 −0.1315 −0.9981] and a refined twin fraction of 0.489. All calculations were performed using the CrystalStructure interface [50]. Selected crystallographic data are presented in Table 4.

Table 4. Selected crystallographic data.

	2.PtCl$_2$	**4.Mo(CO)$_4$**	**5**	**8**
formula	C$_{31}$H$_{34}$Cl$_4$PPtSb	C$_{29}$H$_{30}$MoO$_4$PSb	C$_{27}$H$_{34}$PSb	C$_6$H$_{40}$PSb
fw	896.24	691.22	511.29	505.33
crystal description	Colourless block	Colourless prism	Yellow chip	Colourless chip
crystal size [mm^3]	0.09 × 0.06 × 0.03	0.17 × 0.15 × 0.04	0.10 × 0.08 × 0.06	0.12 × 0.10 × 0.03
temperature [K]	125	173	173	173
space group	*Pna*2$_1$	*P*$\bar{1}$	*P*2$_1$/*n*	*C*2/*c*
a [Å]	14.797(2)	10.1706(3)	9.751(3)	27.3960(13)
b [Å]	18.166(3)	12.6048(5)	13.591(2)	8.6563(4)
c [Å]	11.8407(19)	12.8276(7)	19.091(5)	22.5385(14)
α [°]		76.520(14)		
β [°]		67.547(11)	102.261(8)	109.617(6)
γ [°]		67.048(12)		
vol [Å]^3	3182.8(9)	1392.63(19)	2472.3(11)	5034.7(5)
Z	4	2	4	8
ρ (calc) [g/cm^3]	1.870	1.648	1.374	1.333
μ [mm^{-1}]	5.626	1.507	1.189	1.167
F(000)	1728	688	1048	2096
reflections collected	24,780	24,150	29,608	28,282
independent reflections (R_{int})	6286 (0.0886)	4970 (0.1133)	4528 (0.0604)	22,368 (0.0755)
parameters, restraints	347, 1	330, 0	297, 45	260, 0
GoF on F^2	1.106	1.081	1.109	0.720
R_1 [$I > 2\sigma(I)$]	0.0505	0.0570	0.0620	0.0523
wR_2 (all data)	0.0978	0.1494	0.1308	0.1151
largest diff. peak/hole [e/Å^3]	0.88, −1.30	2.41, −0.72	0.67, −0.58	2.03, −1.39
Flack parameter	0.012(6)	-	-	-

3.11. Computational Methodology

3.11.1. Geometry Optimisations and QTAIM Analysis

Geometry optimisations were performed for models **5** and **8** using coordinates derived from their X-ray crystal structures. These models were geometry optimised without restraints using the ORCA 5.0.4 software package [51] utilising the PBE0 density functional [52] and all-electron ZORA corrected [53] def2-TZVP basis sets [54–57] for all atoms (except Sb), SARC-ZORA-TZVP [58] basis sets for the Sb atoms, along with SARC/J auxiliary basis sets decontracted def2/J up to Kr [59] and SARC auxiliary basis sets beyond Kr. [58,60–62]. Gradient corrections were performed with Grimme's 3rd generation dispersion correction [63,64]. TightSCF and TightOpt convergence criteria were employed, and the location of true minima in these optimisations was confirmed by frequency analysis, which demonstrated that no imaginary vibrations were present.

Extended wavefunction (.wfx) files were generated from these optimisations using Orca 5.0.4 [51]. AIM analysis was performed using MultiWFN 3.8 [65].

3.11.2. EDA-NOCV Analysis

EDA-NOCV calculations were carried out on models of **5** and **8** using coordinates derived from the geometry-optimised structures. The compounds were divided into an Acenap(PiPr$_2$)$^-$ and SbR$_2^+$ fragments (**5** = $^+$Sb(iPr)Ph; **8** = $^+$Sb(nBu)$_2$). Calculations were carried out using the ORCA 5.0.4 software package [51] utilising the PBE0 density functional [52,66] and all-electron ZORA corrected [53] def2-TZVP basis sets [54–57] for all atoms (except Sb), SARC-ZORA-TZVP basis sets [58] for the Sb atoms, along with SARC/J auxiliary basis sets decontracted def2/J up to Kr [59] and SARC auxiliary basis sets beyond Kr [58,60–62]. Gradient corrections were performed with Grimme's 3rd generation dispersion correction [63,64]. VeryTightSCF convergence settings and an integration accuracy value of 6.0 were employed. Deformation density plots were calculated from .cube files of the relevant NOCVs using the MultiWFN 3.8 software package [65] and visualised using Avogadro v1.2.0 [67].

4. Conclusions

The synthetic utility of *peri*-substituted phosphine-chlorostibines **3** and **6** in reactions with carbon nucleophiles has been demonstrated. Reactions with Grignard reagents or nBuLi proceeded rather cleanly and gave alkyl/aryl and alkyl tertiary stibines **4**, **5**, **7** and **8** with very good yields.

The coordination chemistry of the selected tertiary phosphine-stibines has also been probed. Two complexes, **2.PtCl$_2$** and **4.Mo(CO)$_4$**, have been synthesised. Single-crystal X-ray diffraction confirmed that both the phosphine and the stibine groups are attached to the platinum(II) and Mo(0) centres.

In the phosphine-phosphine *peri*-substituted species, such as iPr$_2$P-Ace-PPh$_2$, large magnitude $^{4TS}J_{PP}$ (180 Hz for the above species) were observed due to the forced overlap of the two lone pairs on the phosphorus atoms [68]. As antimony has no spin $\frac{1}{2}$ isotopes, the direct observation of P–Sb couplings (formally $^{4TS}J_{PSb}$) was not possible in the phosphine-stibines reported here. However, long-range $^{5TS}J_{CP}$ couplings of up to 36.8 Hz were observed for carbon atoms attached to Sb atoms in all the phosphine stibines. This indicates the presence of a strong through-space coupling pathway through the phosphorus and antimony atoms (both possessing a lone pair), and the through-bond pathway contribution (^5J) is expected to be negligible [69].

The QTAIM analysis supports the existence of a P$\cdots$Sb interaction in **5** and **8**, which is not purely closed-shell (i.e., Van der Waals), and the visualisation of the deformation densities in an EDA-NOCV analysis supports the view that electron density from the P atom flows towards an apparent Sb–C σ*orbital.

Author Contributions: L.J.T., E.E.L. and B.A.C. carried out the required synthetic steps, collected all data (except X-ray), and analysed the data. A.M.Z.S., D.B.C. and K.S.A.A. collected the X-ray data and solved the structures. L.J.T. performed all computational analysis. P.K. and B.A.C. designed the study. P.K., L.J.T. and D.B.C. wrote the manuscript. P.K. provided the supervision. All authors have contributed to the proof-reading and editing of the manuscript. All authors have read and agreed to the published version of the manuscript.

Funding: This research received no external funding. We are grateful to the University of St Andrews School of Chemistry Undergraduate Project grants.

Institutional Review Board Statement: Not applicable.

Informed Consent Statement: Not applicable.

Data Availability Statement: CCDC 2337824-2337827 contains the supplementary crystallographic data for this paper. These data can be obtained free of charge from The Cambridge Crystallographic Data Centre via www.ccdc.cam.ac.uk/structures. The research data underpinning this publication can be accessed at https://doi.org/10.17630/8f5cd85b-0e47-44ab-9d9a-e27d44a123f6 [70].

Conflicts of Interest: The authors declare no conflicts of interest.

References and Note

1. Champness, N.R.; Levason, W. Coordination chemistry of stibine and bismuthine ligands. *Coord. Chem. Rev.* **1994**, *133*, 115–217. [CrossRef]
2. Levason, W.; Reid, G. Developments in the coordination chemistry of stibine ligands. *Coord. Chem. Rev.* **2006**, *250*, 2565–2594. [CrossRef]
3. Alder, R.W.; Bowman, P.S.; Steele, W.R.S.; Winterman, D.R. The remarkable basicity of 1,8-bis(dimethylamino)naphthalene. *Chem. Commun.* **1968**, 723–724. [CrossRef]
4. Costa, T.; Schmidbaur, H. 1,8-Naphthalindiylbis(dimethylphosphan): Konsequenzen sterischer Hinderung für Methylierung und Borylierung. *Chem. Berichte* **1982**, *115*, 1374–1378. [CrossRef]
5. Karaçar, A.; Thönnessen, H.; Jones, P.G.; Bartsch, R.; Schmutzler, R. 1,8-Bis(phosphino)naphthalenes: Synthesis and molecular structures. *Heteroat. Chem.* **1997**, *8*, 539–550. [CrossRef]
6. Yam, V.W.W.; Chan, C.L.; Choi, S.W.K.; Wong, K.M.C.; Cheng, E.C.C.; Yu, S.C.; Ng, P.-K.; Chan, W.-K.; Cheung, K.-K. Synthesis, photoluminescent and electroluminescent behaviour of four-coordinate tetrahedral gold(i) complexes. X-ray crystal structure of [Au(dppn)$_2$]Cl. *Chem. Commun.* **2000**, 53–54. [CrossRef]
7. Karaçar, A.; Freytag, M.; Jones, P.G.; Bartsch, R.; Schmutzler, R. Platinum(II) Complexes of P-Chiral 1,8-Bis(diorganophosphino) naphthalenes; Crystal Structures of dmfppn, rac-[PtCl$_2$(dmfppn)], and rac-[PtCl$_2$(dtbppn)] (dmfppn = 1,8-di(methyl-pentafluorophenylphosphino)naphthalene and dtbppn = 1,8-di(tert-butylphenylphosphino)naphthalene). *Z. Anorg. Allg. Chem.* **2002**, *628*, 533–544. [CrossRef]
8. Kilian, P.; Knight, F.R.; Woollins, J.D. Synthesis of ligands based on naphthalene peri-substituted by Group 15 and 16 elements and their coordination chemistry. *Coord. Chem. Rev.* **2011**, *255*, 1387–1413. [CrossRef]
9. Di Sipio, L.; Sindellari, L.; Tondello, E.; De Michelis, G.; Oleari, L. NiII complexes with 1,8-naphthalene bis(dimethylarsine). *Coord. Chem. Rev.* **1967**, *2*, 117–128. [CrossRef]
10. Jura, M.; Levason, W.; Reid, G.; Webster, M. Preparation and properties of sterically demanding and chiral distibine ligands. *Dalton Trans.* **2008**, 5774–5782. [CrossRef]
11. Gehlhaar, A.; Wölper, C.; van der Vight, F.; Jansen, G.; Schulz, S. Noncovalent Intra- and Intermolecular Interactions in Peri-Substituted Pnicta Naphthalene and Acenaphthalene Complexes. *Eur. J. Inorg. Chem.* **2022**, *2022*, e202100883. [CrossRef]
12. Gehlhaar, A.; Weinert, H.M.; Wölper, C.; Semleit, N.; Haberhauer, G.; Schulz, S. Bisstibane–distibane conversion via consecutive single-electron oxidation and reduction reaction. *Chem. Commun.* **2022**, *58*, 6682–6685. [CrossRef]
13. Dzialkowski, K.; Gehlhaar, A.; Wolper, C.; Auer, A.A.; Schulz, S. Structure and Reactivity of 1,8-Bis(naphthalenediyl)dipnictanes. *Organometallics* **2019**, *38*, 2927–2942. [CrossRef]
14. Ganesamoorthy, C.; Heimann, S.; Hölscher, S.; Haack, R.; Wölper, C.; Jansen, G.; Schulz, S. Synthesis, structure and dispersion interactions in bis(1,8-naphthalendiyl)distibine. *Dalton Trans.* **2017**, *46*, 9227–9234. [CrossRef] [PubMed]
15. Gehlhaar, A.; Schiavo, E.; Wölper, C.; Schulte, Y.; Auer, A.A.; Schulz, S. Comparing London dispersion pnictogen–π interactions in naphthyl-substituted dipnictanes. *Dalton Trans.* **2022**, *51*, 5016–5023. [CrossRef] [PubMed]
16. Chuit, C.; Corriu, R.J.P.; Monforte, P.; Reyé, C.; Declercq, J.-P.; Dubourg, A. Evidences for intramolecular N → P coordination in (8-dimethylamino-1-naphthyl) diphenylphosphane and derivatives. *J. Organomet. Chem.* **1996**, *511*, 171–175. [CrossRef]
17. Biskup, D.; Bergmann, T.; Schnakenburg, G.; Gomila, R.M.; Frontera, A.; Streubel, R. Synthesis of a 1-aza-2-phospha-acenaphthene complex profiting from coordination enabled chloromethane elimination. *RSC Adv.* **2023**, *13*, 21313–21317. [CrossRef] [PubMed]
18. Chalmers, B.A.; Bühl, M.; Athukorala Arachchige, K.S.; Slawin, A.M.Z.; Kilian, P. Structural, Spectroscopic and Computational Examination of the Dative Interaction in Constrained Phosphine–Stibines and Phosphine–Stiboranes. *Chem.-A Eur. J.* **2015**, *21*, 7520–7531. [CrossRef]
19. Nejman, P.S.; Curzon, T.E.; Bühl, M.; McKay, D.; Woollins, J.D.; Ashbrook, S.E.; Cordes, D.B.; Slawin, A.M.Z.; Kilian, P. Phosphorus–Bismuth Peri-Substituted Acenaphthenes: A Synthetic, Structural, and Computational Study. *Inorg. Chem.* **2020**, *59*, 5616–5625. [CrossRef]
20. Bergsch, J.U.; Slawin, A.M.Z.; Kilian, P.; Chalmers, B.A. Phosphine–Stibine and Phosphine–Stiborane peri-Substituted Donor–Acceptor Complexes. *Molbank* **2023**, *2023*, M1653. [CrossRef]

21. Hupf, E.; Lork, E.; Mebs, S.; Chęcińska, L.; Beckmann, J. Probing Donor–Acceptor Interactions in peri-Substituted Diphenylphosphinoacenaphthyl–Element Dichlorides of Group 13 and 15 Elements. *Organometallics* **2014**, *33*, 7247–7259. [CrossRef]
22. Chalmers, B.A.; Meigh, C.B.E.; Nejman, P.S.; Bühl, M.; Lébl, T.; Woollins, J.D.; Slawin, A.M.Z.; Kilian, P. Geminally Substituted Tris(acenaphthyl) and Bis(acenaphthyl) Arsines, Stibines, and Bismuthine: A Structural and Nuclear Magnetic Resonance Investigation. *Inorg. Chem.* **2016**, *55*, 7117–7125. [CrossRef] [PubMed]
23. Furan, S.; Hupf, E.; Boidol, J.; Brunig, J.; Lork, E.; Mebs, S.; Beckmann, J. Transition metal complexes of antimony centered ligands based upon acenaphthyl scaffolds. Coordination non-innocent or not? *Dalton Trans.* **2019**, *48*, 4504–4513. [CrossRef] [PubMed]
24. Batsanov, S.S. Van der Waals Radii of Elements. *Inorg. Mater.* **2001**, *37*, 871–885. [CrossRef]
25. Cordero, B.; Gomez, V.; Platero-Prats, A.E.; Reves, M.; Echeverria, J.; Cremades, E.; Barragan, F.; Alvarez, S. Covalent radii revisited. *Dalton Trans.* **2008**, 2832–2838. [CrossRef] [PubMed]
26. You, D.; Smith, J.E.; Sen, S.; Gabbaï, F.P. A Stiboranyl Platinum Triflate Complex as an Electrophilic Catalyst. *Organometallics* **2020**, *39*, 4169–4173. [CrossRef]
27. You, D.; Gabbaï, F.P. Unmasking the Catalytic Activity of a Platinum Complex with a Lewis Acidic, Non-innocent Antimony Ligand. *J. Am. Chem. Soc.* **2017**, *139*, 6843–6846. [CrossRef] [PubMed]
28. Yang, H.; Gabbaï, F.P. Solution and Solid-State Photoreductive Elimination of Chlorine by Irradiation of a [PtSb]VII Complex. *J. Am. Chem. Soc.* **2014**, *136*, 10866–10869. [CrossRef] [PubMed]
29. Piesch, M.; Gabbaï, F.P.; Scheer, M. Phosphino-Stibine Ligands for the Synthesis of Heterometallic Complexes. *Z. Anorg. Allg. Chem.* **2021**, *647*, 266–278. [CrossRef]
30. Bader, R.F.W. *Atoms in Molecules: A Quantum Theory*; Clarendon Press: Oxford, UK, 1994.
31. Bader, R.F. A quantum theory of molecular structure and its applications. *Chem. Rev.* **1991**, *91*, 893–928. [CrossRef]
32. Lepetit, C.; Fau, P.; Fajerwerg, K.; Kahn, M.L.; Silvi, B. Topological analysis of the metal-metal bond: A tutorial review. *Coord. Chem. Rev.* **2017**, *345*, 150–181. [CrossRef]
33. Gervasio, G.; Bianchi, R.; Marabello, D. About the topological classification of the metal–metal bond. *Chem. Phys. Lett.* **2004**, *387*, 481–484. [CrossRef]
34. Espinosa, E.; Alkorta, I.; Elguero, J.; Molins, E. From weak to strong interactions: A comprehensive analysis of the topological and energetic properties of the electron density distribution involving X–H···F–Y systems. *J. Chem. Phys.* **2002**, *117*, 5529–5542. [CrossRef]
35. Espinosa, E.; Molins, E.; Lecomte, C. Hydrogen bond strengths revealed by topological analyses of experimentally observed electron densities. *Chem. Phys. Lett.* **1998**, *285*, 170–173. [CrossRef]
36. Mitoraj, M.; Michalak, A. Natural orbitals for chemical valence as descriptors of chemical bonding in transition metal complexes. *J. Mol. Model.* **2007**, *13*, 347–355. [CrossRef] [PubMed]
37. Michalak, A.; Mitoraj, M.; Ziegler, T. Bond Orbitals from Chemical Valence Theory. *J. Phys. Chem. A* **2008**, *112*, 1933–1939. [CrossRef] [PubMed]
38. Mitoraj, M.P.; Michalak, A.; Ziegler, T. A Combined Charge and Energy Decomposition Scheme for Bond Analysis. *J. Chem. Theory Comput.* **2009**, *5*, 962–975. [CrossRef] [PubMed]
39. Mitoraj, M.P.; Michalak, A. σ-Donor and π-Acceptor Properties of Phosphorus Ligands: An Insight from the Natural Orbitals for Chemical Valence. *Inorg. Chem.* **2010**, *49*, 578–582. [CrossRef] [PubMed]
40. In the Orca 5.0.4 implementation of EDA-NOCV, the electrostatic and Pauli terms cannot be separated.
41. Armarego, W.L.F.; Chai, C.L.L. *Purification of Laboratory Chemicals*, 6th ed.; Elsevier: Burlington, MA, USA, 2009.
42. Chalmers, B.A.; Arachchige, K.S.A.; Prentis, J.K.D.; Knight, F.R.; Kilian, P.; Slawin, A.M.Z.; Woollins, J.D. Sterically Encumbered Tin and Phosphorus peri-Substituted Acenaphthenes. *Inorg. Chem.* **2014**, *53*, 8795–8808. [CrossRef]
43. Fuller, A.L.; Scott-Hayward, L.A.S.; Li, Y.; Bühl, M.; Slawin, A.M.Z.; Woollins, J.D. Automated Chemical Crystallography. *J. Am. Chem. Soc.* **2010**, *132*, 5799–5802. [CrossRef]
44. *CrystalClear-SM Expert*; v2.0 and 2.1; Rigaku Americas: The Woodlands, TX, USA, 2010; Rigaku Corporation: Tokyo, Japan, 2015.
45. *CrysAlisPro*; v1.171.42.49; Rigaku Oxford Diffraction, Rigaku Corporation: Oxford, UK, 2022.
46. Sheldrick, G.M. A short history of SHELX. *Acta Crystallogr. Sect. A Found. Crystallogr.* **2008**, *64*, 112–122. [CrossRef]
47. Beurskens, P.T.; Beurskens, G.; de Gelder, R.; Garcia-Granda, S.; Gould, R.O.; Israel, R.; Smits, J.M.M. *DIRDIF-99*; Crystallography Laboratory, University of Nijmegen: Nijmegen, The Netherlands, 1999.
48. Palatinus, L.; Chapuis, G. SUPERFLIP—A computer program for the solution of crystal structures by charge flipping in arbitrary dimensions. *J. Appl. Crystallogr.* **2007**, *40*, 786–790. [CrossRef]
49. Sheldrick, G.M. Crystal structure refinement with SHELXL. *Acta Crystallogr. Sect. C Struct. Chem.* **2015**, *71*, 3–8. [CrossRef]
50. *CrystalStructure*; v4.3.0; Rigaku Americas: The Woodlands, TX, USA; Rigaku Corporation: Tokyo, Japan, 2018.
51. Neese, F. Software update: The ORCA program system—Version 5.0. *WIREs Comput. Mol. Sci.* **2022**, *12*, e1606. [CrossRef]
52. Perdew, J.P.; Burke, K.; Ernzerhof, M. Generalized Gradient Approximation Made Simple. *Phys. Rev. Lett.* **1996**, *77*, 3865–3868. [CrossRef] [PubMed]
53. van Lenthe, E.; Snijders, J.G.; Baerends, E.J. The zero-order regular approximation for relativistic effects: The effect of spin–orbit coupling in closed shell molecules. *J. Chem. Phys.* **1996**, *105*, 6505–6516. [CrossRef]

54. Weigend, F.; Furche, F.; Ahlrichs, R. Gaussian basis sets of quadruple zeta valence quality for atoms H–Kr. *J. Chem. Phys.* **2003**, *119*, 12753–12762. [CrossRef]
55. Schäfer, A.; Horn, H.; Ahlrichs, R. Fully optimized contracted Gaussian basis sets for atoms Li to Kr. *J. Chem. Phys.* **1992**, *97*, 2571–2577. [CrossRef]
56. Schäfer, A.; Huber, C.; Ahlrichs, R. Fully optimized contracted Gaussian basis sets of triple zeta valence quality for atoms Li to Kr. *J. Chem. Phys.* **1994**, *100*, 5829–5835. [CrossRef]
57. Weigend, F.; Ahlrichs, R. Balanced basis sets of split valence, triple zeta valence and quadruple zeta valence quality for H to Rn: Design and assessment of accuracy. *Phys. Chem. Chem. Phys.* **2005**, *7*, 3297–3305. [CrossRef]
58. Pantazis, D.A.; Chen, X.-Y.; Landis, C.R.; Neese, F. All-Electron Scalar Relativistic Basis Sets for Third-Row Transition Metal Atoms. *J. Chem. Theory Comput.* **2008**, *4*, 908–919. [CrossRef] [PubMed]
59. Weigend, F. Accurate Coulomb-fitting basis sets for H to Rn. *Phys. Chem. Chem. Phys.* **2006**, *8*, 1057–1065. [CrossRef] [PubMed]
60. Pantazis, D.A.; Neese, F. All-Electron Scalar Relativistic Basis Sets for the Lanthanides. *J. Chem. Theory Comput.* **2009**, *5*, 2229–2238. [CrossRef] [PubMed]
61. Pantazis, D.A.; Neese, F. All-Electron Scalar Relativistic Basis Sets for the Actinides. *J. Chem. Theory Comput.* **2011**, *7*, 677–684. [CrossRef]
62. Pantazis, D.A.; Neese, F. All-electron scalar relativistic basis sets for the 6p elements. *Theor. Chem. Acc.* **2012**, *131*, 1292. [CrossRef]
63. Grimme, S.; Antony, J.; Ehrlich, S.; Krieg, H. A consistent and accurate ab initio parametrization of density functional dispersion correction (DFT-D) for the 94 elements H-Pu. *J. Chem. Phys.* **2010**, *132*, 154104. [CrossRef]
64. Goerigk, L.; Grimme, S. A thorough benchmark of density functional methods for general main group thermochemistry, kinetics, and noncovalent interactions. *Phys. Chem. Chem. Phys.* **2011**, *13*, 6670–6688. [CrossRef]
65. Lu, T.; Chen, F. Multiwfn: A multifunctional wavefunction analyzer. *J. Comput. Chem.* **2012**, *33*, 580–592. [CrossRef] [PubMed]
66. Adamo, C.; Barone, V. Toward reliable density functional methods without adjustable parameters: The PBE0 model. *J. Chem. Phys.* **1999**, *110*, 6158–6170. [CrossRef]
67. Hanwell, M.D.; Curtis, D.E.; Lonie, D.C.; Vandermeersch, T.; Zurek, E.; Hutchison, G.R. Avogadro: An advanced semantic chemical editor, visualization, and analysis platform. *J. Cheminform.* **2012**, *4*, 17. [CrossRef]
68. Chalmers, B.A.; Nejman, P.S.; Llewellyn, A.V.; Felaar, A.M.; Griffiths, B.L.; Portman, E.I.; Gordon, E.L.; Fan, K.J.H.; Woollins, J.D.; Buhl, M.; et al. A Study of Through-Space and Through-Bond J(PP) Coupling in a Rigid Nonsymmetrical Bis(phosphine) and Its Metal Complexes. *Inorg. Chem.* **2018**, *57*, 3387–3398. [CrossRef] [PubMed]
69. Malkina, O.L.; Hierso, J.-C.; Malkin, V.G. Distinguishing "Through-Space" from "Through-Bonds" Contribution in Indirect Nuclear Spin–Spin Coupling: General Approaches Applied to Complex JPP and JPSe Scalar Couplings. *J. Am. Chem. Soc.* **2022**, *144*, 10768–10784. [CrossRef] [PubMed]
70. Taylor, L.J.; Lawson, E.E.; Cordes, D.B.; Athukorala Arachchige, K.S.; Slawin, A.M.Z.; Chalmers, B.A.; Kilian, P. *Synthesis and Structural Studies of Peri-Substituted Acenaphthenes with Tertiary Phosphine and Stibine Groups Dataset*; University of St Andrews Research Portal: St Andrews, UK, 2024. [CrossRef]

Disclaimer/Publisher's Note: The statements, opinions and data contained in all publications are solely those of the individual author(s) and contributor(s) and not of MDPI and/or the editor(s). MDPI and/or the editor(s) disclaim responsibility for any injury to people or property resulting from any ideas, methods, instructions or products referred to in the content.

Article

Synthesis and Characterization of Novel Cobalt Carbonyl Phosphorus and Arsenic Clusters

Mehdi Elsayed Moussa [†], Susanne Bauer [†], Christian Graßl, Christoph Riesinger, Gábor Balázs and Manfred Scheer *

Department of Inorganic Chemistry, University of Regensburg, 93040 Regensburg, Germany; mehdi.elsayed-mousa@chemie.uni-regensburg.de (M.E.M.); susanne.bauer@chemie.uni-regensburg.de (S.B.); christian.grassl@chemie.uni-regensburg.de (C.G.); christoph.riesinger@chemie.uni-regensburg.de (C.R.); gabor.balazs@chemie.uni-regensburg.de (G.B.)
* Correspondence: manfred.scheer@ur.de; Tel.: +49-(0)941-943-4440; Fax: +49-(0)941-943-4439
[†] These authors contributed equally to this work.

Abstract: Phosphorus- and arsenic-containing cobalt clusters are an interesting class of compounds that continue to provide new structures with captivating bonding patterns. Although the first members of this family were reported 45 years ago, the number of such species is still limited within the broad family of transition metal complexes bearing pnictogen atoms. Herein, we present the reaction of $Co_2(CO)_8$ as a cobalt source with a number of phosphorus- and arsenic-containing compounds under variable reaction conditions. These reactions result in various known and novel cobalt phosphorus and cobalt arsenic clusters in which different nuclearity ratios between P/As and Co exist. All those clusters were characterized by X-ray structural analysis and partly by IR, $^{31}P\{^1H\}$ NMR, EI-MS and elemental analysis. This comprehensive study is the first detailed study in this field that reveals the richness of compounds that could be obtained only by modifying the ratio of used reactants and the involved reaction conditions.

Keywords: phosphorus; arsenic; cobalt clusters; carbonyl; interstitial

1. Introduction

Transition metal complexes that incorporate group 15 elements have attracted increasing interest in the past three decades mainly due to their unprecedented structures [1] and flexible coordination behaviors, which render them versatile building blocks in supramolecular chemistry [1–4]. Within this field, cobalt clusters containing phosphorus and arsenic atoms have emerged as active materials for catalysis [5–7], magnetism [8], and as potential precursors for CoP nanoparticles [9]. In 1969, the Dahl group reported the first examples in this field, which included the tetrahedral arsenic-cobalt carbonyl clusters $As_3\{Co(CO)_3\}$ and $As_2\{Co(CO)_3\}_2$ obtained from the reaction of $Co_2(CO)_8$ with $[AsCH_3]_5$ and $AsCl_3$, respectively [10,11]. Later on, the complex $[\mu_4\text{-}AsCo_3(CO)_9]_3$ [12] was also isolated, which demonstrated the interchangeable roles of As and $Co(CO)_3$ units. Meanwhile, Markó et al. isolated the analogous phosphorus derivative $Co_2(CO)_6P_2$ from the reaction of $Na[Co(CO)_4]$ with PX_3 (X = Cl, Br) [13]. Additionally, the tetrahedral compounds $P_3Co(CO)_3$ and $PCo_3(CO)_9$ were isolated by the Orosz group from the reaction of $Co_2(CO)_8$ with white phosphorus (P_4) and PI_3, respectively [14]. The groups of Seyferth and Nixon isolated the tetrahedral phosphorus cobalt clusters $[(RCP)Co_2(CO)_6)]$ (R = CH_3, tBu, Ph, $SiMe_3$) from the reaction of $Co_2(CO)_8$ with $RCCl_2PCl_2$ [15] and tBuCP [16]. More recently, a number of more complex anionic P- [8,17] and As-containing cobalt clusters [18,19] with higher Co nuclearities were synthesized from the reaction of $Na[Co(CO)_4]$ with variable P- and As-starting materials. Our group contributed to this field by developing new strategies for the synthesis of cobalt clusters incorporating pnictogen atoms. In one approach, the formation of large P_n species was attainable from the reaction of P_4 with the triple-decker complex $[(Cp''' Co)_2(\eta^4:\eta^4-C_7H_8)]$ ($Cp''' = 1,2,4-C_5H_2(tBu)_3$), which dissociates in solution to give 14 VE (valence electron) $Cp'''Co$ moieties [20]. In that respect, a controlled synthesis of P_4, P_8, P_{12}, P_{16} and even P_{24}-containing cobalt complexes proved achievable. This approach was more recently extended to the reaction with As_4 allowing the synthesis of As_4, As_{10} and As_{12} cobalt clusters [21]. In another approach, the sandwich complex $[Cp'''Co(\eta^4-P_4)]$ was found to dimerize in solution forming P_8-containing cobalt complexes [22]. In addition to the aforementioned compounds, only a few other clusters of this type are known and are thus still relatively limited within the broad family of transition metal compounds containing phosphorus atoms [1,23]. Therefore, further investigations to enrich the library of this family of compounds with new candidates, as well as to understand reaction conditions that allow us to obtain them selectively, are still of current interest. Herein, we present the reaction of $Co(CO)_8$ with the phosphorus and arsenic sources: "$[Cr(CO)_{4+n}(PH_3)_{2-n}]$ ($n = 0$, 1), $[Fe(CO)_4(PH_3)]$, $P(SiMe_3)_3$, P_4, As_4, $[W(CO)_5(AsH_3)]$ and $As(SiMe_3)_3$". These reactions afforded a variety of known as well as novel cobalt carbonyl clusters incorporating P and As atoms. Interestingly, most of the known compounds are obtained in much better yields via our novel synthetic strategies presented herein, thus allowing the completion of their characterizations, including X-ray structure analysis.

2. Results and Discussion

*2.1. Synthesis and X-ray Structures of Compound **1** and the Cobalt Clusters $[\{M(CO)_n\}\{Co(CO)_3\}_3E]$ ($n = 5$, $M = Cr$, $E = P$ (**2**), $M = W$, $E = As$ (**3**); $n = 4$, $M = Fe$, $E = P$ (**4**))*

$[W(CO)_5(AsH_3)]$ (**1**) is synthesized in good yields from the reaction of $[W(CO)_6]$ with $As(SiMe_3)_3$ under UV irradiation with subsequent methanolysis of the reaction mixture. This compound was already reported by Fischer et al. [24], who used, however, AsH_3 as an arsenic source. Herein, we completed the analytical data by $^{13}C\{^1H\}$ NMR as well as single crystal structure analysis (for further information see ESI). The reaction of $[Cr(CO)_5(PH_3)]$, $[W(CO)_5(AsH_3)]$ (**1**) or $[Fe(CO)_4(PH_3)]$ with one equivalent of $[Co_2(CO)_8]$ in toluene at room temperature allowed for the synthesis of cobalt clusters with the general formula $[\{M(CO)_n\}\{Co(CO)_3\}_3E]$ ($n = 5$, $M = Cr$, $E = P$ (**2**), $M = W$, $E = As$ (**3**); $n = 4$, $M = Fe$, $E = P$ (**4**)) (Scheme 1, Equations (1) and (2)). Vahrenkamp et al. synthesized **2** using similar starting materials but in benzene instead of toluene [25]. They obtained compound **2** in 22% yield while the yield could be improved to 44% under our reaction conditions. Additionally, **2** was only characterized by IR spectroscopy and elemental analysis. Herein, we completed its analytical data (solution and solid-state IR; ^1H, $^{13}C\{^1H\}$ and $^{31}P\{^1H\}$ NMR; EI-MS; EA) and characterized it by X-ray structural analysis. In the same study of Vahrenkamp, the P-analog of cluster **3** was synthesized from $[W(CO)_5(PH_3)]$ and $[Co_2(CO)_8]$. Vizi-Orosz reported cluster **4** by the reaction of $[Fe_2(CO)_9]$ with $[Co_3P(CO)_9]$ but in very low yields (ca. 1%) [14] and characterized it only by EA. In our case, it was possible to prepare **4** in a different way (Equation (2)), resulting in improved yields (24%) and obtaining complete analytic data. In the $^{31}P\{^1H\}$ NMR spectrum of **2**, a singlet at 697.9 ppm ($\omega_{1/2} = 113$ Hz) was detected while a broad signal at 677.5 ppm ($\omega_{1/2} = 175$ Hz) was observed in that of **4**. The $^{13}C\{^1H\}$ NMR spectra of **2** and **3** show typical broad signals at 197.5 ppm ($\omega_{1/2} = 43$ Hz) (**2**) and 197.0 ppm ($\omega_{1/2} = 35$ Hz) (**3**) for the carbonyl carbon atoms, which belong to the Co_3E tetrahedra (E = P (**2**), As (**3**)). Additionally, a doublet at 214.7 ppm with a C–P coupling constant (13.7 Hz) (**2**) and a singlet at 195.4 ppm (**3**) were observed. In the $^{13}C\{^1H\}$ NMR spectrum of **4**, a doublet at 212.0 ppm was detected with a C–P coupling constant of 18.8 Hz belonging to the cis carbonyl ligands. The signal for the trans carbonyl ligand could not be observed, while a broad signal at 197.0 ppm ($\omega_{1/2} = 46$ Hz) for the carbonyl carbon atoms at the Co_3P tetrahedron was detected. The broad signals in the multinuclear NMR spectra of compounds **2**–**4** originate from the coupling with the ^{59}Co nucleus with a spin of 7/2 and 100% abundance. The EI mass spectra of **3**–**5** exhibit the molecular ion peak as well as peaks showing the successive loss of all carbonyl ligands. Furthermore, the loss of a chromium atom as well as a cobalt atom (**2**), two cobalt atoms (**3**) and one iron atom (**4**) were detected (for further information see Materials and Methods section and ESI).

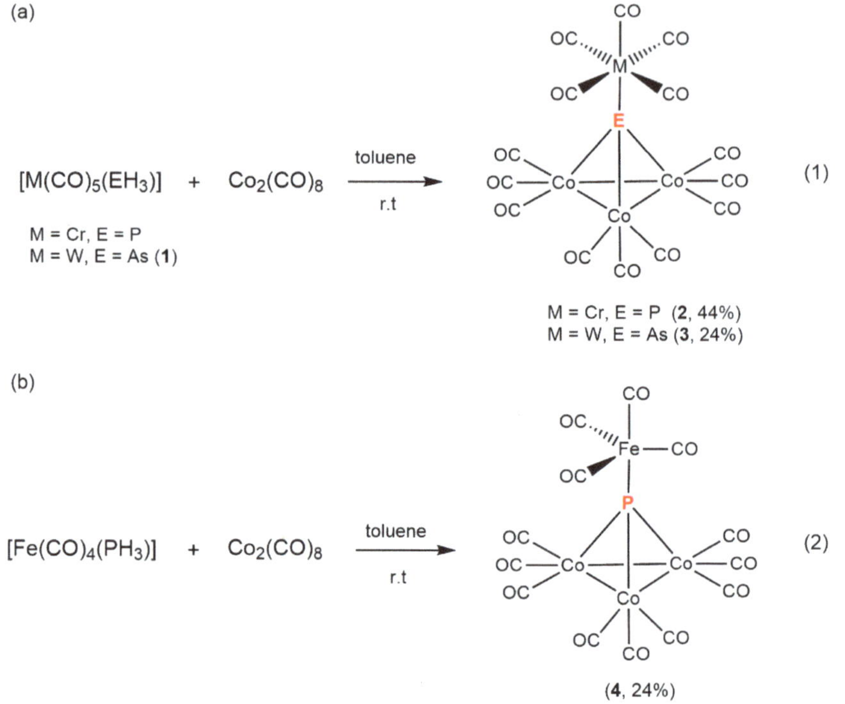

(a)

[M(CO)$_5$(EH$_3$)] + Co$_2$(CO)$_8$ $\xrightarrow{\text{toluene}}_{\text{r.t}}$ [spiked tetrahedral Co$_3$E–M(CO)$_5$ cluster] (1)

M = Cr, E = P
M = W, E = As (**1**)

M = Cr, E = P (**2**, 44%)
M = W, E = As (**3**, 24%)

(b)

[Fe(CO)$_4$(PH$_3$)] + Co$_2$(CO)$_8$ $\xrightarrow{\text{toluene}}_{\text{r.t}}$ [spiked tetrahedral Co$_3$P–Fe(CO)$_4$ cluster] (2)

(**4**, 24%)

Scheme 1. Synthesis of the tetrahedral Co clusters (**a**) **2**–**3** (Equation (1)) and (**b**) **4** (Equation (2)) from the reaction of Co$_2$(CO)$_8$ with [Cr(CO)$_5$(PH$_3$)], [W(CO)$_5$(AsH$_3$)] and [Fe(CO)$_4$(PH$_3$)], respectively, in toluene at room temperature.

Compound **1** is obtained through sublimation as colorless crystals. It crystallizes in the monoclinic space group $P2_1/c$. The central tungsten atom possesses an octahedral coordination sphere with five CO ligands and one As atom. Crystals of compounds **2**–**4** were obtained from concentrated n-hexane solutions stored at −25 °C. They crystallize in the monoclinic space groups Cc (**2**) and $P2_1/n$ (**4**) and the triclinic space group $P\bar{1}$ (**3**). Their molecular structures reveal spiked tetrahedral molecules in which the central structural motifs "Co$_3$P (**2**, **4**) or a Co$_3$As (**3**)" are slightly distorted tetrahedranes that coordinate to the corresponding transition metal carbonyl fragment ([Cr(CO)$_5$] (**2**), [W(CO)$_5$] (**3**), [Fe(CO)$_4$] (**4**)) via the lone pairs of the pnictogen atoms (Figure 1). The Co–Co distances in **2**–**4** (2.536(1)–2.569(1) Å) are in the range of Co–Co single bonds reported, e.g., for [CpCo(µ-PPh)]$_2$ (2.56 Å) [26]. Accordingly, the tetrahedral cores of **2**–**4** each possess 48 cluster valence electrons (CVE) and can be described as closo compounds according to the Wade–Mingos rules.

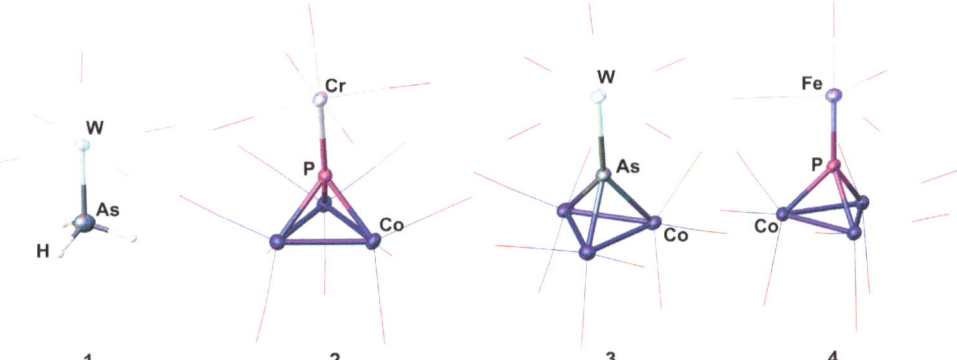

Figure 1. Molecular structure of compound **1** and the tetrahedral Co clusters **2–4** in the solid state (anisotropic displacement parameters (ADPs) are given at the 50% probability level). CO ligands are depicted in the wireframe model for clarity.

2.2. Synthesis and X-ray Structures of the Cobalt Clusters $[Co_8(CO)_{16}(\mu\text{-}CO)_4P]$ (5) and $[\{Co_4(CO)_{11}\}\{Co(CO)_3\}_3P]$ (6)

By using $[Cr(CO)_4(PH_3)_2]$ as a phosphorus source in the reaction with four equivalents of $[Co_2(CO)_8]$ in toluene at room temperature, a new cobalt carbonyl cluster with an interstitial phosphorus atom $[Co_8(CO)_{16}(\mu\text{-}CO)_4P]$ (**5**) is obtained in 6% yield (Scheme 2, Equation (3)). In addition to compound **5** and $[Co_4(CO)_{12}]$, another cobalt phosphorus cluster $[Co_8(CO)_{18}(\mu\text{-}CO)(P)_2]$ (**A**) is formed, previously obtained by our group from the reaction of $[Co_2(CO)_8]$ with $[W(CO)_4(PH_3)_2]$ [27]. Another way to synthesize **5** in a slightly better yield (9%) is the reaction of P_4 with an excess of $[Co_2(CO)_8]$ (1:32 stoichiometry, n-hexane, Equation (4)). However, when the stoichiometry of P_4 and $[Co_2(CO)_8]$ is changed to 1:24 under slightly different reaction conditions, the cluster $[\{Co_4(CO)_{11}\}\{Co(CO)_3\}_3P]$ (**6**) is formed instead of **5** (Scheme 2, Equation (5)).

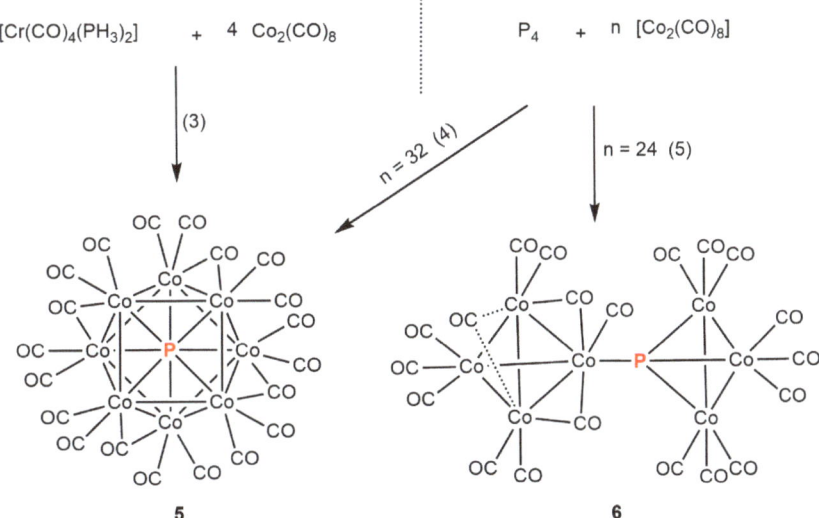

Scheme 2. Synthesis of compound **5** from the reaction of $Co_2(CO)_8$ with $Cr(CO)_4(PH_3)_2]$ in toluene at room temperature (Equation (3)) and with P_4 in hexane at −40 °C (Equation (4)). Synthesis of compound **6** from a similar reaction using 24 eq. of $Co_2(CO)_8$ in toluene at −100 °C (Equation (5)).

Single crystals of **5** were obtained from a dichloromethane solution or an n-hexane solution stored at −25 °C. It crystallizes in the monoclinic space group $P2_1/n$. The central structural feature of **5** can be described as a quadratic antiprism formed by eight Co atoms and an interstitial P atom at the center with a total of 117 CVEs (Figure 2). Each Co atom in **5** is bound to four other Co atoms and to the central P atom with angles between the various Co—Co and Co–P bonds being close to 60 and 90°. Additionally, each Co atom is coordinated by two terminal and one bridging CO ligand. Accordingly, four bridging and sixteen terminal CO ligands exist in **5**. The overall average Co–CO bond distances for bridging and terminal CO ligands are 1.961(8) and 1.810(8) Å, respectively. Similarly, two classes of Co–Co bonds can be distinguished: (a) shorter Co–Co bonds between Co atoms connected via bridging CO ligands range between 2.592(2) and 2.613(2) Å, and (b) longer Co–Co bonds between neighboring Co atoms with no bridging CO ligands ranging between 2.651(1) and 2.787(2) Å. The Co–P lengths are between 2.216(2) and 2.245(2) Å. DFT calculations at the r^{2}SCAN-3c level reproduce the experimental geometry determined by X-ray diffractions well in both doublet and quartet sextet spin states. The spin states do not have a considerable influence on the geometry, which suggests that the spin density is located in a nonbonding Co orbital. Indeed, the spin density in the doublet spin state is evenly distributed on four cobalt atoms, with only small contributions from the other atoms (Figure S12). Energetically, the doublet spin state is 46 kJ·mol^{-1} more stable than the quartet spin state. This is not unexpected since the strong field ligands, for instance, CO, prefer the low spin configurations on the metal centers.

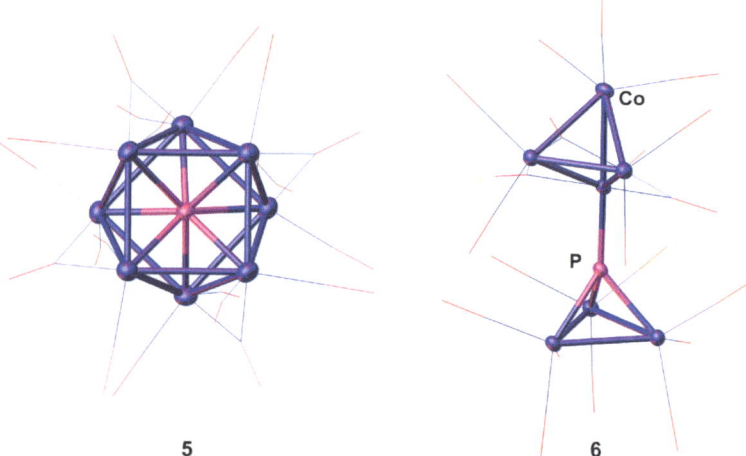

Figure 2. Molecular structure of the cluster compounds **5** and **6** in the solid state (atoms are given at 50% probability level). CO ligands are depicted in the wireframe model.

Compound **6** was obtained as black blocks from a concentrated toluene solution at −25 °C and crystallized in the triclinic space group $P\bar{1}$. Cluster **6** is composed of two slightly distorted tetrahedra, [P{Co(CO)$_3$}$_3$] and [Co$_4$(CO)$_{11}$], which are linked together via the coordination of the lone pair of the phosphorus atom in the former to a cobalt atom in the latter (Figure 2). Interestingly, the [P{Co(CO)$_3$}$_3$] tetrahedron contains only terminal CO ligands (three per cobalt atom) while the [Co$_4$(CO)$_{11}$] fragment contains eight terminal and three bridging CO ligands. The [P{Co(CO)$_3$}$_3$] tetrahedron in **6** shows a structural motif similar to that of compound [{W(CO)$_5$}{Co(CO)$_3$}$_3$P] reported by Vahrenkamp et al. [25], and the structure of compound **6** as a whole is comparable to that of the clusters [{Co$_3$(CO)$_9$}(μ_4-P){Co$_3$(μ-CO)$_3$(CO)$_5$}(μ_3-CR)] (R = CH$_3$, *t*Bu, *i*Pr) recently reported by our group from the reaction of the phopsphaalkynes RCP (R = CH$_3$, *t*Bu, *i*Pr) with Co$_2$(CO)$_8$ [28]. The Co–Co (2.548(6)–2.560(6) Å) bond distances in the [P{Co(CO)$_3$}$_3$] moiety are longer than those reported for [{W(CO)$_5$}{Co(CO)$_3$}$_3$P] and comparable to those found in clusters **2–4**. The Co–Co (2.444(6)–2.529(6) Å) bond lengths in the [Co$_4$(CO)$_{11}$] fragment are, as expected, shorter due to the presence of bridging CO ligands. The Co–P bond lengths (2.171(9)–2.183(9) Å) in the [P{Co(CO)$_3$}$_3$] fragment are slightly shorter than those found in **2** (2.177(3)–2.191(3) Å) and longer than those found in **4** (2.162(2)–2.169(2) Å). The Co atoms in **6** are bound through metal–metal bonds and fulfill the 18-valence electron rule. Thus, the {Co(CO)$_3$}$_3$P part in **6** can be regarded as a closo cluster with 48 CVE with the lone pair at P engaging in dative bonding to the [Co$_4$(CO)$_{11}$] tetrahedron, which is best described as a nido cluster (tetrahedron derived from trigonal bipyramid) with 60 CVE (4 × 9 (Co) + 11 × 2 (CO) + 2 (P{Co(CO)$_3$}$_3$)).

*2.3. Synthesis and X-ray Structures of the Cobalt Clusters [Co$_9$(CO)$_{24}$(μ_4-P)$_3$] (**7**) and [Co$_9$(CO)$_{21}$(μ_5-P)$_3$] (**8**)*

Subsequently, we focused on the reaction of [Co$_2$(CO)$_8$] with P$_4$ using various ratios and reaction conditions. Reactions using eight equivalents of [Co$_2$(CO)$_8$] in toluene (Scheme 3, Equation (6)) and six equivalents of [Co$_2$(CO)$_8$] in hexane (Scheme 3, Equation (7)) at room temperature resulted in the compounds [Co$_9$(CO)$_{24}$(μ_4-P)$_3$] (**7**) and [Co$_9$(CO)$_{21}$(μ_5-P)$_3$] (**8**), respectively. Both compounds consist of three [Co$_3$P(CO)$_9$] fragments that have lost three (**7**) or six (**8**) CO ligands with subsequent formation of new Co–Co and Co–P bonds. Markó et al. [29] isolated a compound with a molecular formula similar to **7** and proposed a cyclic structure as in **7** based on a similar reaction protocol used to obtain the analogous cyclic As–Co trimer. Herein, the formation of compound **7** is evidenced by X-ray crystallography. As for compound **8**, however, the rather poor crystal quality only allowed for the collection of an incomplete data set proving the structure of **8**, but not for structural analysis in detail.

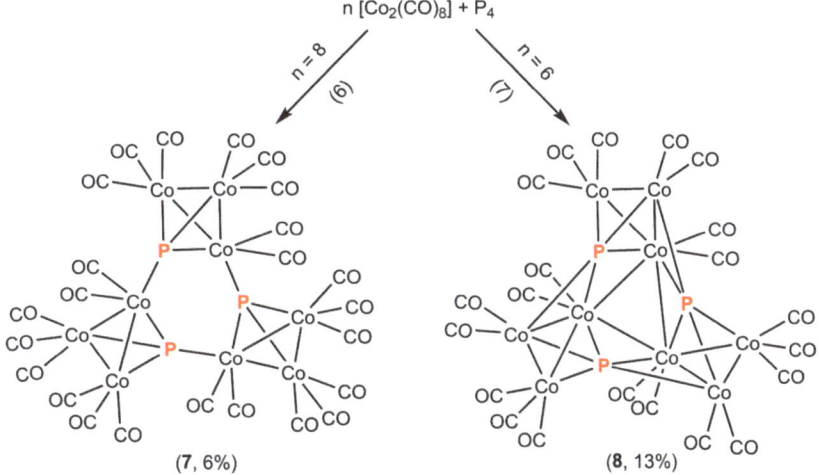

Scheme 3. Synthesis of the cyclic clusters **7** and **8** from the reaction of P$_4$ with eight eq. of Co$_2$(CO)$_8$ in toluene at room temperature (Equation (6)) and six eq. of Co$_2$(CO)$_8$ in hexane at room temperature (Equation (7)), respectively.

Compounds **7** and **8** are isolated as dark violet (**7**) and black block-shaped (**8**) crystals from n-hexane solutions stored at −25 °C and 8 °C, respectively. Cluster **7** crystallizes in the monoclinic space group $P2_1/n$ while **8** crystallizes in the orthorhombic space group $P2_12_12_1$. Compound **7** consists of three [PCo$_3$(CO)$_8$] tetrahedranes that are connected together via dative Co–P bonds, thus forming a cyclic trimer with a six-membered P$_3$Co$_3$ central ring (Figure 3). Compound **8** consists of three [PCo$_3$(CO)$_7$] units connected to one another by two Co–P bonds and one Co–Co bond. Theoretically, cluster **8** could have been formed from **7** by CO elimination, which could, however, not have been proven experimentally due to the insufficient solubility of **7** in common organic solvents. The Co–Co (2.530(8)–2.584(8) Å) bond lengths in **7** are generally comparable to those of **2**–**4** and the Co–P (2.153(1)–2.200(1) Å) bond lengths are comparable to those of **2**. All CO ligands in **7** and **8** are terminal ones with Co–CO lengths ranging between 1.775(7) and 1.898(1) Å, e.g., for **7**. Overall, compound **7** amounts to 144 CVEs from three [PCo$_3$(CO)$_8$] closo tetrahedra each possessing 48 CVEs. In compound **8**, the overall CVEs amount to 138 (9 × 9 (Co) + 21 × 2 (CO) + 3 × 5 (P)).

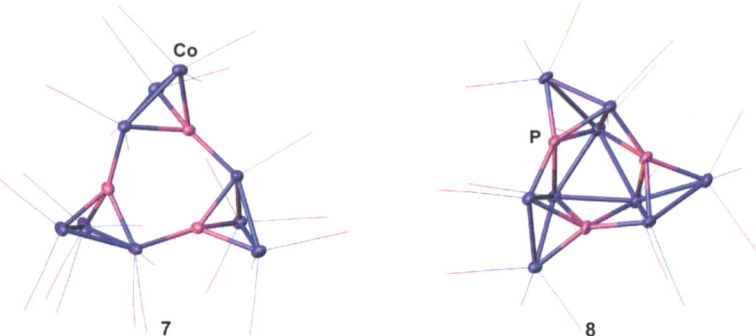

Figure 3. Molecular structures of the cluster compounds **7** and **8** in the solid state (ADPs are given at the 50% probability level). CO ligands are depicted in the wireframe model for clarity.

*2.4. Synthesis and X-ray Structures of the Cobalt Clusters [Co$_{10}$(CO)$_{24}$(μ_3-P)$_2$(μ_6-P$_2$)(μ-CO)$_2$] (**9**) and [Co$_{15}$(μ_6-P)$_6$(μ_{12}-Co)(CO)$_{30}$] (**10**)*

When a solution of five equivalents of [Co$_2$(CO)$_8$] in toluene is layered with a solution of one equivalent P$_4$ in n-hexane, black crystals of [Co$_{10}$(CO)$_{24}$(μ_3-P)$_2$(μ_5-P)$_2$(μ-CO)$_2$] (**9**) are obtained in excellent yields of 75% (Scheme 4, Equation (8)). When the components are instead employed in stirring reactions, using two equivalents of [Co$_2$(CO)$_8$] with one equivalent of P$_4$ in toluene at room temperature, cluster **9** is also obtained but in very low yields (1%). Besides **9**, another cluster [Co$_{15}$(μ_6-P)$_6$(μ_{12}-Co)(CO)$_{30}$] (**10**) can be isolated from the same reaction mixture after a long crystallization time in very low yields of 1% (Scheme 4, Equation (9)). Interestingly, compound **9** can be further refluxed in toluene to give the cluster [Co$_8$(CO)$_{18}$(μ-CO)(P)$_2$] (**A**) [27] in good yields (22%) as a thermally stable cluster via a new synthetic pathway (Scheme 4, Equation (10)).

Compound **9** is isolated as black plates and crystallizes in the monoclinic space group $P2_1/c$. Also, black rods of **9**·C$_6$H$_{14}$ can be obtained from a concentrated n-hexane solution crystallizing in the tetragonal space group $I4_1/a$. The cluster core in **9** is constructed of ten Co and four P atoms with 24 terminal and two bridging CO ligands being coordinated to the Co atoms (Figure 4). It can also be described as two Co$_5$P fragments that are connected together by a P$_2$ moiety located at its inversion center with a P–P bond length of 2.265(1) Å, which is slightly longer but still at the upper limit of a single P–P bond (2.212(2) Å) [30]. Each Co atom in **9** fulfills the 18-valence electron rule with a total of 162 CVEs (10 × 9 (Co) + 26 × 2 (CO) + 4 × 5 (P)).

Crystals of **10**•2C$_7$H$_8$ crystallize in the monoclinic space group $C2/c$. This cluster is composed of 16 Co and 6 P atoms (Figure 4). Besides the central cobalt atom, which is located in the middle of the cluster (Co4), every Co atom is coordinated by one, two or three terminal CO ligands from a total of thirty CO ligands present in **10**. Each P atom in **10** is surrounded by six cobalt atoms, revealing the coordination number six. A simplification of the structural details of **10** is depicted in Figure 5. The core of this compound is composed of three distorted Co$_4$P$_2$ octahedra with Co4 being the center of their intersection (one of them for example is composed of these atoms: Co4, Co5′, Co8′, P2′, P1′; Co7′, Figure 5 left). This central octahedral structural motif is similar to the central structural motif in the reported clusters **A** and [Co$_{10}$(CO)$_{18}$(μ-CO)$_6$P$_2$] (**B**) [27]. When considering the Co–Co bonds between the atoms Co3 and Co5, Co3′ and Co5′ and Co8 and Co8′, another central structural motif becomes apparent. This motif forms a hexagonal antiprism, consisting of six cobalt and six phosphorus atoms with an interstitial cobalt atom (Figure 5 middle). Finally, the addition of the rest of the Co atoms completes the molecular structure of **10** (Figure 5 right). Within **10**, various Co$_3$P tetrahedra are found (e.g., Co7, Co8, Co9, P1) with Co–Co bond lengths ranging between 2.5029(8) and 2.6735(8) Å and, therefore, lie in the range of those discussed for clusters **2**–**9**. For each Co atom in **10**, the 18-valence electron rule is fulfilled with a total of 234 CVEs for the core (16 × 9 (Co) + 30 × 2 (CO) + 6 × 5 (P)).

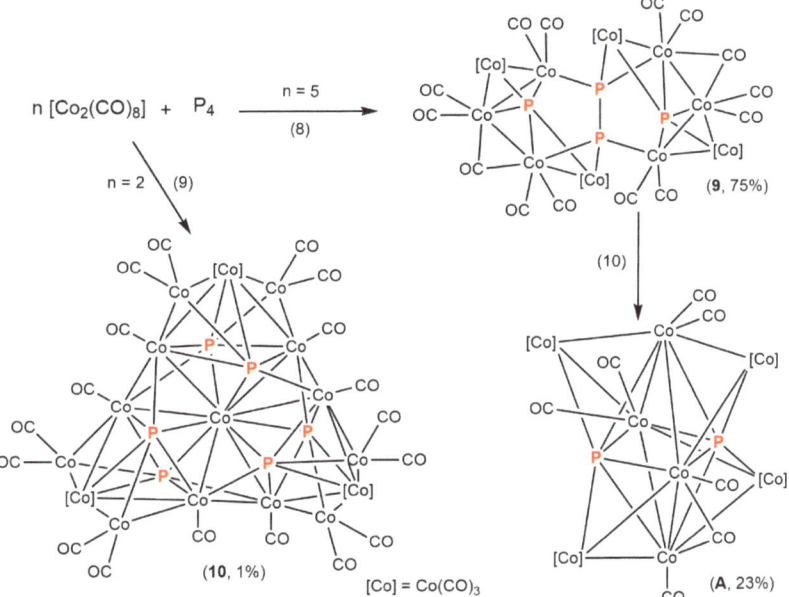

Scheme 4. Synthesis of compounds **9** and **10** from the reaction of P$_4$ with five eq. of Co$_2$(CO)$_8$ from toluene/n-hexane layering at room temperature (Equation (8)) and two eq. of Co$_2$(CO)$_8$ in toluene at room temperature (Equation (9)), respectively. Thermal transformation of **9** to cluster **A** in boiling toluene (Equation (10)).

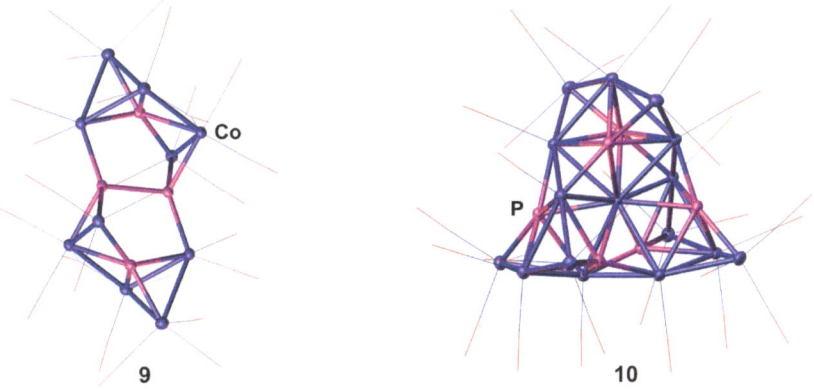

Figure 4. Molecular structure of the cluster compounds **9** and **10** in the solid state (ADPs are given at the 50% probability level). CO ligands are depicted in the wireframe model for clarity.

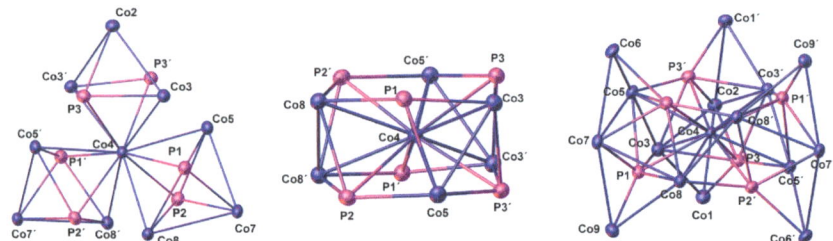

Figure 5. Molecular structure details of the core of cluster **10** without CO ligands. (**Left**): central distorted Co$_4$P$_2$ octahedra; (**middle**): central hexagonal antiprism-like structural motif; (**right**): whole core structure.

2.5. New Synthetic Protocol for the Synthesis of the Cobalt Clusters [Co$_9$(CO)$_{24}$(μ_4-As)$_3$] (11) and [{Co$_4$(CO)$_{11}$}{Co(CO)$_3$}$_3$As] (12)

Finally, [Co$_2$(CO)$_8$] was reacted with the arsenic sources yellow arsenic As$_4$ (Scheme 5, Equation (11)) and As(SiMe$_3$)$_3$ (Scheme 5, Equation (12)), respectively. The first reaction afforded the cluster [Co$_9$(CO)$_{24}$(μ_4-As)$_3$] (**11**) good yields (39%). Markó et al. [29] and the groups of Mackay and Nicholson [12] reported the synthesis of **11** in 8% and 17% yields, respectively. The former described it as a "cyclic trimer" of the trigonal pyramidal cluster [AsCo$_3$(CO)$_9$]. This compound was obtained together with AsCo$_3$(CO)$_9$ and As$_2$Co$_2$(CO)$_6$ from the reaction of Na[Co(CO)$_4$] with AsCl$_3$. Using our new strategy, **11** was obtained in better yields, and its full analytical characterization was performed. From the reaction with As(SiMe$_3$)$_3$, besides product **11**, it was possible to isolate the cobalt arsenic cluster [{Co$_4$(CO)$_{11}$}{Co(CO)$_3$}$_3$As] (**12**) in low yields (1%), which had already been reported by Huttner et al. from the reaction of [Co$_2$(CO)$_8$] with [Cr(CO)$_5$(AsPhH$_2$)] [31]. Interestingly, we were able to obtain compounds **11** and **12** as single crystals but in polymorphs differing from those of the initially reported ones.

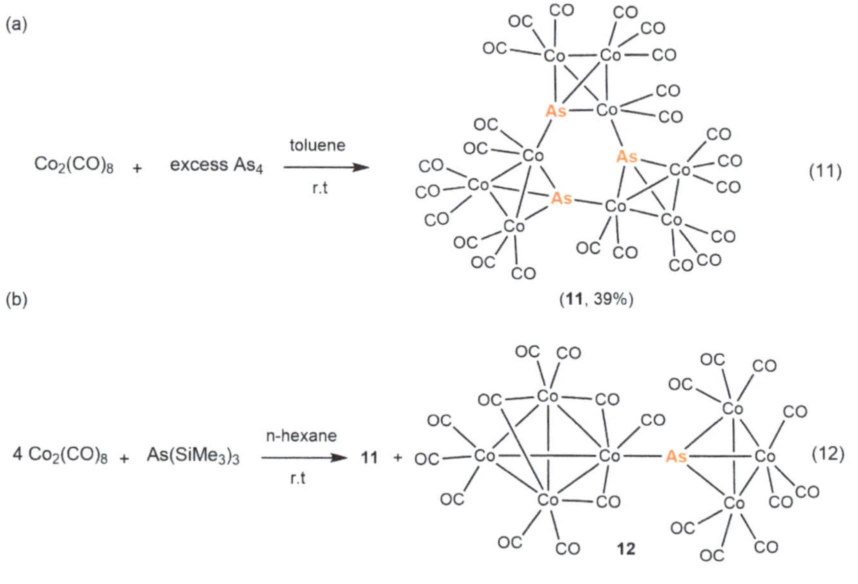

Scheme 5. Synthesis of the cyclic clusters (**a**) **11** from the reaction of Co$_2$(CO)$_8$ with As$_4$ in toluene at room temperature (Equation (11)), and (**b**) **12** from the reaction of Co$_2$(CO)$_8$ with As(SiMe$_3$)$_3$ in n-hexane at room temperature (Equation (12)).

Compounds **6, 7, 8, 11** and **12** are soluble in toluene and THF while compounds **5, 9** and **10** are nearly insoluble in common organic solvents. In the EI mass spectra of **6, 7, 8, 11** and **12**, the molecular ion peak as well as the peaks showing the successive loss of all carbonyl ligands are detected (for further information see Section 3 and ESI). Thus, the fragments at m/z = 443.6 for **6** and 487.6 for **12** can be observed, which shows the remaining cluster cores "Co$_7$P" or "Co$_7$As". For compounds **7, 8** and **11**, peaks contributing to the cluster cores "Co$_9$P$_3$" at m/z = 623.4 or "Co$_9$As$_3$" at m/z = 755.2 are detected. The ^{31}P{^{1}H} NMR spectra of **6, 7**, and **8** show broad signals at 666.4 ($\omega_{1/2}$ = 114 Hz), 667.3 ($\omega_{1/2}$ = 362 Hz) or 658.2 ppm ($\omega_{1/2}$ = 205 Hz). In the ^{31}P{^{1}H} MAS NMR spectrum of **9**, a broad signal at 684 ppm ($\omega_{1/2}$ = 2800 Hz) can be detected. Due to the insolubility of some compounds and the high sensitivity of most of the clusters, it was not possible to characterize all of them completely. As the cluster cores of compounds **2–12** are mainly surrounded by CO ligands that can only be readily released according to the mass spectra, clusters **2–12** could be used as potential single source precursors for the synthesis of Co$_x$P$_y$ and/or Co$_x$As$_y$ nanoparticles with variable ratios of the elements.

3. Materials and Methods

3.1. General Information

All manipulations were carried out under a dry argon or dinitrogen atmosphere using glovebox or standard Schlenk techniques. The solvents were dried using standard procedures and were freshly distilled prior to use. The starting materials [Cr(CO)$_5$(PH$_3$)] [32], [Fe(CO)$_4$(PH$_3$)] [33], [Cr(CO)$_4$(PH$_3$)$_2$] [34], As$_4$ [35] and As(SiMe$_3$)$_3$ [36] were synthesized according to literature procedures. [Cr(CO)$_6$], [W(CO)$_6$] and [Co$_2$(CO)$_8$] were purchased from Merck (Darmstadt, Germany) and used without further purification. The NMR spectra were recorded on a Bruker Advance 400 (Billerica, MA, USA), ^{1}H, 400.132 MHz; ^{31}P, 161.975 MHz, ^{13}C, 100.613 MHz) referenced to external SiMe$_4$ (^{1}H) or H$_3$PO$_4$ (^{31}P), respectively. ^{31}P{^{1}H} MAS NMR spectra were recorded with a Bruker Advance 300 spectrometer equipped with a double resonance 2.5 mm MAS probe. The spectra were acquired at MAS rotation frequencies of 30 and 34 kHz, a 90° pulse length of 2.3 µs, and with relaxation delays between 120 and 450 s. Mass spectra were recorded on a Finnnigan MAT SSQ 710 A (EI) spectrometer (Scientific Instrument Services, Palmer, MA, USA). IR spectra were measured with a Varian FTS-800 spectrometer (Palo Alto, CA, USA). Elemental analyses (CHN) were determined on a Vario EL III instrument (Elementar Analysensysteme GmbH, Langenselbold, Germany).

3.2. Synthesis and Characterization of Clusters **2–12**

3.2.1. Synthesis and Characterization of Clusters **2** and **3**

To a stirred solution of [Cr(CO)$_5$(PH$_3$)] (69.0 mg, 0.3 mmol) or [W(CO)$_5$(AsH$_3$)] (**1**) (121 mg, 0.3 mmol) in toluene, a solution of [Co$_2$(CO)$_8$] (103 mg, 0.3 mmol) in toluene was added. The crude mixture was then stirred for 24 h at room temperature. The solvent was removed under reduced pressure, and the dark residue was extracted with 10 mL n-hexane and stored at −25 °C. After one day, dark red crystals of **2** and violet black crystals of **3**, respectively, were obtained. Yield of **2**: (1.11 g, 57%). Yield of **3**: 86.2 mg (44%) (**2**); 59.6 mg (24%) (**3**). IR (KBr): $\tilde{\upsilon}$/cm^{-1} = ν_{CO}: 2113 (s), 2085 (vs sh), 2053 (vs), 2031 (s), 2000 (w), 1990 (m), 1963 (s sh), 1944 (vs) (**2**); IR (toluene): $\tilde{\upsilon}$/cm^{-1} = ν_{CO}: 2111 (w), 2062 (vs), 2043 (vw), 2034 (w), 1977 (sh), 1956 (m) (**2**); IR (KBr): $\tilde{\upsilon}$/cm^{-1} = ν_{CO}: 2110 (w), 2079 (m), 2057 (vs), 2038 (m), 2032 (m), 2018 (w), 1992 (w), 1937 (vs) (**3**). ^{31}P{^{1}H} NMR (161.975 MHz, C$_6$D$_6$): δ[ppm] = 697.9 (br s, $\omega_{1/2}$ = 113 Hz). ^{13}C{^{1}H} NMR (100.613 MHz, C$_6$D$_6$): δ = 197.5 (br s, $\omega_{1/2}$ = 43 Hz, (Co(CO)$_3$)$_3$), 214.7 (d, $^2J_{CP}$ = 13.7 Hz, Cr(CO)$_4$) (**2**); 197.0 (br s, $\omega_{1/2}$ = 35 Hz, (Co(CO)$_3$)$_3$), 195.4 (s, W(CO)$_4$) (**3**). EI-MS (70 eV): m/z (%) = 651.7 (25) [M$^+$], 623.6 (21) [M$^+$–CO], 567.6 (6) [M$^+$–3 CO], 539.7 (17) [M$^+$–4 CO], 511.6 (42) [M$^+$–5 CO], 483.7 (95) [M$^+$–6 CO], 455.7 (68) [M$^+$–7 CO], 427.7 (52) [M$^+$–8 CO], 399.7 (58) [M$^+$–9 CO], 371.7 (59) [M$^+$–10 CO], 343.7 (59) [M$^+$–11 CO], 315.7 (59) [M$^+$–12 CO], 287.7 (53) [M$^+$–13 CO], 259.7 (100) [M$^+$–14 CO], 207.7 (53) [M$^+$–14 CO–Cr], 148.8 (38) [M$^+$–14 CO–Cr–Co] (**2**); 827.8 (24)

[M$^+$], 799.9 (34) [M$^+$–CO], 743.8 (4) [M$^+$–3 CO], 715.8 (13) [M$^+$–4 CO], 687.8 (69) [M$^+$–5 CO], 659.9 (100) [M$^+$–6 CO], 603.8 (57) [M$^+$–8 CO], 575.9 (80) [M$^+$–9 CO], 547.8 (49) [M$^+$–10 CO], 519.8 (55) [M$^+$–11 CO], 491.8 (63) [M$^+$–12 CO], 463.8 (63) [M$^+$–13 CO], 435.9 (99) [M$^+$–14 CO], 376.8 (76) [M$^+$–14 CO–Co], 317.9 (28) [M$^+$–14 CO–2 Co] (**3**). Elemental analysis, calcd. for Co$_3$PCr(CO)$_{14}$ (651.64 g/mol): C, 25.79. Found C, 25.72 (**2**). Elemental analysis, calcd. for Co$_3$AsW(CO)$_{14}$ (827.60 g/mol): C, 20.32. Found C, 20.22 (**3**).

3.2.2. Synthesis and Characterization of Cluster 4

A solution of [Fe(CO)$_4$(PH$_3$)] (60.6 mg, 0.3 mmol) in toluene was added to a solution of [Co$_2$(CO)$_8$] (103 mg, 0.3 mmol) in toluene and stirred for 18 h. After removing the solvent under reduced pressure, the black residue was dissolved in 10 mL hexane and filtrated. Brown blocks of **4** were obtained within a few hours by storing the hexane solution at −25 °C. Yield: 45.2 mg (24%). IR (KBr): $\tilde{v}$/cm^{-1} = v_{CO}: 2115 (m), 2080 (s sh), 2055 (vs), 2033 (s), 2020 (m), 1984 (m), 1952 (m), 1941 (s); IR (hexane): $\tilde{v}$/cm^{-1} = v_{CO}: 2112 (w), 2089 (w br), 2069 (vs), 2058 (s), 2039 (w), 1986 (m br), 1960 (w), 1932 (w br). ^{31}P{^{1}H} NMR (161.975 MHz, C$_6$D$_6$): δ [ppm] = 677.5 (br s, $\omega_{1/2}$ = 175 Hz). ^{13}C{^{1}H} NMR (100.613 MHz, C$_6$D$_6$): δ = 197.0 (br s, $\omega_{1/2}$ = 46 Hz, Co(CO)$_3$)$_3$), 212.0 (d, $^2J_{CP}$ = 18.8 Hz, Fe(CO)$_3$). EI-MS (70 eV): m/z (%) = 627.7 (22) [M$^+$], 599.8 (20) [M$^+$–CO], 571.8 (39) [M$^+$–2 CO], 543.8 (19) [M$^+$–3 CO], 515.7 (9) [M$^+$–4 CO], 487.8 (21) [M$^+$–5 CO], 459.8 (100) [M$^+$–6 CO], 431.8 (75) [M$^+$–7 CO], 403.8 (50) [M$^+$–8 CO], 375.8 (55) [M$^+$–9 CO], 347.8 (50) [M$^+$–10 CO], 319.8 (44) [M$^+$–11 CO], 291.8 (42) [M$^+$–12 CO], 263.9 (92) [M$^+$–13 CO], 207.9 (36) [M$^+$–13 CO–Fe]. Elemental analysis, calcd. for Co$_3$PFe(CO)$_{13}$ (627.64 g/mol): C, 24.87. Found C, 24.96.

3.2.3. Synthesis and Characterization of Cluster 5

Method 1: A solution of [Cr(CO)$_4$(PH$_3$)$_2$] (232 mg, 1 mmol) in 50 mL toluene was added to a solution of [Co$_2$(CO)$_8$] (1368 mg, 4 mmol) in 50 mL toluene and stirred for seven days at room temperature. The solvent was then completely removed under reduced pressure, and the crude product was dissolved in hexane and filtrated. This filtrate contained [Co$_4$(CO)$_{12}$]. The residue in the frit was extracted with dichloromethane and found to contain the product [Co$_8$(CO)$_{18}$(μ-CO)$_4$P] (**5**). The remaining dark red residue was collected and dissolved in dichloromethane overnight. After filtration and reducing the solution under reduced pressure up to 30 mL, crystals of **5** were obtained upon storing at −25 °C after two weeks. Yield: 63.7 mg (6%).

Method 2: A stirred solution of P$_4$ (3 mg, 0.025 mmol) and [Co$_2$(CO)$_8$] (274 mg, 0.8 mmol) in 30 mL cold hexane at −40 °C was warmed up to room temperature under stirring. After further stirring for two days, the reaction mixture was filtrated and stored at room temperature. Crystals of **5** were obtained within two weeks. Yield: 19.0 mg (9%).

3.2.4. Synthesis and Characterization of Cluster 6

P$_4$ (4 mg, 0.03 mmol) and [Co$_2$(CO)$_8$] (274 mg, 0.8 mmol) were dissolved in 40 mL toluene, cooled to −100 °C and stirred for 30 min. The reaction mixture was then warmed up to room temperature and further stirred for six days. The reaction mixture was filtrated and stored at −25 °C for three months from which black blocks of **6** were obtained. Yield: ~1%. IR (KBr): $\tilde{v}$/cm^{-1} = v_{CO}: 2053 (vs), 2037 (vs), 1896 (m), 1848 (s). ^{31}P{^{1}H} NMR (161.975 MHz, C$_6$D$_6$): δ [ppm] = 666.4 (br s, $\omega_{1/2}$ = 114 Hz). EI-MS (70 eV): m/z (%) = 1003.4 (3) [M$^+$], 947.5 (27) [M$^+$–2 CO], 919.4 (35) [M$^+$–3 CO], 891.4 (21) [M$^+$–4 CO], 863.3 (17) [M$^+$–5 CO], 835.3 (65) [M$^+$–6 CO], 807.4 (79) [M$^+$–7 CO], 779.4 (54) [M$^+$–8 CO], 751.4 (57) [M$^+$–9 CO], 723.9 (49) [M$^+$–10 CO], 695.5 (43) [M$^+$–11 CO], 667.4 (59) [M$^+$–12 CO], 639.5 (59) [M$^+$–13 CO], 611.5 (57) [M$^+$–14 CO], 583.6 (55) [M$^+$–15 CO], 555.4 (65) [M$^+$–16 CO], 527.5 (58) [M$^+$–17 CO], 499.6 (1) [M$^+$–18 CO], 471.6 (1) [M$^+$–19 CO], 443.6 (3) [M$^+$–20 CO].

3.2.5. Synthesis and Characterization of Cluster 7

A solution of P$_4$ (47 mg, 0.38 mmol) and [Co$_2$(CO)$_8$] (1026 mg, 3.0 mmol) in 40 mL toluene was stirred for ten days at room temperature. After removing the solvent under reduced pressure, the residue was dissolved in hexane and filtrated. Black blocks of **7** were obtained after storage at 8 °C. Yield: 36.0 mg (6%). IR (KBr): $\tilde{v}/\text{cm}^{-1} = v_{CO}$: 2034 (s br). ^{31}P{^{1}H} NMR (161.975 MHz, C$_6$D$_6$): δ = 658.2 (br s, $\omega_{1/2}$ = 205 Hz). EI-MS (70 eV): m/z (%) = 1211.2 (6) [M$^+$], 1183.2 (23) [M$^+$–CO], 1127.3 (11) [M$^+$–3 CO], 1099.2 (3) [M$^+$–4 CO], 1071.3 (16) [M$^+$–5 CO], 1043.3 (36) [M$^+$–6 CO], 1015.3 (42) [M$^+$–7 CO], 987.3 (29) [M$^+$–8 CO], 959.3 (24) [M$^+$–9 CO], 931.4 (23) [M$^+$–10 CO], 903.3 (24) [M$^+$–11 CO], 875.3 (25) [M$^+$–12 CO], 847.4 (25) [M$^+$–13 CO], 819.4 (26) [M$^+$–14 CO], 791.4 (23) [M$^+$–15 CO], 763.3 (32) [M$^+$–16 CO], 735.4 (27) [M$^+$–17 CO], 707.4 (23) [M$^+$–18 CO], 679.4 (28) [M$^+$–19 CO], 651.4 (28) [M$^+$–20 CO], 623.4 (100) [M$^+$–21 CO]. Elemental analysis, calcd. for Co$_9$P$_3$C$_{21}$O$_{21}$ (1211.21 g/mol): C, 20.82. Found C, 21.06.

3.2.6. Synthesis and Characterization of Cluster **8**

A solution of P$_4$ (12 mg, 0.1 mmol) and [Co$_2$(CO)$_8$] (205 mg, 0.6 mmol) in 40 mL cold hexane at −40 °C was warmed up to room temperature under stirring. After further stirring for two days the reaction mixture was filtrated and stored at −25 °C from which dark violet crystals of **8** were obtained. Yield: 67.0 mg (13%). IR (KBr): $\tilde{v}/\text{cm}^{-1} = v_{CO}$: 2050 (s br), 2038 (s sh). ^{31}P{^{1}H} NMR (161.975 MHz, C$_6$D$_6$): δ = 667.3 (br s, $\omega_{1/2}$ = 362 Hz). EI-MS (70 eV): m/z (%) = 1267.7 (8) [M$^+$–CO], 1239.6 (3) [M$^+$–2 CO], 1211.8 (19) [M$^+$–3 CO], 1183.8 (15) [M$^+$–4 CO], 1155.8 (2) [M$^+$–5 CO], 1127.8 (16) [M$^+$–6 CO], 1071.1 (18) [M$^+$ –8 CO], 1043.2 (63) [M$^+$–9 CO], 1015.2 (77) [M$^+$–10 CO], 987.2 (44) [M$^+$–11 CO], 959.2 (40) [M$^+$–12 CO], 931.2 (41) [M$^+$–13 CO], 903.2 (39) [M$^+$–14 CO], 875.2 (42) [M$^+$–15 CO], 847.2 (38) [M$^+$–16 CO], 819.2 (40) [M$^+$–17 CO], 791.2 (40) [M$^+$–18 CO], 763.3 (41) [M$^+$–19 CO], 679.3 (2) [M$^+$–22 CO], 651.3 (3) [M$^+$–23 CO], 623.3 (4) [M$^+$–24 CO].

3.2.7. Synthesis and Characterization of Cluster **9**

A solution of [Co$_2$(CO)$_8$] (274 mg, 0.8 mmol) in 25 mL toluene was slowly layered with a solution of P$_4$ (20 mg, 0.16 mmol) in 25 mL hexane. Within five days, a few black crystals of **10** were obtained. The complete crystallization needed a further five weeks. Yield: 175 mg (75%). IR (KBr): $\tilde{v}/\text{cm}^{-1} = v_{CO}$: 2110 (vw), 2064 (s sh), 2054 (s sh), 2043 (vs br), 2032 (s sh), 2025 (s sh), 2001 (m sh), 1958 (w sh). ^{31}P{^{1}H} NMR (161.975 MHz, C$_6$D$_6$): δ = 684.0 (br s, $\omega_{1/2}$ = 2768 Hz). Elemental analysis, calcd. for Co$_{10}$P$_4$C$_{26}$O$_{26}$ (1441.44 g/mol): C, 21.66. Found C, 21.64.

3.2.8. Synthesis and Characterization of Cluster **10**

A solution of P$_4$ (25 mg, 0.2 mmol) in 20 mL toluene was added to a solution of [Co$_2$(CO)$_8$] (137 mg, 0.4 mmol) in 20 mL toluene and stirred for two days at room temperature. On days three and four, a little vacuum was applied over the reaction mixture to remove the evolved CO from the reaction atmosphere, and the solution mixture was stirred for two more days. The reaction mixture was then filtrated and stored at −25 °C. After six months, a few crystals of **9** and **10** were obtained. Yield: ~1%. IR (KBr): $\tilde{v}/\text{cm}^{-1} = v_{CO}$: 2080 (m sh), 2060 (vs), 2039 (s sh), 1994 (m).

3.2.9. Synthesis and Characterization of Cluster **11**

Method 1: A freshly prepared solution of As$_4$ (17 mg, 55 µmol, 3.67 mmol L^{-1}) in toluene was added to a solution of [Co$_2$(CO)$_8$] (274 mg, 0.8 mmol) in 30 mL toluene under light exclusion and stirred for four days at room temperature. The reaction mixture was filtrated and concentrated to 25 mL under reduced pressure. The mixture was then stored at −25 °C from which black blocks of **11** were obtained within three weeks. Yield 24 mg (39%).

Method 2: As(SiMe$_3$)$_3$ (0.03 mL, 0.1 mmol) was added to a solution of [Co$_2$(CO)$_8$] (137 mg, 0.4 mmol) in hexane and stirred for two days at room temperature. The dark red solution was filtrated and stored at −25 °C. After one month, black rods of **11** were obtained. Yield: ~1%. IR (KBr): $\tilde{v}/\text{cm}^{-1} = v_{CO}$: 2089 (s), 2070 (vs), 2026 (vs), 2055 (s sh), 2008 (s). EI-MS (70 eV): m/z (%) = 1427.1 (7) [M$^+$], 1399.2 (35) [M$^+$–CO], 1343.2 (11) [M$^+$–3 CO], 1315.3 (17) [M$^+$–4 CO], 1287.4 (3) [M$^+$–5 CO], 1259.4 (12) [M$^+$–6 CO], 1231.4 (47) [M$^+$–7

CO], 1203.4 (100) [M$^+$–8 CO], 1175.4 (80) [M$^+$–9 CO], 1147.5 (57) [M$^+$–10 CO], 1119.4 (74) [M$^+$–11 CO], 1063.4 (76) [M$^+$–13 CO], 1035.0 (23) [M$^+$–14 CO], 1007.0 (23) [M$^+$–15 CO], 979.0 (24) [M$^+$–16 CO], 951.1 (28) [M$^+$–17 CO], 923.0 (21) [M$^+$–18 CO], 895.1 (21) [M$^+$–19 CO], 867.1 (21) [M$^+$–20 CO], 839.2 (21) [M$^+$–21 CO], 811.1 (22) [M$^+$–22 CO], 783.2 (21) [M$^+$–23 CO], 755.2 (80) [M$^+$–24 CO]. Elemental analysis, calcd. for $Co_9As_3C_{24}O_{214}$ (1427.04 g/mol): C, 20.19. Found C, 20.30.

3.2.10. Synthesis and Characterization of Cluster 12

A solution of $As(SiMe_3)_3$ (0.03 mL, 0.1 mmol) and $[Co_2(CO)_8]$ (137 mg, 0.4 mmol) in 20 mL hexane was stirred for two days at room temperature. The solution was filtered and stored at −25 °C from which black rods of **12** were isolated besides crystals of **11**, which could be manually separated from each other in a glove box. Yield: ~1%. EI-MS (70 eV): *m/z* (%) = 1047.4 (14) [M$^+$], 991.3 (61) [M$^+$–2 CO], 963.4 (20) [M$^+$–3 CO], 935.3 (19) [M$^+$–4 CO], 907.4 (20) [M$^+$–5 CO], 879.4 (60) [M$^+$–6 CO], 851.4 (72) [M$^+$–7 CO], 823.4 (86) [M$^+$–8 CO], 795.5 (56) [M$^+$–9 CO], 767.5 (60) [M$^+$–10 CO], 739.6 (70) [M$^+$–11 CO], 711.5 (72) [M$^+$–12 CO], 683.6 (64) [M$^+$–13 CO], 655.5 (4) [M$^+$–14 CO], 627.6 (4) [M$^+$–15 CO], 599.6 (5) [M$^+$–16 CO], 543.8 (16) [M$^+$–18 CO], 515.8 (8) [M$^+$–19 CO], 487.6 (14) [M$^+$–20 CO].

3.2.11. Synthesis and Characterization of Cluster A

The cluster **9** (100 mg, 0.07 mmol) was dissolved in 20 mL toluene and refluxed for three hours. The suspension was filtrated and stored at −25 °C. After one month, black crystals of **A** were isolated. Yield: 20 mg (21.7%). IR (KBr): $\tilde{v}/cm^{-1}$ = ν_{CO}: 2108 (m), 2088 (vs), 2052 (vs), 2033 (vs), 2023 (vs), 2000 (vs), 1991 (s), 1979 (vs), 1966 (vs), 1826 (m), 1800 (s). $^{31}P\{^1H\}$ NMR (161.975 MHz, C_6D_6): δ [ppm] = 474.6 (br s, $\omega_{1/2}$ = 738 Hz). EI-MS (70 eV): *m/z* (%) = 1065.3 (26) [M$^+$], 1037.2 (27) [M$^+$–CO], 1009.2 (9) [M$^+$–2 CO], 981.2 (5) [M$^+$–3 CO], 953.2 (25) [M$^+$–4 CO], 925.2 (63) [M$^+$–5 CO], 897.4 (57) [M$^+$–6 CO], 869.4 (50) [M$^+$–7 CO], 841.4 (40) [M$^+$–8 CO], 813.4 (57) [M$^+$–9 CO], 785.4 (36) [M$^+$–10 CO], 757.4 (40) [M$^+$–11 CO], 729.5 (38) [M$^+$–12 CO], 701.5 (45) [M$^+$–13 CO], 673.5 (40) [M$^+$–14 CO], 645.5 (46) [M$^+$–15 CO], 617.6 (48) [M$^+$–16 CO], 589.6 (33) [M$^+$–17 CO], 561.5 (23) [M$^+$–18 CO], 533.6 (100) [M$^+$–19 CO]. Elemental analysis, calcd. for $Co_8P_2C_{19}O_{19}$ (1065.32 g/mol): C, 21.42. Found C, 21.30.

4. Conclusions

In the present study, we have demonstrated the high potential of $[Co_2(CO)_8]$ to form cobalt clusters embedding P and As atoms in their core upon its reaction with various P and As sources, such as white phosphorus and yellow arsenic. Accordingly, a large number of novel, as well as reported, clusters were formed depending on the reaction conditions involved. Those include stoichiometry of the reactants, temperature, reaction time, method of crystallization, and solvent. The formed clusters are surrounded by carbonyl ligands. However, EI mass spectra reveal the successive loss of those CO ligands and the possibility of isolating the substituent-free metal-P or metal-As cluster cores. Our current efforts in this field focus on enlarging the family of these valuable candidates further and investigating their potential as single-source precursors for the synthesis of Co_xP_y or Co_xAs_y nanoparticles with varied metal-to-main group element ratios.

Supplementary Materials: The supplementary materials are available online at https://www.mdpi.com/article/10.3390/molecules29092025/s1. Figure S1. 1H NMR spectrum of **1**. Figure S2. 13C{1H} NMR spectrum of **1**. Figure S3. 31P{1H} NMR spectrum of **2**. Figure S4. 13C{1H} NMR spectrum of **2** (above) and **3** (below). Figure S5. 31P{1H} NMR spectrum of **4**. Figure S6. 13C{1H} NMR spectrum of **4**. Figure S7. 31P{1H} NMR spectrum of **6**. Figure S8. 31P{1H} NMR spectrum of **7**. Figure S9. 31P{1H} NMR spectrum of **8**. Figure S10. 31P{1H} MAS NMR spectrum of **9**. Figure S11. 31P{1H} NMR spectrum of **A**. Figure S12. Spin density distribution (left) and singly occupied natural orbital in [Co8(CO)16(μ-CO)4P] (**5**) in the doublet ground state, calculated on the r2SCAN-3c level (right). Figure S13. View of the asymmetric unit of **1**. Figure S14. View of the asymmetric unit of **2**. Figure S15. View of the asymmetric unit of **3**. Figure S16. View of the asymmetric unit of **4**.

Figure S17. View of the asymmetric unit of **5**. Figure S18. View of the asymmetric unit of **6**. Figure S19. View of the asymmetric unit of **7**. Figure S20. View of the asymmetric unit of **8**. Figure S21. (a) View of the asymmetric unit of **9**; (b) Molecular structure of compound **9** in the solid state. Figure S22. (a) View of the asymmetric unit of **10**; (b) Molecular structure of compound **10** in the solid state. Figure S23. View of the asymmetric unit of **11**. Figure S24. View of the asymmetric unit of **12**. Table S1. Thermodynamic parameters (Hartree) calculated for **5** in different spin states at the r2SCAN-3c level of theory. Table S2. Crystallographic data for compounds **1–4**. Table S3. Crystallographic data for compounds **5–8**. Table S4. Crystallographic data for compounds **9–12**. References [29,31,37–47] are cited in the Supplementary Materials.

Author Contributions: Literature study and manuscript preparation, M.E.M.; synthesis of all compounds and their characterizations, S.B.; helping S.B. in the synthesis of all compounds, C.G.; preparation of CIF files for all compounds and calculating CVE wherever mentioned in the manuscript, C.R.; DFT computations, G.B.; reviewing and editing the manuscript, supervision of the whole project and funding acquisition, M.S. All authors have read and agreed to the published version of the manuscript.

Funding: This work was financially supported by the Deutsche Forschungsgemeinschaft within the projects Sche 384/40-1.

Institutional Review Board Statement: Not applicable.

Informed Consent Statement: Not applicable.

Data Availability Statement: Data are contained within the article and Supplementary Materials.

Acknowledgments: C.R. is grateful to the Studienstiftung des Deutschen Volkes e.V. for a PhD fellowship.

Conflicts of Interest: The authors declare no conflicts of interest.

References

1. Whitmire, K.H. Transition metal complexes of the naked pnictide elements. *Coord. Chem. Rev.* **2018**, *376*, 114–195. [CrossRef]
2. Shelyganov, P.A.; Elsayed Moussa, M.; Seidl, M.; Scheer, M. Diantimony Complexes [Cp^R$_2$Mo$_2$(CO)$_4$(μ,η^2-Sb$_2$)] (Cp^R = C$_5$H$_5$, C$_5$H$_4$^{t}Bu) as Unexpected Ligands Stabilizing Silver(I)$_n$ (n = 1–4) Monomers, Dimers and Chains. *Angew. Chem. Int. Ed.* **2023**, *62*, e202215650. [CrossRef] [PubMed]
3. Peresypkina, E.; Virovets, A.; Scheer, M. Organometallic polyphosphorus complexes as diversified building blocks in coordination chemistry. *Coord. Chem. Rev.* **2021**, *446*, 213995–214038. [CrossRef]
4. Bai, J.; Virovets, A.V.; Scheer, M. Synthesis of Inorganic Fullerene-Like Molecules. *Science* **2003**, *300*, 781–782. [CrossRef]
5. Cesari, C.; Shon, J.-H.; Zacchini, S.; Berben, L.A. Metal carbonyl clusters of groups 8–10: Synthesis and catalysis. *Chem. Soc. Rev.* **2021**, *50*, 9503–9539. [CrossRef]
6. Gauthier, J.A.; King, L.A.; Stults, F.T.; Flores, R.A.; Kibsgaard, J.; Regmi, Y.N.; Chan, K.; Jaramillo, T.F. Transition Metal Arsenide Catalysts for the Hydrogen Evolution Reaction. *J. Phys. Chem. C* **2019**, *123*, 24007–24012. [CrossRef]
7. Chen, W.-C.; Wang, X.-L.; Qin, C.; Shao, K.-Z.; Su, Z.-M.; Wang, E.-B. A carbon-free polyoxometalate molecular catalyst with a cobalt-arsenic core for visible light-driven water oxidation. *Chem. Commun.* **2016**, *52*, 9514–9517. [CrossRef]
8. Hong, C.S.; Berben, L.A.; Long, J.R. Synthesis and characterization of a decacobalt carbonyl cluster with two semi-intertitial phosphorus atoms. *Dalton Trans.* **2003**, 2119–2120. [CrossRef]
9. Buchwalter, P.; Rosé, J.; Lebeau, B.; Ersen, O.; Girleanu, M.; Rabu, P.; Braunstein, P.; Paillaud, J.-L. Characterization of cobalt phosphide nanoparticles derived from molecular clusters in mesoporous silica. *J. Nanopart. Res.* **2013**, *15*, 2132–2152. [CrossRef]
10. Foust, A.S.; Foster, M.S.; Dahl, L.F. Organometallic pnictogen complexes. III. Preparation and structural characterization of the triarsenic-cobalt Atom cluster system As$_3$Co(CO)$_3$, The first known X3-transition metal analog of group VA tetrahedral X$_4$ molecules. *J. Am. Chem. Soc.* **1969**, *91*, 5631–5633. [CrossRef]
11. Foust, A.S.; Foster, M.S.; Dahl, L.F. Organometallic pnictogen complexes. IV. Synthesis, structure, and bonding of new organometallic arsenic-metal atom clusters containing a metal-bridged multiply bonded As$_2$ ligand: Co$_2$(CO)$_6$As$_2$ and Co$_2${(CO)$_5$P(C$_6$H$_5$)$_3$}As$_2$. *J. Am. Chem. Soc.* **1969**, *91*, 5633–5635. [CrossRef]
12. Arnold, L.J.; Mackay, K.M.; Nicholson, B.K. Reaction of arsane with cobalt or iron carbonyls, and the X-ray crystal structures of [Fe$_2$(CO)$_8$(μ$_4$-As)]$_2$[Fe$_2$(CO)$_6$] and [μ$_4$-AsCo$_3$(CO)$_8$]$_3$. *J. Organomet. Chem.* **1990**, *387*, 197–207. [CrossRef]
13. Vizi-Orosz, A.; Pályi, G.; Markó, L. Phosphido cobalt carbonyl clusters: Co$_2$(CO)$_6$P$_2$ and Co$_3$(CO)$_9$PS. *J. Organomet. Chem.* **1973**, *60*, C25–C26. [CrossRef]
14. Vizi-Orosz, A. Phosphido cobalt carbonyl clusters P$_n$[Co(CO)$_3$]$_{4-n}$ (n = 1, 2, 3). *J. Organomet. Chem.* **1976**, *111*, 61–64. [CrossRef]
15. Seyferth, D.; Henderson, R.S. phosphaacetylenehexacarbonyldicobalt complexes: New cluster lewis bases. *J. Organomet. Chem.* **1978**, *162*, C35–C38. [CrossRef]

16. Burckett-St. Laurent, J.C.T.R.; Hitchcock, P.B.; Kroto, H.W.; Nixon, J.F. Novel transition metal phospha-alkyne complexes. X-Ray crystal and molecular structure of a side-bonded tBuC≡P complex of zerovalent platinum, Pt(PPh$_3$)$_2$(tBuCP). *J. Chem. Soc. Chem. Commun.* **1981**, *21*, 1141–1143. [CrossRef]
17. Ciani, G.; Sironi, A.; Martinengo, S.; Garlaschelli, L.; Pergola, R.D.; Zanello, P.; Laschi, F.; Masciocchi, N. Synthesis and X-ray characterization of the phosphide-carbonyl cluster anions [Co$_9$(μ_8,P)(CO)$_{21}$]$^{2-}$ and [Co$_{10}$(μ_8,P)(CO)$_{22}$]$^{3-}$. *Inorg. Chem.* **2001**, *40*, 3905–3911. [CrossRef]
18. Della Pergola, R.; Sironi, A.; Colombo, V.; Garlaschelli, L.; Racioppi, S.; Sironi, A.; Macchi, P. Periodical trends in [Co$_6$E(CO)$_{16}$]$^-$ clusters: Structural, synthetic and energy changes produced by substitution of P with As. *J. Organomet. Chem.* **2017**, *849–850*, 130–136. [CrossRef]
19. Della Pergola, R.; Garlaschelli, L.; Macchi, P.; Facchinetti, I.; Ruffo, R.; Racioppi, S.; Sironi, A. From small metal clusters to molecular nanoarchitectures with a core-shell structure: The synthesis, redox fingerprint, theoretical analysis, and solid-state structure of [Co$_{38}$As$_{12}$(CO)$_{50}$]$^{4-}$. *Inorg. Chem.* **2022**, *61*, 26, 9886–9896. [CrossRef] [PubMed]
20. Dielmann, F.; Sierka, M.; Virovets, A.V.; Scheer, M. Access to extended polyphosphorus frameworks. *Angew. Chem. Int. Ed.* **2010**, *49*, 6860–6864. [CrossRef] [PubMed]
21. Graßl, C.; Bodensteiner, M.; Zabel, M.; Scheer, M. Synthesis of arsenic-rich As$_n$ ligand complexes from yellow arsenic. *Chem. Sci.* **2015**, *6*, 1379–1382. [CrossRef]
22. Dielmann, F.; Timoshkin, A.; Piesch, M.; Balázs, G.; Scheer, M. The cobalt cyclo-P$_4$ sandwich complex and its role in the formation of polyphosphorus compounds. *Angew. Chem. Int. Ed.* **2017**, *56*, 1671–1675. [CrossRef]
23. Del Mar Conejo, M.; Pastor, A.; Montilla, F.; Galindo, A. P atom as ligand in transition metal chemistry: Structural aspects. *Coord. Chem. Rev.* **2021**, *434*, 213730–213773. [CrossRef]
24. Fischer, E.O.; Bathelt, W.; Müller, J. Arsine pentacarbonyl complexes of chromium(0), Molybdenum(0#9 and Tungsten(0). *Chem. Ber.* **1970**, *103*, 1815–1821. [CrossRef]
25. Lal De, R.; Vahrenkamp, H. Polynuclear complexes from PH$_3$ and RPH$_2$ complexes with Co$_2$(CO)$_8$. *Z. Naturforsch.* **1985**, *40b*, 1250–1257. [CrossRef]
26. Coleman, J.M.; Dahl, L.F. Molecular structures of [(C$_6$H$_5$)$_2$PCoC$_5$H$_5$]$_2$ and [(C$_6$H$_5$)$_2$PNiC$_5$H$_5$]$_2$. An assessment of the influence of a metal-metal bond on the molecular geometry of an organometallic ligand-bridged complex. *J. Am. Chem. Soc.* **1967**, *89*, 542–552. [CrossRef]
27. Dreher, C.; Zabel, M.; Bodensteiner, M.; Scheer, M. [(CO)$_4$W(PH$_3$)$_2$] as a source of semi-interstitial phosphorus ligands in cobalt carbonyl clusters. *Organometallics* **2010**, *29*, 5187–5191. [CrossRef]
28. Elsayed Moussa, M.; Rummel, E.-M.; Eckhardt, M.; Riesinger, C.; Scheer, M. Unusual cleavage of phosphaalkynes triple bond in the coordination sphere of transition metals. *Dalton Trans.* **2023**, *52*, 15656–15659. [CrossRef]
29. Vizi-Orosz, A.; Galamb, V.; Pályi, G.; Markó, L.; Bor, G.; Natile, G. AsCo$_3$(CO)$_9$, its cyclic trimer, As$_3$Co$_9$(CO)$_{24}$ and the phosphorus-containing analog As$_3$Co$_9$(CO)$_{24}$. *J. Organomet. Chem.* **1976**, *107*, 235–240. [CrossRef]
30. Maxwell, S.B.H.L.R.; Mosley, V.M. Electron diffraction by gases. *J. Chem. Phys.* **1935**, *3*, 699–709. [CrossRef]
31. Lang, H.; Huttner, G.; Sigwarth, B.; Jibril, I.; Zsolnai, L.; Orama, O. μ_3-P und μ_3-As-verbrückte cluster als liganden. *J. Organomet. Chem.* **1986**, *304*, 137–155. [CrossRef]
32. Vogel, U.; Scheer, M. Zur oxidativen addition von komplexierten phosphanen bzw. arsanen an platin(0)-komplexen. *Anorg. Allg. Chem.* **2001**, *627*, 1593–1598. [CrossRef]
33. Hunger, C.; Ojo, W.-S.; Bauer, S.; Xu, S.; Zabel, M.; Chaudret, B.; Lacroix, L.-M.; Scheer, M.; Nayral, C.; Delpech, F. Stoichiometry-controlled FeP nanoparticles synthesized from a single source precursor. *Chem. Commun.* **2013**, *49*, 11788–11790. [CrossRef] [PubMed]
34. Dreher, C. Darstellung und Reaktivität von P–H-funktionellen Übergangsmetallcarbonylkomplexen. PhD Thesis, Universität Regensburg, Regensburg, Germany, 2009.
35. Seidl, M.; Balázs, G.; Scheer, M. The Chemistry of Yellow Arsenic. *Chem. Rev.* **2019**, *119*, 8406–8434. [CrossRef]
36. Becker, G.; Gutekunst, G.; Wessely, H.J. Trimethylsilylverbindungen der Vb-Elemente. I Synthese und Eigenschaften von Trimethylsilylarsanen. *Z. Anorg. Allg. Chem.* **1980**, *462*, 113–129. [CrossRef]
37. Kendall, R.A.; Früchtl, H.A. The impact of the Resolution of the Identity approximate integral method on modern ab initio algorithm development. *Theor. Chim. Acta* **1997**, *97*, 158–163. [CrossRef]
38. Neese, F. The ORCA program system. *WIREs Comput. Mol. Sci.* **2012**, *2*, 73–78. [CrossRef]
39. Neese, F. Software update: The ORCA program system, version 4.0. *WIREs Comput. Mol. Sci.* **2018**, *8*, e1327. [CrossRef]
40. Grimme, S.; Hansen, A.; Ehlert, S.; Mewes, J.-M. r^{2}SCAN-3c: A "Swiss army knife" composite electronic-structure method. *J. Chem. Phys.* **2021**, *154*, 064103. [CrossRef]
41. Kruse, H.; Grimme, S. A geometrical correction for the inter- and intra-molecular basis set superposition error in Hartree-Fock and density functional theory calculations for large systems. *J. Chem. Phys.* **2012**, *136*, 154101. [CrossRef] [PubMed]
42. Caldeweyher, E.; Bannwarth, C.; Grimme, S. Extension of the D3 dispersion coefficient model. *J. Chem. Phys.* **2017**, *147*, 034112. [CrossRef] [PubMed]
43. Agilent Technologies Ltd. *CrysAlis PRO*; Agilent Technologies Ltd.: Oxfordshire, UK, 2014.
44. Dolomanov, O.V.; Bourhis, L.J.; Gildea, R.J.; Howard, J.A.K.; Puschmann, H. OLEX2: A Complete Structure Solution, Refinement and Analysis Program. *J. Appl. Crystallogr.* **2009**, *42*, 339–341. [CrossRef]

45. Sheldrick, G.M. SHELXT–Integrated space-group and crystal-structure determination. *Acta Cryst. A* **2015**, *71*, 3–8. [CrossRef] [PubMed]
46. Sheldrick, G.M. A short history of SHELX. *Acta Cryst. A* **2008**, *64*, 112–122. [CrossRef]
47. Sheldrick, G.M. Crystal Structure Refinement with SHELXL. *Acta Cryst. C* **2015**, *71*, 3–8. [CrossRef]

Disclaimer/Publisher's Note: The statements, opinions and data contained in all publications are solely those of the individual author(s) and contributor(s) and not of MDPI and/or the editor(s). MDPI and/or the editor(s) disclaim responsibility for any injury to people or property resulting from any ideas, methods, instructions or products referred to in the content.

Review

Monofluorophos–Metal Complexes: Ripe for Future Discoveries in Homogeneous Catalysis

Alexandra M. Miles-Hobbs [1], Paul G. Pringle [1,*], J. Derek Woollins [2] and Daniel Good [1]

[1] The School of Chemistry, University of Bristol, Cantock's Close, Bristol BS8 1TS, UK
[2] Department of Chemistry Khalifa University, Abu Dhabi P.O. Box 127788, United Arab Emirates; jdw3@st-andrews.ac.uk
* Correspondence: paul.pringle@bristol.ac.uk

Abstract: The discovery that cyclic (ArO)$_2$PF can support Rh-catalysts for hydroformylation with significant advantages in tuning regioselectivity transformed the study of metal complexes of monofluorophos ligands from one of primarily academic interest to one with potentially important applications in catalysis. In this review, the syntheses of monofluorophosphites, (RO)$_2$PF, and monofluorophosphines, R$_2$PF, are discussed and the factors that control the kinetic stability of these ligands to hydrolysis and disproportionation are set out. A survey of the coordination chemistry of these two classes of monofluorophos ligands with d-block metals is presented, emphasising the bonding of the fluorophos to d-block metals, predominantly in low oxidation states. The application of monofluorophos ligands in homogeneous catalysis (especially hydroformylation and hydrocyanation) is discussed, and it is argued that there is great potential for monofluorophos complexes in future catalytic applications.

Keywords: fluorophosphites; fluorophosphines; coordination chemistry; homogeneous catalysis

Citation: Miles-Hobbs, A.M.; Pringle, P.G.; Woollins, J.D.; Good, D. Monofluorophos–Metal Complexes: Ripe for Future Discoveries in Homogeneous Catalysis. *Molecules* **2024**, *29*, 2368. https://doi.org/10.3390/molecules29102368

Academic Editor: Graham Saunders

Received: 23 April 2024
Revised: 8 May 2024
Accepted: 14 May 2024
Published: 17 May 2024

Copyright: © 2024 by the authors. Licensee MDPI, Basel, Switzerland. This article is an open access article distributed under the terms and conditions of the Creative Commons Attribution (CC BY) license (https://creativecommons.org/licenses/by/4.0/).

1. Introduction

Phosphorus ligands containing P–C, P–N, and P–O bonds are ubiquitous in homogeneous catalysis. By contrast, fluorophos ligands (those containing a P–F bond) have attracted relatively little attention in catalysis, despite the extensive fluorophos coordination chemistry of late transition metals that has been developed and the industrial interest in the application of monofluorophosphite **L1** (Figure 1) in Rh-catalysed hydroformylation dating back to 1998 [1]. In other contexts, **L1** (commercial name: Ethanox 398) has been employed as an antioxidant [2] and as a flame retardant [3].

The extreme electronegativity of fluorine means that it can withdraw electron density from any atom it is bonded to, contributing to its reputation as the *Tyrannosaurus Rex* of chemistry [4]. It should be noted that the electron-withdrawing power of F is a σ-inductive effect and, in some cases, this is offset by an electron-donating π-resonance effect (see later) [5]. This property, combined with the diminutive size of P–F (only P–H is smaller), makes the steric and electronic properties of an F substituent of particular academic interest. The high electronegativity of F would be expected to enhance the π-acceptor capacity of ligands containing P–F bonds compared to analogous ligands containing P–O bonds. Since one of the reasons cited for the success of phosphites such as **L2–4** (Figure 1) as ligands in Rh-catalysed hydroformylation is their strong π-acceptor capacity, it is understandable why monofluorophosphite **L1** performs well in hydroformylation [6–11].

The simplest fluorophos ligand, PF$_3$, has a special place in coordination and organometallic chemistry as a ligand that has π-acceptor properties on par with, or surpassing, those of CO [12]. The volatility of some PF$_3$ complexes has made them attractive for applications in chemical vapour deposition [13–15] and recently, a PF$_3$ complex, identified as [Co$_2$(μ-CO)$_2$(CO)$_2$(PF$_3$)$_4$], was reported to be a catalyst precursor for 1-hexene hydroformylation [16]. However, progress in the application of PF$_3$ as an ancillary ligand is hampered by it being an odourless gas with toxicity similar to phosgene [17], and it is not amenable to chemical modification.

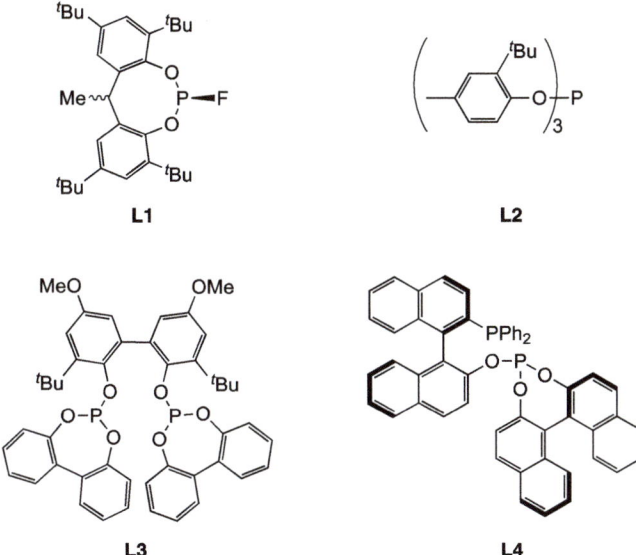

Figure 1. Ethanox-398 (**L1**) and some other landmark phosphite ligands **L2**–**L4** for Rh-catalysed hydroformylation.

There are no such disadvantages for the collage of P–F ligands, depicted in Figure 2, which have C-, O-, or N-substituents. These substituted fluorophos ligands have the advantages of being systematically modifiable via R substituents and they are generally straightforward to synthesise.

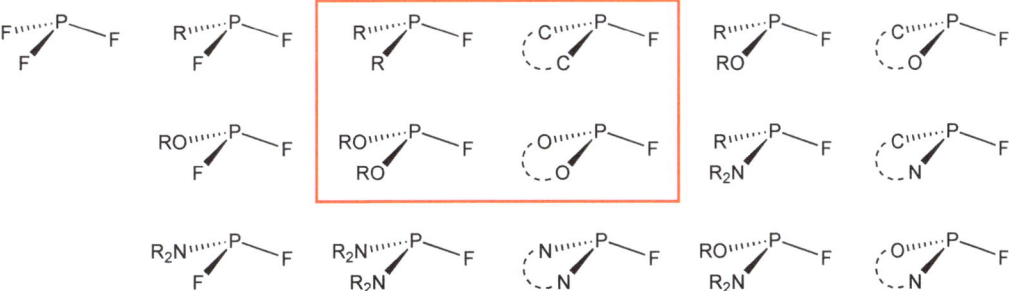

Figure 2. A selection of P–F containing monophos ligands (R = alkyl or aryl group) including P-heterocycles showing the diversity of ligands that are potentially available. The structures in the red box are the subject of this review.

The focus of this review is acyclic and cyclic monofluorophos ligands of the type $(RO)_2P–F$ and $R_2P–F$, since these are amongst the simplest achiral P^{III} compounds that contain a P–F bond. Both of these classes of P-ligand have attracted considerable academic and industrial interest since the 1960s, including in the area of homogeneous catalysis. To the best of our knowledge, there has not previously been a review of monofluorophos ligands, although difluorophos ligands have been reviewed [18]. The topics covered in this review include (1) the synthetic routes to monofluorophosphites and monofluorophosphines;

(2) the factors controlling the stability of monofluorophos ligands that limit their applications; (3) the transition metal coordination chemistry of monofluorophos ligands that may be pertinent to an understanding of their role in homogeneous catalysis; (4) the homogeneous catalysis that has been reported with metal–monofluorophos complexes. This review is not comprehensive and there is a bias to more recent developments that build upon the early foundational work reported by the groups of Schmutzler and Nixon. The main conclusion that is drawn from this review is that the tunability of the steric and electronic effects in monofluorophosphites and monofluorophosphines augurs well for future applications of these and related classes of P–F ligands in homogeneous catalysis.

2. Monofluorophosphites
2.1. Synthesis and Hydrolytic Stability of Monofluorophosphites

Cyclic and acyclic monofluorophosphites are most readily prepared from the corresponding chlorophosphite, $PCl(OR)_2$, and a source of fluoride, such as CsF or SbF_3. The precursor chlorophosphites are prepared from PCl_3 and the appropriate phenol/alcohol, or a siloxy derivative (as exemplified in Scheme 1) [19]. Monofluorophosphites have also been made from PCl_2F, but this precursor is not readily accessible [20,21].

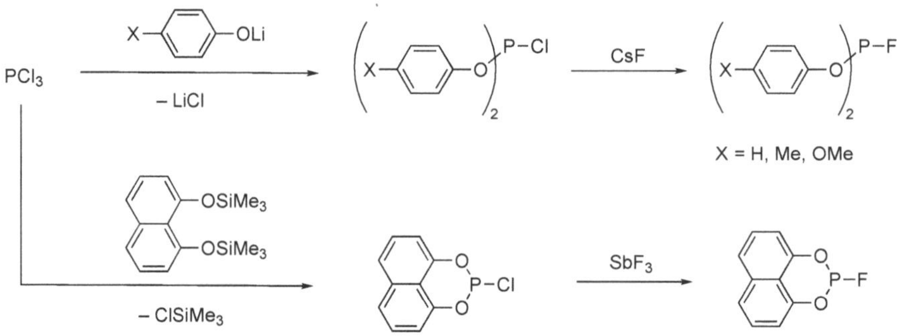

Scheme 1. Typical examples of the synthesis of acyclic and cyclic monofluorophosphites.

At first sight, the prospects for using any halophos ligand of the type Z_2P–Hal (Z = alkyl, aryl, OR, OAr, NR_2; Hal = F, Cl, Br, I) in catalysis may appear bleak because of the reactivity of P-Hal bonds. For example, chlorophos compounds (Z_2P–Cl) are normally viewed as useful intermediates rather than ligands because they react readily with a wide range of C-, O-, or N-nucleophiles [22]; this reactivity makes chlorophos ligands incompatible with many reactive functional groups. Moreover, chlorophos compounds commonly fume in air because of their high susceptibility to hydrolysis, during which HCl is produced (Equation (1), X = Cl).

$$Z_2P-X + H_2O \longrightarrow Z_2P(=O)H + HX \quad (1)$$

The favourable thermodynamics of P–Cl hydrolysis are largely driven by the P=O bond formation in the P-containing product (Equation (1), X = Cl). However, the thermodynamics of P–OAr hydrolysis (Equation (1), X = OAr) are at least as favourable as those of P–Cl hydrolysis and yet ligands containing P-OAr groups are widely used in coordination chemistry and catalysis. It can therefore be surmised that the high reactivity of chlorophosphites is primarily due to their high kinetic lability. Indeed, chlorophosphites that are remarkably stable to moisture have also been developed and some have been applied in catalysis [23,24].

It has been shown that phosphite P–O bonds can be stabilised to hydrolysis by integrating them into cyclic structures and/or incorporating bulky hydrophobic groups into the ligand framework, as in aryl phosphite ligands **L2–4** (Figure 1). Indeed, diphosphite **L3** and its derivatives have been successfully applied in large scale industrial hydroformylation processes [7]. It is of no surprise, therefore, that the Eastman monofluorophos ligand **L1** is a phosphadioxacycle which contains bulky *t*-butyl substituents that shroud the P–F moiety [1].

While **L1** is reportedly stable to hydrolysis [25], the hydrolytic stability of the related cyclic monofluorophosphites **L5–8** (Figure 3) in aqueous methanol depends on ring size: the half-lives increase in the order **L5 < L7 ~ L8 < L6** [19].

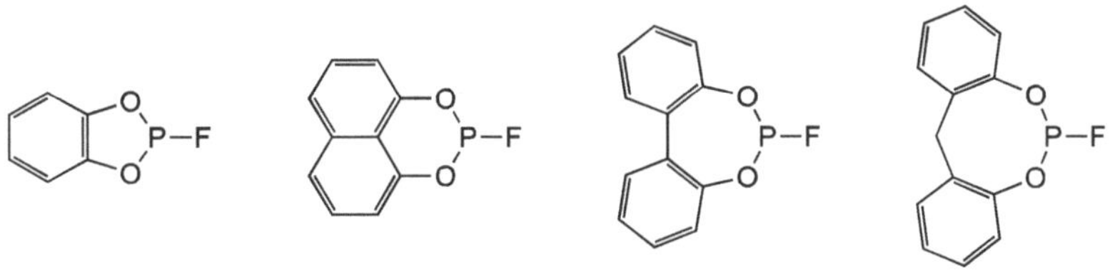

Figure 3. Cyclic monofluorophosphites **L5–L8** with ring sizes of 5–8, respectively.

2.2. Coordination Chemistry of Monofluorophosphites

Metal complexes of monofluorophosphites have been produced by the two routes shown in Scheme 2: (a) by substitution of a labile, neutral ligand (A) by a monofluorophosphite; (b) by methanolysis of a coordinated PF_3 or by addition of an equivalent of HF to a coordinated $P(OR)_3$.

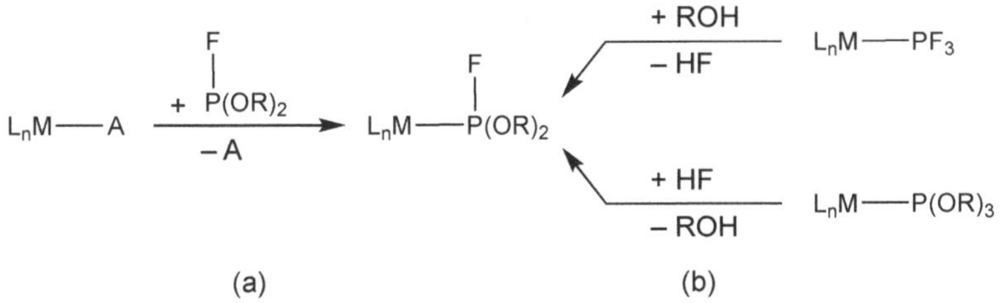

Scheme 2. Routes to monofluorophosphite complexes: (**a**) conventional substitution at the metal of neutral ligand A by monofluorophos ligand; (**b**) substitution at the phosphorus of a coordinated P-ligand.

2.2.1. Group 6 Metal Complexes of Monofluorophosphites

The range of Group 6 metal(0) complexes of monofluorophosphites that have been prepared is summarised in Scheme 3 [20,26,27]. UV photolysis of each of the metal hexacarbonyls in the presence of $(MeO)_2PF$ (**L9**) gave the homoleptic complexes **1–3** [27]. $[Cr(CO)_6]$ reacts with **L5** to give the trisubstituted **4** while the molybdenum analogue **5** is formed when **L5** reacts with [(cycloheptatriene)$Mo(CO)_3$] [26].

The *cis*-disubstituted Mo complexes **6–9** were prepared by substitution of the norbornadiene ligand in [Mo(nbd)$(CO)_4$] with the cyclic monofluorophosphites **L5–L8** and the products were fully characterised, including by X-ray crystallography. The IR data for **6–9** are consistent with the π-acceptor capacities of **L5–L8** lying between those of PF_3 and $P(OPh)_3$. The ν_{CO} values for the highest frequency band increases in the order $P(OPh)_3$ < **L8** ~ **L7** < **L6** < **L5** < PF_3, which is consistent with the π-acceptor capacity of the cyclic phosphites increasing as the ring size decreases [19].

2.2.2. Group 8 Metal Complexes of Monofluorophosphites

The synthesis of the iron(0)–monofluorophosphite complexes **10–12** is summarised in Scheme 4. Complex **10** is formed by addition of ligand **L7** to [Fe$_2$(CO)$_9$] (Scheme 4, route (a)) [20]. Complex **11** is produced by two routes: (1) addition of ligand **L10** to [Fe$_2$(CO)$_9$] (Scheme 4, route (a)); (2) treatment of the Fe–PFCl$_2$ precursor complex **13** with the sodium alkoxide nucleophile shown in Scheme 4 route (b) [28].

Complex **12** has been identified by IR spectroscopy as a product of the methanolysis of the PF$_3$ complex **14** in a detailed study of the alcoholysis of [Fe(PF$_3$)$_x$(CO)$_{5-x}$] (x = 1–4) species [29].

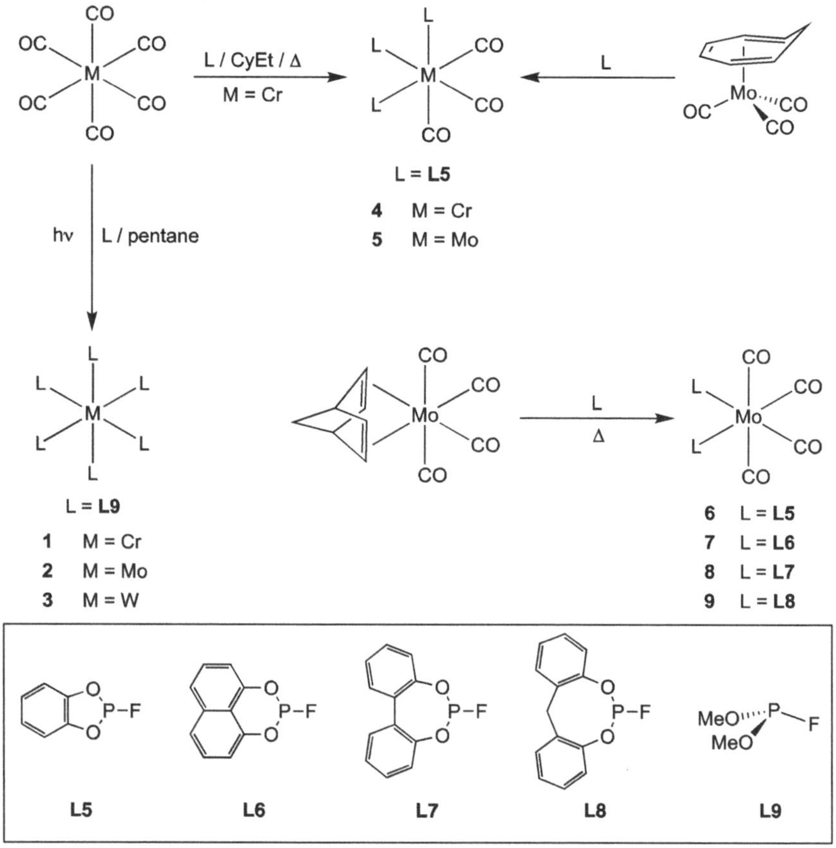

Scheme 3. Group 6 metal complexes of monofluorophosphites.

The equilibrium proportions of equatorial (*e*) and apical (*a*) isomers of [Fe(CO)$_4$L] can be determined by IR spectroscopy; sterically demanding and good π-acceptor ligands prefer to bind at the equatorial sites [29]. As shown in Scheme 4 (d), for complex **14**, the predominant isomer has the PF$_3$ equatorial, although the *e*:*a* ratio is close to the statistical 60:40 ratio, reflecting the similarity of PF$_3$ and CO as ligands. For complex **12**, only the apical isomer was detected, consistent with **L9** being small and a poorer π-acceptor than PF$_3$. For complex **11**, a higher proportion of equatorial isomer was present than even in the PF$_3$ complex **14**, as expected for the bulky **L10**. The ν$_{CO}$ values for the complexes **14** and **11** are very similar, showing that PF$_3$ and **L10** have similar π-acceptor properties. This demonstrates that the steric and electronic effects of monofluorophosphite ligands can be controlled via the phosphorus alkoxy substituents.

The ruthenium(II) phosphite complexes *trans*-[(dppe)$_2$Ru(H){P(OR)$_3$}]$^+$ react with HBF$_4$ to give the homologous series of monofluorophosphite complexes **15–17** (Scheme 5); the HBF$_4$ is providing the source of HF in these reactions. The coordinated monofluorophosphite ligands **L9**, **L11**, and **L12** are readily displaced by a H$_2$ to give the η^2-H$_2$ complex **18** (Scheme 5) [30].

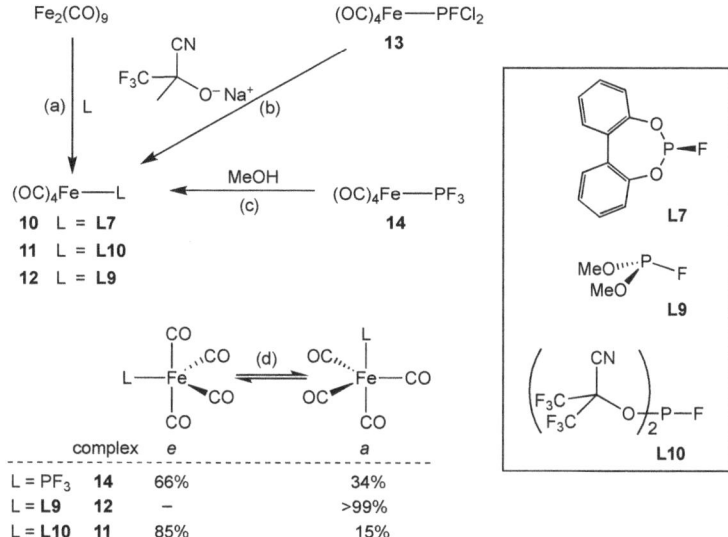

Scheme 4. Iron(0) complexes of monofluorophosphites. Routes (a)-(c) and equilibrium (d) are referred to in the text.

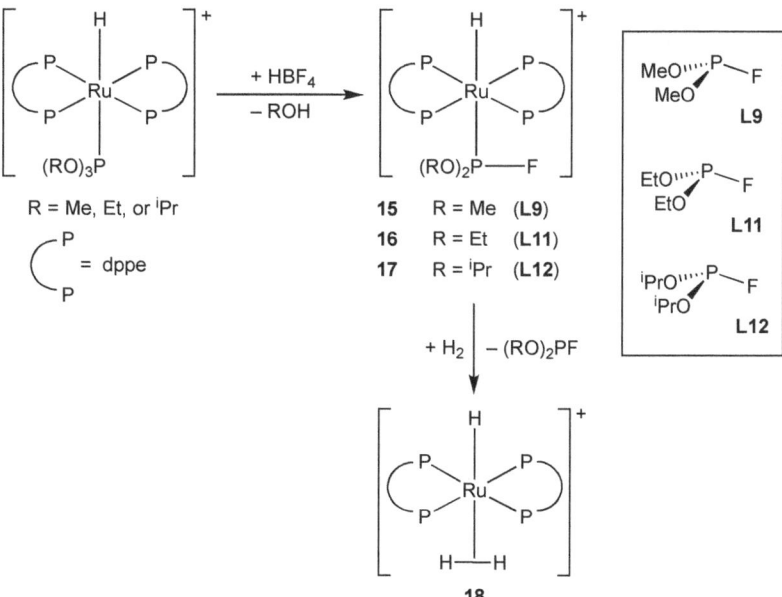

Scheme 5. In situ generation of monofluorophosphite ligands on ruthenium(II).

2.2.3. Group 9 Metal Complexes of Monofluorophosphites

The tetrahedral cobalt complexes **19** and **20** containing the coordinated **L9** have been separated by preparative GLC from the mixtures obtained by methanolysis of the corresponding PF_3 complexes (Scheme 6) [31]. The IR spectra of the complexes showed that the ν_{CO} and ν_{NO} stretching bands are both shifted to significantly lower wavenumber in the monofluorophosphite complexes **19** and **20** with respect to their PF_3 precursors, consistent with **L9** being a poorer π-acceptor ligand than PF_3.

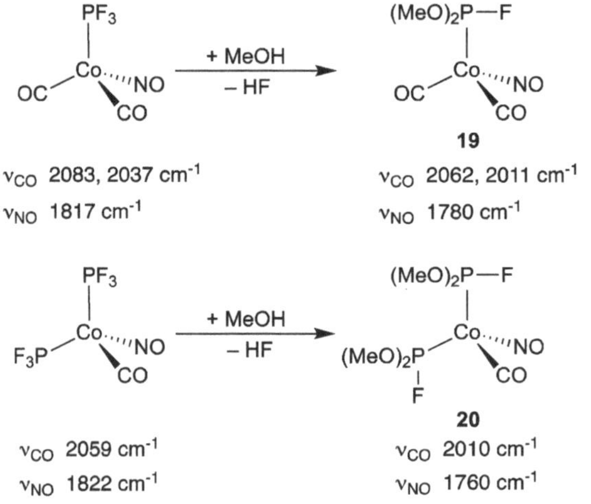

ν_{CO} 2083, 2037 cm^{-1} ν_{CO} 2062, 2011 cm^{-1}
ν_{NO} 1817 cm^{-1} ν_{NO} 1780 cm^{-1}

19

ν_{CO} 2059 cm^{-1} ν_{CO} 2010 cm^{-1}
ν_{NO} 1822 cm^{-1} ν_{NO} 1760 cm^{-1}

20

Scheme 6. Monofluorophosphite derivatives of nitosylcobalt(−I) complexes.

The rhodium(I) chemistry with the cyclic monofluorophosphites **L5–L8** is summarised in Scheme 7 [19]. Treatment of $[Rh_2Cl_2(CO)_4]$ with **L5–L8** gave the three products **21–23** in the proportions shown in Scheme 7. These products were characterised by multinuclear NMR spectroscopy and comparison of the spectra with the products exclusively formed from $[Rh_2Cl_2(diene)_2]$ (diene = 1,5-hexadiene or 1,5-cyclooctadiene) and $[Rh(cod)_2][BF_4]$. There is a consistent trend of increasing proportion of binuclear complex **21** formed with decreasing ring size; indeed, with **L5**, binuclear **21d** is exclusively formed. It is significant that PF_3 is the only other monophos ligand that selectively forms the binuclear product **21e** [32,33]. The interpretation of these observations is that **L5** and PF_3 are sufficiently good π-acceptors to displace the CO from the Rh.

The trend of increasing PF_3-like behaviour with decreasing size of phosphacycle in relation to the reactions of **L8–L5** with $[Rh_2Cl_2(CO)_4]$ parallels the trend observed in the spectroscopic properties of cis-$[Mo(CO)_4(L)_2]$ (see above) [19].

2.2.4. Group 10 Metal Complexes of Monofluorophosphites

The homoleptic nickel(0) and platinum(0) complexes **24–27** containing monofluorophosphites **L5** or **L13** were prepared (Scheme 8) [34,35] and their ^{31}P and ^{19}F NMR spectra were analysed extensively because they are rare examples of $[AX]_4$ spin systems [35,36]. It was noted that the $^2J_{P,P}$ values for the Ni(0) complexes **24** and **25** (ca. 20 Hz) are significantly smaller than for the analogous Pt(0) complexes **26** and **27** (ca. 100 Hz), although no rationale was given for this large difference [35]. The nickel(0) complexes **24** and **25** were originally prepared from $[Ni(CO)_4]$ [26,34] but it was shown that complexes **24–27** can be conveniently prepared from the corresponding $[M(cod)_2]$ (Scheme 8) [35].

The trans-palladium(II) and cis-platinum(II) complexes **28** and **29**, containing the cyclic monofluorophosphite **L5**, were prepared by cleavage of the corresponding binuclear complex (Scheme 9) [37]. The phosphacycle **L14**, which can be viewed as a saturated analogue of **L5**, forms the cis-platinum(II) complex **30**; comparison of the ^{31}P NMR parameters for

29 and 30 shows that they are similar, e.g., $J_{Pt,P}$ = 5600 and 5490 Hz, respectively. The platinum(0) complex 31 contains monofluorophosphite L15, a saturated analogue of L6 (Scheme 9) [37].

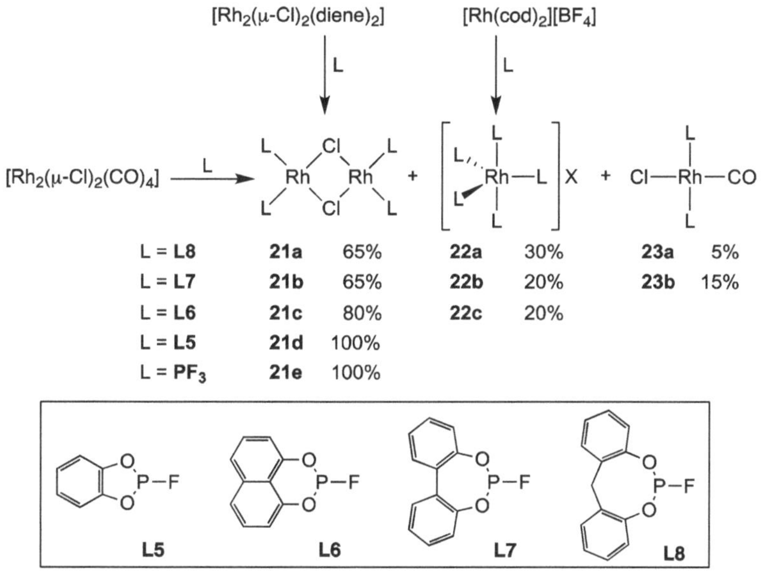

Scheme 7. Cyclic monofluorophosphite chemistry of rhodium(I).

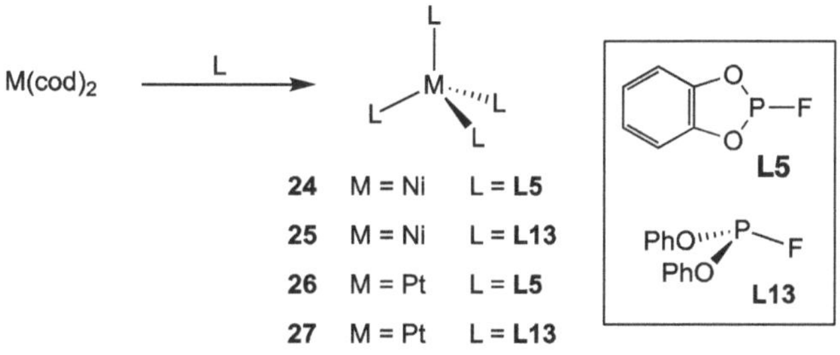

Scheme 8. Nickel(0) and platinum(0) chemistry of monofluorophosphites.

The tetrahedral platinum(0) complexes 32a–d are readily formed by the addition of 4 equiv. of L5–L8 to [Pt(nbe)₃] (nbe = norbornene). Complex 32b crystallised from solution even when a sub-stoichiometric amount of L6 was added (Scheme 10). However, the addition of 2 equiv. of L5, L7 or L8 to [Pt(nbe)₃] in THF gave mixtures of [Pt(L)₄] (32a,c,d) [Pt(L)₂(nbe)] (33a,c,d), and [Pt(L)(nbe)₂] (34a,c,d), identified from their characteristic ^{31}P and ^{195}Pt NMR signals (Scheme 10) [19]. The ratios of complexes observed at equilibrium (Scheme 10) were rationalised to be the result of the competing steric and electronic factors for the nbe and monofluorophosphite ligands; for example, while [Pt(L)₄] is more sterically crowded than [Pt(L)₂(nbe)], the greater π-acceptor properties of monofluorophosphites makes them better than norbornene at stabilising Pt(0) [19].

2.3. Catalysis with Complexes of Monofluorophosphites

2.3.1. Hydroformylation Catalysis with Rhodium Complexes of Monofluorophosphites

The most notable example of the application of monofluorophos ligands in homogeneous catalysis is the use of cyclic monofluorophosphites such as L1 in the Rh-catalysed hydroformylation reactions, reported by Eastman and shown in Scheme 11 [1,25]. Initially, the application of monofluorophosphite ligands in catalysis was approached with scepticism, as it was suspected that monofluorophosphites may be thermally unstable, and be prone to hydrolysis, especially at elevated temperatures, generating hydrogen fluoride (HF), which is a known catalyst poison [25,38,39]. However, it was demonstrated that **L1** is stable to degradation at temperatures up to 350 °C and stable to hydrolysis even in refluxing aqueous isopropanol, with no free fluoride ions detected [40]. While acidic conditions promote the degradation of monofluorophosphites, it has been shown that the catalyst system can be stabilised by the addition of an epoxide or a complex such as [Co(acac)$_3$] [41,42].

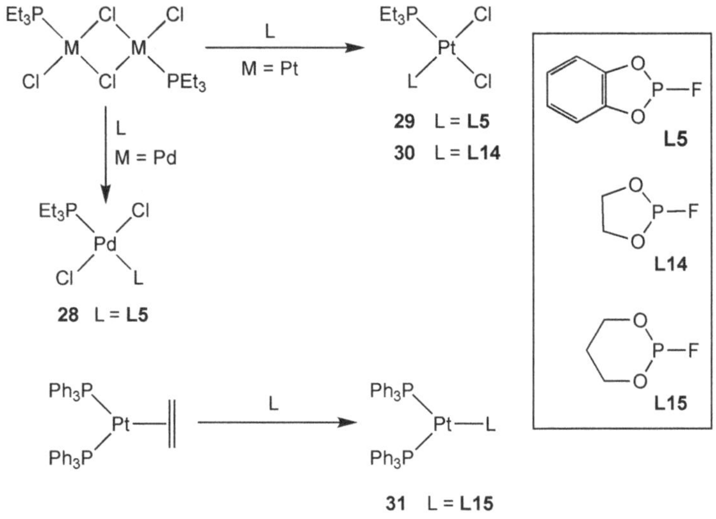

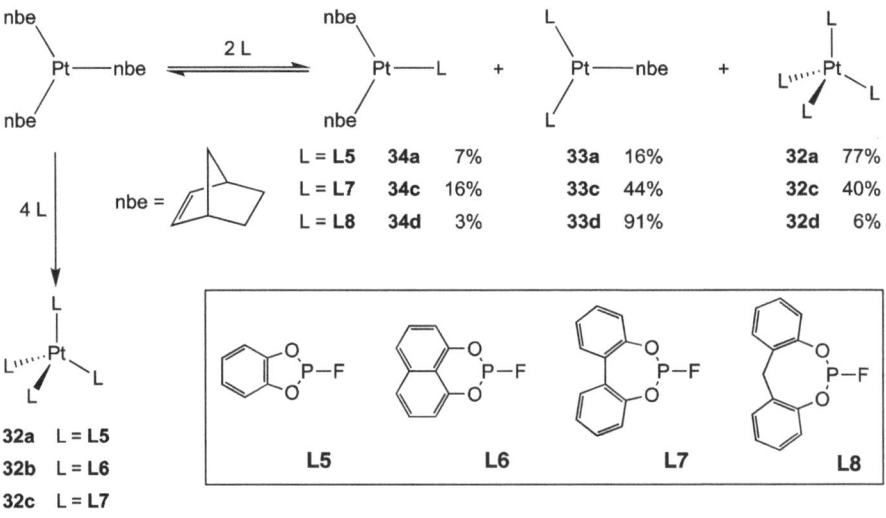

Scheme 9. Platinum(II) and palladium(II) chemistry of monofluorophosphites.

Scheme 10. Platinum(0) chemistry of monofluorophosphites.

The striking stability of **L1** is attributed to the 8-membered phosphacycle which entropically stabilises the ligand to P–O cleavage and to the ^{t}Bu substituents which sterically shield the P atom and provide a hydrophobic environment in the vicinity of the P–F bond.

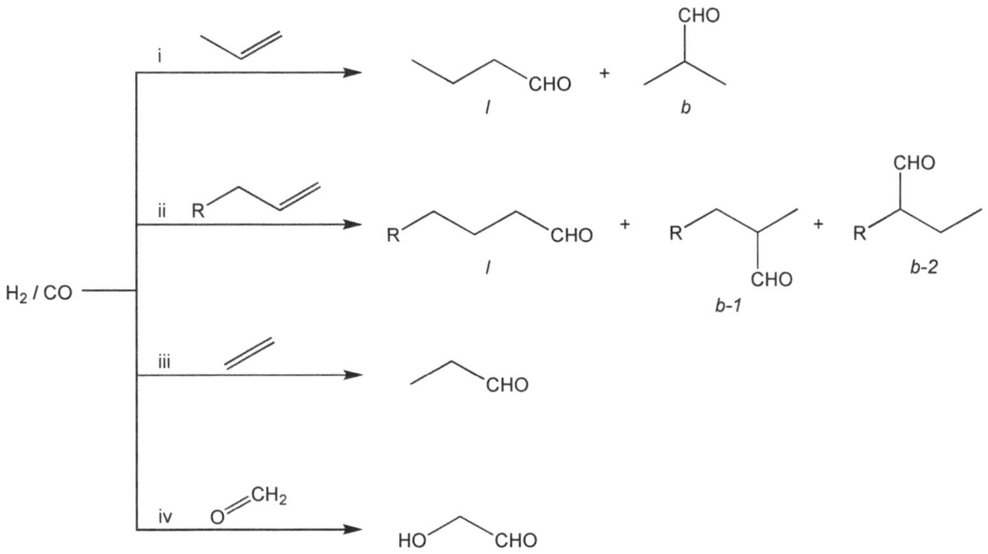

Scheme 11. Rh-monofluorophosphite catalysed hydroformylation of alkenes (reactions (**i**–**iii**)) and formaldehyde (reaction (**iv**)).

Ligand **L1** exists as two geometric isomers, labelled *cis*-**L1** and *trans*-**L1** in Figure 4, associated with the relative stereochemistry of the F substituent on P and the Me substituent on the CH of the ligand backbone. The isomers of **L1** have been separated, and it was shown by ^{31}P NMR spectroscopy that, when [Rh(CO)$_2$(acac)] was treated with 2 equiv. of *cis*-**L1**, a mono-ligated RhL$_1$ species was produced whereas with 2 equiv. of *trans*-**L1** a bis-ligated RhL$_2$ species was the product. Furthermore, *trans*-**L1** readily displaced *cis*-**L1** from its Rh(acac) complex, showing that *trans*-**L1** has a greater affinity for the Rh(I) centre than *cis*-**L1** [43,44]. These differences in coordination chemistry are likely due to the 8-membered heterocycle having to adopt a more strained ring conformation in *cis*-**L1** than in *trans*-**L1** in order to accommodate the bulky metal moiety being bound at a pseudo-equatorial site. The observed coordination chemistry differences of the isomers of **L1** may be the source of the differences in hydroformylation activity and selectivity that are observed with the various mixtures of isomers of **L1** [43,44].

The alkene substrates employed in Rh/**L1** catalysed hydroformylations include terminal alkenes (1-propene and 1-octene), and internal alkenes (isomeric nonenes) [1,38]. As a consequence of their unsymmetrical nature, alkenes other than ethene give linear (*l*) and branched (*b*) aldehydes. For propene, two isomeric aldehydes (one linear and one branched) are formed (reaction i in Scheme 11), while for longer chain alkenes, alkene isomerisation is a competing reaction which can lead to several branched aldehyde products, e.g., for 1-hexene, there are two branched isomers (see reaction ii where R = ^{n}Pr in Scheme 11). The *l*:*b* ratio of products is affected by a wide array of factors, including temperature, syngas pressure, ligand–metal (L:Rh) ratio, and the nature of the ligands [7–9,25,45]. With monofluorophosphite ligands, it has been shown that the impact of the L:Rh ratio on the alkene hydroformylation activity is strongly dependent on the structure of the ligand. Increasing the L:Rh ratio (L = P-donor ligand) normally decreases catalytic activity, and this is indeed observed with monofluorophosphite **L1**. However, with the bulkier cyclic monofluorophosphite **L16**, increasing the L:Rh ratio increased catalytic activity. The cyclic structure of **L16** appears to be critical for this unusual concentration effect on rate, since the conventional decrease in activity with increase in L:Rh is observed with **L17**, an acyclic analogue of **L16** [46].

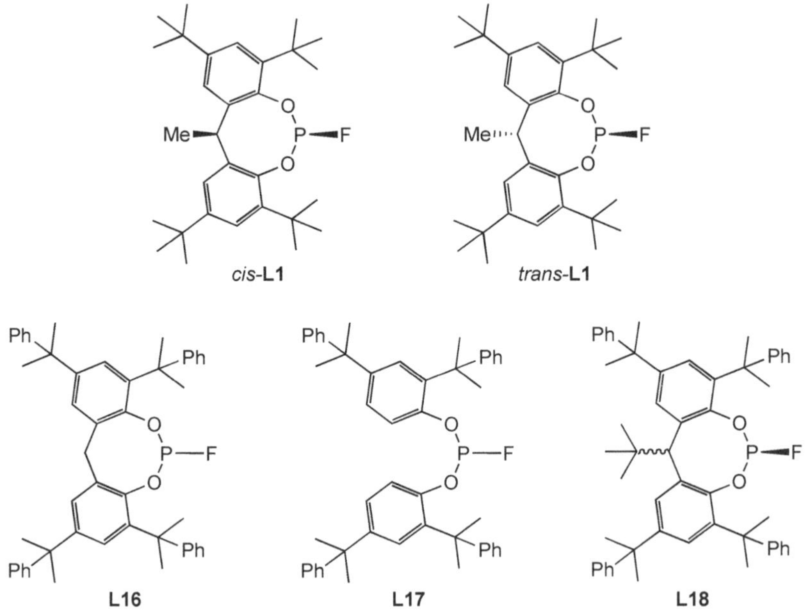

Figure 4. Some of the Eastman fluorophophites used as ligands in hydroformylation.

A thorough study of the alkene hydroformylation catalytic properties of Rh complexes of monofluorophosphite **L18** has been reported, which includes in-flow and batch hydroformylation of propene, 1-octene, and 2-octene [47]. High activities, with TOF up to 75,000 mol(RCHO) mol(Rh)$^{-1}$ h^{-1}, have been observed and outstanding control of the aldehyde *l:b* ratio can be achieved by modulating the temperature, PCO, PH_2, time of reaction, the pre-activation of the catalyst, and Rh:**L18** ratio; for example, for 1-octene, the *l:b* ratio can be 'tuned' from 0.27 to 15 (corresponding to selectivity ranging from 78% branched to 94% linear). The higher the concentration of **L18**, the more the linear aldehyde is favoured, and this has been rationalised by postulating two mechanisms are operating in parallel: one based on RhL$_2$(CO) species, favouring linear aldehyde formation, and the other based on the less bulky RhL(CO)$_2$ moiety, favouring branched aldehyde formation [47].

The hydroformylation of ethylene to produce propionaldehyde (Scheme 11, reaction iii) is a potentially useful transformation but acetylene, typically present in ethylene feedstocks in small quantities, acts as a reversible poison towards Rh-based catalysts [25]. The activity of ethylene hydroformylation using a Rh–PPh$_3$ catalyst suffered greatly when subjected to ethylene containing 1000 ppm of acetylene. By contrast, the Rh–**L1** catalyst system was shown to be remarkably acetylene-tolerant under the same conditions; the activity of the Rh–**L1** catalyst eventually deteriorated upon increasing the concentration of acetylene to 10,000 ppm [48].

The hydroformylation of formaldehyde (in the form of paraformaldehyde) is potentially a valuable route to produce glycolaldehyde (Scheme 11, reaction iv) which can then be hydrogenated to ethylene glycol. It has been shown that a Rh–**L1** catalyst is more active and selective than a Rh-PPh$_3$ catalyst under the same conditions [49].

2.3.2. Other Catalytic Reactions with Monofluorophosphite Ligands

The bulky, optically active monofluorophosphite BIFOP-F (**L19**), derived from fenchol, has been employed in the intramolecular Pd-catalysed cross-coupling reaction shown in Scheme 12 [50]. A library of 12 related fenchol-derived BIFOP-X ligands were screened for catalysis and complex **35**, derived from **L19,** was the most enantioselective (64% *ee*) and gave good yields (88%).

An attempt to use the same ligand **L19** in a Cu-catalysed 1,4-addition of R_2Zn or RMgBr (R = Me, Et) to enones was unsuccessful; it was suggested that **L19** was unstable under the reaction conditions used [51].

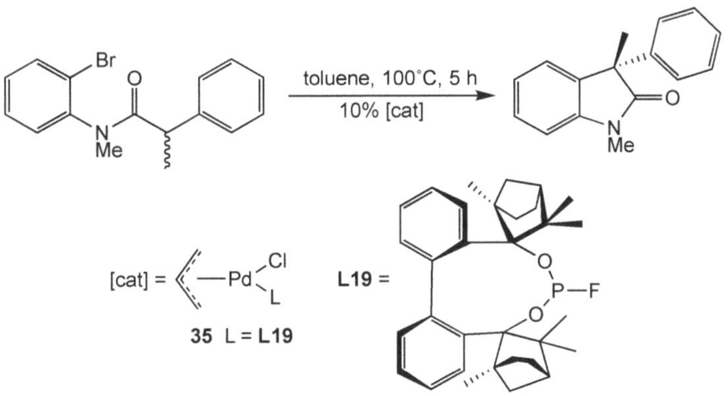

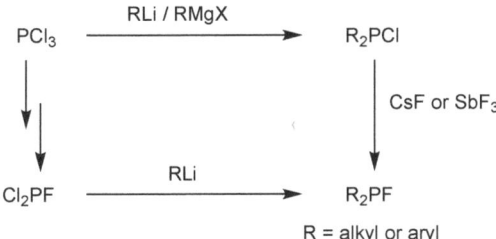

Scheme 12. Palladium-BIFOP-F catalysts.

3. Monofluorophosphines

3.1. Synthesis and Stability of Monofluorophosphines

Two general routes to R_2PF where R = alkyl or aryl are shown in Scheme 13. The R_2PCl route has the advantage of the ready availability of chlorophosphines from PCl_3 but the Cl_2PF route can provide access to R_2PF for which the corresponding R_2PCl is unknown, as demonstrated for $(PhC\equiv C)_2PF$ [52].

Scheme 13. Routes to monofluorophosphines.

Simple R_2PF (which are P^{III} species) are generally unstable with respect to the disproportionation to the P^V in R_2PF_3 and P^{II} in R_2P-PR_2, as shown in Scheme 14 [53,54]. The pathway shown in Scheme 14, involving the intermediates **A** and **B**, has been proposed for the disproportionation; examples of $P^{III}-P^V$ species **A** have been isolated and characterised spectroscopically [55,56]. This chemistry would militate against the application of monofluorophosphines as ligands in homogeneous catalysis unless, under the catalytic reaction conditions, the equilibrium in Scheme 14 lies in favour of the R_2PF, or the equilibrium is rapidly reversible, such that it can be entrained via metal complexation.

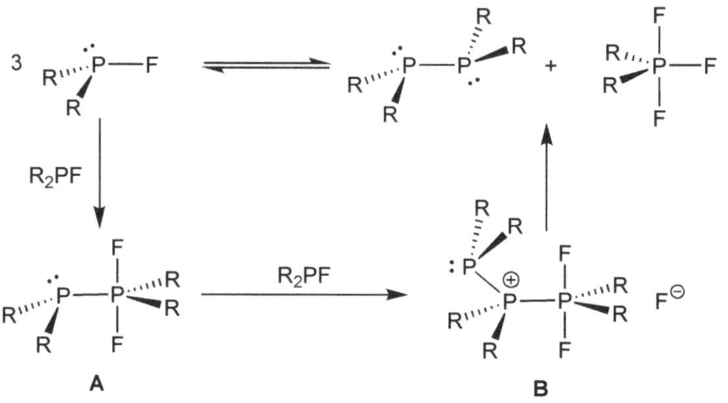

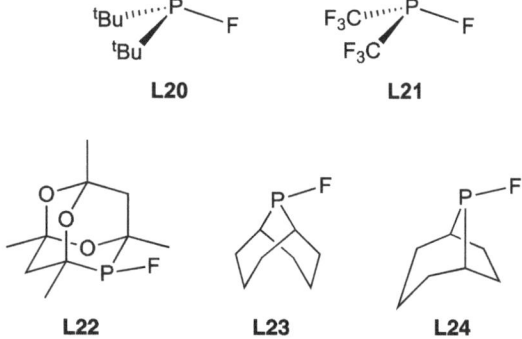

Scheme 14. Disproportionation of monofluorophosphines R$_2$PF.

The following generalisations on the stability of R$_2$PF to disproportionation (Scheme 14) have been established from extensive studies:

(1) Many common R$_2$PF (e.g., R = Ph, Me, ^{n}Bu) readily disproportionate [54,57,58];
(2) Bulky substituents and electron-withdrawing substituents stabilise R$_2$PF with respect to disproportionation [59,60];
(3) Cyclic monofluorophosphines with constrained C–P–C bonds are more stable with respect to disproportionation than acyclic analogues [61].

The stabilising effects of the P-substituents noted in generalisation (2) accounts for the dominance of ^{t}Bu$_2$PF (**L20**) and (CF$_3$)$_2$PF (**L21**) in the early literature concerning the coordination chemistry of monofluorophosphines (Figure 5). A simple rationale for the R$_2$PF-stabilising effect of bulky and electron-withdrawing substituents is that these substituents raise the energy of the disproportionation diphosphane product, R$_2$P–PR$_2$ because (a) bulky R groups maximise 1,2-steric repulsions in the relatively crowded diphosphane—^{t}Bu$_2$P–P^{t}Bu$_2$ has been calculated to have a weak P-P bond [62]; (b) electron-withdrawing groups destabilise the P–P bond due to electrostatic repulsion between the resulting δ+ charges on each of the P atoms—it has been reported that (CF$_3$)$_2$P–P(CF$_3$)$_2$ has an elongated P-P bond [63]. A mechanism for disproportionation involving sterically crowded intermediates **A** and **B**, which would also be disfavoured by electron-withdrawing substituents [54–56], has been proposed (Scheme 14).

Figure 5. Stable monofluorophosphine ligands.

The monofluorophosphines CgPF (**L22**), containing a phospha-adamantane cage, and the PhobPF species **L23** and **L24**, containing a phospha-bicycle (Figure 5), are remarkably stable to disproportionation [61]. The CgP and PhobP moieties are rigid and bulky, and so the stability of **L22–L24** may be, at least in part, explained using similar steric congestion arguments to those used above for the stability of **L20** [64–67]. In addition, it has been argued that the constrained C–P–C angles in **L22–L24** also contribute to their observed stability to disproportionation (generalisation (3) above) using the following reasoning [61]. The two geometric isomers of R_2PF_3 have diapical–equatorial (*aae*) or apical–diequatorial (*aee*) F groups, with the high apicophilicity of F leading to the *aae* isomer being preferred for R_2PF_3 [68]. Therefore, the favoured isomer has the two R substituents occupying two equatorial sites with a 120° angle between them, as depicted in Scheme 14. X-ray crystallography has shown that the C–P–C angles are close to 90° in multiple compounds containing either the CgP or PhobP moieties [64–67]. Consequently, the observed stability to disproportionation of **L22–L24** can be partly attributed to the high degree of C–P–C ring strain in R_2PF_3 that would be incurred by the 2 C substituents occupying equatorial sites; if, instead, the *eea* isomer were adopted, there would be an unfavourable cost in the P–F bond energies associated with two of the F substituents occupying equatorial sites [68].

3.2. Coordination Chemistry of Monofluorophosphines

In general, monofluorophosphine (R_2PF) complexes are made just like many other P-ligand complexes: by the substitution of a labile ligand on a precursor complex. In metal complexes of monofluorophosphines, the coordinated R_2PF is not susceptible to disproportionation. Consequently, ligated Ph_2PF (which is unstable as the free ligand) has been generated within a Cr, Mo, or W coordination sphere by fluoride substitution of a labile X group on a precursor R_2PX complex [69–71].

3.2.1. Group 6 Metal Complexes of Monofluorophosphines

The Group 6 complexes **36–44** of monofluorophosphines **L20** and **L21** are shown in Scheme 15 [72–74]. The $[ML(CO)_5]$ complexes **36–40** were made by photolysis of a mixture of $[M(CO)_6]$ and ligand in THF (for **L20**) or CH_2Cl_2 (for **L21**) [72,73]. The *cis*-disubstituted complexes **41** and **42** were formed by stirring $[Mo(norbornadiene)(CO)_4]$ with the ligand at ambient temperatures for several hours [72,74]. The $[MoL_3(CO)_3]$ complexes **43** and **44** were both prepared from $[Mo(cycloheptatriene)(CO)_3]$, but the products were assigned different geometries (*fac* in **43** and *mer* in **44**, respectively) based on the unambiguous IR and ^{19}F NMR spectra for the C_{2v} and C_{3v} isomers. Extensive NMR (^{31}P and ^{19}F) and IR spectroscopic studies have been carried out on all complexes **36–44**. It was shown that the trend in the position of the highest energy ν_{CO} band in the IR spectra of **36** and its analogues are consistent with the expected π-acidities being in the order: $^tBu_3P$ (2067 cm^{-1}) < $^tBu_2PF$ (2076 cm^{-1}) < $^tBuPF_2$ (2088 cm^{-1}) < PF_3 (2104 cm^{-1}) [74].

A notable conclusion drawn on the basis of the IR spectra of *cis*-$[MoL_2(CO)_4]$ and *mer*-$[MoL_3(CO)_3]$ is that $(CF_3)_2PF$ and CF_3PF_2 are stronger π-acceptors than PF_3, notwithstanding the greater electronegativity of F than that of CF_3 (χ of 4.0 and 3.3, respectively, on the Pauling Scale). It has been suggested [74] that an explanation for this apparent anomaly lies in the π component present in the P–F bond that involves a HOMO (lone pair) orbital on F and the LUMO (σ*) on P which has π symmetry. This is the same orbital on P that is involved in the π backbonding from the metal. Thus, in a M–P–F fragment, the M competes with F for the π acceptor orbital on P (Figure 6(i)); this competition is not present in a M–P–CF_3 fragment which would explain the greater π acceptor capacity of $(CF_3)_2PF$ than PF_3 [74]. This explanation in terms of π interactions between the LUMO (σ*) on P and a HOMO with π symmetry on a P-substituent is reminiscent of the arguments used by Woollins et al. to explain why $P^tBu(pyrrolyl)_2$ is a stronger σ donor than $P(pyrrolyl)_3$ (see Figure 6(ii)) [75].

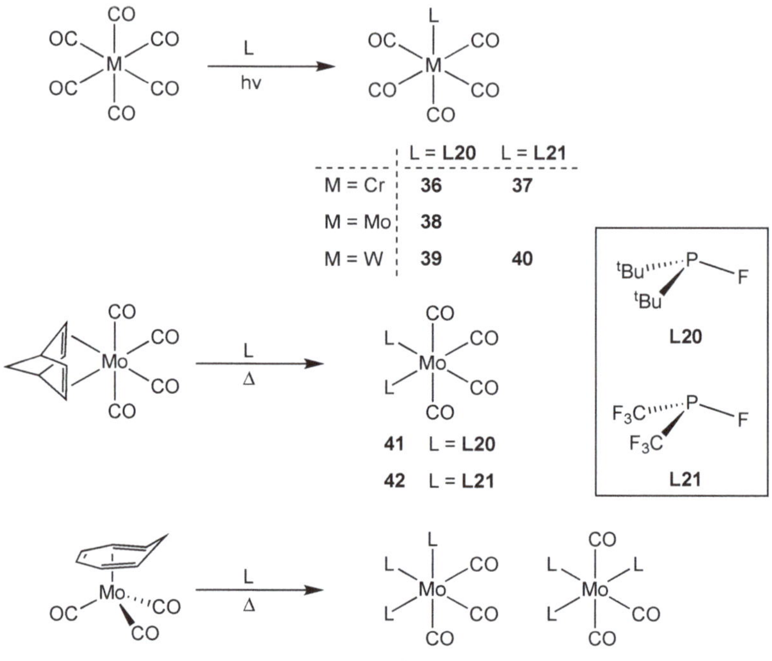

Scheme 15. Monofluorophosphine complexes of Group 6 metals.

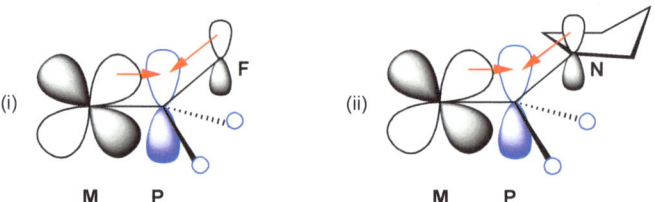

Figure 6. MO pictures of the π-bonding involved in (i) fluorophos and (ii) pyrrolylphos ligands.

3.2.2. Group 7 Metal Complexes of Monofluorophosphines

The only reported Group 7 metal complexes containing a monofluorophosphine ligand are the isomeric hydridomanganese(I) complexes **45** and **46**, formed as a 3:1 mixture by the reaction of [HMn(CO)$_5$] with **L21** (Scheme 16) [76]. The ^2J(HP) values for **46** (72 Hz) and **47** (4 Hz) are consistent with the assignment of their respective *trans* and *cis* geometries.

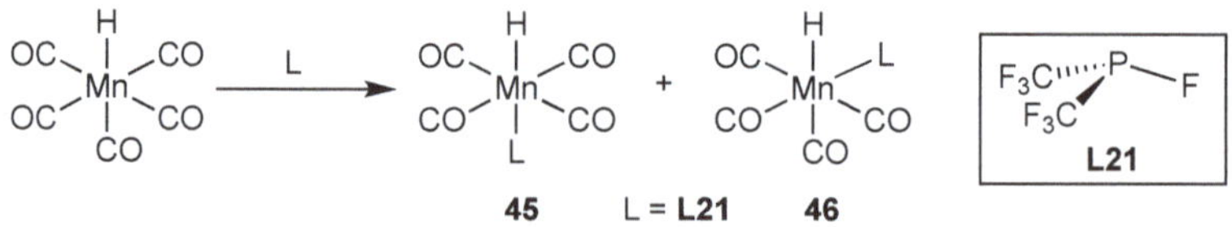

Scheme 16. Monofluorophosphine-manganese(I) complexes.

3.2.3. Group 8 Metal Complexes of Monofluorophosphines

The tetracarbonyliron complex **47** can be generated in situ by photolysis of a mixture of **L21** and [Fe(CO)$_5$] and the IR spectrum suggests that **47** is predominantly the equatorial isomer (Scheme 17). This is consistent with **L21** being bulkier than PF_3 and of comparable π acceptor capacity to it [29].

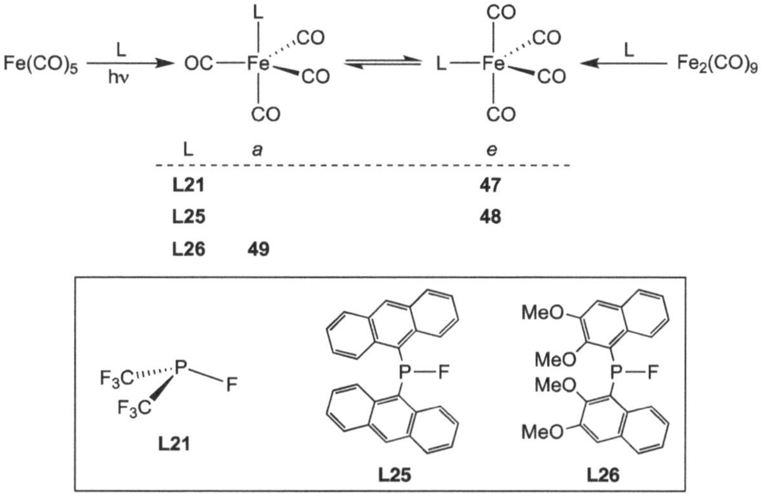

Scheme 17. Monofluorophosphine complexes of iron(0).

The anthracene-derived monofluorophosphine **L25** and the naphthalene-derived monofluorophosphine **L26** were prepared from Cl_2PF (Scheme 13) [52]. Ligand **L25** was purified by distillation and showed no tendency to undergo disproportionation presumably because it is stabilised by its bulky substituents. Reaction of **L25** with $[Fe_2(CO)_9]$ gave complex **48**, whose IR spectrum was consistent with C_{2v} symmetry and was therefore assigned to the equatorial isomer. Ligand **L26** was not obtained in pure form but the impure material was reacted with $[Fe_2(CO)_9]$ to produce the iron complex **49**, the IR spectrum of which was consistent with C_{3v} symmetry and was therefore assigned to the apical isomer [52]. The different geometries assigned to **48** and **49** may be rationalised by **L25** being larger and more electron poor (making it a better π-acceptor) than **L26**.

The unusual monofluorophosphine **50** has been prepared by treatment of its anionic precursor with *N*-fluoropyridinium tetrafluoroborate which acts as an electrophilic source of F^+ (see Scheme 18) [77]. The P–F bond in **50** was shown to be covalent in the solid state by single-crystal X-ray diffraction (d_{P-F} = 1.658(4) Å), and in solution by ^{31}P and ^{19}F NMR spectroscopy, which showed that $^1J_{PF}$ = 918 Hz). The data for **50** are comparable to values for conventional R_2PF compounds: d_{P-F} = 1.619(7) Å for $^tBu_2PF$ [78]; $^1J_{PF}$ = 905 Hz for Ph_2PF [57]. In principle, **50** could act as an monofluorophos ligand, but this has not been reported to date.

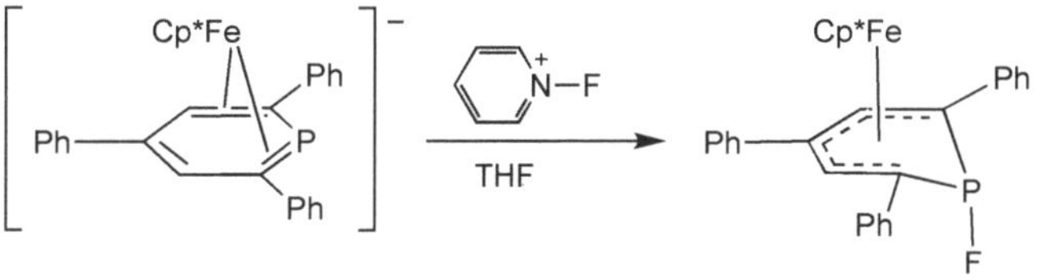

Scheme 18. Formation of metalla-monofluorophosphine. Cp* = η^5-C_5Me_5.

The osmium cluster complexes **51** and **52** were readily formed by the addition of an excess of the bulky monofluorophosphine **L20** to the corresponding labile MeCN complex precursors (Scheme 19) [79].

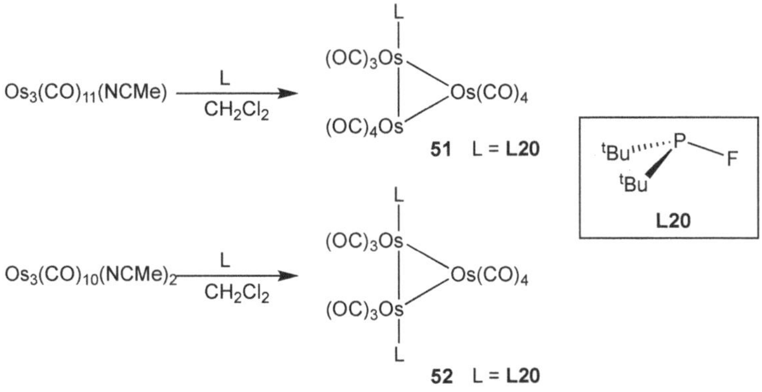

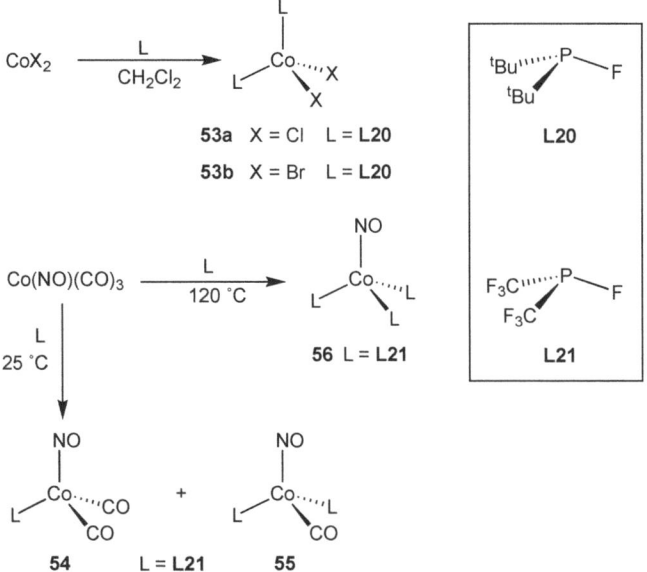

Scheme 19. Osmium cluster complexes of monofluorophosphines.

3.2.4. Group 9 Metal Complexes of Monofluorophosphines

The paramagnetic cobalt complexes **53a** and **53b** were prepared by stirring a suspension of CoX_2 in CH_2Cl_2 with **L20**. The highly coloured **53a** (blue) and **53b** (blue-green) had electronic spectra, IR spectra, and magnetic moments ($\mu \approx 4.5$ BM) consistent with the tetrahedral geometry depicted in Scheme 20 [80].

Scheme 20. Monofluorophosphine-cobalt complexes.

Reaction of 1 equiv. of **L21** with [Co(CO)₃(NO)] at ambient temperatures over 7 days yielded a mixture of monosubstituted and disubstituted complexes **54** and **55**, which were separated by fractional distillation [81]. The trisubstituted complex **56** was obtained by heating a mixture of [Co(CO)₃(NO)] and an excess of **L21** to 120 °C (Scheme 20). The position of the ν_{NO} band in the IR spectra of **54** (1832 cm^{-1}), **55** (1842 cm^{-1}), and **56** (1854 cm^{-1}) are consistent with **L21** being a better π-acceptor than CO.

Mononuclear rhodium complexes **57–61** are formed rapidly upon reaction between [Rh$_2$Cl$_2$(CO)$_4$] and the appropriate monofluorophosphine in CH$_2$Cl$_2$ (Scheme 21). The ν_{CO} values given in Scheme 21 show that the cage monofluorophosphine **L22** is the strongest π-acceptor followed by the dimethoxynaphthalene ligand **L26** and then the *sym* and *asym* isomers of the bicyclic fluorophobanes **L23** and **L24** straddle the bulky **L20** [52,61].

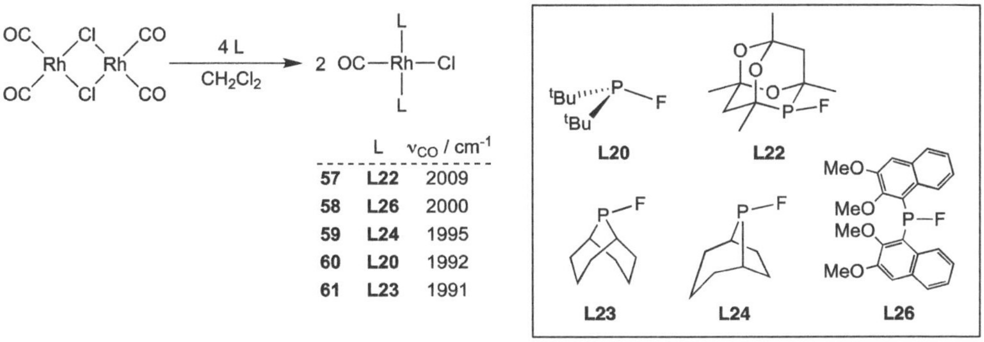

Scheme 21. Monofluorophosphine–rhodium complexes.

The fluoro analogue of Wilkinson's Catalyst, [RhF(PPh$_3$)$_3$], undergoes the rearrangement shown in Scheme 22 to generate complex **62**, which contains a 'trapped' Ph$_2$P–F ligated to Rh [82]. This remarkable isomerisation occurs under mild conditions and is reversible. Several examples are known where late transition metal fluoro complexes with PR$_3$ ancillary ligands undergo related P–C/M–F rearrangements to generate coordinated R$_2$P–F ligands as products or transient intermediates [82].

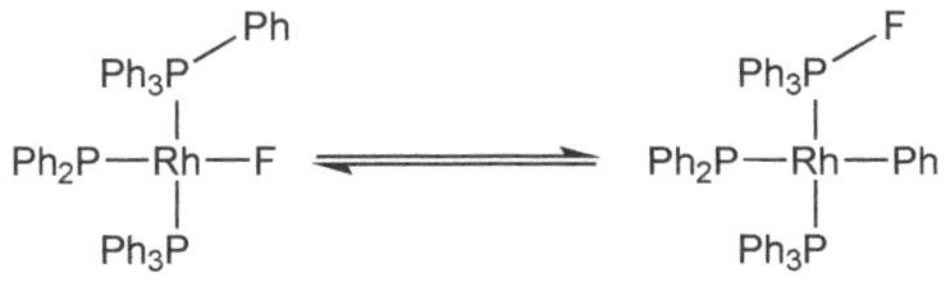

Scheme 22. Rearrangement leading to in situ formation of Ph$_2$PF complex.

3.2.5. Group 10 Metal Complexes of Monofluorophosphines

Treatment of nickel tetracarbonyl with an excess of **L21** at 25 °C gives predominantly monocarbonyl **63** with traces of dicarbonyl **64**, which can be separated by fractional distillation. The fully substituted complex **65** is produced under more forcing conditions (95 °C, 24 h), but the product is contaminated with traces of **63** (Scheme 22) [83]. The volatile, air-stable nickel(0) complex **65** can be more readily prepared by mixing **L21** with nickelocene [84] or by reaction of **L21** with metallic nickel, generated by thermolysis of nickel oxalate at 60 °C (Scheme 23) [85]. The reaction between [Ni(cod)$_2$] and the phosphacage flurophosphine **L22** was reported to give complex **66** (Scheme 23), identified in solution on the basis of the stoichiometry used and the characteristic AA'XX' pattern observed in the ^{31}P NMR spectrum [61].

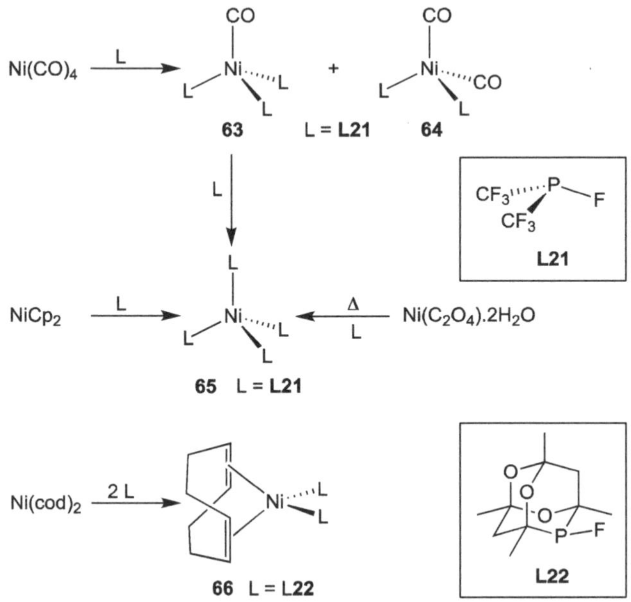

Scheme 23. Routes to nickel(0)–monofluorophosphine complexes.

Diamagnetic nickel(II) complexes **67a–c** are formed when suspensions of NiX$_2$ in acetone or toluene are treated with **L20** (Scheme 24) [80]. The *trans* geometry of **57a** was established from the large $^2J_{PP}$ of 425 Hz and the crystal structure of **57b** confirms its *trans* geometry in the solid state [86].

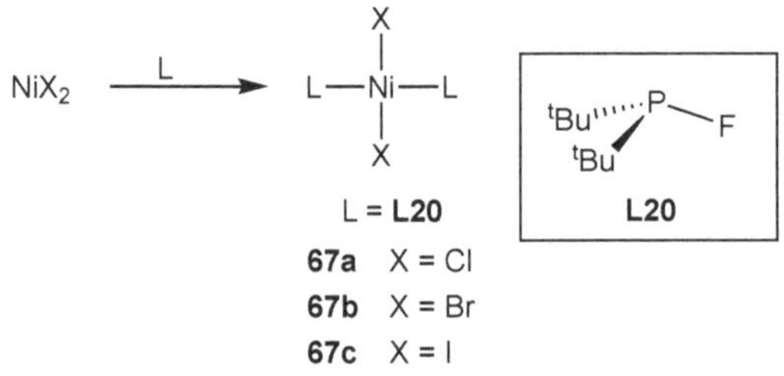

Scheme 24. Nickel(II)–monofluorophosphine complexes.

Chiral monofluorophosphine **L27** disproportionates (Scheme 14) over a period of 16 h, but the rate of the disproportionation for dilute solutions of **L27** in benzene was slow enough to measure its optical purity [87]. Reaction of a racemic mixture of **L27** with the optically pure dipalladium complex shown in Scheme 25 gave a diastereomeric mixture of complexes **68** and **69**. Pure complex **68** was obtained selectively by repeated crystallisation from diethyl ether and the absolute configuration at P was determined by X-ray crystallography. Enantiomerically pure *S*-**L27** was then displaced from complex **68** by addition of a chelating diphosphine. It was shown by polarimetry that *S*-**L27** racemised in benzene over a period of 6 h [87].

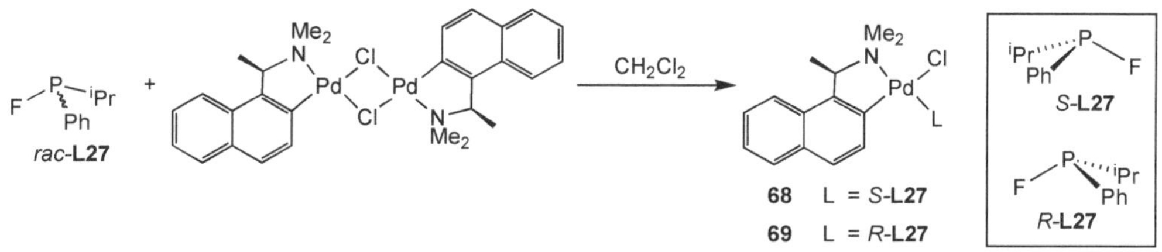

Scheme 25. Resolution of optically active monofluorophosphine.

The platinum(0) complex **70** was prepared by heating $K_2[PtCl_4]$ (or $PtCl_2$) with a large excess of **L21** followed by prolonged shaking at ambient temperature (Scheme 26); the P^V by-product $(CF_3)_2PFCl_2$ was identified, consistent with **L21** acting as the reducing agent [88]. Complex **70** was inert to the addition of MeI, HCl, C_2H_4, or CS_2, even upon prolonged heating, in contrast to the triphenylphosphine analogue $[Pt(PPh_3)_4]$. This behaviour likely reflects the greater π-acceptor properties of **L21** stabilising Pt(0) and reducing its nucleophilicity, coupled with the greater steric bulk of PPh_3 promoting the formation of reactive, coordinatively unsaturated PtL_3 species [88].

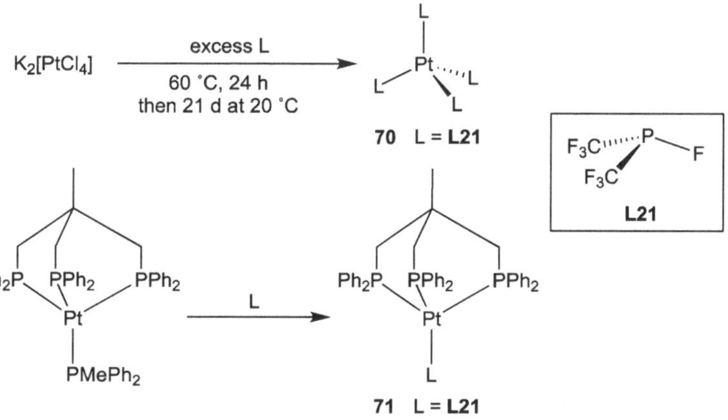

Scheme 26. Platinum(0) complexes of monofluorophosphine.

The insoluble platinum(0) complex **71** was prepared by the replacement of $PMePh_2$ by **L21** in the reaction shown in Scheme 26, a reaction presumably driven by the greater π-acceptor properties of **L21** than $PMePh_2$ [89].

The substituted diarylfluorophosphines **L28**, **L29**, and **L30** form the platinum(II) complexes **72**, **73**, and **74** by the routes shown in Scheme 27. The *cis* geometry of **72** and **73** was confirmed by their X-ray crystal structures [52,90], and the *trans*-configuration of **74** was confirmed by the large value of $^2J_{P,P}$ = 567 Hz [91].

3.3. Catalysis with Complexes of Monofluorophosphines

3.3.1. Hydroformylation Catalysis with Rhodium Complexes of Monofluorophosphines

The first step in the homologation of 1-heptene to 1-octene is the hydroformylation shown in Scheme 28 [92]. Rhodium complexes of monofluorophos ligands **L20**, **L22**, **L23**, and **L24** all showed catalytic activity comparable to the commercialised Rh—PPh_3 catalyst. The *l:b* ratio of 3.9 obtained for the Rh–**L22** catalyst compares favourably with the *l:b* ratio of 2.2 for the Rh-PPh_3 catalyst under the same conditions. The ^{31}P NMR spectrum of the exit solutions for the Rh–**L22** catalysis showed the presence of Rh–monofluorophos complexes, indicating that the coordinated **L22** had survived the reaction conditions [61].

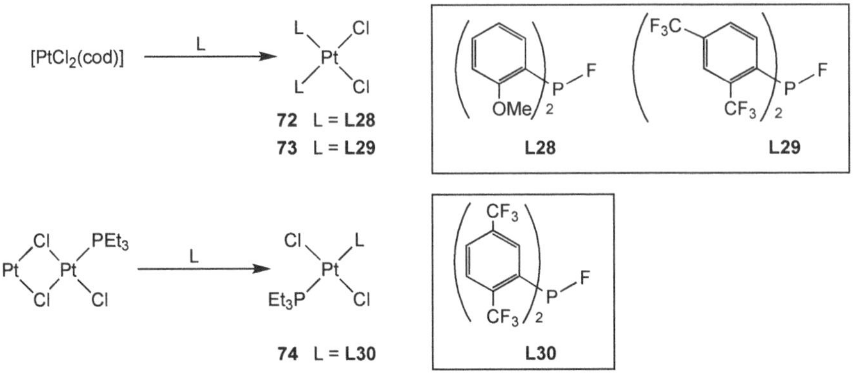

Scheme 27. Platinum(II) complexes of monofluorophosphines.

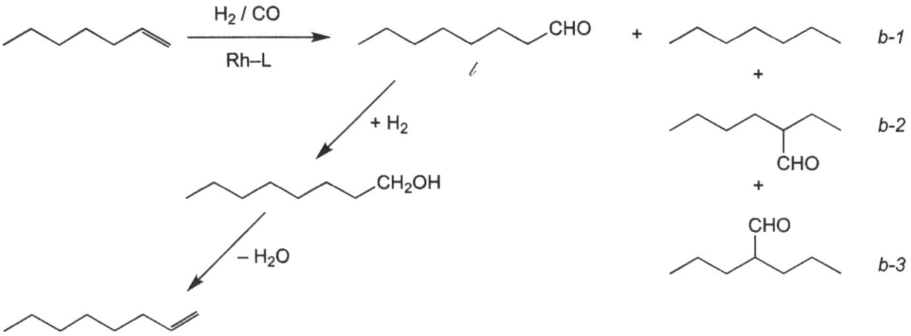

Scheme 28. Hydroformylation of 1-heptene.

3.3.2. Hydrocyanation Catalysis with Nickel Complexes of Monofluorophosphines

Catalysts derived from nickel complexes of **L20**, **L22**, **L23**, and **L24** with a Lewis acid ($ZnCl_2$ or Ph_2BOBPh_2) co-catalyst were tested for the Ni-catalysed isomerisation-hydrocyanation of 3-pentenenitrile (3-PN) to give adiponitrile (ADN) via 4-pentenenitrile (4-PN), as shown in Scheme 29. Nickel complexes of **L24** showed essentially no activity (only traces of ADN detected). Compared with the commercialised catalyst based on Ni–P(OTol)$_3$, the Ni–**L20** and Ni–**L23** catalysts were modestly active and selective but Ni–**L22** system showed good activity and selectivity [61,93,94]. The fluorine substituent in CgP–F (**L22**) was critical to the success of the hydrocyanation catalyst (Scheme 29), since attempts to use CgP–Br or CgP–Ph as ligands gave only traces of ADN.

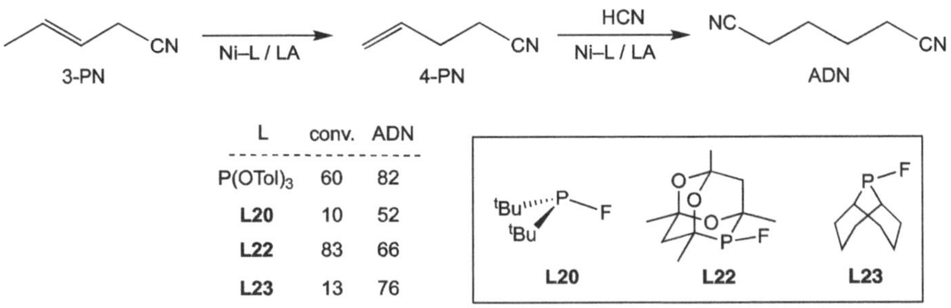

L	conv.	ADN
P(OTol)$_3$	60	82
L20	10	52
L22	83	66
L23	13	76

Scheme 29. Hydrocyanation to give ADN catalysed by Ni(0)-monofluorophosphine complex.

4. Conclusions and Prospective Applications of Monofluorophos Ligands in Coordination Chemistry and Catalysis

The combination of the extreme electronegativity and smallness of F has made ligands containing a P–F bond of academic interest for many years. The strength of the P–F bond at 490 kJ mol^{-1} dwarfs other P–X single bonds (cf. P–C, 264 kJ mol^{-1}; P–O, 335 kJ mol^{-1}) and is the source of the thermodynamically stability of P–F compounds. PF_3 is often characterised as the ultimate π-acceptor, outstripping even CO in its capacity to stabilise electron-rich, low oxidation state metal complexes. What has attracted particular attention to substituted monofluorophos ligands is their capacity to be 'tunable' analogues of PF_3 and indeed to make ligands such as $(CF_3)_2PF$ which are more powerful π-acceptors.

The focus of this review has been on the coordination chemistry of monofluorophosphites, $(RO)_2PF$, and monofluorophosphines, R_2PF, and the successful applications of monofluorophos–metal complexes in homogeneous catalysis. At the outset, the prospects for applications of monofluorophos ligands in homogeneous catalysis appeared to be inauspicious because of two fundamental instabilities: (1) notwithstanding the great P–F bond strength, monofluorophos compounds are generally susceptible to hydrolysis, a reaction driven by the formation of the even stronger bonds, H–F (565 kJ mol^{-1}) and P=O (544 kJ mol^{-1}); (2) the propensity of F to stabilise high oxidation states explains the observation that many P^{III}–F compounds readily decompose by disproportionation into P^{V}–F compounds and P^{II} species containing P–P bonds.

The 1998 report by Puckette and coworkers at Eastmann of the application of the cyclic monofluorophosphite **L1** in Rh-catalysed hydroformylation under commercially viable conditions and the impressive advantages of this catalyst (including its tunable regioselectivity) emphatically established that monofluorophos ligands have great potential as ligands for catalysis. It was shown that **L1** has structural features that make it resistant to both hydrolysis and disproportionation. These features were borrowed from diphosphites such as **L3** which are: the PO_2 heterocycle and the bulky hydrophobic t-butyl groups that protect the P–F group and kinetically stabilise the monofluorophosphite.

Early studies (in the 1970s and 1980s) demonstrated that monofluorophosphines **L21** and **L22** were stable to disproportionation and this was rationalised in terms of the great steric bulk and strong electron-withdrawing properties of the substituents. It was later shown that constraining the C–P–C angle in bicyclic or tricyclic monofluorophos ligands such as **L22** also led to greater stability with respect to disproportionation. Ligands such as **L22** have been shown to be effective not only in hydroformylation but also in hydrocyanation under commercially viable conditions.

In view of the observed powerful stabilising effects of P-substituents on monofluorophos ligands, and the demonstrated capacity of monofluorophos ligands to support homogeneous catalysis, it is surprising to us that, to date, the area of monofluorophos chemistry remains so underdeveloped and it is our contention that there are a plethora of opportunities in the areas of ligand design, fundamental coordination chemistry studies, and catalyst discovery based on ligands containing a P–F bond.

It is clear from this review that, firstly, a P–F group confers unusual donor properties on the P^{III} ligand, but there are striking 'holes' in our knowledge due to the paucity of information on monofluorophos coordination chemistry of many d-block metals; for instance, to the best of our knowledge, there are no examples of monofluorophos complexes of Re or Au. Secondly, the few catalytic studies on monofluorophos–metal complexes that have been reported have led to impressive discoveries. Some suggestions for potentially fruitful lines of enquiry that build on the results presented in this review are outlined below.

The monofluorophosphites, denoted {O,O}PF, and monofluorophosphines, denoted {C,C}PF, that are the subject of this review represent only a minor portion of the monofluorophos landscape that is available (Figure 2). There are many related {N,N}PF as well as mixed {C,O}PF, {C,N}PF, and {N,O}PF ligands waiting to be developed. Indeed, a series of acyclic and cyclic {N,O}PF ligands, (see Figure 7) of general structure **L31** (R = alkyl) [95] and **L32** (R = aryl or alkyl) [96], have been reported. Ligand **L32** generates Rh catalysts for alkene hydroformylation with *l:b* ratios ranging from 0.41 to 12.8 depending on ligand concentration and the nature of R [96].

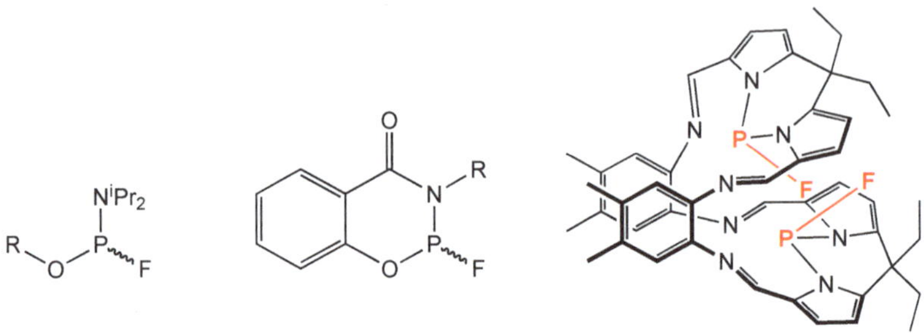

Figure 7. Monofluorophos ligands worthy of future study for catalysis.

Chelating bis(monofluorophos) ligands would be an exciting avenue to explore and an example of a bis{N,N}PF ligand was recently described: the "Pacman" fluorophos ligand **L33** (see Figure 7) [97].

Hydroformylation and hydrocyanation catalysis have been successfully demonstrated with monofluorophos ligands. These observations are consistent with the monofluorophos ligands behaving like other P-donors that are relatively electron-poor, such as phosphites. Monofluorophos–metal catalysts should be capable of catalysing other reactions that are catalysed by metal-phosphites and related ligands such as alkene isomerisation, hydrogenation, and C-C coupling reactions.

It was discovered that the optically active monofluorophosphite **L19** was an effective ligand for the enantioselective Pd-catalysed intramolecular C–C coupling reaction. It would certainly be of interest to develop other optically active monofluorophos ligands (including bidentates) and investigate their efficacy in asymmetric catalysis. All of the {X,Y}PF heterocycles shown in Figure 2 have a stereogenic P-centre, and it should be possible to resolve these molecules and investigate the application of their complexes in asymmetric catalysis.

The overarching conclusion is that there is great scope to design new fluorophos ligands containing a PF group and expand the range of steric and electronic effects such ligands can have. There are good reasons to believe that new catalysts will emerge.

Funding: We would like to thank Khalifa University for a Visiting Scholar Grant (to PGP). This work was also supported by the Engineering and Physical Sciences Research Council with the award of PhD studentships to AMH and, via the Centre for Doctoral Training in Catalysis [grant number EP/L016443], to DG.

Institutional Review Board Statement: Not applicable.

Informed Consent Statement: Not applicable.

Data Availability Statement: Not applicable.

Conflicts of Interest: The authors declare no conflict of interest.

References

1. Puckette, T.A.; Struck, G.E. Hydroformylation Process Using Novel Phosphite-Metal Catalyst System. US 5840647, 24 November 1998.
2. Burton, L.P.J. Antioxidant Aromatic Fluorophosphites. European Patent EP0280938B1, 15 June 1994.
3. Kaprinidis, N.; Chandrika, G.; Zingg, J. Flame Retardant Compositions. World Patent WO 2004/031286 A1, 15 April 2004.
4. Chemjobber blog. Fluorine: The T rex of the periodic table. *Chemistry World.* 30 July 2019. Available online: https://www.chemistryworld.com/opinion/fluorine-the-t-rex-of-the-periodic-table/3010748.article (accessed on 16 May 2024).

5. Hansch, C.; Leo, A.; Taft, R.W. A Survey of Hammett Substituent Constants and Resonance and Field Parameters. *Chem. Rev.* **1991**, *91*, 165–195. [CrossRef]
6. Tricas, H.; Diebolt, O.; van Leeuwen, P.W.N.M. Bulky Monophosphite Ligands for Ethene Hydroformylation. *J. Catal.* **2013**, *298*, 198–205. [CrossRef]
7. Billig, E.; Abatjoglou, A.G.; Bryant, D.R. Transition Metal Complex Catalysed Processes. US 4769498, 6 September 1988.
8. van Leeuwen, P.W.N.M.; Claver, C. (Eds.) *Rhodium Catalyzed Hydroformylation*; Springer: Berlin/Heidelberg, Germany, 2002.
9. Börner, A.; Franke, R. *Hydroformylation: Fundamentals, Processes, and Applications in Organic Synthesis*; Wiley-VCH GmbH: Weinheim, Germany, 2016.
10. Chakrabortty, S.; Almasalma, A.A.; de Vries, J.G. Recent Developments in Asymmetric Hydroformylation. *Catal. Sci. Technol.* **2021**, *11*, 5388–5411. [CrossRef]
11. Tazawa, T.; Phanopoulos, A.; Nozaki, K. *Enantioselective Hydroformylation*; Wiley Online Library: Hoboken, NJ, USA, 2021.
12. Nixon, J.F. Trifluorophosphine Complexes of Transition Metals. *Adv. Inorg. Chem. Radiochem.* **1985**, *29*, 41–141.
13. Vargas Garcia, J.R.; Goto, T. Chemical Vapor Deposition of Iridium, Platinum, Rhodium and Palladium. *Mater. Trans.* **2003**, *44*, 1717–1728. [CrossRef]
14. Tran, P.D.; Doppelt, P. Gold CVD Using Trifluorophosphine Gold(I) Chloride Precursor and Its Toluene Solutions. *J. Electrochem. Soc.* **2007**, *154*, D520–D525. [CrossRef]
15. Utke, L.; Swiderek, P.; Höflich, K.; Madajska, K.; Jurczyk, J.; Martinović, P.; Szymańska, I.B. Coordination and Organometallic Precursors of Group 10 and 11: Focused Electron Beam Induced Deposition of Metals and Insight Gained from Chemical Vapour Deposition, Atomic Layer Deposition, and Fundamental Surface and Gas Phase Studies. *Coord. Chem. Rev.* **2022**, *458*, 213851. [CrossRef]
16. Carpenter, A.E.; Singleton, D.G.; Kheir, S.A. Hydroformylation Catalysts Comprising Fluorophosphine Ligands and Precursors Thereof. World Patent WO 2021/202225 A1, 7 October 2021.
17. Kloprogge, T.; Ponce, C.P.; Loomis, T. *The Periodic Table: Nature's Building Blocks: An Introduction to the Naturally Occurring Elements, Their Origins and Their Uses*; Elsevier: Amsterdam, The Netherlands, 2021.
18. Heuer, L. Fluorophosphine Complexes of the Platinum Group Metals. *Platin. Met. Rev.* **1991**, *35*, 86–93. [CrossRef]
19. Miles-Hobbs, A.M.; Hunt, E.; Pringle, P.G.; Sparkes, H.A. Ring Size Effects in Cyclic Fluorophosphites: Ligands That Span the Bonding Space between Phosphites and PF_3. *Dalton Trans.* **2019**, *48*, 9712–9724. [CrossRef]
20. Meyer, T.G.; Fischer, A.; Jones, P.G.; Schmutzler, R. Darstellung und Einkristall-Röntgenstrukturanalyse Einiger Fluorphosphite Und Phosphitester. *Z. Naturforsch. B* **1993**, *48*, 659–671. [CrossRef]
21. Albers, W.; Krüger, W.; Storzer, W.; Schmutzler, R. Improved Synthesis of Halo-Phosphorus(III) Fluorides. *Synth. React. Inorg. Met.-Org. Chem.* **1985**, *15*, 187–195. [CrossRef]
22. Quin, L.D. *A Guide to Organophosphorus Chemistry*; Wiley-Interscience: New York, NY, USA, 2000.
23. Tolleson, G.S.; Puckette, T.A. Hydroformylation Process Using Chlorophosphite-Metal Catalyst System. EP 1133356 B1, 17 March 2004.
24. Trillo, R.B.; Neudörfl, J.M.; Goldfuss, B. An unusually stable chlorophosphite: What makes BIFOP–Cl so robust against hydrolysis? *Bellstein J. Org. Chem.* **2015**, *11*, 313–322. [CrossRef] [PubMed]
25. Puckette, T.A. Hydroformylation Catalysis at Eastman Chemical: Generations of Catalysts. *Top. Catal.* **2012**, *55*, 421–425. [CrossRef]
26. Schmutzler, R. Complexes of Organophosphorus Fluorides with Zerovalent Transition Metals. US 3242171, 22 March 1966.
27. Mathieu, R.; Poilblanc, R. New Penta- and Hexasubstituted Derivatives of Group VIb Metal Hexacarbonyls. *Inorg. Chem.* **1972**, *11*, 1858–1861. [CrossRef]
28. Bauer, D.P.; Ruff, J.K. Novel Iron Tetracarbonyl Fluorophosphine Complexes. *Inorg. Chem.* **1983**, *22*, 1686–1689. [CrossRef]
29. Udovich, C.A.; Clark, R.J.; Haas, H. Stereochemical Nonrigidity in Iron Carbonyl Fluorophosphine Compounds. *Inorg. Chem.* **1969**, *8*, 1066–1072. [CrossRef]
30. Mathew, N.; Jagirdar, B.R. Influence of the Cone Angles and the π-Acceptor Properties of Phosphorus-Containing Ligands in the Chemistry of Dihydrogen Complexes of Ruthenium. *Organometallics* **2000**, *19*, 4506–4517. [CrossRef]
31. Clark, R.J.; Morgan, K.A. Methanol Solvolysis of Metal Trifluorophosphine Complexes. *Inorg. Chim. Acta* **1968**, *2*, 93–96. [CrossRef]
32. Nixon, J.F.; Swain, J.R. Trifluorophosphine Complexes of Rhodium(1): Syntheses and Ligand-exchange Studies. *J. Chem. Soc. Dalton Trans.* **1972**, *10*, 1044–1048. [CrossRef]
33. Hitchcock, P.B.; Morton, S.; Nixon, J.F. Fluorophosphine Complexes of Rhodium(I) and Iridium(I): Towards the Design of Systems with Extended Metal-Metal Interactions. The Crystal Structure of [{IrCl(PF_3)$_2$}$_2$]. *J. Chem. Soc. Dalton Trans.* **1985**, *7*, 1295–1301. [CrossRef]
34. Reddy, G.S.; Schmutzler, R. Phosphorus-Fluorine Chemistry. XVIII. Nuclear Magnetic Resonance Studies on Coordination Compounds Involving Fluorine-Containing Phosphine Ligands. *Inorg. Chem.* **1967**, *6*, 823–830. [CrossRef]
35. Crocker, C.; Goodfellow, R.J. Heteronuclear INDOR Spectra of Some Tetrakis(Fluorophosphine)-Nickel(0) and -Platinum(0) Complexes Having the $[AX]_4$ (T_d) Spin System. *J. Chem. Soc. Dalton Trans.* **1977**, *17*, 1687–1689. [CrossRef]
36. Lynden-Bell, R.M. The $[AX]_4$ Nuclear Spin System with Tetrahedral Symmetry. *Mol. Phys.* **1968**, *15*, 523–531. [CrossRef]

37. Matos, R.M.; da Costa, R.F.F.; Knupp, V.F.; Silva, J.A.D.; Passos, B.F.T. Syntheses and ^{31}P NMR Studies of Transition Metal Complexes Containing Derivatives of Dioxaphospholane and Dioxaphosphorinane. *J. Braz. Chem. Soc.* **2000**, *11*, 311–316. [CrossRef]
38. Puckette, T.A. Halophosphite Ligands for the Rhodium Catalyzed Low-Pressure Hydroformylation Reaction. In *Catalysis of Organic Reactions*; Schmidt, S.R., Ed.; CRC Press: Boca Raton, FL, USA, 2006; Chapter 4.
39. Tau, K.D. Production of 2-Methylbutanal. US 4605781, 12 August 1986.
40. Klender, G.J.; Gatto, V.J.; Jones, K.R.; Calhoun, C.W. Further Developments in the Study of Fluorophosphonite Stabilizers. In *Polymer Preprints*; American Chemical Society: Washington, DC, USA, 1993; Volume 34, pp. 156–157.
41. Puckette, T.A.; Tolleson, G.S.; Devon, T.J.; Stavinoha, J.L. Epoxide Stabilization of Fluorophosphite-Metal Catalyst System in a Hydroformylation Process. WO 02/098825 A2, 12 December 2002.
42. Puckette, T.A.; Tolleson, G.S. Stabilization of Fluorophosphite-Containing Catalysts. US 6831035 B2, 14 December 2004.
43. Puckette, T.A.; Shan, X.; Rogers, J.L.; Green, B.E. Hydroformylation Catalyst. US 9550179 B1, 24 January 2017.
44. Puckette, T.A.; Shan, X.; Rogers, J.L.; Green, B.E. Hydroformylation Catalyst Containing Isomerically Enriched Halophosphite. WO 2017/044277 A1, 16 March 2017.
45. Zuidema, E.; Daura-Oller, E.; Carbó, J.J.; Bo, C.; van Leeuwen, P.W.N.M. Electronic Ligand Effects on the Regioselectivity of the Rhodium-Diphosphine-Catalyzed Hydroformylation of Propene. *Organometallics* **2007**, *26*, 2234–2242. [CrossRef]
46. Liu, Y.-S.; Rodgers, J.L. Fluorophosphite Containing Catalysts for Hydroformylation Processes. US 7872156 B2, 18 January 2011.
47. Ibrahim, M.Y.S.; Bennett, J.A.; Mason, D.; Rodgers, J.; Abolhasani, M. Flexible Homogeneous Hydroformylation: On-Demand Tuning of Aldehyde Branching with a Cyclic Fluorophosphite Ligand. *J. Catal.* **2022**, *409*, 105–117. [CrossRef]
48. Puckette, T.A. Acetylene Tolerant Hydroformylation Catalysts. US 2010/0069679 A1, 18 March 2010.
49. Puckette, T.A. Process for the Preparation of Glycolaldehyde. US 7301054 B1, 27 November 2007.
50. Trillo, R.B.; Leven, M.; Neudörfl, J.M.; Goldfuss, B. Electronegativity Governs Enantioselectivity: Alkyl-Aryl Cross-Coupling with Fenchol-Based Palladium-Phosphorus Halide Catalysts. *Adv. Synth. Catal.* **2012**, *354*, 1451–1465. [CrossRef]
51. Brüllingen, E.; Neudörfl, J.M.; Goldfuss, B. Enantioselective Cu-Catalyzed 1,4-Additions of Organozinc and Grignard Reagents to Enones: Exceptional Performance of the Hydrido-Phosphite-Ligand BIFOP-H. *New J. Chem.* **2019**, *43*, 4787–4799. [CrossRef]
52. Meyer, T.G.; Jones, P.G.; Schmutzler, R. Darstellung Neuer Monofluorphosphine Und Einiger Ihrer Übergangsmetallkomplexe; Einkristall-Röntgenstrukturanalyse Eines Platin(II)-Komplexes. *Z. Naturforsch. B* **1993**, *48*, 875–885. [CrossRef]
53. Seel, F.; Rudolph, K.; Gombler, W. Dimethylfluorophosphine. *Angew. Chem. Int. Ed.* **1967**, *6*, 708. [CrossRef]
54. Seel, F.; Rudolph, K. Uber Dimethylfluorophosphin und Dimethyldifluorophosphoran. *Z. Anorg. Allg. Chem.* **1968**, *363*, 233–244. [CrossRef]
55. Riesel, L.; Haenel, J.; Ohms, G. Zur Disproportionierung der Phenylfluorphosphane $(C_6H_5)_2$PF Und (C_6H_5)PF$_2$. *J. Fluor. Chem.* **1988**, *38*, 335–340. [CrossRef]
56. Haenel, J.; Ohms, G.; Riesel, L. Die Dimerisierung yon Di(n-buty1)fluorphosphan und seine Reaktion mit Benzaldehyd. *Z. Anorg. Allg. Chem.* **1992**, *607*, 161–163. [CrossRef]
57. Brown, C.; Murray, M.; Schmutzler, R. Fluorodiphenylphosphine. *J. Chem. Soc. C* **1970**, 878–881. [CrossRef]
58. Schmutzler, R.; Stelzer, O.; Liebman, J.F. Catalytic and Autocatalytic Disproportionation Reactions of Fluorophosphines and Related Lower Valence Nonmetal Fluorides. *J. Fluor. Chem.* **1984**, *25*, 289–299. [CrossRef]
59. Fild, M.; Schmutzler, R. Phosphorus-Fluorine Chemistry. Part XXIII. t-Butyl-Fluorophosphines and -Fluorophosphoranes and Their Derivatives. *J. Chem. Soc. A* **1970**, 2359–2364. [CrossRef]
60. Fild, M.; Schmutzler, R. Phosphorus-Fluorine Chemistry. Part XXI. Pentafluorophenylfluorophosphines and Pentafluorophenylfluorophosphoranes. *J. Chem. Soc. A* **1969**, 840–843. [CrossRef]
61. Fey, N.; Garland, M.; Hopewell, J.P.; McMullin, C.L.; Mastroianni, S.; Orpen, A.G.; Pringle, P.G. Stable Fluorophosphines: Predicted and Realized Ligands for Catalysis. *Angew. Chem. Int. Ed.* **2012**, *51*, 118–122. [CrossRef] [PubMed]
62. Szynkiewicz, N.; Ponikiewski, L.; Grubba, R. Symmetrical and unsymmetrical diphosphanes with diversified alkyl, aryl, and amino substituents. *Dalton Trans.* **2018**, *47*, 16885–16894. [CrossRef] [PubMed]
63. Becker, G.; Golla, W.; Grobe, J.; Klinkhammer, K.W. Element-Element Bonds. IX. Structures of Tetrakis(trifluoromethyl)diphosphane and -diarsane: Experimental and Theoretical Investigations. *Inorg. Chem.* **1999**, *38*, 1099–1107. [CrossRef] [PubMed]
64. Dodds, D.L.; Floure, J.; Garland, M.; Haddow, M.F.; Leonard, T.R.; McMullin, C.L.; Orpen, A.G.; Pringle, P.G. Diphosphanes derived from phobane and phosphatrioxa-adamantane: Similarities, differences and anomalies. *Dalton Trans.* **2011**, *40*, 7137–7146. [CrossRef] [PubMed]
65. Downing, J.H.; Floure, J.; Heslop, K.; Haddow, M.F.; Hopewell, J.; Lusi, M.; Hirahataya Phetmung, H.; Orpen, A.G.; Pringle, P.G.; Pugh, R.I.; et al. General Routes to Alkyl Phosphatrioxaadamantane Ligands. *Organometallics* **2008**, *27*, 3216–3224. [CrossRef]
66. Carreira, M.; Charernsuk, M.; Eberhard, M.; Fey, N.; van Ginkel, R.; Hamilton, A.; Mul, W.P.; Orpen, A.G.; Phetmung, H.; Pringle, P.G. Anatomy of Phobanes. Diastereoselective Synthesis of the Three Isomers of n-Butylphobane and a Comparison of their Donor Properties. *J. Am. Chem. Soc.* **2009**, *131*, 3078–30929. [CrossRef] [PubMed]
67. Lister, J.M.; Carreira, M.; Haddow, M.F.; Hamilton, A.; McMullin, C.L.; Orpen, A.G.; Pringle, P.G.; Stennett, T.E. Unexpectedly High Barriers to M–P Rotation in Tertiary Phobane Complexes: PhobPR Behavior That Is Commensurate with ^{t}Bu$_2$PR. *Organometallics* **2014**, *33*, 702–714. [CrossRef]

68. Deiters, J.A.; Holmes, R.R.; Holmes, J.M. Fluorine and Chlorine Apicophilicities in Five-Coordinated Phosphorus and Silicon Compounds via Molecular Orbital Calculations, A Model for Nucleophilic Substitution. *J. Am. Chem. Soc.* **1988**, *110*, 7672–7681. [CrossRef]
69. Hófler, M.; Stubenrauch, M.; Rlcharz, E. A New Method for the Preparation of Monofluorophosphane Complexes. Isolation of [(CO)$_5$CrP(NEt$_2$Me)Et$_2$][BF$_4$]. *Organometallics* **1987**, *6*, 198–199. [CrossRef]
70. Yih, K.-H. Syntheses and Characterization of Molybdenum Complexes with the (1,3-Dithioliumyl)diphenylphosphine Containing Ligands. *J. Chin. Chem. Soc.* **1999**, *46*, 535–538. [CrossRef]
71. Yih, K.-H.; Lee, G.-H.; Wang, Y. Syntheses and Crystal Structures of Tungsten Complexes with Various Ligands Containing (1,3-Dithioliumyl)diphenylphosphine. *Organometallics* **2001**, *20*, 2604–2610. [CrossRef]
72. Stelzer, O.; Schmutzler, R. Phosphorus-Fluorine Chemistry. Part XXVIII. Fluorophosphines with Bulky Substituents as Ligands in Transition Metal Carbonyl Complexes. *J. Chem. Soc. A* **1971**, 2867–2873. [CrossRef]
73. Grobe, J.; Le Van, D.; Meyring, W. Chrom- und Wolframpentacarbonylkomplexe von Bis(Trifluormethyl)Phosphanen des Typs (F$_3$C)$_2$PX' (X' = H, F, Cl, Br, I, NEt$_2$). *Z. Anorg. Allg. Chem.* **1990**, *586*, 149–158. [CrossRef]
74. Barlow, C.G.; Nixon, J.F.; Webster, M. The Chemistry of Phosphorus-Fluorine Compounds. Part IX. Preparation and Spectroscopic Studies of Fluorophosphine-Molybdenum Carbonyl Complexes. *J. Chem. Soc. A* **1968**, 2216–2223. [CrossRef]
75. Clarke, M.L.; Holliday, G.L.; Slawin, A.M.Z.; Woollins, J.D. Highly electron rich alkyl- and dialkyl-N-pyrrolidinyl phosphines: An evaluation of their electronic and structural properties. *J. Chem. Soc. Dalton Trans.* **2002**, *6*, 1093–1103. [CrossRef]
76. Dobbie, R.C. Action of Bistrifluoromethylphosphino-Compounds on Pentacarbonyl-Manganese Hydride. *J. Chem. Soc. A* **1971**, 230–233. [CrossRef]
77. Hoidn, C.M.; Leitl, J.; Ziegler, C.G.P.; Shenderovich, I.G.; Wolf, R. Halide-Substituted Phosphacyclohexadienyl Iron Complexes: Covalent Structures vs. Ion Pairs. *Eur. J. Inorg. Chem.* **2019**, *2019*, 1567–1574. [CrossRef]
78. Oberhammer, H.; Schmutzler, R.; Stelzer, O. Molecular Structures of Phosphorus Compounds. 6. An Electron Diffraction Study of tert-Butylfluorophosphines Bu^t_nPF$_{3-n}$ (n = 1, 2, 3). *Inorg. Chem.* **1978**, *17*, 1254–1258. [CrossRef]
79. Heuer, L.; Schomburg, D. ^{t}Bu$_2$PF as a Ligand in Tri-osmium Clusters. *J. Organomet. Chem.* **1995**, *495*, 53–59. [CrossRef]
80. Stelzer, O.; Unger, E. Alkyl- and Aryl-Fluorophosphines as Ligands in Transition-Metal Complexes with Metals in Positive Oxidation States. Part I. Nickel(II) and Cobalt(II) Halide Complexes of Di-(t-Butyl)Fluorophosphine. *J. Chem. Soc. Dalton Trans.* **1973**, *17*, 1783–1788. [CrossRef]
81. Dobbie, R.C.; Morton, S. Trifluoromethylphosphine Complexes of Tricarbonylnitrosylcobalt. *J. Chem. Soc. Dalton Trans.* **1976**, *14*, 1421–1423. [CrossRef]
82. Macgregor, S.A.; Roe, D.C.; Marshall, W.J.; Bloch, K.M. The F/Ph Rearrangement Reaction of [(Ph$_3$P)$_3$RhF]. *J. Am. Chem. Soc.* **2005**, *127*, 15304–15321. [CrossRef] [PubMed]
83. Burg, A.B.; Street, G.B. Perfluoromethylphosphine-Nickel Compounds, Including a New Volatile Heterocycle. *Inorg. Chem.* **1966**, *5*, 1532–1537. [CrossRef]
84. Nixon, J.F. The Chemistry of Phosphorus-Fluorine Compounds. Part VIII. Synthesis and Nuclear Magnetic Resonance Spectra of Tetrakisfluorophosphine Derivatives of Zerovalent Nickel. *J. Chem. Soc. A* **1967**, 1136–1139. [CrossRef]
85. Nixon, J.F.; Sexton, M.D. Phosphorus-Fluorine Compounds. Part XIV. Direct Syntheses of Tetrakis(Fluorophosphine) Complexes of Zerovalent Nickel. *J. Chem. Soc. A* **1969**, 1089–1091. [CrossRef]
86. Sheldrick, W.S.; Stelzer, O. Preparation, Crystal and Molecular Structure of trans-Dibromobis[di(t-butyl)fluorophosphine]nickel(II). *J. Chem. Soc. Dalton* **1973**, *9*, 926–929. [CrossRef]
87. Pabel, M.; Willis, A.C.; Wild, S.B. First Resolution of a Free Fluorophosphine Chiral at Phosphorus. Resolution and Reactions of Free and Coordinated (±)-Fluorophenylisopropylphosphine. *Inorg. Chem.* **1996**, *35*, 1244–1249. [CrossRef]
88. Nixon, J.F.; Sexton, M.D. Phosphorus-Fluorine Compounds. Part XVII. Fluorophosphine Complexes of Zerovalent Platinum. *J. Chem. Soc. A* **1970**, 321–323. [CrossRef]
89. Al-Ohaly, A.-R.; Nixon, J.F. ^{31}P Nuclear Magnetic Resonance Spectroscopic Studies on Some Zerovalent Platinum Phosphine Complexes. *Inorg. Chim. Acta* **1980**, *47*, 105–109. [CrossRef]
90. Heuer, L.; Jones, P.G.; Schmutzler, R. Preparation of Bis[2,4-bis(trifluoromethyl)phenyl]fluorophosphine and 2,4-Bis(trifluoromethyl)phenyl-[2,6-bis(trifluoromethyl)phenyl]fluorophosphine—Two Distillable Monofluorophosphines. *J. Fluor. Chem.* **1990**, *46*, 243–254. [CrossRef]
91. Capel, V.L.; Dillon, K.B.; Goeta, A.E.; Howard, J.A.K.; Monks, P.K.; Probert, M.R.; Shepherd, H.J.; Zorina, N.V. Stereochemically Inactive Lone Pairs in Phosphorus(III) Compounds: The Characterisation of Some Derivatives with the 2,5-(CF$_3$)$_2$C$_6$H$_3$ (Ar) Substituent and Their Complexation Behaviour towards Pt(II) Species. *Dalton Trans.* **2011**, *40*, 1808–1816. [CrossRef] [PubMed]
92. van Leeuwen, P.W.N.M.; Nicolas, D.; Cleément, N.D.; Mathieu, J.-L.; Tschan, M.J.-L. New processes for the selective production of 1-octene. *Coord. Chem. Rev.* **2011**, *255*, 1499–1517. [CrossRef]
93. Mastroianni, S.; Pringle, P.; Garland, M.; Hopewell, J. Method for the Production of Nitrile Compounds from Ethylenically-Unsaturated Compounds. WO 2010/145960 A1, 23 December 2010.
94. Mastroianni, S.; Pringle, P.; Hopewell, J.; Garland, M. Method for Producing Nitrile Compounds from Ethylenically Unsaturated Compounds. WO 2013/045524 A1, 4 April 2013.
95. Michalski, J.; Wojciech Dabkowski, W. New chemistry and stereochemistry of tricoordinate phosphorus esters containing phosphorus–fluorine bond. *Comptes Rendus. Chim.* **2004**, *7*, 901–907. [CrossRef]

96. Puckette, T.A. Amido-Fluorophosphite Compounds and Catalysts. US 8492593 B2, 23 July 2013.
97. Eickhoff, L.; Kramer, P.; Bresien, J.; Michalik, D.; Villinger, A.; Schulz, A. On the Dynamic Behaviour of Pacman Phosphanes—A Case of Cooperativity and Redox Isomerism. *Inorg. Chem.* **2023**, *62*, 6768–6778. [CrossRef] [PubMed]

Disclaimer/Publisher's Note: The statements, opinions and data contained in all publications are solely those of the individual author(s) and contributor(s) and not of MDPI and/or the editor(s). MDPI and/or the editor(s) disclaim responsibility for any injury to people or property resulting from any ideas, methods, instructions or products referred to in the content.

MDPI AG
Grosspeteranlage 5
4052 Basel
Switzerland
Tel.: +41 61 683 77 34

Molecules Editorial Office
E-mail: molecules@mdpi.com
www.mdpi.com/journal/molecules

Disclaimer/Publisher's Note: The title and front matter of this reprint are at the discretion of the Guest Editor. The publisher is not responsible for their content or any associated concerns. The statements, opinions and data contained in all individual articles are solely those of the individual Editor and contributors and not of MDPI. MDPI disclaims responsibility for any injury to people or property resulting from any ideas, methods, instructions or products referred to in the content.

www.ingramcontent.com/pod-product-compliance
Lightning Source LLC
LaVergne TN
LVHW072325090526
838202LV00019B/2352